DISCRETE SYSTEMS AND INTEGRABILITY

This first introductory text to discrete integrable systems introduces key notions of integrability from the vantage point of discrete systems, also making connections with the continuous theory where relevant. While treating the material at an elementary level, the book also highlights many recent developments. Topics include: Darboux and Bäcklund transformations, difference equations and special functions, multidimensional consistency of integrable lattice equations, associated linear problems (Lax pairs), connections with Padé approximants and convergence algorithms, singularities and geometry, Hirota's bilinear formalism for lattices, intriguing properties of discrete Painlevé equations and the novel theory of Lagrangian multiforms. The book builds the material in an organic way, emphasizing interconnections between the various approaches, while the exposition is mostly done through explicit computations on key examples. Written by respected experts in the field, the book's numerous exercises and thorough list of references will benefit both upper-level undergraduate and beginning graduate students as well as researchers from other disciplines.

JARMO HIETARINTA is Professor Emeritus of Theoretical Physics at the University of Turku, Finland. His work has focused on the search for integrable systems of various forms, including Hamiltonian mechanics, Hirota bilinear form, Yang–Baxter and tetrahedron equations as well as lattice equations. He was instrumental in the setting up of the nlin.SI category in arxiv.org and created the web pages for the SIDE (Symmetries and Integrability of Difference Equations) conference series: http://side-conferences.net

NALINI JOSHI is Professor of Applied Mathematics at the University of Sydney. She is best known for her work on the Painlevé equations and works at the leading edge of international efforts to analyze discrete and continuous integrable systems in the geometric setting of their initial-value spaces, constructed by resolving singularities in complex projective space. She was elected as a Fellow of the Australian Academy of Science in 2008, holds a Georgina Sweet Australian Laureate Fellowship and was awarded the special Hardy Fellowship of the London Mathematical Society in 2015.

FRANK NIJHOFF is Professor of Mathematical Physics in the School of Mathematics of the University of Leeds. His research focuses on nonlinear difference and differential equations, symmetries and integrability of discrete systems, variational calculus, quantum integrable systems and linear and nonlinear special functions. He was the principal organizer of the 2009 six-month programme on Discrete Integrable Systems at the Isaac Newton Institute, and a Royal Society Leverhulme Trust Senior Research Fellow in 2011.

Cambridge Texts in Applied Mathematics

All titles listed below can be obtained from good booksellers or from Cambridge University Press.
For a complete series listing, visit www.cambridge.org/mathematics.

Nonlinear Dispersive Waves
MARK J. ABLOWITZ

Flow, Deformation and Fracture
G. I. BARENBLATT

Hydrodynamic Instabilities
FRANÇOIS CHARRU

The Mathematics of Signal Processing
STEVEN B. DAMELIN & WILLARD MILLER, JR

An Introduction to Stochastic Dynamics
JINQIAO DUAN

Singularities: Formation, Structure and Propagation
J. EGGERS & M. A. FONTELOS

A First Course in Continuum Mechanics
OSCAR GONZALEZ & ANDREW M. STUART

A Physical Introduction to Suspension Dynamics
ÉLISABETH GUAZZELLI & JEFFREY F. MORRIS

Applied Solid Mechanics
PETER HOWELL, GREGORY KOZYREFF & JOHN OCKENDON

A First Course in the Numerical Analysis of Differential Equations (2nd Edition)
ARIEH ISERLES

Iterative Methods in Combinatorial Optimization
LAP CHI LAU, R. RAVI & MOHIT SINGH

An Introduction to Polynomial and Semi-Algebraic Optimization
JEAN BERNARD LASSERRE

An Introduction to Computational Stochastic PDEs
GABRIEL J. LORD, CATHERINE E. POWELL & TONY SHARDLOW

DISCRETE SYSTEMS AND INTEGRABILITY

J. HIETARINTA
University of Turku

N. JOSHI
University of Sydney

F. W. NIJHOFF
University of Leeds

CAMBRIDGE
UNIVERSITY PRESS

CAMBRIDGE
UNIVERSITY PRESS

Shaftesbury Road, Cambridge CB2 8EA, United Kingdom

One Liberty Plaza, 20th Floor, New York, NY 10006, USA

477 Williamstown Road, Port Melbourne, VIC 3207, Australia

314–321, 3rd Floor, Plot 3, Splendor Forum, Jasola District Centre, New Delhi – 110025, India

103 Penang Road, #05–06/07, Visioncrest Commercial, Singapore 238467

Cambridge University Press is part of Cambridge University Press & Assessment, a department of the University of Cambridge.

We share the University's mission to contribute to society through the pursuit of education, learning and research at the highest international levels of excellence.

www.cambridge.org
Information on this title: www.cambridge.org/9781107042728

First published 2016

A catalogue record for this publication is available from the British Library

Library of Congress Cataloging-in-Publication data
Names: Hietarinta, J. (Jarmo), author. | Joshi, Nalini, author. | Nijhoff, Frank W., author.
Title: Discrete systems and integrability / J. Hietarinta (University of Turku), N. Joshi (University of Sydney), F.W. Nijhoff (University of Leeds).
Other titles: Cambridge texts in applied mathematics.
Description: Cambridge, United Kingdom ; New York, NY : Cambridge University Press, 2016. | © 2016 | Series: Cambridge texts in applied mathematics | Includes bibliographical references and index.
Identifiers: LCCN 2016005109 | ISBN 9781107042728 (hardback) | ISBN 1107042720 (hardback) | ISBN 9781107669482 (paperback) | ISBN 1107669480 (paperback)
Subjects: LCSH: Integral equations. | Mathematical physics.
Classification: LCC QC20.7.I58 H54 2016 | DDC 511/.1–dc23
LC record available at http://lccn.loc.gov/2016005109

ISBN 978-1-107-04272-8 Hardback
ISBN 978-1-107-66948-2 Paperback

To Marja, Robert and Alain

Contents

Preface

There has been a surge of interest in discrete integrable systems in the last two decades. The term "discrete integrable systems" (DIS) combines two aspects: discreteness and integrability. The subtle concept of integrability touches on global existence and regularity of solutions, explicit solvability, as well as compatibility and consistency – fundamental features, which form the recurrent themes of this book. On the other hand, by discrete systems we mean mathematical models that involve *finite* (as opposed to *infinitesimal*) operations. In a sense, discrete systems are essential to an understanding of integrability, and this book serves to provide an introduction to integrability from the perspective of discrete systems.

Discrete integrable systems include many types of equations, such as recurrence relations, difference equations and dynamical mappings as well as equations that contain a mixture of derivative and difference operators. Integrable systems have appeared throughout the history of mathematics without being recognized as integrable. An example is the equation that arises from the geometric collinearity theorem of Menelaus of Alexandria in the second century. Other examples are found in the defining equations of classical and nonclassical special functions. Important physical models, such as the equations of motion of the Euler top, are also integrable. The elliptic billiard is a classic example of a DIS, and the corresponding geometric result is the *Poncelet's porism*. It is, however, only in the second half of the twentieth century that the subject of integrable systems has grown as a discipline in its own right, and only in the last roughly two decades that the field of DIS has come to prominence as an area in which numerous breakthroughs have taken place, inspiring new developments in other areas of mathematics.

The number of integrable systems is large and growing, and the list of them includes a large number of examples that have application to physics and other scientific fields. In many cases, it turns out that one and the same discrete equation may be interpreted in different ways: as a dynamical map, as a difference equation and as an addition formula. This means that in the study of a given DIS a large number of branches of mathematics come together. Moreover, integrable equations are not necessarily isolated objects, but have many mathematical interconnections between each other. Highlighting these interconnections is another aim of this book.

Integrable systems are both rare and universal. Like prime numbers, their rarity is due to the intricate mathematical structures underlying them, while at the same time these structures explain their universality. This book aims to describe their underlying structures, but there are still many that remain to be discovered.

This book grew from lectures given by the authors in courses on DIS for students at the advanced undergraduate or beginning graduate level, at the University of Leeds, the University of Turku and the University of Sydney. The presentation of the book is pitched at a level suitable for students as well as researchers who are seeking an introduction to the subject.

The exposition of the book will be led by exploring key examples. We will show how the major features of integrability arise in an organic way by revealing them through explicit calculations. Each chapter contains several exercises, which will further illustrate the methods and expand on the main text. The initial chapters contain the basic material. Additional specialized or advanced topics are provided in the later chapters. Some important background material on difference calculus, elliptic functions and determinantal identities are provided in the appendices.

We start in Chapter 1 with an overview of the types of difference equations we will encounter, highlighting integrable systems as examples. In Chapter 2 we show how discrete systems, in the form of difference equations, arise naturally from known continuous equations in two different settings: the linear case (with difference equations arising from the recurrence structure of families of special functions) and the nonlinear case (addition rules for elliptic functions and Bäcklund–Schlesinger transformations for Painlevé transcendents). In the multivariable case we show how integrable partial difference equations arise as permutability conditions from Bäcklund transforms of soliton equations. This emergence of fully discrete equations (also called *lattice equations*) leads naturally to a characterization of integrability by *multidimensional consistency*. This property is explored in depth in Chapter 3, where we study its consequences and also discuss additional sources of DIS. Chapter 4 provides an interlude from the main theory by illustrating the links between integrable systems and applications in the theory of numerical algorithms. The circle is closed in Chapter 5 by providing continuum limits that lead back to continuous systems from the discrete ones. In the remaining chapters we will delve deeper into the theory and explore reductions and solutions, as well as other characteristics of integrability. In Chapter 6, we shift the focus to ordinary difference equations and dynamical mappings and provide various points of view to study them. The problem of detecting and identifying integrable difference equations by methods based on their singularity structure and growth is discussed in Chapter 7. Chapters 8 and 9 discuss two different approaches to the construction of special solutions of the partial difference equations encountered in the earlier chapters. First, in Chapter 8, we present and explain Hirota's bilinear method: in particular, how it can be used to obtain soliton solutions. In Chapter 9, another approach to soliton solutions is provided by the Cauchy matrix approach, which allows us to see that many seemingly different systems are closely interrelated. Other types of solutions are obtained in Chapter 10 through similarity reductions of partial difference equations. These are transcendental solutions obtained as solutions of a class of nonautonomous nonlinear ordinary difference equations, called the discrete Painlevé equations. The discrete Painlevé equations have very deep mathematical properties, which are described in Chapter 11. The final chapter, Chapter 12, provides insight into the geometrical aspects of integrable systems through their Lagrangian structure and the corresponding new variational approach suitable for integrable systems.

The book is by no means intended to be an exhaustive treatment of all topics related to discrete integrability, but rather reflects the authors' interests and experience. For example, we do not give a comprehensive treatment of Lie symmetries and conservation laws of discrete systems; for that, we refer to Hydon (2014). Other topics we omit are ultra-discrete equations and tropical geometry, as well as discrete quantum integrable systems. Currently, there exist only a few monographs on DIS: Suris (2003) focuses primarily on integrable discretizations from a Hamiltonian perspective, while Bobenko and Suris (2008) consider differential geometric aspects; Duistermaat (2010) deals mainly with the QRT (Quispel–Roberts–Thompson) dynamical mapping and the algebraic geometry of the associated elliptic surfaces.

We would like to draw the reader's attention also to the biennial conference series SIDE, devoted to *Symmetries and Integrability of Difference Equations*, the proceedings of which provide a good account of the state of the art of the subject. For further details, see `http://side-conferences.net`. We would also like to mention some collections of lectures on DIS; for example, Grammaticos et al. (2004) and Levi et al. (2011).

Many colleagues and students have contributed to the evolution of this book in various ways. We would in particular like to thank (in alphabetical order) James Atkinson, Sam Butler, Neslihan Delice, Chris Field, Wei Fu, Basil Grammaticos, Mike Hay, Anthony Henderson, Phil Howes, Paul Jennings, Pavlos Kassotakis, Sotiris Konstantinou-Rizos, Sarah Lobb, Nobutaka Nakazono, Maciej Nieszporski, Chris Ormerod, Vassilis Papageorgiou, Reinout Quispel, Alfred Ramani, John Roberts, Yang Shi, Ying Shi, Paul Spicer, Tomoyuki Takenawa, Tasos Tongas, Dinh Tran, Peter van der Kamp, Claude Viallet, Pavlos Xenitidis, Sikarin Yoo-Kong, Da-jun Zhang and Songlin Zhao. Special thanks go to Da-jun Zhang for his careful and detailed reading of the galley proofs.

1

Introduction to difference equations

This chapter serves to introduce the various concepts of discretization and types of equations that we will encounter in the course of this book. We start by recalling basic properties of differential equations and show how some of them translate to difference equations. An important equation that we will meet several times is the discrete Riccati equation, which we discuss here in some detail. Partial difference equations form a major theme of this book, and we introduce linear partial difference equations, which are discrete versions of canonical types that are important in mathematical physics.

For basic methods and tools of the elementary theory of difference calculus and difference equations, we refer the reader to Appendix A.

1.1 A first look at discrete equations

At the most fundamental level, differential equations can be divided into *ordinary differential equations* (ODE) and *partial differential equations* (PDE), depending on the number of independent variables. Their respective theories are quite different in nature. A similar division can be made for difference equations, distinguishing between *ordinary difference equations* (OΔEs) and *partial difference equations* (PΔEs), depending on whether there is one or more than one independent (discrete) variable(s).

1.1.1 Ordinary differential equations (ODEs)

Let us first recall some basic notions about ODEs. The general form of an ODE is

$$\mathfrak{F}\Big(\, y(x), y'(x), y''(x), \dots y^{(n)}(x); x \,\Big) = 0, \tag{1.1}$$

in which $\mathfrak{F}$ is some given function of its arguments. Here x is the **independent variable**, while $y(x)$ is the **dependent variable** since it depends on x. The primes in (1.1) denote differentiation with respect to x; if there are higher derivatives we use a superscript in parentheses:

$$y'(x) = \frac{dy}{dx}, \quad y''(x) = \frac{d^2y}{dx^2}, \quad \dots \quad y^{(n)} = \frac{d^ny}{dx^n},$$

where the number of primes gives the **order** of the derivative and the order of the equation is given by the order of the highest derivative appearing in the equation, while the **degree** of the equation is the power of the highest derivative. The domain and range of the function y needs to be specified; usually $y : \mathbb{R} \to \mathbb{R}$ or $y : \mathbb{C} \to \mathbb{C}$. The dependent variable can be multi-component, e.g. $y : \mathbb{R} \to \mathbb{R}^M$, in which case there can also be several equations. The equation may depend also on various **parameters**, i.e. unspecified constants (quantities independent of x). If $\mathfrak{F}$ does not depend explicitly on x, i.e. x enters only through the function y and its derivatives, then the ODE is called **autonomous**; otherwise it is called **nonautonomous**.

Sometimes we can separate the highest derivative:

$$y^{(n)}(x) = F(y, y', \ldots, y^{(n-1)}; x). \tag{1.2}$$

Such a higher-order equation can be written as a first-order multi-component equation

$$\boldsymbol{y}' = \boldsymbol{F}(\boldsymbol{y}; x), \tag{1.3}$$

by defining (here the superscript T stands for transpose)

$$\boldsymbol{y} := (y, y', \ldots, y^{(n-1)})^T, \quad \boldsymbol{F} := \left(y', y'', \ldots, y^{(n-1)}, F(y, y', \ldots, y^{(n-1)}; x)\right)^T.$$

Usually the aim is to solve for y as a function of x, wherever possible, from (1.1) or (1.2) or (1.3). The solution, if we can find it at all, is not unique, and needs further data in order to render it unique. This can be done by imposing, in addition to the ODE itself, a number of *initial data*; by fixing at a given value of x, say at $x = x_0$, the values of y, y', y'', …, $y^{(n-1)}$; or by fixing the data at several boundary points. This leads to solutions $y(x)$ that are functions defined over a domain that is determined by the initial or boundary data. The idea that this process may lead to functions $y(x)$ that have never before been described was very influential in the development of the theory that led to Painlevé transcendents around the turn of the twentieth century. This idea will arise again when we consider functions defined as solutions of difference equations.

Example 1.1.1 Here are some examples of famous differential equations that have been studied widely in the literature and will play a role in the later chapters. The Verhulst equation (or logistic growth model) is given by

$$\frac{dy}{dx} = a\, y(1 - y), \tag{1.4}$$

where a is a constant parameter, and is first order and first degree. Weierstrass' elliptic function $\wp$ satisfies an equation of first order and second degree

$$(\wp')^2 = 4\wp^3 - g_2\wp - g_3, \tag{1.5}$$

where g_2 and g_3 are constant parameters. The first Painlevé equation $\mathrm{P_I}$ is given by

$$y''(x) = 6y^2(x) + x, \tag{1.6}$$

and is second order and nonautonomous, while the Chazy equation

$$y''' - 2yy'' + 3y'^2 = 0 \tag{1.7}$$

is third order and autonomous. We will later consider discrete versions of some of these equations.

In the examples given above, the equations contain parameters, which are usually fixed at a given value. However, there is often merit in considering these parameters as *free* (i.e. unspecified) so that the solutions of the differential equations can be studied as functions of both the parameters and the independent variable.

1.1.2 The difference operator

Given a continuous system, an important question is how to find a corresponding discrete equation, which has qualitatively the same properties (such as conserved quantities). This question arises as a core problem in numerical analysis, where there is a vast literature on appropriate ways of discretizing a given differential equation.

The starting point of discretizing an ODE is to recall that the derivative can be considered as a limit of a difference

$$\frac{dy}{dx} = \lim_{\delta \to 0} \frac{y(x+\delta) - y(x)}{\delta}. \tag{1.8}$$

The discretization procedure effectively amounts to going one step back, namely to the object we had before the limit is taken, which is given by the **difference operator** Δ_δ defined by

$$\Delta_\delta y(x) := \frac{y(x+\delta) - y(x)}{\delta}. \tag{1.9}$$

Replacing a derivative by Δ_δ (or by similar discrete objects) often leads to a good approximation to the differential equation, the solution of which can be iterated on a computer. In numerical analysis, more subtle approaches are often needed to capture the correct qualitative behavior of the solution (which may also involve replacing the nonlinear terms by suitably chosen expressions).

Higher derivatives also need to be considered. Since these are also defined by limits, for example

$$\frac{d^2y}{dx^2} = \lim_{\delta \to 0} \frac{y(x+\delta) - 2y(x) + y(x-\delta)}{\delta^2}, \tag{1.10}$$

their discretization requires more points where the function $y(x)$ is evaluated. These operations can be expressed in terms of the **(lattice) shift operator**:

$$T_\delta y(x) = y(x+\delta). \tag{1.11}$$

It is easy to see that the difference operator Δ_δ and its powers can be simply expressed in terms of the shifts T_δ through the following formulae

$$\begin{aligned} \Delta_\delta y(x) &= \frac{1}{\delta}\,(T_\delta - \mathrm{id})\, y(x)\,, \\ \Delta_\delta^2 y(x) &= \frac{1}{\delta^2}\left(T_\delta^2 - 2T_\delta + \mathrm{id}\right) y(x), \\ &\ \vdots \\ \Delta_\delta^n y(x) &= \frac{(-1)^n}{\delta^n} \sum_{j=0}^{n} \binom{n}{j} (-T_\delta)^{n-j}\, y(x) \end{aligned} \tag{1.12}$$

(here and later "id" denotes the identity operator) and thus the nth-order difference operator acting on $y(x)$ can be expressed in terms of the shifted variables $y(x), y(x+\delta), \ldots, y(x+n\delta)$.

One way to obtain discrete dynamics from continuous ones is to take stroboscopic pictures of the flow. We assume that the underlying dynamics is given by an equation such as (1.3), and consider a sequence of values of the solution $y : \mathbb{Z} \to \mathbb{R}^N$ at regularly spaced points, $x + n\delta$, by defining

$$y_n := y(x_0 + n\delta) = y(x). \tag{1.13}$$

Another circumstance in which discrete maps arise is by taking a **Poincaré section** of the orbit of a continuous system; see Figure 1.1. If the orbit revisits the same region of the space of the dependent variable repeatedly, the flow will intersect again and again with some given surface. The intersections give rise to a discrete sequence y_n, where n labels the crossings of the given surface. (Sometimes it is best to keep only those crossings that come from the same side of the surface.)

1.1.3 Ordinary difference equations

In analogy to (1.1) we can, by replacing derivatives with differences, define an equation of the form:

$$\mathcal{F}(y(x), \Delta_\delta y(x), \Delta_\delta^2 y(x), \ldots; x) = 0.$$

Using the replacements (1.12) this can also be rewritten in the form

$$\overline{\mathcal{F}}(y(x), y(x+\delta), y(x+2\delta), \ldots; x) = 0\,, \tag{1.14}$$

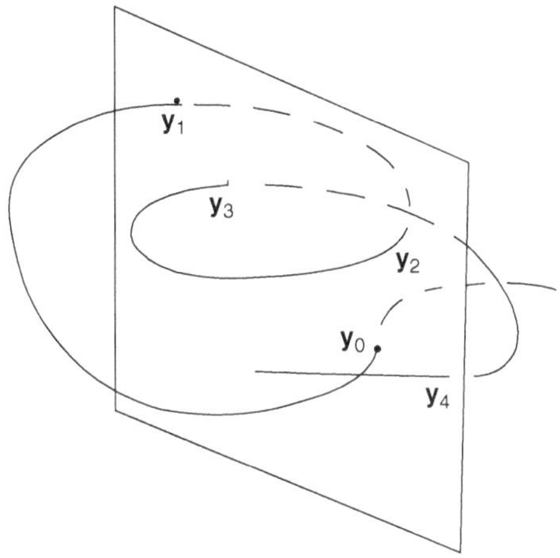

Figure 1.1 Dynamical mapping arising from the Poincaré section of an orbit.

Figure 1.2 An illustration of the points involved in the difference equation (1.15) of order $k + l$.

where the expression $\overline{\mathcal{F}}$ can straightforwardly be obtained from $\mathcal{F}$ by using the above-mentioned substitutions.

At this point, we should ask ourselves what is actually meant by an equation of the form (1.14). In other words, what range of values can the independent variable x take and what would represent a solution of this equation? These are deep questions. For the time being, we could distinguish between the cases where x takes values in a discrete set or where it takes values in a continuous domain. We call the former type of equation a **finite-difference equation** and the latter an **analytic difference equation**.

In the latter case, we can rewrite the equation in different ways; by shifting x, for example

$$\overline{\mathcal{F}}\,(y(x-k\delta), y(x-k\delta+\delta), \ldots, y(x+l\delta-\delta), y(x+l\delta); x) = 0, \tag{1.15}$$

where k, l are fixed integers with $k+l > 0$ (see Figure 1.2). Thus the equation only involves integer multiples of δ in the arguments of the dependent variable $y(x)$, and if $n\delta = x$ and then the point named $n + s$ corresponds to the value $x + s\delta$ of the independent variable.

Analytic difference equations

If the independent variable x is a *continuous* variable (typically x in subdomains of $\mathbb{R}$ or $\mathbb{C}$), solving the difference equation (1.15) would amount to finding a *function* $y(x)$, for x in some appropriate domain. We may want to specify the function class or space to which

y belongs, for example, it may be a meromorphic function of x. However, often we will leave this open, depending on the type of problem we wish to solve.

Clearly, there is also indeterminacy in this problem because a solution is needed for the whole domain, whereas only values at discrete intervals are used in the equation itself. This implies that the initial data has to be given over an entire interval on the real line, and consequently the solution of the difference equations is determined only up to periodic functions, i.e. functions $\pi(x)$ obeying

$$\pi(x+\delta) = \pi(x),$$

which here play the role of "integration constants".

Finite-difference equations

In contrast to the analytic difference case, we may assume that x starts from a fixed base point value (such as 0), in which case it is natural to consider y as a function of the integers $y : \mathbb{Z} \to \mathbb{R}$ or $\mathbb{C}$. We then use the shorthand notation $y_n := y(x + n\delta)$. In this case, we have equations containing the discrete variable y_n and its various shifts $y_{n-k}, y_{n-k+1}, \ldots,$ $y_{n+l-1}, \ldots\ y_{n+l}$. Thus the canonical form of a difference equation is

$$\mathcal{F}(y_{n-k}, y_{n-k+1}, \ldots, y_{n+l-1}, y_{n+l}; n) = 0, \quad \forall n \in \mathbb{Z}. \tag{1.16}$$

This is variously called a *recurrence relation* or *iterative scheme* from which we wish to solve y at discrete points n only. Thus, giving initial values at a sufficient number of points, generically $y_0, y_1, \ldots, y_{k+l-1}$, we can hope to iterate the equation and calculate y_{k+l} and subsequent values step by step.

If we can solve (1.16) for y_{n+l} then the equation provides a **map**

$$(y_{n-k}, y_{n-k+1}, \ldots, y_{n+l-1}) \mapsto y_{n+l},$$

and if we can also solve for y_{n-k} then we have a **reversible map**. If we cannot uniquely solve for the first or last term from equation (1.16) then it is called a *correspondence* and the possibility of multiple ways to proceed must be studied carefully.

The **order** of equation (1.16) is $k+l$, which is the minimal number of initial data required to uniquely define the evolution of y_n as a function of integer n. We will sometimes refer to an equation of the form (1.16) also as a $(k+l+1)$-**point map**, where it is understood that the map is iterated by composing the map with itself. (The iteration scheme could break down if the composition of the map becomes ill-defined. That may occur if the original difference equation becomes noninvertible at certain points in the domain of the independent variable.)

Almost all maps we consider in this book are **rational** maps; that is, the latest iterate is a rational function of previous iterates. We will see that integrable maps turn out to be **birational**; that is, the map is rational in both directions.

Example 1.1.2 A simple example of a finite-difference equation is

$$y_{n+1} = a\, y_n, \tag{1.17}$$

where a is a constant parameter. The explicit general solution is

$$y_n = y_0\, a^n,$$

where y_0 is a free constant. On the other hand, if it is an analytic difference equation

$$y(x+1) = a(x)\, y(x), \tag{1.18}$$

where $a(x+1) = a(x)$, we get the general solution

$$y(x) = \pi_0(x)\, a(x)^x,$$

where $\pi_0(x)$ is a periodic function $\pi_0(x+1) = \pi_0(x)$ but is otherwise free.

The periodic functions that arise in the solution of analytic difference equations may be subject to further conditions that may depend on the context in which the equation arises, such as in physical models.

As before, if $\mathfrak{F}$ is explicitly independent of n (i.e. n does not appear in the equation other than through the dependence of y on n), then the equation is called *autonomous*. Otherwise, it is called *nonautonomous*.

Example 1.1.3 The McMillan map is defined by the difference equation

$$x_{n+1} + x_{n-1} = \frac{2ax_n}{1 - x_n^2}, \tag{1.19}$$

which is an autonomous birational equation.

Example 1.1.4 The first Painlevé equation (1.6) can be discretized as

$$y_{n+1} + y_n + y_{n-1} = \frac{\alpha + \beta n}{y_n} + \gamma, \tag{1.20}$$

which is called "discrete Painlevé I" ($\mathrm{dP_I}$). The "alternate" $\mathrm{dP_I}$ is given by

$$\frac{n+1}{y_{n+1} + y_n} + \frac{n}{y_n + y_{n-1}} = n + a + by_n^2. \tag{1.21}$$

Both maps are clearly birational (i.e. rational in both directions) and nonautonomous.

Example 1.1.5 The recurrence relation for Hermite polynomials is given by

$$H_{n+1}(x) = 2x H_n(x) - 2n H_{n-1}(x), \tag{1.22}$$

which can be interpreted as a (nonautonomous) difference equation with independent variable n and parameter x. Hermite polynomials also solve a differential equation with x as the independent variable and n as a parameter

$$H_n'' - 2x H_n' + 2n H_n = 0, \tag{1.23}$$

where the prime refers to derivation with respect to x.

Finite-difference equations of higher order can also be viewed as dynamical mappings by converting them to multicomponent first-order maps in the same way as is done with ODEs. Assuming that we can solve y_{n+l} uniquely from (1.16), leading to an expression of the form

$$y_{n+l} = F(y_{n-k}, \ldots, y_{n+l-1}),$$

we can write it as a dynamical map

$$\boldsymbol{y}_n \quad \mapsto \quad \boldsymbol{y}_{n+1} = \boldsymbol{F}(\boldsymbol{y}_n),$$

where we have introduced the $(k+l)$-component vectors

$$\boldsymbol{y}_n := (y_{n-k}, \ldots, y_{n+l-1})^t, \quad \boldsymbol{F}(\boldsymbol{y}) := (y_{n-k+1}, \ldots, y_{n+l-1}, F(y_{n-k}, \ldots, y_{n+l-1})).$$

This is one of many ways in which a scalar difference equation, of order $N \geq 2$, can be transformed to a system of N first-order equations. (Note that the converse statement, i.e. that a system of N first-order equations leads to a scalar equation of order N, is not always true.)

1.1.4 Continuum limits

In many circumstances, difference equations arise as discretizations of differential equations, such as in numerical analysis. It makes sense in this context to ask how to recover the differential equation from the difference equation. This can be done by taking a **continuum limit** of the difference equation. Even for those cases where we study the difference equation in its own right, its continuum limit can be useful for identifying the equation (e.g. as a physical model) and to get further insight into the behavior of solutions by comparing the continuous and discrete equations.

Before we can compute a continuum limit we must connect y_n and $y(x)$. We usually think of x as generic point on the real line and n as a generic point in the lattice; see

Figure 1.2. They are connected by taking x to be n steps of length δ from some fixed point x_0, i.e. $x = x_0 + \delta n$. Thus we have the identifications

$$y(x) = y(x_0 + \delta n) = y_n, \tag{1.24a}$$

and as a consequence,

$$y_{n+1} = y(x_0 + \delta(n+1)) = y(x+\delta), \text{ etc.} \tag{1.24b}$$

Given a difference equation of the form (1.15) or (1.16), asking for its continuum limit amounts to making an approximation in which a parameter related to the spacing between successive points becomes infinitesimally small. Thus for $y_{n+k} = y(x + k\delta)$, the limit $\delta \to 0$ leads to a continuum limit of the difference equation. Performing the limit involves expanding $y(x + k\delta)$ as a Taylor series

$$y(x+k\delta) = y(x) + k\delta y'(x) + \tfrac{1}{2}(k\delta)^2 y''(x) + \ldots$$

taking into account the possibility that the parameters of the equation, and the dependent variable y itself for that matter, may explicitly depend on δ or that they may depend on δ in a hidden way. (The latter case corresponds to the point of view where the parameters are yet to be specified.)

Example 1.1.6 (i) Consider the difference equation $y_{n+1} = \alpha y_n(1 - y_n)$. Letting $y_n = y(x)$ we expand $y_{n+1} = y(x+\delta) = y + \delta y' + \ldots$. Then for w, defined by $y(x) = a\delta w(x)$, and with α depending on the expansion parameter δ by $\alpha = 1 + \delta a$, we get

$$\begin{aligned} a\delta w + a\delta^2 w' + \ldots &= (1 + \delta a)a\delta w(1 - a\delta w) \\ &= a\delta w + a\delta^2(aw(1-w)). \end{aligned}$$

Now the first-order terms cancel and equation (1.4) is recovered for w at order δ^2.

(ii) For dP_I (1.20), the continuum limit is more involved. We need to take $y_n = \frac{\gamma}{6}(1 - 2\delta^2 w(x_0 + n\,\delta))$, $\alpha + \beta n = -\frac{\gamma^2}{36}(3 + 2\delta^4 x)$, $(x = x_0 + n\,\delta)$. Then at order δ^4 we get $w'' = 6\,w^2 + x$, which is (1.6) for $w(x)$.

1.1.5 Functional equations

Another way to look at equations of the type (1.14) is to consider the step-size δ no longer to be fixed, but to consider the equation as a problem posed for arbitrary values of x and δ. This point of view leads to **functional equations**, which are equations that relate the value of a function (or more than one function) at one point to its values at any other point. These are equations of the form

$$\mathfrak{F}(f(x), g(x), f(y), \ldots, f(x+y), \ldots, g(x+y+z), \ldots; x, y, z, \ldots) = 0\,, \tag{1.25}$$

where x, y, z, ... take all values in a given domain. This is not just a superficial change in perspective; it changes the problem fundamentally from that of a difference equation. In fact, under general assumptions on these functions (such as continuity, differentiability, etc.), the requirement that the equation holds for all values of the arguments of the functions is often sufficient to almost uniquely fix the functions that solve the functional equation. The notion of initial values is irrelevant in this context.

Example 1.1.7 Consider the functional equation

$$\mathcal{F}(f(x), f(y), f(x+y)) \equiv f(x+y) - f(x)f(y) = 0.$$

Evaluating the equation at $y = 0$ we find $f(x) = f(x)f(0)$ and therefore $f(0) = 1$. Let us denote $f'(0) = \alpha \neq 0$. Computing the y derivative and then evaluating the result at $y = 0$ yields $f'(x) = \alpha f(x)$ so that $f = e^{\alpha x}$.

Example 1.1.8 Consider the functional equation

$$\mathcal{F}(f(x), f(y), g(x), g(y), f(x+y)) \equiv f(x+y) - f(x)f(y) + g(x)g(y) = 0.$$

Under the assumption that f, g are both differentiable for all real values of their arguments, and that f is an *even* function of its argument, one can show that either f, g are trivial or $f(x) = \cos(\alpha x)$ and $g(x) = \sin(\alpha x)$. This is left as an exercise.

Many integrable difference equations can also be interpreted as functional equations. An example is provided by the addition formulas for elliptic functions, which are discussed further in Section 2.2.2. The discrete Painlevé equations have also been interpreted as functional equations; see e.g. Gromak and Tsegel'nik (1994).

1.1.6 Delay-difference equations

Equations that involve iterates in one independent variable and derivatives in another are referred to as differential-difference equations. However, there are dynamical processes that depend on finite time-steps and instantaneous changes in the *same* variable. (This happens, for example, in population models where the population growth in one generation may depend on the numbers in a previous generation as well as the birth rate of the current population.) The resulting equations contain iterates and derivatives of a function in the same variable.

When such a model only contains iterates from a previous time-step, the resulting equation is called a *delay*-difference equation, or delay equation for short. In general, equations may contain both backward and forward iterates as well as derivatives in the same variable. For simplicity, we also describe such equations as "delay" equations, but they can also be called ordinary differential-difference equations. As a first example we mention (Quispel et al., 1992)

$$\partial_\eta u(\eta) = u(\eta)[u(\eta+1) - u(\eta-1) + a].$$

Another example is the coupled system for $H(\eta)$, $G(\eta)$ (Joshi, 2009)

$$-c_0 H(\eta) + \partial_\eta H(\eta) = H(\eta)[G(\eta+1) - G(\eta)], \tag{1.26a}$$

$$p_0 - c_0 G(\eta) + \partial_\eta G(\eta) = 2[H(\eta)^2 - H(\eta-1)^2], \tag{1.26b}$$

where c_0, p_0 are parameters. Both of these equations can be called delay-P_I equations because they have continuum limits to equation (1.6). Other examples of delay-difference equations were given in Levi and Winternitz (1993), and Grammaticos et al. (1993). See also Levi and Winternitz (2006).

1.1.7 Difference equations iterated on curves

So far we have discretized the derivative on a regularly spaced grid, such as in Figure 1.2. There are still other ways to discretize a derivative using more general grids. One way is by using the **q-derivative** or the **Euler–Jackson q-difference operator**:

$$D_q y(z) := \frac{y(qz) - y(z)}{qz - z}. \tag{1.27}$$

Thus, in this case the shift is *multiplicative* as opposed to *additive* and it is easy to see that

$$\frac{y(qz) - y(z)}{qz - z} \to \frac{dy(z)}{dz}, \quad \text{as} \quad q \to 1.$$

Corresponding to the operator D_q one can also define the q-shift operator by

$$T_q y(z) = y(qz), \tag{1.28}$$

and develop a full q-difference calculus.[1]

Example 1.1.9 The first Painlevé equation (1.6) has a q-difference version

$$y(qz)\, y(q^{-1}z) = \frac{-2\,z}{y(z)} + \frac{\mu\, z^2}{y(z)^2}, \tag{1.29}$$

where μ is a parameter. Transforming variables to $w(z) = z^{-1/2}\, y(z)$, then taking $w(z) = 1 + \epsilon^2\, W(t)$, $t = n\,\epsilon$, $\mu = -3$, $q = (1 - \epsilon^5/4)^2$, we arrive at the first Painlevé equation governing $W(t)$, with t as independent variable, in the limit as $\epsilon \to 0$. This example illustrates how the same differential equation (such as (1.6)) may arise as the continuum limit of more than one difference equation (in this case, (1.20), (1.21), as well as (1.29)).

[1] A correspondence with the difference case can be made by setting $f(x) = F(\ln(x))$, $q = e^h \Rightarrow f(qx) = F(\ln(x) + h)$ but this affects the analytic nature of the solutions in a dramatic way.

The q-shift operator relates points on an exponential grid. This can be further generalized to arbitrary grids, such as

$$(\mathcal{D}f)(z) := \frac{f(\psi(z)) - f(\phi(z))}{\psi(z) - \phi(z)}, \tag{1.30}$$

where the functions ψ, ϕ are defined through some algebraic relations. The previous cases are included with $\psi(z) = z + \delta$, $\phi(z) = z$ and $\psi(z) = qz$, $\phi(z) = z$. (For modern developments, see Magnus (2009); Spiridonov and Zhedanov (2007)).

Returning to the q-difference case, the general q-difference equation would be of the form

$$\mathcal{F}(y(z), y(qz), \ldots, y(q^k z); q, z) = 0, \tag{1.31}$$

which if needed could be rewritten in terms of the q-difference operator D_q. Here, both q and z can be complex numbers, although q is given and fixed, while z ranges in a suitable domain in $\mathbb{C}$.

Formally, a q-difference equation can be transformed into an analytic difference equation (1.15) by taking $y_n := y(q^n z_0)$ (cf. (1.13)). However, analytic continuation may markedly alter the types of domains that arise in each case. (For example, if the independent variable of the analytic difference equation lies in a disk, iteration of the equation leads to a linear sequence of possibly overlapping disks, while in the case of (1.31), the iteration leads to a spiral of such disks.) For an autonomous equation, there is no real distinction between a q-difference equation and an analytic difference equation except for the domains on which they are posed. When the equation is nonautonomous, a q-difference interpretation is natural if the equation contains coefficients that can be expressed in terms of q^n.

Sometimes the case when $|q| = 1$ deserves special attention. In this case, the iterates of the independent variable lie on a circle of constant modulus $|z|$.

1.1.8 Partial differential equations

Partial differential equations (PDEs) govern functions of more than one variable. For example, if we have two independent variables, say x and t, a PDE takes on the general form:

$$\mathcal{F}(u, u_x, u_t, u_{xx}, u_{xt}, u_{tt}, \ldots; x, t) = 0, \tag{1.32}$$

where $\mathcal{F}$ is a function of $u(x, t)$ (the dependent variable which depends on both independent variables), and its partial derivatives

$$u_x = \frac{\partial u}{\partial x}, \quad u_t = \frac{\partial u}{\partial t}, \quad u_{xx} = \frac{\partial^2 u}{\partial x^2}, \quad u_{xt} = \frac{\partial^2 u}{\partial x\, \partial t}, \quad u_{tt} = \frac{\partial^2 u}{\partial t^2}, \quad \ldots$$

The general solutions of PDEs in n-independent variables involve functions of $n-1$ variables and, therefore, are often described as infinite-dimensional systems. These "functions of integration" are determined by auxiliary conditions, such as initial or boundary data (or a mix of the two), depending on the type of the PDE.

The PDEs that we consider later in this book are either linear or belong to a class referred to as "soliton equations," which, as we see later, will admit special classes of solitary wave solutions called solitons. The theory of such equations is well developed and the equations have many interesting mathematical properties; see e.g. Ablowitz and Segur (1981); Calogero and Degasperis (1982); Newell (1985); Drazin and Johnson (1989); Ablowitz and Clarkson (1991); Dickey (2003); Hirota (2004) .

Example 1.1.10 Some of the integrable PDEs that we will meet later include the Korteweg–de Vries (KdV) equation

$$u_t = u_{xxx} + 6uu_x, \tag{1.33}$$

the modified Korteweg–de Vries (mKdV) equation

$$v_t = v_{xxx} - 6\,v^2\,v_x, \tag{1.34}$$

the sine-Gordon (sG) equation

$$\theta_{xt} = \sin\,\theta, \tag{1.35}$$

the nonlinear Schrödinger equation (here ψ is complex)

$$i\psi_t = \psi_{xx} + 2\,\sigma|\psi|^2\psi, \tag{1.36}$$

and the Kadomtsev–Petviashvili (KP) equation

$$\partial_x\,(u_{xxx} + 6u_x u - 4u_t) + 3\sigma u_{yy} = 0, \tag{1.37}$$

where $\sigma = \pm 1$, which is a soliton equation in $2+1$ dimensions.

1.1.9 Semi-discrete equations

In this section, we consider differential-difference equations. These are equations in which both a derivative in one variable and a difference in another variable occur. An example is the semi-discrete potential KdV equation (Nijhoff et al., 1983a: Equation (17) with $A = B = -1$)

$$\frac{\partial u_n}{\partial t} = 1 - \frac{2p}{2p + u_{n-1} - u_{n+1}}, \qquad n \in \mathbb{Z}. \tag{1.38}$$

Another example is given by the Toda lattice equation (Toda, 1967, 1989)

$$\frac{\partial^2 q_j}{\partial t^2} = e^{-(q_j - q_{j-1})} - e^{-(q_{j+1} - q_j)}. \tag{1.39}$$

Both of these equations are integrable as semi-discrete soliton equations. Such differential-difference equations have been considered widely in the literature; see e.g. Yamilov (2006).

Rather than considering an equation of the type (1.38) as an equation in terms of two independent variables (namely the variables t and n), we can also think of the discrete variable as a label enumerating components of a multi-component dependent variable u_n. From the latter point of view, we can consider these components as coordinates of particles in the extreme limit that the number of particles goes to infinity. Having made that connection, we could just as well take the number of particles to be finite, in which case the equation becomes a system of ordinary differential equations.

Many such models arise from the physical picture of a chain of particles on a line, interacting pairwise. Thus let $q_j(t)$ be the (time-dependent) position of the jth particle having mass m_j, and assume that the particles have pairwise interaction determined by a potential depending on the distance between the particles, i.e. $V(q_j - q_k)$. If every particle interacts with all particles, then V must be symmetric and Newton's equations of motion are given by

$$m_j \frac{\partial^2 q_j}{\partial t^2} = -\sum_{k \neq j} V'(q_j - q_k). \tag{1.40}$$

If the particles only interact with nearest neighbors, then the equations are given by

$$m_j \frac{\partial^2 q_j}{\partial t^2} = V'(q_{j+1} - q_j) - V'(q_j - q_{j-1}). \tag{1.41}$$

We can think of these as a dynamical systems in $1+1$ dimensions: one dimension relates to the continuous time t and the other to the discrete particle label j.

We would like to highlight two very important equations of the above type: the Calogero–Moser (CM) equation (Calogero, 1969; Moser, 1975) is of the type (1.40), usually with inverse square potential $V(q) = \alpha/q^2$ and identical masses:

$$\frac{\partial^2 q_j}{\partial t^2} = -\sum_{k \neq j} \frac{2\alpha}{(q_j - q_k)^3}. \tag{1.42}$$

Choosing the potential $V(q) = e^{-q}$ for the chain (1.41) yields the Toda chain (1.39). Both of these models can also be generalized in several ways while preserving integrability (Olshanetsky and Perelomov, 1981) and even quantized.

The range of the particle label j above can be bounded or unbounded. In the bounded case, the particles at the ends of the chain can be free (i.e. of "molecule type") or have fixed, given positions. The particles could also form a closed chain, i.e. there is a periodicity condition of the type $q_{j+N} = q_j$, and so the particles can be viewed as forming a *periodic chain* of length N.

We will discuss semi-discrete equations further in the context of continuum limits of fully discrete equations in Chapter 5.

1.1.10 Partial difference equations

As before, we distinguish PΔEs by the nature of their (now multiple) independent variables. In the case of partial *analytic* difference equations, the independent variables could, for example, be $x, t \in \mathbb{R}$, but the dependent variable $u(x,t) : \mathbb{R}^2 \to \mathbb{R}$ would be evaluated at discrete shifts $u(x+n\delta, t+m\varepsilon)$. The equations would take the form

$$\mathcal{F}(u(x-k_1\delta, t-k_2\varepsilon), \ldots, u(x+l_1\delta, t+l_2\varepsilon)\,; x, t) = 0, \quad \forall x, t, \tag{1.43}$$

where k_1, k_2, l_1, l_2 are fixed integers.

In the finite-difference case the independent variables only take integer values, $n, m \in \mathbb{Z}$, $u : \mathbb{Z}^2 \to \mathbb{R}$ and the equation would be of the form

$$\mathcal{F}(u_{n-k_1,m-k_2}, \ldots, u_{n+l_1,m+l_2}\,; n, m) = 0, \quad \forall n, m. \tag{1.44}$$

Formally these two cases are related by

$$u_{n,m} = u(x_0 + n\delta, t_0 + m\varepsilon) = u(x,t)\ ,$$

but obviously the problems are quite different in nature, since in the analytic case we want to solve for $u(x,t)$ as a function of a continuous range of values of its arguments.

We can think of the independent variables (n,m) as integer points in $\mathbb{R}^2$, forming a rectangular lattice or *Cartesian lattice*, and the dependent variable as being defined (only) on those lattice points. Later we will consider higher-dimensional generalization, but it is also possible to formulate dynamics in other kinds of lattices or even on arbitrary planar graphs (Adler, 2001).

Example 1.1.11 The discrete KdV is given by (Hirota, 1977a)

$$Q_{n,m+1} - Q_{n+1,m} = (p^2 - q^2)\left(\frac{1}{Q_{n,m}} - \frac{1}{Q_{n+1,m+1}}\right) \tag{1.45}$$

and in its potential form by

$$(p - q + u_{n,m+1} - u_{n+1,m})(p + q + u_{n,m} - u_{n+1,m+1}) = p^2 - q^2. \tag{1.46}$$

The equation of "similarity constraint" for KdV is given by

$$\left(\lambda(-1)^{n+m} + 1/2\right)u_{n,m} + \frac{np^2}{2p + u_{n-1,m} - u_{n+1,m}} + \frac{mq^2}{2q + u_{n,m-1} - u_{n,m+1}} = 0, \tag{1.47}$$

which is not integrable by itself, but appears in the context of reductions of integrable quadrilateral equations; see Section 10.4.1.

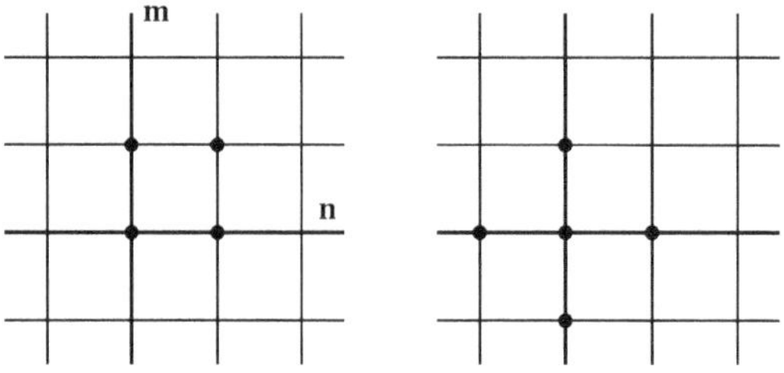

Figure 1.3 A difference equation may involve a different configuration of points in the lattice. Case (a) corresponds to (1.45), (1.46) and case (b) to (1.47).

The configuration of points that are involved in the above two-dimensional difference equations (often called the "stencil" of the equation) are given in Figure 1.3.

Example 1.1.12 Hirota's DAGTE-equation (Hirota, 1981) is the three-dimensional equation

$$\alpha\, \tau_{n+1,m,k}\tau_{n,m+1,k+1} + \beta\, \tau_{n,m+1,k}\tau_{n+1,m,k+1} + \gamma\, \tau_{n,m,k+1}\tau_{n+1,m+1,k} = 0, \tag{1.48}$$

where the dependent variable is τ and there are three independent variables n, m, k. In the above, $p, q, \lambda, \alpha, \beta, \gamma$ can be considered as constant parameters.

In Section 1.3 we will discuss the classification PΔEs and their initial/boundary value problems.

1.2 The Riccati equation

Sometimes a nonlinear difference equation can be *linearized*, i.e. rendered into a linear equation by a suitable change of variables, or can be solved by applying other tricks. A case where this happens is the important example of the *discrete Riccati equation*, i.e. a difference analogue of the famous Riccati differential equation, named after the Count Jacopo Francesco Riccati (1676–1754). Since we will often encounter both the discrete as well as the continuous Riccati equation in the context of integrable systems, we will present here a few properties of both equations.

Figure 1.4 Count J. F. Riccati

1.2.1 Continuous Riccati equation

The usual Riccati differential equation (see Reid, 1972) is a first-order nonlinear differential equation of the form:

$$\frac{dy}{dx} = a(x)y^2 + b(x)y + c(x)\,, \tag{1.49}$$

i.e. it possesses a quadratic nonlinearity. The coefficients $a(x)$, $b(x)$ and $c(x)$ can be arbitrary functions of x. In the case that they are constants, the equation (1.49) can be solved simply by separation of variables.

In the general case, when the coefficients are not necessarily constant, separation of variables will not always work, but instead we can rely on a number of special properties, the most important of which we will now briefly summarize.

First, the Riccati equation (1.49) preserves its form under **fractional linear transformations** also referred to as **Möbius transformations**, namely under transformations of the type:

$$y(x) \mapsto \widetilde{y}(x) = \frac{\alpha y(x) + \beta}{\gamma y(x) + \delta} \quad \Leftrightarrow \quad y(x) = \frac{\delta \widetilde{y}(x) - \beta}{-\gamma \widetilde{y}(x) + \alpha}\,, \tag{1.50}$$

where $\alpha, \beta, \gamma, \delta$ are constants such that $\alpha\delta - \beta\gamma \neq 0$. In fact, implementing the transformation (1.50) turns (1.49) into an equation for $\widetilde{y}$ of the form

$$\frac{d\widetilde{y}}{dx} = \widetilde{a}(x)\widetilde{y}^2 + \widetilde{b}(x)\widetilde{y} + \widetilde{c}(x)\,.$$

with new coefficients $\widetilde{a}(x)$, $\widetilde{b}(x)$, $\widetilde{c}(x)$.

Second, the Riccati equation can be **linearized** by setting: $y(x) = f(x)/g(x)$, substitution of which leads to

$$gf' - fg' = af^2 + bfg + cg^2.$$

Since y can be written in an infinite number of ways as a ratio of the form f/g, we have the freedom to fix one of the two variables as we choose and then adapt the other variable accordingly. This allows us to split the latter relation into two linear equations, leading to the system:

$$\begin{cases} f' = \lambda f + cg, \\ g' = -af + (\lambda - b)g, \end{cases} \tag{1.51}$$

where λ can be chosen arbitrarily. Choosing λ in appropriate way, e.g. as a constant, and eliminating either g or f from the system (1.51), we can derive a second-order linear ODE for the remaining function f or g respectively.

Third, we note that any four solutions $y_i(x)$, $(i = 1, 2, 3, 4)$ of the same Riccati equation, i.e. having the same coefficient functions $a(x)$, $b(x)$ and $c(x)$, are related through the **cross-ratio relation**

$$\mathrm{CR}[y_1, y_2, y_3, y_4] := \frac{(y_1 - y_2)(y_3 - y_4)}{(y_1 - y_3)(y_2 - y_4)} = \text{constant}, \tag{1.52}$$

where CR stands for **cross-ratio** of four points in the complex plane. In fact, for any two solutions y_i, y_j of the Riccati equation, it is easy to derive:

$$\frac{d}{dx} \ln(y_i - y_j) = a(y_i + y_j) + b \,,$$

and, hence, the combination

$$\frac{d}{dx} \ln[(y_1 - y_2)(y_3 - y_4)]$$

is invariant under permutations of the indices.

Last, given two independent solutions $y_1(x)$, $y_2(x)$ of the same Riccati equation (1.49), a new solution can be found through the **interpolation formula**:

$$y(x) = \frac{y_1 + \rho y_2}{1 + \rho} \quad \text{provided} \quad \rho(x) \text{ solves } \frac{d\rho}{dx} = a(y_1 - y_2)\rho \,. \tag{1.53}$$

Thus, given any two special solutions of the Riccati equation, the general solution can be found by interpolating between those solutions through the general solution of the above linear equation for ρ. (This is a nonlinear analogue of the superposition principle for constructing the *general solution* of a second-order homogeneous linear differential equation from two independent special solutions.)

1.2.2 Discrete Riccati equation

The Riccati equation (1.49) has a natural discrete analogue, the **discrete Riccati equation** given by the first-order difference equation

$$y_n y_{n+1} + a_n y_{n+1} + b_n y_n + c_n = 0, \tag{1.54}$$

where the coefficients a_n, b_n, c_n are given functions of n. In fact, it is sometimes more natural to introduce the form

$$R(y_n, y_{n+1}) = C_n y_n y_{n+1} + D_n y_{n+1} - A_n y_n - B_n = (-1, y_{n+1}) \begin{pmatrix} A_n & B_n \\ C_n & D_n \end{pmatrix} \begin{pmatrix} y_n \\ 1 \end{pmatrix}, \tag{1.55}$$

and denote by $R(y_n, y_{n+1}) = 0$ the discrete Riccati equation. It follows immediately from this form that the discrete Riccati equation can be linearized as

$$\begin{pmatrix} y_{n+1} \\ 1 \end{pmatrix} = \lambda \begin{pmatrix} A_n & B_n \\ C_n & D_n \end{pmatrix} \begin{pmatrix} y_n \\ 1 \end{pmatrix}, \tag{1.56}$$

where $\lambda = (C_n y_n + D_n)^{-1}$. Furthermore, solving for y_{n+1}, we obtain the fractional linear form:

$$y_{n+1} = \frac{A_n y_n + B_n}{C_n y_n + D_n}, \tag{1.57}$$

which reminds us of the Möbius transformation (1.50). In other words, the Möbius transformation, interpreted as a map $y_n \mapsto y_{n+1}$, is equivalent to the iteration of solutions of the discrete Riccati equation.

We recover the differential Riccati equation (1.49) from the discrete one of the form (1.57) by performing a continuum limit as described in Section 1.1.4. We use (1.24) and expand in δ, but at the same time we have to fix the δ dependence of the coefficients A_n, B_n, C_n, D_n as follows

$$\frac{C_n}{D_n} = -a(x)\delta + O(\delta^2), \quad \frac{A_n - D_n}{D_n} = b(x)\delta + O(\delta^2), \quad \frac{B_n}{D_n} = c(x)\delta + O(\delta^2).$$

One can then show that from the limit described above one indeed recovers, at order δ^1, the continuous Riccati equation in the form (1.49).

Let us return to the discrete Riccati equation in the form (1.54). Since it occurs often in later chapters of this book, we collect some important observations about this equation here. These are analogues of the properties given before for the continuous Riccati equation.

The equation is form-invariant under the Möbius transformation. That is, under the transformation

$$y_n \mapsto \widetilde{y}_n = \frac{\alpha y_n + \beta}{\gamma y_n + \delta}, \quad \alpha\delta - \beta\gamma \neq 0,$$

the equation (1.54) goes over into an equation for $\widetilde{y}_n$ of the same form with coefficients $\widetilde{a}_n$, $\widetilde{b}_n$, $\widetilde{c}_n$.

By setting $y_n = f_n/g_n$ in (1.54) the equation can be *linearized* leading to the linear matrix equation:

$$\begin{pmatrix} f_{n+1} \\ g_{n+1} \end{pmatrix} = \kappa \begin{pmatrix} b_n & c_n \\ -1 & -a_n \end{pmatrix} \begin{pmatrix} f_n \\ g_n \end{pmatrix}, \tag{1.58}$$

in which κ can be chosen arbitrarily. It is easy to derive from this matrix equation a second-order linear difference equation for f_n or g_n by eliminating the other variable.

For any two different solutions $y_n^{(i)}$ and $y_n^{(j)}$ we have the following relation

$$\frac{y_{n+1}^{(i)} - y_{n+1}^{(j)}}{y_n^{(i)} - y_n^{(j)}} = \frac{c_n - a_n b_n}{(a_n + y_n^{(i)})(a_n + y_n^{(j)})},$$

and hence we conclude that

$$\frac{(y_{n+1}^{(1)} - y_{n+1}^{(2)})(y_{n+1}^{(3)} - y_{n+1}^{(4)})}{(y_n^{(1)} - y_n^{(2)})(y_n^{(3)} - y_n^{(4)})} = \frac{(c_n - a_n b_n)^2}{(a_n + y_n^{(1)})(a_n + y_n^{(2)})(a_n + y_n^{(3)})(a_n + y_n^{(4)})},$$

for any four solutions $y_n^{(1)}$, $y_n^{(2)}$, $y_n^{(3)}$, $y_n^{(4)}$. Consequently, by permuting the (2) and (3) superscripts we infer that any *four* solutions $y_n^{(1)}$, $y_n^{(2)}$, $y_n^{(3)}$, $y_n^{(4)}$ of the discrete Riccati equation satisfy the cross-ratio equation

$$\mathrm{CR}[y_n^{(1)}, y_n^{(2)}, y_n^{(3)}, y_n^{(4)}] = \frac{(y_n^{(1)} - y_n^{(2)})(y_n^{(3)} - y_n^{(4)})}{(y_n^{(1)} - y_n^{(3)})(y_n^{(2)} - y_n^{(4)})} = \text{constant}\,. \tag{1.59}$$

Given two independent solutions $y_n^{(1)}$, $y_n^{(2)}$ the combination

$$y_n = \frac{y_n^{(1)} + \rho_n y_n^{(2)}}{1 + \rho_n}$$

is again a solution of the discrete Riccati equation provided that ρ_n solves the linear difference equation

$$(y_n^{(2)} y_{n+1}^{(1)} + a_n y_{n+1}^{(1)} + b_n y_n^{(2)} + c_n)\rho_n + (y_n^{(1)} y_{n+1}^{(2)} + a_n y_{n+1}^{(2)} + b_n y_n^{(1)} + c_n)\rho_{n+1} = 0\,.$$

By using again the discrete Riccati equation for $y_n^{(1)}$, $y_n^{(2)}$, we can deduce from this that this linear equation can be written in either of the two equivalent forms:

$$\rho_{n+1} = -\frac{R(y_n^{(2)}, y_{n+1}^{(1)})}{R(y_n^{(1)}, y_{n+1}^{(2)})} = \frac{y_n^{(2)} + a_n}{y_n^{(1)} + a_n}\rho_n = \frac{y_{n+1}^{(1)} + b_n}{y_{n+1}^{(2)} + b_n}\rho_n\,.$$

Example 1.2.1 An alternative, more direct way to linearize the Riccati equation (1.54) is by rewriting the equation as

$$(y_n + a_n)(y_{n+1} + b_n) + c_n - a_n b_n = 0\,, \tag{1.60}$$

which suggests a substitution of the form

$$y_n + a_n = \frac{w_n}{w_{n-1}} \quad \Rightarrow \quad w_{n+1} + (b_n - a_{n+1})w_n + (c_n - a_n b_n)w_{n-1} = 0\,,$$

which brings us directly to a linear second-order difference equation for w_n.

1.3 Partial difference equations

Partial difference equations (PΔE) arise, e.g. in the context of numerical methods for PDEs, where they correspond to finite-difference schemes on lattices; cf. the monographs by Garabedian (1964) and Hildebrand (1968). Such problems were in fact already considered by Courant et al. (1928) as approximation techniques for equations arising in mathematical physics. In this section we discuss how PΔEs can be used to approximate well-known PDEs, as well as consider aspects of their initial and boundary value problems.

There exists a classification of linear PDEs that separates hyperbolic, parabolic and elliptic equations, which leads to well-known canonical forms of second-order equations, such as the wave equation, heat equation and Laplace's equation. Below, we show how to discretize these canonical forms to arrive at corresponding PΔEs. However, we do not know of a classification of linear PΔEs in full generality.

Related to the question of classification, there is the issue of whether solutions of a given PΔE arise from boundary value problems or from initial value problems. A PΔE of the form

$$F(u_{n,m}, u_{n+1,m}, u_{n,m+1}, u_{n+1,m+1}) = 0, \tag{1.61}$$

which relates the values of u at the corners of a quadrilateral has features reminiscent of *hyperbolic* PDEs. If from equation (1.61) we can solve uniquely one value of u in terms of the other three (say, the value at $(n+1, m+1)$ in terms of the values at (n, m), $(n+1, m)$, $(n, m+1)$), then on a Cartesian lattice we may pose an *initial value problem* by giving the values of u on a "sawtooth" or a "ladder," such as given in Figure 1.5, or on a corner.

An example of the form (1.61) is equation (1.45), whose solution can be thought of as evolving by iteration from the initial values on a sawtooth, in a way very similar to solutions of equations such as the KdV equation (1.33) for which the solution evolves from initial data $u(x, 0)$ given on a line $t = 0$.

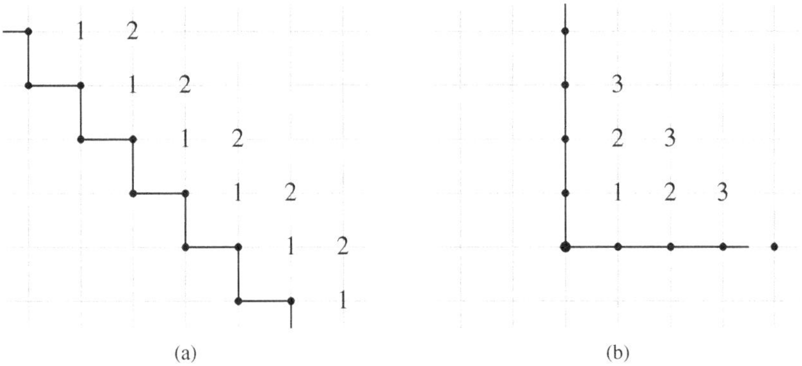

Figure 1.5 (a) Evolution from initial values given on the staircase. The initial data is given on the black points after which the values at "1" can be computed using the equation, then values at "2," etc. (b) Evolution from initial values given on a corner.

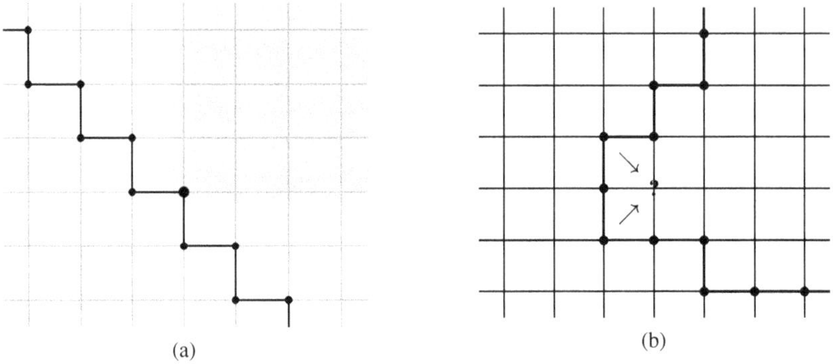

Figure 1.6 (a) The influence propagating from a given initial value for an equation of the type (1.61) forms a "light-cone." (b) If initial values are given in each other's light cone, trouble follows.

The initial data being given on a sawtooth rather than on a straight line arises from a subtle distinction between the continuous and discrete cases. In fact, if we give an initial value problem on a straight line, such as $u_{n,0}$, for all n, for equation (1.61), this leads to a *nonlocal* problem: to calculate any value $u_{n,1}$, say, for given n, we need three previous values of u, i.e. $u_{n,0}$, $u_{n-1,0}$ as well as $u_{n-1,1}$. The last of these values requires in turn the values of u further to the left of n, i.e. $u_{n-1,0}$, $u_{n-2,0}$, $u_{n-2,1}$ and so on. This means the iteration of $u_{n,m}$ to the line $m = 1$ would need to involve all initial values $u_{n',0}$ with $n' < n$ and some information on the asymptotic value $u_{n,1}$ as $n \to \infty$. In contrast, on a sawtooth given by two parallel lines $n - m = 0$, $n - m = 1$ in the lattice (see Figure 1.5), the evolution is a local process: the values of u at the next parallel line $n - m = 2$ are obtained by using three initial values of u on an elementary square. The same holds whenever the initial values are given on a line where steps in the horizontal and vertical directions alternate, without reversing direction.

Hyperbolic PDEs have another feature that is shared with their discrete analogues. The domain on dependence of the solution at any point (x, t) encapsulates a "memory" of the initial values that contributed to the evaluation of the solution u at that point. In the discrete case and for an equation like (1.61), defined on a quadrilateral, the influence of a given initial value can be inferred from Figure 1.6a: a change in the initial value at the lower left corner of the shadowed region influences by equation (1.61) all points in the shadowed quadrant, which in the present context is the "light-cone". This also means that giving initial values inside the light-cone of some other initial values leads to trouble; see Figure 1.6b.

A discussion of admissible configurations and initial value problems for equations of the type (1.61) can be found in Adler and Veselov (2004), van der Kamp and Quispel (2010) and van der Kamp (2015), and for lattice equations on more general stencils in van der Kamp (2009).

1.3.1 Classification of linear partial difference equations

In numerical analysis, PΔEs arise as difference approximations which are based on difference quotients used to represent partial derivatives of smooth functions. Suppose we are given a point (x, t) and a function $u(x, t)$ that has a convergent Taylor series expansion around that point. Then, for sufficiently small δ and ε, we have

$$\begin{aligned} u(x+\delta, t+\varepsilon) = u(x,t) &+ \delta\, u_x(x,t) + \varepsilon\, u_t(x,t) \\ &+ \frac{\delta^2}{2} u_{xx}(x,t) + \delta\,\varepsilon\, u_{xt}(x,t) + \frac{\varepsilon^2}{2} u_{tt}(x,t) \\ &+ \mathcal{O}\big(|\delta|^3 + |\varepsilon|^3\big), \end{aligned}$$

where the meaning of the big-$\mathcal{O}$ notation is standard ($f = \mathcal{O}(g)$, in the limit $g \to 0$, if $\lim_{g\to 0} |f/g|$ is bounded). Below, we will drop absolute values from the errors in statements where the meaning is clear.

From the Taylor expansion, we have the following results for the derivatives of u in terms of u at incremented arguments:

$$u_x(x,t) = \frac{1}{\delta}\,(u(x+\delta,t) - u(x,t)) + \mathcal{O}(\delta) \tag{1.62a}$$

$$u_t(x,t) = \frac{1}{\varepsilon}\,(u(x,t+\varepsilon) - u(x,t)) + \mathcal{O}(\varepsilon) \tag{1.62b}$$

$$u_{xx}(x,t) = \frac{1}{\delta^2}\,(u(x+\delta,t) - 2\,u(x,t) + u(x-\delta,t)) + \mathcal{O}(\delta^2) \tag{1.62c}$$

$$u_{tt}(x,t) = \frac{1}{\varepsilon^2}\,(u(x,t+\varepsilon) - 2\,u(x,t) + u(x,t-\varepsilon)) + \mathcal{O}(\varepsilon^2) \tag{1.62d}$$

$$\begin{aligned} u_{xt}(x,t) = \frac{1}{4\,\delta\,\varepsilon}\,&(u(x+\delta,t+\varepsilon) - u(x+\delta,t-\varepsilon) \\ &+ u(x-\delta,t-\varepsilon) - u(x-\delta,t+\varepsilon)) + \mathcal{O}(\delta^2+\varepsilon^2). \end{aligned} \tag{1.62e}$$

We use these identities to find partial difference equations that are "close" to the canonical ones that arise from classification of linear PDEs. To do so, we will replace x and t by $x = n\,\delta$, $t = m\,\varepsilon$ and $u(x, t)$ by $u_{n,m} = u(n\delta, m\varepsilon)$ on a lattice.

Hyperbolic equations

Consider the wave equation

$$u_{xx} - u_{tt} = 0, \tag{1.63}$$

with initial values

$$\begin{aligned} u(x,0) &= f(x) \\ u_t(x,0) &= g(x). \end{aligned}$$

Assume that f and g are differentiable as many times as needed.

Using the replacements suggested by Taylor expansions (as explained above), we get the "discrete wave equation"

$$\frac{1}{\delta^2}\left(u_{n+1,m}-2\,u_{n,m}+u_{n-1,m}\right)-\frac{1}{\varepsilon^2}\left(u_{n,m+1}-2\,u_{n,m}+u_{n,m-1}\right)=0,$$

or

$$u_{n,m+1}=2\,u_{n,m}-u_{n,m-1}+\frac{\varepsilon^2}{\delta^2}\left(u_{n+1,m}-2\,u_{n,m}+u_{n-1,m}\right). \tag{1.64}$$

Consider the initial values. We replace $f(x)$ by $f(n\delta)=f_n$ and $g(x)$ by $g(n\delta)=g_n$. We can show (see Exercises) that the appropriate discretization of the initial values are given by

$$u_{n,0}=f_n, \tag{1.65a}$$

$$u_{n,1}=f_n+g_n\varepsilon+\frac{\varepsilon^2}{2}f_n'', \tag{1.65b}$$

for all n. Equation (1.64) is often referred to as a "five-point" equation, because it relates the values of u at five neighboring points on the lattice. (See Figure 1.7b.)

Example 1.3.1 Consider the square lattice given by $\delta=\varepsilon$ and assume the initial conditions in (1.65) are given by $f_n=1$ and $g_n=\left(1+(-1)^n\right)/2$, for all m. Then using the equation

$$u_{n,m+1}=u_{n+1,m}-u_{n,m-1}+u_{n-1,m}$$

we can find the values of u along the line $n=2$ by using the initial values given along the lines $n=0$ and $n=1$ on the right-hand side of this equation. For example, $u_{3,2}=u_{4,1}-u_{3,0}+u_{2,1}=2-1+2=3$. We can then use the values of u along $n=2$ and those along $n=1$ to get the values of u along the next line $n=3$, etc.

Parabolic equations

Consider the heat equation

$$u_{xx}-u_t=0. \tag{1.66}$$

By using equations (1.62c) and (1.62b), we get the discrete analogue

$$\frac{1}{\delta^2}\left(u_{n+1,m}-2\,u_{n,m}+u_{n-1,m}\right)-\frac{1}{\varepsilon}\left(u_{n,m+1}-u_{n,m}\right)=0.$$

That is, we find the discrete heat equation

$$u_{n,m+1}=u_{n,m}+\frac{\varepsilon}{\delta^2}\left(u_{n+1,m}-2\,u_{n,m}+u_{n-1,m}\right). \tag{1.67}$$

This discrete equation is a four-point scheme illustrated in Figure 1.7a.

Example 1.3.2 Suppose $\varepsilon = \delta^2$. Then we have

$$u_{n,m+1} = -u_{n,m} + u_{n+1,m} + u_{n-1,m}.$$

Assume the initial value is $u_{n,0} = f_n = \left(1 + (-1)^n\right)/2$, then the equation gives the values of $u_{n,1} = \left(1 - 3\,(-1)^n\right)/2$, $u_{n,2} = \left(1 + 9\,(-1)^n\right)/2$ and so on. Notice that the absolute values of u are increasing fast as m increases.

Elliptic equations

Consider Laplace's equation

$$u_{xx} + u_{tt} = 0. \tag{1.68}$$

By using equations (1.62c) and (1.62d), we get the discrete analogue

$$\frac{1}{\delta^2}\left(u_{n+1,m} - 2\,u_{n,m} + u_{n-1,m}\right) + \frac{1}{\varepsilon^2}\left(u_{n,m+1} - 2\,u_{n,m} + u_{n,m-1}\right) = 0.$$

Note that this implies the "discrete Laplace's equation"

$$2\left(\frac{1}{\delta^2} + \frac{1}{\varepsilon^2}\right) u_{n,m} = \frac{1}{\delta^2}\left(u_{n+1,m} + u_{n-1,m}\right) + \frac{1}{\varepsilon^2}\left(u_{n,m+1} + u_{n,m-1}\right). \tag{1.69}$$

Remark: On a square grid, where $\varepsilon = \delta$, we have

$$u_{n,m} = \frac{1}{4}\left(u_{n+1,m} + u_{n-1,m} + u_{n,m+1} + u_{n,m-1}\right). \tag{1.70}$$

So the values of u at a grid point coincide with the average of the values at the four neighboring grid points indicated in Figure 1.7b. This automatically implies that the values $u_{n,m+1}$, $u_{n+1,m}$, $u_{n,m-1}$, $u_{n-1,m}$ cannot all be greater or smaller than $u_{n,m}$. That is, $u_{n,m}$ cannot be a maximum or minimum in the interior of the lattice.

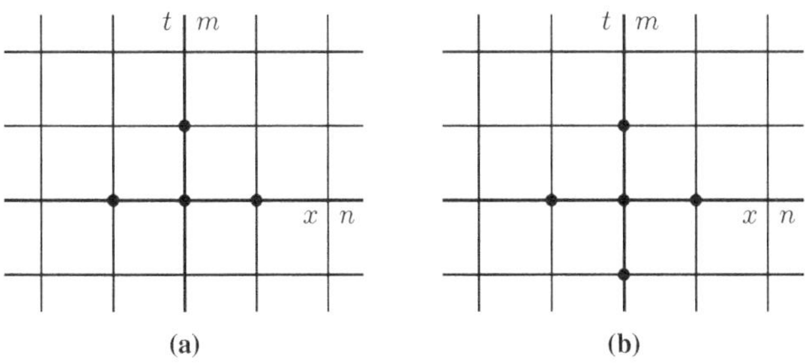

Figure 1.7 Points involved (a) in equation (1.67) and (b) in (1.64) and (1.69).

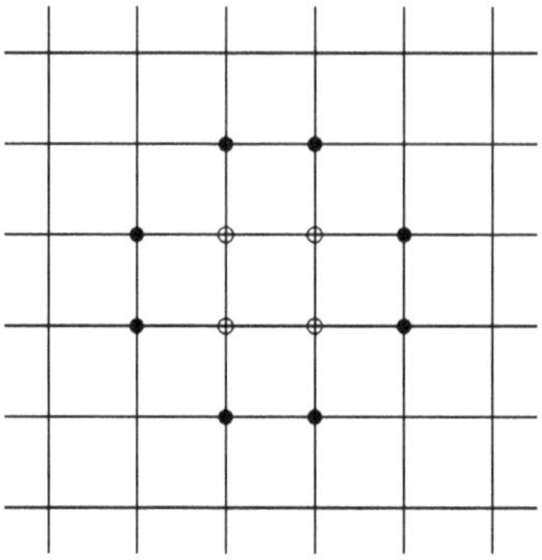

Figure 1.8 A boundary value (or Dirichlet) problem: values at black disks are given and the values at the open circles are to be determined by four equations centered at the open circles.

We have called equation (1.64) the discrete wave equation, (1.67) the discrete heat equation and (1.69) the discrete Laplace equation by analogy with the PDEs for which they provide discretizations. But it is worth noting that, in general, there is no classification of PΔEs as hyperbolic, parabolic or elliptic equations. Equations (1.64) and (1.69) both involve five points on a lattice arranged in a cross-like formation; however, the nature of the associated conditions for which they can be generically solved are quite different: for (1.64), these are initial values (1.65), while for (1.69), they are boundary values. A typical acceptable boundary value problem for an elliptic equation is illustrated in Figure 1.8.

1.3.2 Separation of variables

For practical calculations it is often useful to introduce two shift operators,

$$Ty_{n,m} = y_{n+1,m}, \qquad Sy_{n,m} = y_{n,m+1}.$$

For example the discrete heat equation (1.67) can then be written as

$$Sy_{n,m} = (\alpha T + \beta + \alpha T^{-1})y_{n,m}.$$

If we have now the initial value given as $y_{n,0} = f_n$ then we find the solution in the form

$$y_{n,m} = (\alpha T + \beta + \alpha T^{-1})^m f_n.$$

For linear constant-coefficient PDEs the ansatz $y(x,t) = f(ax + bt)$ often leads to a solution if the parameters a, b are chosen suitably. If the terms in the equation have the same total order we get a characteristic equation for the parameters, while the function f can be arbitrary; in other cases we can reduce the equation to an ODE.

In the discrete case the above works much less frequently. It is useful if the points involved fall on a line; we can then change to new variables and reduce the problem to an OΔE.

Example 1.3.3 Consider the equation

$$y_{n,m} = r\, y_{n+1,m-2} + s\, y_{n+2,m-4}.$$

If we now define new variables as follows:

$$k = n, \quad l = 2n + m, \quad y_{n,m} = w_{k,l},$$

the equation becomes

$$w_{k,l} = r\, w_{k+1,l} + s\, w_{k+2,l},$$

which can be considered as an OΔE in k. Let now $f_i(k)$, $\mathrm{i} = 1, 2$, be the two solutions of this equation (as discussed in Section A.2.1); then the solution of the original equation is

$$y_{n,m} = h_1(2n + m) f_1(n) + h_2(2n + m) f_2(n),$$

where the h_i, $\mathrm{i} = 1, 2$, are arbitrary functions.

Other classes of PΔEs will be considered in subsequent chapters, motivated by the context of integrable systems.

The continuum limit is also much more complicated for PΔEs than for OΔEs. We return to this problem later; at this point it suffices to note that sometimes we may have to take the continuum limit along a diagonal of the lattice.

1.4 Notes

The basic theory of difference calculus, including the elementary theory of difference equations, is outlined in Appendix A. The classical texts on difference equations are Nörlund (1954) and Milne-Thomson (2000). Modern approaches include Garabedian (1964), Hildebrand (1968), Bender and Orszag (1978) and Kelley and Peterson (2001). A useful source of examples is Spiegel (1971).

The theory of linear *analytic difference equations* was developed in the beginning of the twentieth century by the school of George D. Birkhoff and his collaborators W. J. Trjitzinsky, R. Carmichael, J. LeCaine; see Carmichael (1912); Birkhoff (1913); Trjitzinsky (1933); LeCaine (1943). This school developed systematic methods for the Fuchsian theory of linear difference equations (some of which is represented in the monograph by Nörlund (1954)). Unfortunately, the work in this direction was interrupted by the Second World War, but it was revived in recent times by J.-P. Ramis and his students (Sauloy, 1999; Di Vizio et al., 2003; Ramis et al., 2013).

An alternative approach to difference equations is provided by difference algebras, which goes back to the works of J. F. Ritt (Ritt and Doob 1933; Ritt 1934). Through the work of E. R. Kolchin (1973) this was further developed in the 1960s to difference Galois theory. For a modern account of these developments, see the monographs of van der Put and Singer (1997) and Levin (2008).

The Riccati equation has a long history (cf. e.g. Reid (1972)) and has close connections to the classical theory of special functions and differential equations in the complex domain (Hille, 1976). A group theoretical picture closely related to the theory of integrable PΔEs was developed in Atkinson (2008b). The discrete Riccati equation was already considered by Wallenberg (1917, 1918).

The history of partial difference equations was initially mostly motivated by the problem of creating finite-difference schemes for numerical approximations to PDEs; cf. e.g. Garabedian (1964); Hildebrand (1968). Setting up such finite-difference schemes for integrable PDEs led to the pioneering discoveries of PΔEs which possess integrable characteristics themselves (Ablowitz and Ladik 1976, 1977; Taha 1991), and they will be discussed at length in later chapters. The problem of classification of linear difference operators according to the parallel of hyperbolic, parabolic and elliptic classes of the continuous case was discussed e.g. in Nieszporski and Santini (2005); Nieszporski (2007), where it was noted that, in spite of some similarities, such distinctions in the discrete domain are rather more subtle and not quite as clear-cut as in the continuous situation.

Exercises

1.1 (a) Solve the equation (1.4) by the standard method of separation of variables. (Solution: $y(t) = y_0/(y_0 + (1 - y_0)e^{-\alpha(t-t_0)}$.)

(b) Linearize the Verhulst equation with the substitution $y = 1/(1+w)$, which leads to $w' + \alpha w = 0$.

(c) Show that the "naive" discretization $y' \to (y(t+h) - y(t))/h$ leads to a version of the discrete equation $y(t + h) = \alpha y(t)(\beta - y(t))$, also known as the logistic map, which is known to have chaotic behavior.

(d) Consider the case $\alpha = 2$ and $\beta = 1$ in part (c) and solve it by substituting $y(t) = (1 - u(t))/2$.

(e) Discretize the derivative in the linearized equation the "naive" way, and w itself by $\lambda w(t + h) + (1 - \lambda)w(t)$, and solve for $w(t + h)$. Find then the general solution. (Answer: $w(t + nh) = w(t)\left(\frac{1+\alpha h(\lambda-1)}{1+\alpha h\lambda}\right)^n$.)

(f) It seems that $\lambda = 1/2$ gives a particularly symmetric result. Starting with the previous results, return to y and show that this choice corresponds to the discretization

$$\frac{y(t+h) - y(t)}{h} = \alpha\left[\tfrac{1}{2}(y(t+h) + y(t)) - y(t+h)y(t)\right],$$

with the solution

$$y_n = \frac{y_0}{y_0 + (1 - y_0)\left(\frac{1-\alpha h/2}{1+\alpha h/2}\right)^n}.$$

(g) Show that the continuum limit (where $h \to 0$, $n \to \infty$ with $hn = 2t$ constant) of the above solution agrees with the solution in point (a), above.

1.2 Complete the example 1.1.7. Furthermore, follow similar methods to solve the following examples.

(a) Determine the general solution of the following functional equation

$$f(xy) = f(x) + f(y),$$

in which $f(x)$ is a real-valued differentiable function for all x real positive.

(b) Consider the functional equation

$$f(x + y) = f(x)f(y)g(xy)$$

and determine the real-valued differentiable functions f and g as functions of x real.

(c) Determine the general solution of the functional equation

$$f(x + y) = \frac{f(x) + f(y)}{1 - f(x)f(y)},$$

in which $f(x)$ is a real-valued function of real x which is differentiable around $x = 0$.

1.3 Consider the general discrete Riccati equation

$$n(n-1)y_{n+1}y_n + n^2 y_{n+1} + (n-1)y_n + n(7 - 6n) = 0,$$

in which y_n is the unknown function of the discrete variable n. By making the substitution $y_n + n/(n-1) =: w_n/w_{n-1}$ reduce this equation to a second-order linear differential equation and find the general solution of the latter equation. Show that the solution depends effectively on only one integration constant.

1.4 (a) Show that any three known solutions, $y_n^{(1)}$, $y_n^{(2)}$, $y_n^{(3)}$, of the discrete Riccati equation (1.54) (or in the alternative form (1.55)) determine the dynamical map (1.57) uniquely. In fact, show that this map can be given as

$$\begin{vmatrix} y_n^{(1)}y_{n+1}^{(1)} & y_{n+1}^{(1)} & y_n^{(1)} & 1 \\ y_n^{(2)}y_{n+1}^{(2)} & y_{n+1}^{(2)} & y_n^{(2)} & 1 \\ y_n^{(3)}y_{n+1}^{(3)} & y_{n+1}^{(3)} & y_n^{(3)} & 1 \\ y_n y_{n+1} & y_{n+1} & y_n & 1 \end{vmatrix} =$$

$$= \left(y_n^{(2)} - y_n^{(3)}\right)\left(y_n^{(1)} - y_n\right)\left(y_{n+1}^{(2)} - y_{n+1}\right)\left(y_{n+1}^{(3)} - y_{n+1}^{(1)}\right)$$
$$- \left(y_n^{(2)} - y_n\right)\left(y_n^{(1)} - y_n^{(3)}\right)\left(y_{n+1}^{(3)} - y_{n+1}^{(2)}\right)\left(y_{n+1}^{(1)} - y_{n+1}\right) = 0\,.$$

Derive from this result a fractional linear form of the Riccati equation in terms of the given solutions $y_n^{(i)}$, $i = 1, 2, 3$. Furthermore, conclude from this result that the canonical cross-ratio of any four solutions of the Riccati equation is constant.

(b) Deduce from (1.56) that the *fixed-point solutions* of the discrete Riccati equation (1.55) correspond to the eigenvectors of the given matrix. Let $w_n^{(1)}$ and $w_n^{(2)}$ be two independent fixed-point solutions of (1.55), i.e. $R(w_n^{(i)}, w_n^{(i)}) = 0, i = 1, 2,$ (note that this does not mean that these functions are constant), then use this fact to derive the following form of the Riccati map (1.57) in terms of its fixed points:

$$\frac{y_{n+1} - w_n^{(2)}}{y_{n+1} - w_n^{(1)}} = \frac{C_n w_n^{(1)} + D_n}{C_n w_n^{(2)} + D_n} \frac{y_n - w_n^{(2)}}{y_n - w_n^{(1)}}.$$

1.5 Show that the appropriate discretization of the initial value problem $u(x, 0) = f(x)$ and $u_t(x, 0) = g(x)$ is given by (1.65).
Hint: note that equation (1.62b) yields $g_n = u_t(nh, 0) = \frac{1}{k}(u(nh, k) - u(nh, 0)) - \frac{k}{2}u_{tt}(nh, 0) + \mathcal{O}(k^2)$.

1.6 Assume that the discrete Laplace's equation is given together with boundary conditions

$$\begin{aligned} u_{m,0} &= 1,\ 0 \le m \le 3 \\ u_{m,3} &= 0,\ 0 \le m \le 3 \\ u_{0,n} &= 0,\ 1 \le n \le 2 \\ u_{3,n} &= 0,\ 1 \le n \le 2. \end{aligned}$$

Solve this system of four linear equations for the four unknowns $u_{1,1}, u_{1,2}, u_{2,1}, u_{2,2}$ to find the solution at the interior four points:

$$u_{1,1} = 3/8, u_{1,2} = 1/8, u_{2,1} = 3/8, u_{2,2} = 1/8.$$

2

Discrete equations from transformations of continuous equations

In this chapter we discuss how difference equations arise from transformations applied to differential equations. Thus we find that some functions can be defined both by a differential equation and by a difference equation. The independent variables of these two types of equations are not necessarily the same, but nevertheless we often find an interesting duality between parameters and independent variables.

Often functions defined by a differential equation possess special transformations. One particularly important class of transformations is the one that acts on the differential equation by *changing the values of its parameters*. When such transformations are iterated, we obtain a sequence of differential equations (and often a sequence of solutions) that are strung together by the changing sequence of parameter values.

As a particularly simple example consider the differential equation

$$x\, w' = \alpha\, w. \tag{2.1}$$

Here x is the independent (continuous) variable; the prime stands for x-derivative ($w' = \frac{dw}{dx}$) and α is a parameter of the equation. It is easy to find the general solution of this equation; it is

$$w = c\, x^{\alpha} =: w_{\alpha}(x), \tag{2.2}$$

where c is a constant of integration. From the form of the solution (2.2) we can see that the simple action of multiplying the function w_α by x changes α, that is

$$w_{\alpha+1} = x\, w_{\alpha}. \tag{2.3}$$

Equation (2.3) can now be interpreted as a discrete or difference equation, where α is the (discrete) independent variable and x is a parameter. Thus we have found that the two equations (2.1) and (2.3) describe from two different points of view the same function $w_\alpha(x)$ of (2.2).

Another way of looking at this is to view the ordinary difference equation (OΔE) (2.3) as being **compatible** with the differential equation (2.1). In fact, we find

$$\frac{d}{dx}(T_\alpha w) = \frac{d}{dx}(x\, w_\alpha) = w_\alpha + x\, w'_\alpha = (1+\alpha)w_\alpha,$$
$$T_\alpha(\frac{d}{dx}w) = T_\alpha(\frac{\alpha}{x}w_\alpha) = (1+\alpha)\frac{1}{x}w_{\alpha+1} = (1+\alpha)w_\alpha,$$

and thus the operators T_α and $\frac{d}{dx}$ commute. The concept of compatibility of different operations, acting on solution of systems of equations, forms an enduring theme in the subject of integrability.

In this chapter we will first discuss how the above point of view works for ordinary differential equations (ODEs), using as examples the equations satisfied by certain classical *special functions*, such as the Weber, Bessel and hypergeometric functions. These can be shown to satisfy not only linear ODEs, but also linear OΔEs in terms of certain parameters. We will also take a look at some *nonlinear special functions*, such as the elliptic functions and the Painlevé transcendents, that satisfy nonlinear differential equations, and to which similar ideas apply. Finally, we will move from ODEs to partial differential equations (PDEs), and treat briefly some basic theory of completely integrable PDEs (also known as *soliton equations*), and show how the application of similar ideas leads to the construction of integrable partial difference equations (PΔEs).

2.1 Special functions and linear equations

Special functions such as the Bessel, parabolic cylinder and hypergeometric functions (Whittaker and Watson, 1902; Abramowitz and Stegun, 1970; Andrews et al., 1999; DLMF, 2010; Olver et al., 2010) satisfy ODEs with independent variable x and at least one parameter α. They also satisfy *recurrence relations* in which α is shifted. In this section we show how to derive their corresponding parameter-changing transformations and find related difference equations.

2.1.1 Weber functions and Hermite polynomials

Consider the differential equation

$$w'' + \left(\alpha + \tfrac{1}{2} - \tfrac{1}{4}x^2\right) w = 0, \tag{2.4}$$

where the primes denote differentiation in x. Since the equation is second order it has two linearly independent solutions. One of these can be chosen to be the *parabolic cylinder* or *Weber function*, denoted by $D_\alpha(x)$, uniquely specified by the asymptotic behavior

$$D_\alpha(x) = x^\alpha\, e^{-x^2/4}\left(1 - \frac{\alpha(\alpha-1)}{2x^2} + \mathcal{O}\left(\frac{1}{x^4}\right)\right), \quad x \to +\infty. \tag{2.5}$$

The simplest special case arises when α is an integer n: this yields the Hermite polynomials $H_n(x)$, defined by

$$D_n(x) = 2^{-n/2}e^{-x^2/4}\, H_n(z), \quad z = x/\sqrt{2}. \tag{2.6}$$

Instead of thinking of equation (2.4) as one equation specified by one fixed value of α, it is more productive to think of it as an infinite sequence of equations, each of which is specified by successive values of α differing by unity. This alternative perspective has a wonderful consequence: we can generate new solutions of each successive equation by knowing solutions of an earlier equation in the sequence. To see how to do this, note that the differential operator in equation (2.4) factorizes:

$$\left(\partial_x - \tfrac{1}{2}x\right)\left(\partial_x + \tfrac{1}{2}x\right)w = -\alpha\, w. \tag{2.7}$$

Assume that we know a solution of this equation for a particular value of α, denoted by $w = w_\alpha$. Let us define

$$\widetilde{w} := \left(\partial_x + \tfrac{1}{2}x\right)w. \tag{2.8}$$

This and equation (2.7) form a system of two equations

$$\begin{cases} \left(\partial_x + \tfrac{1}{2}x\right)w = \widetilde{w}, \\ \left(\partial_x - \tfrac{1}{2}x\right)\widetilde{w} = -\alpha\, w. \end{cases} \tag{2.9}$$

If we use the first equation to eliminate $\widetilde{w}$ in the second equation, we immediately get (2.7), while using the second equation to eliminate w from the first one leads to

$$\left(\partial_x + \tfrac{1}{2}x\right)\left(\partial_x - \tfrac{1}{2}x\right)\widetilde{w} = -\alpha\, \widetilde{w}, \tag{2.10}$$

or after expanding

$$\widetilde{w}'' + \left((\alpha - 1) + \tfrac{1}{2} - \tfrac{1}{4}x^2\right)\widetilde{w} = 0, \tag{2.11}$$

which is the same equation as (2.4) except that the parameter α has been replaced by $\tilde{\alpha} = \alpha - 1$. The solution $\widetilde{w}$ can therefore be used to define $w_{\alpha-1}(x)$, and thus we have constructed a new solution of (2.4) corresponding to a different parameter value $\alpha - 1$. This can be repeated recursively, yielding a sequence of functions $w_{\alpha-n}(x)$.

In the above argument, we assumed knowledge of *some* solution corresponding to a given value of α. However, we could have started with a particular solution corresponding to a given value of α, namely the one given by the Weber function D_α (see (2.5)). We now start with this function and apply the above procedure to find $\widetilde{w}$, but it is not a priori clear that it yields a Weber function again for the new value of α.[1] By explicit computation, we find

$$\left(\partial_x + \tfrac{1}{2}x\right)D_\alpha(x) = \alpha\, x^{\alpha-1}\, e^{-x^2/4}\left(1 + \mathcal{O}(1/x^2)\right) = \alpha D_{\alpha-1},$$

[1] Since the general solution depends on two integration constants, $\widetilde{w}$ could be a linear combination of $D_{\alpha-1}$ and another linearly independent solution.

and therefore comparing this with (2.8) we can identify

$$w(x) = D_\alpha(x), \quad \widetilde{w}(x) = \alpha D_{\alpha-1}(x). \tag{2.12}$$

Thus, in order to get the sequence of functions $\{D_\alpha\}$ we must introduce a proper normalization in the definition (2.8).

Substituting (2.12) into (2.9) we obtain

$$\begin{cases} D_{\alpha-1}(x) = \frac{1}{\alpha}[D'_\alpha(x) + \frac{1}{2}x D_\alpha(x)], \\ D_\alpha(x) = -D'_{\alpha-1}(x) + \frac{1}{2}x D_{\alpha-1}(x), \end{cases} \tag{2.13}$$

which can also be written as

$$\begin{cases} D'_\alpha(x) = -\frac{1}{2}x D_\alpha(x) + \alpha D_{\alpha-1}(x), \\ D'_{\alpha-1}(x) = \frac{1}{2}x D_{\alpha-1}(x) - D_\alpha(x). \end{cases} \tag{2.14}$$

The first pair (2.13) allows us to travel up and down in the chain of functions $D_\alpha(x)$ by changing α by integer steps. The second form (2.14) allows us to derive a fully discrete equation: substituting $\alpha \to \alpha + 1$ into the second equation of (2.14) and then subtracting the two equations we get

$$D_{\alpha+1}(x) - x\, D_\alpha(x) + \alpha\, D_{\alpha-1}(x) = 0. \tag{2.15}$$

This is a (linear homogeneous) ordinary difference equation, where α is the independent variable and x plays the role of a parameter. Thus the function $D_\alpha(x)$ satisfies two equations: the difference equation (2.15) and the differential equation (2.4). In fact, these two equations are compatible in the sense that the shift in α and derivative in x are commutative. In the next section, we see how this scenario arises as a special case of application of the Darboux transformation.

Remark: Equation (2.4) has a form similar to the stationary Schrödinger equation of quantum mechanics for a particle confined by a harmonic potential, namely

$$-\frac{\hbar^2}{2m}\psi'' + \frac{1}{2}m\omega^2 x^2 \psi = E\psi, \tag{2.16}$$

so the discussion above has a quantum mechanical interpretation. Comparing (2.16) with (2.4) we see that α, which we used to label the solution D_α, is related to the energy, $E \propto \alpha + \frac{1}{2}$, while the analysis of asymptotic behavior forces α to be an integer. Furthermore, from (2.13) we can see that $\partial_x + \frac{1}{2}x$ is a lowering operator since it changes D_α to $D_{\alpha-1}$ and that $-\partial_x + \frac{1}{2}x$ is a raising operator. These matters are further elaborated in introductory books on quantum mechanics; see e.g. Schiff (1968) Sections 13 and 25, or Griffiths (2005) Section 2.3.

2.1.2 Darboux and Bäcklund transformations

Jean Gaston Darboux (1842–1917) was one of the pioneers of classical differential geometry. Furthermore, he made important contributions to many other fields of mathematics, including differential equations. In his book (Darboux, 1914, p. 210, Section 408) there is a theorem, which has been revived in the modern theory of integrable systems, leading to what in the modern literature (see e.g. Matveev and Salle (1991), Rogers and Schief (2002)) is called a **Darboux transformation**. Related to it is the notion of a so-called Bäcklund transformation, which we describe briefly below and in the next section in more detail (for the case of PDEs). Whereas Darboux's theorem applies primarily to linear equations, the Bäcklund transformation, as we shall see, is most relevant to nonlinear equations.

Figure 2.1 J. G. Darboux

Darboux' theorem can be stated as follows:

Theorem 2.1.1 *(Darboux transformation) Consider the differential equation*

$$y'' = [\phi(x) + h]y, \tag{2.17}$$

with parameter h. Suppose $f(x)$ is a particular solution of this equation for some specific value h_1 of h. Let us define a new function $\widetilde{y}$ by

$$\widetilde{y} := [\partial_x - (\log f)']y. \tag{2.18}$$

Then $\widetilde{y}$ solves the equation

$$\widetilde{y}'' = [\widetilde{\phi}(x) + h]\widetilde{y}, \quad \textit{where} \quad \widetilde{\phi} := \phi - 2(\log f)''. \tag{2.19}$$

We say that $\widetilde{y}$ and $\widetilde{\phi}$ are the Darboux transforms of the wave-function and the potential, respectively.

This can be proved by a factorization procedure similar to the one used in Section 2.1.1. We start by assuming that $f(x) = y(x, h_1)$ is a solution corresponding to parameter h_1 and express its second derivative by using the following identity:

$$\frac{f''}{f} = (\log f)'' + \left((\log f)'\right)^2.$$

Then, noting that the potential $\phi(x)$ in equation (2.17) is independent of the spectral parameter h, we obtain

$$\phi(x) = -h_1 + (\log f)'' + \big((\log f)'\big)^2,$$

and hence equation (2.17) for general h can be rewritten as:

$$y'' = (h - h_1 + v' + v^2)y, \quad \text{where} \quad v = \partial_x \log f.$$

This can be factorized as follows:

$$(\partial_x + v)(\partial_x - v)y = (h - h_1)y,$$

and setting $\widetilde{y} = (\partial_x - v)y$, we obtain by interchanging the factors:

$$(\partial_x - v)(\partial_x + v)\widetilde{y} = (h - h_1)\widetilde{y}\,,$$

which is the equation (2.19) with a new potential $\widetilde{\phi}(x) = \phi(x) - 2v'$.

Note that the transformation of Weber functions found in the previous subsection can be viewed as an application of Darboux's theorem: in comparison with (2.4) we see that we should choose $\phi = x^2/4$, $f = e^{x^2/4}$, $h_1 = 1/2$. Then we find that $\widetilde{\phi} = \phi - 1$, in agreement with (2.11).

Notice that the pair of simultaneous equations (2.13), which we deduced for Weber functions, involves D_α, $D_{\alpha-1}$ and their first derivatives. In Section 2.1.1, we saw how to obtain the discrete equation (2.15) relating D_α, $D_{\alpha-1}$ and $D_{\alpha+1}$, by eliminating the first derivatives. However, we could instead have eliminated D_α, which leads to a second-order ODE for $D_{\alpha-1}$, while eliminating $D_{\alpha-1}$ leads to a copy of the same second-order equation for D_α. This transformation between two copies of the Weber equation is a special case of a transformation well known in geometry, called a Bäcklund transformation.

Albert Victor Bäcklund (1845–1922) worked on transformations of surfaces in differential geometry (Bäcklund, 1883). In the modern era, the connection between the latter subject and the theory of differential equations has become quite prominent. In fact, there is a close relationship between transformations of special surfaces (in terms of coordinates on these surfaces) and transformations between solutions of (linear and nonlinear) differential equations. As we shall see in Section 2.5.5, these transformations form the basis of the construction of *exact discretizations* of those very same differential equations. We defer the precise definition of a Bäcklund transformation to a later section when we deal with PDEs, but present here a loose definition of this concept as follows:

Figure 2.2 A. Bäcklund

Definition 2.1.1 (Bäcklund transformation) *Suppose we have a pair of equations satisfied by two dependent variables u and v and possibly on their (partial) derivatives with respect to an independent variable x:*

$$\begin{cases} F(u, u_x, \ldots, v, v_x, \ldots) = 0, \\ G(u, u_x, \ldots, v, v_x, \ldots) = 0. \end{cases} \tag{2.20}$$

If upon eliminating v we obtain the equation $R(u, u_x, u_{xx} \ldots) = 0$ *and upon eliminating u we obtain* $S(v, v_x, v_{xx} \ldots) = 0$ *then* (2.20) *is called a* ***Bäcklund transformation (BT)*** *between the equations* $R = 0$ *and* $S = 0$. *If R and S coincide, then the transformation is called an* ***auto-Bäcklund transformation***. *There are also situations when the equations* $R = 0$ *and* $S = 0$ *belong to a family of differential equations depending on an unspecified parameter. The term auto-Bäcklund transformation is also used for mappings between members of such a parameter family of equations, where they map solutions of an equation with a specific value of the parameter to solutions of the equation with a different value of the parameter.*

In the situation where a Bäcklund transformation maps between different equations (which do not belong to the same parameterized family), it is sometimes also called a hetero-Bäcklund transformation.

In the following subsections, we consider examples for which Theorem 2.1.1 is not directly applicable at first glance, for example, because the separation between the potential $\phi(x)$ and parameter h may not be as clear cut as in (2.17). Nevertheless, we show that the underlying ideas of factorization and transformations that are linear in the first derivative (such as in (2.19)) still apply and can be used to find transformations between solutions corresponding to different values of parameters. We will continue to refer to such transformations as Darboux transformations even when the ODE of interest is not strictly in the form of equation (2.17).

2.1.3 Bessel functions

Bessel functions are defined as solutions of

$$x^2 w'' + x w' + (x^2 - \nu^2) w = 0, \tag{2.21}$$

where ν is a parameter, with specific asymptotic behaviors. The standard *Bessel function of the first kind* is defined by[2]

$$J_\nu(x) = \left(\frac{x}{2}\right)^\nu \sum_{k=0}^{\infty} \frac{\left(-x^2/4\right)^k}{k!\,\Gamma(\nu + k + 1)}. \tag{2.22}$$

[2] Recall the definition of the Γ-function, which is used in this formula, as the function obeying the difference equation $\Gamma(z+1) = z\Gamma(z)$.

These functions occur in many applications, e.g. in the study of waves in circular domains.

We will consider Bessel's equation as an infinite sequence of equations in the space of the parameter value ν. To allow us to iterate in this parameter space, we need recurrence relations for Bessel functions. To find Darboux transformations from which recurrence relations follow, we start by transforming equation (2.21) into the so-called Sturm–Liouville form, i.e. without the first derivative term.

Transforming the dependent variable w by setting $w(x) = p(x)y(x)$, and substituting this into equation (2.21), we can derive a second-order ODE for y without a term containing y', by choosing $2x\,p'(x) + p(x) = 0$, which implies $p(x) = 1/\sqrt{x}$. Thus by setting $w(x;\nu) = y(x;\nu)/\sqrt{x}$, we get

$$y'' + \left(1 - \frac{\nu^2 - 1/4}{x^2}\right) y = 0. \tag{2.23}$$

Note that this equation is not in the form (2.17) assumed in Darboux's theorem, because the coefficient function $\phi(x)$ here also involves the parameter ν, playing a role analogous to h, which will be changed by the transformation. Nevertheless, the equation can be written in factorized form by observing that

$$\left[\partial_x - (\nu - \tfrac{1}{2})\tfrac{1}{x}\right]\left[\partial_x + (\nu - \tfrac{1}{2})\tfrac{1}{x}\right] = \partial_x^2 - (\nu^2 - \tfrac{1}{4})\tfrac{1}{x^2}. \tag{2.24}$$

From the two factors in (2.24) we obtain the BT

$$\begin{cases} \left[\partial_x + (\nu - \frac{1}{2})/x\right] y = \widetilde{y}, \\ \left[\partial_x - (\nu - \frac{1}{2})/x\right] \widetilde{y} = -y. \end{cases} \tag{2.25}$$

Explicit calculation shows that eliminating $\tilde{y}$ (by using the first equation to replace $\widetilde{y}$ in the second equation) yields (2.23) for y, while eliminating y (by using the second equation to replace y in the first one) yields

$$\tilde{y}'' + \left(1 - \frac{(\nu-1)^2 - 1/4}{x^2}\right) \tilde{y} = 0;$$

that is, we obtain a copy of equation (2.23) with ν replaced by $\nu - 1$. By starting with $y(x) = J_\nu(x)/\sqrt{x}$, and expanding the first equation to find $\widetilde{y}$ near $x = 0$, we find that $\widetilde{y} = J_{\nu-1}(x)/\sqrt{x}$.

Returning to the original variables with $y = \sqrt{x}\,w$, the above equations become

$$\begin{cases} \left[\partial_x + \nu/x\right] w = \widetilde{w}, \\ \left[\partial_x - (\nu - 1)/x\right] \widetilde{w} = -w. \end{cases} \tag{2.26}$$

Now identifying $w(x) = J_\nu(x)$, $\widetilde{w}(x) = J_{\nu-1}(x)$, and shifting $\nu \to \nu + 1$ in the second equation we obtain

$$\begin{cases} x J_\nu' = -\nu\, J_\nu + x\, J_{\nu-1}, \\ x J_\nu' = \nu\, J_\nu - x\, J_{\nu+1}, \end{cases} \tag{2.27}$$

where primes denote x-derivatives. The x-derivative J_ν' can be eliminated by subtracting the second equation from the first, which leads to

$$x\, J_{\nu+1} - 2\nu\, J_\nu + x\, J_{\nu-1} = 0, \tag{2.28}$$

which is a difference equation in ν, with x playing the role of a parameter.

In Exercise 2.1, we show that there exists a transformation of Bessel's equation to which Darboux's theorem 2.1.1 can be applied directly to yield the above results.

2.1.4 Hypergeometric functions

Hypergeometric functions form a very rich class of functions that have intricate connections to geometry and complex analysis. Gauss' hypergeometric function ${}_2F_1(a, b; c; x)$ is defined by the series

$${}_2F_1(a, b; c; x) \equiv {}_2F_1\left(\begin{matrix} a, b \\ c \end{matrix} ; x \right) = \sum_{n=0}^{\infty} \alpha_n \frac{x^n}{n!} = \sum_{n=0}^{\infty} \frac{(a)_n\, (b)_n}{(c)_n} \frac{x^n}{n!} \tag{2.29}$$

for $|x| < 1$. Here $(a)_n$ is the raising factorial defined in (A.11). If we let the coefficients in the series (2.29) be denoted by α_n, we get[3]

$$\frac{\alpha_{n+1}}{\alpha_n} = \frac{(n + a)\,(n + b)}{n + c}.$$

We will derive recurrence relations for ${}_2F_1(a, b; c; x)$ by the method of Darboux transformations. These relations are called *contiguity relations* because they relate two hypergeometric functions having neighboring (i.e. contiguous) parameter values. Note that there are now several parameters and it is possible to have different transformations changing each parameter independently.

As a function of x, ${}_2F_1(a, b; c; x)$ satisfies the differential equation

$$x\,(1 - x)\, y'' + (c - (a + b + 1)\, x)\; y' - a\, b\, y = 0. \tag{2.30}$$

[3] A hypergeometric series is a series $\sum c_n$ such that c_{n+1}/c_n is a rational function of n (see e.g. Andrews et al. (1999), Section 2.1). The notation ${}_pF_q$ refers to such a series in which the ratio of successive coefficients is a degree p polynomial divided by a degree q polynomial.

Our previous experience suggests that a Bäcklund transformation of the form

$$\begin{cases} \left[A(x)\partial_x + B(x)\right] w = \tilde{w} \\ \left[C(x)\partial_x + D(x)\right] \tilde{w} = w \end{cases}$$

should work. However this leads to rather complicated equations for the functions a, b, c, d. Let us instead consider the ansatz

$$\begin{cases} x^{\varphi_1}(1-x)^{\varphi_2}\left(\partial_x - \dfrac{\mu\, x + \nu}{2\,x\,(x-1)}\right) w = \widetilde{w} \\ \kappa\, x^{\varphi_3}(1-x)^{\varphi_4}\left(\partial_x + \dfrac{\rho\, x + \sigma}{2\,x\,(x-1)}\right) \widetilde{w} = w, \end{cases} \tag{2.31}$$

which is sufficiently general to allow us to derive several contiguity relations.

In order to recover (2.30) after eliminating $\widetilde{w}$ from (2.31) we must set

$$\varphi_1 = c+\tfrac{1}{2}(\sigma-\nu),\ \varphi_2 = a+b+\tfrac{1}{2}(\mu-\rho)-\varphi_1+1,\ \varphi_3 = -\varphi_1+2-\omega_1,\ \varphi_4 = -\varphi_2+2-\omega_2,$$

where ω_i are two integers satisfying $0 \leq \omega_1 + \omega_2 \leq 2$. With different choices of ω_i we can derive different relations. We will furthermore restrict ourselves to cases where the exponents φ_i are all integers.

For example, if $\omega_1 = 0, \omega_2 = 1$ we can obtain, among others, Bäcklund transformation pairs that only change a (or b). If we take $\mu = -2a,\ \nu = 2a,\ \rho = 2b, \sigma = 2(a-c+1)$, $\kappa = -1/[a(a-c+1)]$ we get the pair

$$\begin{cases} (x\partial_x + a)\, y(a,b,c;x) = a y(a+1,b,c;x) \\ \frac{1}{a-c+1}\left[x(x-1)\partial_x + a - c + 1 + bx\right]\, y(a+1,b,c;x) = y(a,b,c;x). \end{cases} \tag{2.32}$$

(The normalization $\widetilde{w} = a\, y(a+1,b,c;x)$ can be derived from the first terms of the power series expansion (2.29).) After shifting $a \mapsto a-1$ in the second equation and then eliminating $y'(a,b,c;x)$ we get one of Gauss' relations for contiguous functions (Abramowitz and Stegun (1970), Equation (15.2.10); Olver et al. (2010); DLMF (2010), equation (15.5.11))

$$a(x-1)\,{}_2F_1(a+1,b,c;x) + [(b-a)x+2a-c]\,{}_2F_1(a,b,c;x) \\ + (c-a)\,{}_2F_1(a-1,b,c;x) = 0. \tag{2.33}$$

Another transformation is obtained with the choice $\omega_1 = 1, \omega_2 = 0,\ \mu = -2a$, $\nu = 0,\ \rho = 2b,\ \sigma = -2c, \kappa = 1/[a(b-c)]$:

$$\begin{cases} \left[(1-x)\partial_x - a\right] y(a,b,c;x) = \frac{a(b-c)}{c} y(a+1,b,c+1;x), \\ \frac{1}{c}\left[x(1-x)\partial_x + (c-bx)\right] y(a+1,b,c+1;x) = y(a,b,c;x). \end{cases} \tag{2.34}$$

After eliminating the derivative terms, we get the difference equation

$$a(b-c)x\,{}_2F_1(a+1,b;c+1;x)+c[x(a-b)+c-1]\,{}_2F_1(a,b;c;x)$$
$$+c(1-c)\,{}_2F_1(a-1,b;c-1;x)=0. \tag{2.35}$$

The difference equations (2.33, 2.35) involve three terms on a line in the three-dimensional space of parameters a, b, c. One can also have corner relations. For example, by taking the first equations of the pairs (2.32, 2.34) and eliminating the derivative between them one obtains (Abramowitz and Stegun (1970), 15.2.20; Olver et al. (2010); DLMF (2010), 15.5.16 with $a \to a+1$)

$$(b-c)x\,{}_2F_1(a+1,b;c+1;x)+c(x-1)\,{}_2F_1(a+1,b;c;x)$$
$$+c\,{}_2F_1(a,b;c;x)=0. \tag{2.36}$$

Numerous other contiguity relations can be derived this way; see Andrews et al. (1999); DLMF (2010); Olver et al. (2010).

2.2 Addition formulae

Many classical special functions, such as those we have so far investigated, are defined by *linear* ODEs and possess *linear* recurrence relations. However, the existence of recurrence relations is not restricted to the linear case. As an intermediate step toward nonlinear equations, in this section, we show how to deduce recurrences and discrete equations for the Jacobi elliptic functions.

2.2.1 Trigonometric functions

To motivate the derivation of transformations, recurrences and discrete equations for Jacobi elliptic functions, it is instructive to consider trigonometric functions; for example, the sine function, for which we have the well-known identities

$$\begin{cases} \sin(a+b)=\sin(a)\cos(b)+\cos(a)\sin(b), \\ \sin(a-b)=\sin(a)\cos(b)-\cos(a)\sin(b). \end{cases} \tag{2.37}$$

Adding these two equations together leads to

$$\sin(a+b)+\sin(a-b)=2\sin(a)\cos(b).$$

Consider a as a discrete independent variable and b as its shift, i.e. $a=a_0+bn$, and introduce the function y_n by

$$y_n := \sin(a_0+bn).$$

Then we have obtained

$$y_{n+1}+y_{n-1}=2\cos(b)\,y_n, \tag{2.38}$$

i.e. a linear difference equation.

The identities (2.37) can be interpreted in terms of Darboux transformations. To make a parallel with previous derivations we introduce the notation $\widetilde{y} = y_{n+1}$ and omit n in $y = y_n$. We just note that $\sin(x)$ satisfies $y'' + y = 0$, and from the first equation in (2.37) we also get the Darboux transformation $\widetilde{y} = \sin(b)\, y' + \cos(b)\, y$. The pair

$$\begin{cases} \widetilde{y} = \sin(b)\, y' + \cos(b)\, y, \\ y = -\sin(b)\, \widetilde{y}' + \cos(b)\, \widetilde{y}, \end{cases} \tag{2.39}$$

which provides the corresponding Bäcklund transformation, namely from equations (2.39) if follows both that $y'' + y = 0$ and $\widetilde{y}'' + \widetilde{y} = 0$. And if we eliminate the derivatives we recover (2.38).

2.2.2 *Elliptic functions*

In a sense the *Jacobi elliptic functions* can be viewed as generalizations of the trigonometric functions. The basic properties of elliptic functions are discussed in Appendix B; see also Abramowitz and Stegun (1970), Chapter 16; Olver et al. (2010); DLMF (2010), Chapter 22.

Jacobi elliptic functions satisfy the first-order differential equation

$$s'^2 = (1 - s^2)(1 - k^2 s^2), \tag{2.40}$$

and its differential consequence

$$\frac{d^2 s}{du^2} = 2k^2 s^3 - \left(1 + k^2\right) s, \tag{2.41}$$

where the parameter k is called the *modulus*. We focus on the solution denoted by $s(u) = \mathrm{sn}(u; k)$, which is an *odd* function of u. In the limit $k \to 0$ the sn-function reduces to the sine function. We can interpret the relation (2.40) also as defining an algebraic curve (in analogue to the circle for the trigonometric functions) given by

$$Y^2 = (1 - X^2)(1 - k^2 X^2),$$

in the complex (X, Y) plane.

Associated with the Jacobi sn-function we have two conjugate functions, denoted by $\mathrm{cn}(u; k)$, $\mathrm{dn}(u; k)$, both *even* functions of u; they are related to sn through

$$\mathrm{cn}^2(u) + \mathrm{sn}^2(u) = 1, \tag{2.42a}$$

$$\mathrm{dn}^2(u) + k^2\mathrm{sn}^2(u) = 1. \tag{2.42b}$$

Using such relations, we can show from (2.40) that

$$\mathrm{sn}'(u) = \mathrm{cn}(u)\mathrm{dn}(u). \tag{2.43}$$

The addition formula for the Jacobi sn-function is given by

$$\operatorname{sn}(u+v)=\frac{\operatorname{sn}(u)\operatorname{cn}(v)\operatorname{dn}(v)+\operatorname{sn}(v)\operatorname{cn}(u)\operatorname{dn}(u)}{1-k^2\operatorname{sn}^2(u)\operatorname{sn}^2(v)}, \tag{2.44a}$$

and by the antisymmetry of sn

$$\operatorname{sn}(u-v)=\frac{\operatorname{sn}(u)\operatorname{cn}(v)\operatorname{dn}(v)-\operatorname{sn}(v)\operatorname{cn}(u)\operatorname{dn}(u)}{1-k^2\operatorname{sn}^2(u)\operatorname{sn}^2(v)}. \tag{2.44b}$$

Adding these two equations together, we can show that the general solution of the McMillan map (1.19) can be expressed in terms of the Jacobi sn function.

If, in the above, we consider u as the independent variable and v as the parameter, let $y(u) = \operatorname{sn}(u)$, $\widetilde{y}(u) = \operatorname{sn}(u+v)$, and also recall the derivative rule (2.43), then we can write (2.44a, 2.44b) in the form of a Darboux–Bäcklund transformation as follows:

$$\begin{cases} \frac{1}{1-k^2\operatorname{sn}^2(v)\,y^2(u)}[\operatorname{sn}(v)\partial_u+\operatorname{cn}(v)\operatorname{dn}(v)]y(u)=\widetilde{y}(u),\\ \frac{1}{1-k^2\operatorname{sn}^2(v)\,\widetilde{y}^2(u)}[\operatorname{sn}(v)\partial_u-\operatorname{cn}(v)\operatorname{dn}(v)]\widetilde{y}(u)=-y(u). \end{cases} \tag{2.45}$$

Indeed, if one now solves for $\widetilde{y}$ from the first and substitutes to the second, the result is proportional to $\partial_u[(2.40)/(1-k^2\ \operatorname{sn}^2(v)\,y^2(u))]$, and vice versa after eliminating y. (To verify this it is also necessary to use the relations (2.42a) and (2.42b).)

Note that the transformations (2.45) do not change the parameter k of the equation. The Darboux approach we have followed here can be extended to find transformations that do change the modulus k, i.e. Landen transforms (see Appendix B.3); however, this extension lies outside the scope of this chapter and we will not go into it here.

We will next show that the well-known addition formulae for the *Weierstrass elliptic functions* can be obtained from a BT of the form (2.20). Consider the BT between functions u, v given by

$$F(u,u',v,v')\equiv(v'+b)(u-a)+(u'-b)(v-a)=0, \tag{2.46a}$$

$$G(u,u',v,v')\equiv\frac{1}{4}\left(\frac{u'-b}{u-a}\right)^2-u-v-a=0, \tag{2.46b}$$

where u, v are functions of x and the prime denotes derivative with respect to x, and where we assume that the parameters a, b are related by

$$b^2=4a^3-g_2a-g_3. \tag{2.47}$$

Recall the idea behind the BT: by compatibility it leads on the one hand to an equation for u, on the other hand to an equation for v. In fact, solving v from the equation $G=0$ and inserting this into $F=0$ we obtain

$$\partial_x\left[(u'^2-4\,u^3+g_2\,u+g_3)/(u-a)\right]=0. \tag{2.48}$$

Similarly, after first eliminating u' from (2.46b) using (2.46a) we can solve for u from (2.46b) and then (2.46a) leads to the same equation (2.48) for v. A particular solution of (2.48) is to take

$$u'^2 = 4\,u^3 - g_2\,u - g_3. \tag{2.49}$$

Thus we have found that the pairs (a, b), (u, u') and (v, v') can be assumed to satisfy the same relation

$$Y^2 = 4\,X^3 - g_2\,X - g_3.$$

On the (X, Y)-plane, this defines a curve called the Weierstrass elliptic curve, and we have just found that the points (a, b), (u, u') and (v, v') all lie on the same curve. Hence the BT maps a point on the Weierstrass curve to another point on the same curve. Since both u and v are functions of x, they can be identified with $u = \wp(x - x_0)$, $v = \wp(x - x_0 + \beta)$, where x_0 can be taken as arbitrary, but where β will depend on the parameters a, b. Setting $x = x_0$ we find: $a = \wp(\beta)$, $b = \wp'(\beta)$, where we have used the fact that a, b are assumed to be related through the curve. Thus we have found the well-known addition formula for the Weierstrass $\wp$-function

$$\wp(x) + \wp(y) + \wp(x + y) = \frac{1}{4}\left(\frac{\wp'(x) - \wp'(y)}{\wp(x) - \wp(y)}\right)^2. \tag{2.50}$$

Addition formulae like (2.50) and (2.44) play an important role in the study of DIS. Here we have demonstrated how they can be found from BTs of the nonlinear differential equations for functions such as the $\wp$-function.

2.3 The Painlevé equations

In this section we illustrate how Darboux's approach can be applied to nonlinear ordinary differential equations called the Painlevé equations.

Figure 2.3 P. Painlevé

They form an important class of classical second-order nonlinear ODEs whose solutions satisfy a certain property that is now commonly referred to as the *Painlevé property* (Conte, 1999). (A description of this property, the list of continuous Painlevé equations and a brief outline of their history is given in Appendix C.) The Painlevé property allows the solutions of all of these equations to be globally defined in the complex plane, a property that is now recognized as very important and useful for integrability. Due to the regularity of the solutions, the equations themselves have many interesting mathematical properties, which we will use in this section to derive associated difference equations.

2.3.1 P_{II} and P_{34}

In this subsection, we show how Darboux's approach can be followed to find a transformation between solutions of two nonlinear ODEs, namely the second Painlevé equation (P_{II}) and the equation known as P_{34}:[4]

$$w'' = \frac{w'^2}{2\,w} + 4\,\alpha\,w^2 - x\,w - \frac{1}{2\,w} \tag{2.51}$$

where $w = w(x)$ and α is a constant parameter. Following Darboux's approach, we consider the transformation

$$w(x) = A\big(y\big)\,y' + B\big(y, x\big) \tag{2.52}$$

where we will assume that $y(x)$ satisfies an ODE of the form

$$y'' = R(y, x)$$

where $R(y, x)$ is rational in its arguments (note that it is taken to be independent of y'). Although we do not assume an explicit form of $R(y, x)$ at this stage, we find below that our assumption (2.52) imposes conditions on this function, which lead to a nontrivial ODE for $y(x)$.

Differentiating equation (2.52) twice and using equation (2.51), we find an equation of the form:

$$F_4\,y'^4 + F_3\,y^3 + F_2\,y'^2 + F_1\,y' + F_0 = 0 \tag{2.53}$$

where F_j, $j = 0, 1, 2, 3, 4$ are functions of y and x. Since y and y' can be considered as independent variables in the initial value space of P_{34}, we find that each coefficient F_j must vanish, for $j = 0, \ldots, 4$.

We find

$$F_4 = \big(A_y\big)^2 - 2\,A\,A_{yy} = 0 \;\Rightarrow\; A = (a_0\,y + b_0)^2,$$

where a_0 and b_0 are constants. For simplicity, we take $a_0 \equiv 0$ in what follows. In this case, we find

$$F_3 = 2\,b_0^2\,\big(-B_{yy} + 4\,\alpha\,b_0^4\big) = 0 \;\Rightarrow\; B(y, x) = 2\alpha\,b_0^4\,y^2 + c_0(x)\,y + c_1(x).$$

Again, for simplicity, we take $c_0(x) \equiv 0$ in the following. Now we obtain

$$F_2 = 48\,\alpha^2\,b_0^8\,y^2 + 16\,\alpha\,b_0^4\,c_1(x) - 2\,b_0^4\,\big(R_y + x\big)$$

and the vanishing of F_2 implies that

$$R_y = 8\alpha\,c_1(x) + 24\,\alpha^2\,b_0^4\,y^2 - x \Rightarrow\; R = 8\,\alpha^2\,b_0^4\,y^3 + (8\alpha\,c_1(x) - x)\,y + c_2(x).$$

[4] Due to its appearance as the 34th equation in the list of equations with the Painlevé property in the classification presented by Ince (1956).

Now the vanishing of F_1 and F_0 provides the following results:

$$c_1(x) = \frac{x}{4\,\alpha}, \quad c_2(x) = \frac{\pm 4\,\alpha - 1}{4\,\alpha\, b_0^2}.$$

Notice that α was given in equation (2.51); the freedom of sign in c_2 will be found useful later. These show immediately that the equation satisfied by $y(x)$ is

$$y'' = 8\,\alpha^2\, b_0^4\, y^3 + x\, y + \frac{\pm 4\,\alpha - 1}{4\alpha\, b_0^2},$$

which, upon taking $b_0^2 = 1/(2\,\alpha)$, becomes a more conventional form of $\mathrm{P_{II}}$:

$$\mathrm{P_{II}}: \quad y'' = 2\,y^3 + x\,y \pm 2\,\alpha - 1/2. \tag{2.54}$$

We have therefore found a Darboux-type transformation from $\mathrm{P_{34}}$ to $\mathrm{P_{II}}$:

$$w(x) = \frac{1}{2\alpha}\big(y' + y^2 + x/2\big). \tag{2.55}$$

This transformation is also often called a Miura transformation, because it is a reduced form of the well-known Miura transformation between integrable PDEs, which we will consider in Section 2.4.2 below.

The transformation (2.55) appears at first glance to be a one-way transformation. But in fact it is invertible. To see this, simply differentiate the transformation and use $\mathrm{P_{II}}$ to replace y'', which shows that y is given uniquely by

$$y(x) = \frac{1}{2\,w}\big(w' + 1\big).$$

Auto-Bäcklund transformations that map the solutions of two different copies of $\mathrm{P_{II}}$, with different values of the parameter α, are also known (Fokas and Ablowitz, 1982). We omit the details of their derivation and only state them here:

$$\widehat{y} := -y - \frac{1 + 2\alpha}{2\,y^2 + 2\,y' + x}, \qquad \widehat{\alpha} = \alpha + 1, \tag{2.56a}$$

$$\widetilde{y} := -y - \frac{2\alpha - 1}{2\,y^2 - 2\,y' + x}, \qquad \widetilde{\alpha} = \alpha - 1. \tag{2.56b}$$

Here, the functions $\widehat{y}(x)$ and $\widetilde{y}(x)$ solve $\mathrm{P_{II}}$ again but with parameters $\widehat{\alpha}$ and $\widetilde{\alpha}$, respectively.

Next we want to write these in the form of a difference equation. Let

$$y_n(x) := y\big(x; \alpha_n\big), \quad y_{n+1}(x) := \widehat{y}\big(x; \widehat{\alpha}_n\big), \quad y_{n-1}(x) := \widetilde{y}\big(x; \widetilde{\alpha}_n\big). \tag{2.57}$$

With this notation (2.56) can be written as

$$y_{n+1} = -y_n - \frac{1+2\alpha_n}{2\,y_n^2 + 2y_n{}' + x}, \quad \alpha_{n+1} = \alpha_n + 1, \tag{2.58a}$$

$$y_{n-1} = -y_n - \frac{2\alpha_n - 1}{2\,y_n^2 - 2y_n{}' + x}, \quad \alpha_{n-1} = \alpha_n - 1. \tag{2.58b}$$

Note that the equations governing α_n in each case can be solved explicitly, yielding $\alpha_n = n + \alpha_0$, where α_0 is arbitrary.

These equations allow us to iterate in n forward or backward. And as usual with BTs, if we shift the second equation $n \mapsto n+1$ and then eliminate y_{n+1} (i.e. use one equation to go forward, another to go backward) we get (2.54). The same holds if we eliminate y_n. Thus (2.58) form a normal BT pair, even though the derivative term appears in the denominator.

If we multiply out the denominators in the right sides respectively in equations (2.58) we find

$$2\,y_n^2 + 2y_n{}' + x = -\frac{1+2\alpha_n}{y_n + y_{n+1}},$$

$$2\,y_n^2 - 2y_n{}' + x = -\frac{2\alpha_n - 1}{y_n + y_{n-1}}.$$

Adding these two equations serves to eliminate $y_n{}'(x)$ and gives rise to

$$4\,y_n^2 + 2x = -\frac{1+2\alpha_n}{y_n + y_{n+1}} - \frac{2\alpha_n - 1}{y_n + y_{n-1}}.$$

Here we have $1+2\alpha_n = 1+2\alpha_0+2n$, $2\alpha_n - 1 = 2\alpha_0 - 1 + 2n$. If we denote $\zeta_n = \alpha_0 + n - 1/2$ then we get the difference equation (Jimbo and Miwa (1981), equation (5.3))

$$\frac{\zeta_{n+1}}{y_{n+1} + y_n} + \frac{\zeta_n}{y_{n-1} + y_n} + 2\,y_n^2 + x = 0, \tag{2.59}$$

where, now, y is a function of n, with x a fixed parameter. This equation is often referred to in the literature (Fokas et al., 1993) as the alternate discrete first Painlevé equation or alt-dP$_{\mathrm{I}}$ because its continuum limit yields the first Painlevé equation. Thus from the transformation property of the *second* Painlevé equation (2.54) we can derive a difference equation whose continuum limit is the *first* Painlevé equation (1.6).

2.3.2 P_{IV}

The fourth Painlevé equation P$_{\mathrm{IV}}$ is given by

$$w'' = \frac{(w')^2}{2w} + \frac{3}{2}w^3 + 4t\,w^2 + 2(t^2 - \alpha)w + \frac{\beta}{w}, \tag{2.60}$$

where α and β are (constant) parameters. We search for Darboux-type transformations of P_{IV} by using the hypothesis

$$W(t) = A(w(t), t)\, w'(t) + B(w(t), t)$$

where we assume $w(t; \alpha, \beta)$ satisfies P_{IV} with parameters α, β while $W(t; a, b)$ satisfies the same equation with possibly different respective parameters a, b. For convenience later, we define $\beta = -\gamma^2/2$ and $b = -c^2/2$.

Substituting the hypothesis into P_{IV} with parameters a, c, and using the fact that w satisfies P_{IV} with parameters α, γ, we find that the resulting equation is polynomial in $w'(t)$, of degree four. We demand that the coefficient of each power of $w'(t)$ vanishes. These lead to five partial differential equations for $A(w, t)$ and $B(w, t)$, the first of which is

$$2\,A\,\frac{\partial^2 A}{\partial w^2} - \frac{\partial A}{\partial w}^2 = \frac{A^2}{4\,w^2} - 2\,\frac{A}{w}\,\frac{\partial A}{\partial w} + 3\,A^4.$$

Since w possesses simple poles at arbitrary points, it is useful to study this equation as $w \to \infty$. We consider the rational behavior $A \sim p_0/w^p$ and find that $p = 1$ and $p_0^2 = \frac{1}{4}$. In fact, this is an exact solution.

Consider the choice $p_0 = 1/2$, i.e. $A = 1/(2w)$. The next coefficient equation, i.e. the coefficient of $(w')^3$ then gives

$$\left(w\,B_w\right)_w = \frac{B+t}{w}.$$

We can solve this equation by writing $B = F(w, t)/w$, and using separation of variables, which implies

$$F_{ww} = \frac{t + F_w}{w} \quad \Rightarrow t + F_w = w\,G(t)$$

where $G = G(t)$ is a function of integration in w. Integrating with respect to w one more time, we find

$$F = \frac{1}{2}G(t)\,w^2 - t\,w + H(t),$$

where H is another function of integration. Substitution into the remaining coefficient equations shows that

$$G(t) = -1,\ H(t) = (1 - \alpha - 2\,a)/3, a = \frac{1}{4}\left(2 \pm 3\,\gamma - 2\alpha\right),$$

and

$$a = \frac{1}{4}\left(2 + 3\,\gamma - 2\alpha\right) \Rightarrow c = \pm\left(1 + \gamma/2 + \alpha\right)$$

or

$$a = \frac{1}{4}\left(2 - 3\,\gamma - 2\alpha\right) \Rightarrow c = \pm\left(1 - \gamma/2 + \alpha\right).$$

In summary, these two choices give the following well-known BTs. The first choice is

$$\widetilde{w}^{\pm}(t) := \frac{w' - w^2 - 2t\,w \mp \gamma}{2w}, \tag{2.61a}$$

where $\widetilde{w}^{\pm}$ solves P_{IV} with parameters $\widetilde{\alpha}^{\pm}$ and $\widetilde{\beta}^{\pm}$ which are given by

$$\widetilde{\alpha}^{\pm} = \frac{1}{4}(2 - 2\alpha \pm 3\gamma), \tag{2.61b}$$

$$\widetilde{\beta}^{\pm} = -\frac{1}{2}\left(1 + \alpha \pm \frac{1}{2}\gamma\right)^2. \tag{2.61c}$$

A second BT for P_{IV} is

$$\widehat{w}^{\pm}(t) := -\frac{w' + w^2 + 2tw \mp \gamma}{2w}, \tag{2.62a}$$

where $\widehat{w}^{\pm}$ solves P_{IV} with parameters $\widehat{\alpha}^{\pm}$, $\widehat{\beta}^{\pm}$ which are given by

$$\widehat{\alpha}^{\pm} = -\frac{1}{4}(2 + 2\alpha \pm 3\gamma) \tag{2.62b}$$

$$\widehat{\beta}^{\pm} = -\frac{1}{2}\left(1 - \alpha \pm \frac{1}{2}\gamma\right)^2. \tag{2.62c}$$

Consider the choice $\widetilde{w}^-$ and $\widehat{w}^+$. We take $\widehat{w}^+$ to be a forward iterate and $\widetilde{w}^-$ to be a backward iterate in a discrete parameter n, so that

$$w_{n+1}(t) = -\frac{w_n{}' + w_n^2 + 2t w_n - \gamma_n}{2w_n} \tag{2.63}$$

$$w_{n-1}(t) = \frac{w_n' - w_n^2 - 2t\,w_n + \gamma_n}{2w_n}, \tag{2.64}$$

where we have written $\gamma = \sqrt{-2\beta}$ (with a choice of branch so that γ is real positive for large negative real β), and assumed that n parameterizes α, β, so that their iterates are consistent. We have

$$\alpha_{n+1} = -\frac{1}{4}(2 + 2\alpha_n + 3\gamma_n), \tag{2.65a}$$

$$\gamma_{n+1} = 1 - \alpha_n + \frac{\gamma_n}{2}, \tag{2.65b}$$

$$\alpha_{n-1} = \frac{1}{4}(2 - 2\alpha_n - 3\gamma_n), \tag{2.65c}$$

$$\gamma_{n-1} = 1 + \alpha_n - \frac{\gamma_n}{2}. \tag{2.65d}$$

Considering these as difference equations for α_n and γ_n, we can eliminate γ_n from equations (2.65a) and (2.65c) to get

$$\alpha_{n+1} - \alpha_{n-1} = -1 \quad \Rightarrow \quad \alpha_n = -\frac{n}{2} + c_0 + c_1(-1)^n, \tag{2.66}$$

and then substituting into equation (2.65a) gives

$$\gamma_n = n - 2c_0 + \frac{2c_1}{3}(-1)^n. \tag{2.67}$$

It is straightforward to check that equations (2.65) are all satisfied by these solutions. Now adding equations (2.63) and (2.64) we find that w_n must satisfy the difference equation (Fokas et al., 1993)

$$w_{n+1} + w_n + w_{n-1} = -2t + \frac{\gamma_n}{w_n}, \tag{2.68}$$

with γ_n given by equation (2.67). This difference equation is called the first discrete Painlevé equation, or $\mathrm{dP_I}$.

The process of obtaining difference equations from BTs can be extended to other Painlevé equations. It can, in fact, be extended to any pair of BTs, associated with ODEs, that are able to be consistently parameterized in terms of a discrete variable and in which the derivatives with respect to the independent variable of the ODE can be eliminated.

2.4 Bäcklund transformations for nonlinear PDEs

In this section we will mainly discuss the Korteweg–de Vries (KdV) equation, which is given by the PDE:

$$u_t = u_{xxx} + 6uu_x. \tag{2.69}$$

This so-called *nonlinear evolution equation* was derived in 1895 (Korteweg and de Vries, 1895) by the two people whose names it bears, in a study on shallow water waves, where also an exact "solitary wave" solution was presented. The equation was a key milestone in a big controversy on the nature of waves, following the famous "real life" observation of a solitary wave by John Scott Russell in 1834. It is not the place here to recite the whole history of the soliton, which can be found in several of the existing monographs on solitons and integrable systems; see e.g. Ablowitz and Segur (1981); Calogero and Degasperis (1982); Newell (1985); Drazin and Johnson (1989); Ablowitz and Clarkson (1991); Dickey (2003); Babelon et al. (2003); Hirota (2004). We will just mention that the KdV equation was revived 70 years after Korteweg and de Vries in a seminal

Figure 2.4 M. D. Kruskal

paper by Zabusky and Kruskal (1965). This led to results that were later described as a profound advance of the twentieth century (David Jr. et al., 1984). In a celebrated study by C. Gardner, J. Greene, M. Kruskal and R. Miura, (Gardner et al., 1967) where it was shown that the nonlinear PDE (2.69) can be exactly solved by an ingenious method, which is nowadays referred to as the *inverse scattering transform method*. Although this method is only applicable to very special equations, equations that we refer to as *soliton equations* or *exactly integrable* equations, we now know entire infinite families of such equations to which the method can be applied. (This is in stark contrast with the generic situation in which the nonlinear PDEs cannot be exactly solved and typically we have to resort to either qualitative studies or perturbative and numerical methods to study their solutions.)

2.4.1 Lax pair for KdV

A very important property of equation (2.69) is that there is an underlying over-determined system of *linear* equations governing a function $\psi(x, t; \lambda)$:

$$\psi_{xx} + u\psi = \lambda\psi, \tag{2.70a}$$

$$\psi_t = 4\psi_{xxx} + 6u\psi_x + 3u_x\psi, \tag{2.70b}$$

The first equation (2.70a) has the form of a Schrödinger equation, cf. equation (2.16), except that instead of the harmonic potential $u = \frac{1}{2}x^2$ we now consider a general potential $u = u(x; t)$. Equation (2.70a) can also be written as

$$\mathcal{L}\psi = \lambda\psi, \quad \text{where} \quad \mathcal{L} := \partial_x^2 + u \tag{2.71}$$

is called the (Schrödinger) differential operator. The parameter λ is an eigenvalue of the operator $\mathcal{L}$ and can, in principle, depend on t if u depends on it, however, by definition, it is independent of x.

The second equation (2.70b) describes the (linear) time-evolution of the function $\psi(x, t)$, where again the same u enters in the coefficients. The system (2.70) is over-determined: the two linear equations, holding simultaneously on the function ψ, can only be compatible with each other if additional conditions hold for the coefficients, which are all expressed in terms of u. This leads to the following statement:

Theorem 2.4.1 *Under the assumption of isospectrality, i.e.* $\lambda_t = 0$, *the linear system (2.70) is self-consistent, i.e.* $(\psi_{xx})_t = (\psi_t)_{xx}$, *iff either* $\psi \equiv 0$ *(the trivial case) or the potential* $u = u(x, t)$ *obeys the KdV equation (2.69).*

Proof The proof is by direct computation: First, using (2.70a) rewrite (2.70b) as follows:

$$\psi_t = (4\lambda + 2u)\psi_x - u_x\psi$$

and now consider the cross-derivatives:

$$
\begin{aligned}
(\psi_{xx})_t &= ((\lambda - u)\psi)_t = (\lambda_t - u_t)\psi + (\lambda - u)\,[(4\lambda + 2u)\psi_x - u_x\psi] \\
(\psi_t)_{xx} &= (4\lambda + 2u)\psi_{xxx} + 4u_x\psi_{xx} + 2u_{xx}\psi_x - u_x\psi_{xx} - 2u_{xx}\psi_x - u_{xxx}\psi \\
&= (4\lambda + 2u)\,((\lambda - u)\psi)_x + 3u_x(\lambda - u)\psi - u_{xxx}\psi \\
\Rightarrow \quad & (\psi_t)_{xx} - (\psi_{xx})_t = (u_t - u_{xxx} - 6uu_x)\,\psi - \lambda_t\psi.
\end{aligned}
$$

Hence, under the condition of *isospectrality*, $\lambda_t = 0$, we see that the system is compatible, i.e. $(\psi_t)_{xx} = (\psi_{xx})_t$ provided $u(x,t)$ obeys the KdV equation. □

The existence of a linear system of equations associated with a nonlinear PDE is symptomatic of its integrability through the inverse scattering method. Such a linear system, consisting of a spectral problem and an equation for the time evolution, is called a **Lax pair**, after P. D. Lax who gave a systematic framework for describing such linear problems in his celebrated paper (Lax, 1968). Although for virtually all soliton equations Lax pairs have been found, there is no fully algorithmic method known to produce a Lax pair for a given equation.

The Lax pair is the key ingredient in the *inverse scattering transform method* for solving the KdV equation (for a class of initial values given on the real line).

2.4.2 Miura transformation

The KdV equation possesses a remarkable transformation, called the Miura transformation after its inventor (Miura, 1968), which gives rise to many insights about its solutions, and in particular can be used to derive a Bäcklund transformation for KdV. To find it, consider how we can eliminate the function u from the system (2.70) and obtain a PDE in terms of the "eigenfunction" ψ itself.

From (2.70a) we find

$$u = \lambda - \psi_{xx}/\psi, \tag{2.72}$$

and inserting this into (2.70b) we obtain the following equation for ψ

$$\psi_t = \psi_{xxx} - 3\frac{\psi_x\psi_{xx}}{\psi} + 6\lambda\psi_x. \tag{2.73}$$

Introducing the variable:

$$v := \partial_x \log\, \psi, \tag{2.74}$$

we easily obtain from (2.72) the **Miura transformation**

$$u = \lambda - v_x - v^2. \tag{2.75}$$

Furthermore, after taking derivatives with respect to x on both sides of (2.73) and expressing all terms using v we get a PDE governing v:

$$v_t = v_{xxx} - 6v^2 v_x + 6\lambda v_x. \tag{2.76}$$

The latter equation (for $\lambda = 0$) is known as the **modified KdV equation** (mKdV), and it differs from the KdV equation (2.69) notably in the nonlinear term. The differential substitution (2.75) allows one to find a solution of the KdV equation given a solution of the mKdV equation: if v solves the mKdV (2.76) and u is defined by (2.75), then u solves the KdV equation (2.69).

2.5 Infinite sequence of conservation laws and KdV hierarchy

One important application of the Lax pair (2.70), is that it leads to the construction of an *infinite sequence of conservation laws* for the KdV equation. A conservation law is an identity of the form:

$$\partial_t \mathfrak{U} = \partial_x \mathfrak{F}, \tag{2.77}$$

holding on solutions of the equation. Here $\mathfrak{U}$ and $\mathfrak{F}$ are expressions involving the KdV variable u and its spatial derivatives $u_x, u_{xx}, \ldots$, where $\mathfrak{U}$ is called the *conserved density* and $\mathfrak{F}$ the associated *flux*. Under appropriate boundary conditions as $x \to \pm\infty$ on u and its derivatives, the integral of the right-hand side of (2.77) vanishes,

$$\int_{-\infty}^{\infty} \partial_x \mathfrak{F}\, dx = 0 \quad \Rightarrow \quad \frac{d}{dt}\int_{-\infty}^{\infty} \mathfrak{U}\, dx = 0,$$

leading, thus, to a conserved quantity given by the integral of $\mathfrak{U}$.

The construction of an infinite sequence of conservation laws can be done (Miura et al., 1968) by using a generalization of the Miura transform (2.75)

$$u = v - \epsilon v_x - \epsilon^2 v^2. \tag{2.78}$$

When this is substituted into the KdV equation (2.69) the result can be written as

$$-u_t + u_{xxx} + 6uu_x = (1 - \epsilon\partial_x - 2\epsilon^2 v)(-v_t + v_{xxx} + 6(v + \epsilon^2 v^2)v_x).$$

From this we conclude that if v solves

$$v_t = v_{xxx} + 6(v + \epsilon^2 v^2)v_x, \tag{2.79}$$

then u constructed by (2.78) solves (2.69). But since (2.69) does not depend on ϵ, we can use it as an expansion parameter

$$v(x,t;\epsilon) = \sum_{n=0}^{\infty} v_n(x,t)\,\epsilon^n,$$

and applying this term by term we can find v_n in terms of u and its derivatives, $v_0 = u$, $v_1 = u_x$ and thereafter recursively

$$v_{n+1} = \partial_x v_n + \sum_{m=0}^{n-1} v_m v_{n-1-m}, \quad n = 0, 1, 2, \ldots . \tag{2.80}$$

Some of the first terms are

$$\begin{aligned}
v_0 &= u, \\
v_1 &= u_x, \\
v_2 &= u_{xx} + u^2, \\
v_3 &= u_{xxx} + 4uu_x, \\
v_4 &= u_{xxxx} + 6u_{xx}u + 5u_x^2 + 2u^3, \\
&\vdots
\end{aligned}$$

Since the equation (2.79) is in the conservation law form (2.77), we find that integral of $v(x,t;\epsilon)$ is conserved and therefore each term in the ϵ expansion. However, since the odd v_n are total derivatives we find conserved quantities only from the even ones:

$$\begin{aligned}
v_0 &= u, \\
v_2 &= u_{xx} + u^2, \\
v_4 &= u_{xxxx} + 6u_{xx}u + 5u_x^2 + 2u^3, \\
&\vdots
\end{aligned}$$

Associated with these conservation laws comes another remarkable fact about the KdV equation, namely that the PDE (2.69) admits a *hierarchy* of compatible equations. This means that we can allow the function $u(x,t)$ to depend on a number of additional variables (which may be hidden in the integration constants of the solutions). The equations can be written in terms of evolution equations containing higher-time variables t_i, $(i = 0, 1, 2, \ldots)$ and take the form:

$$\partial_{t_j} u = 2\partial_x \frac{\delta v_{2j+1}}{\delta u}, \quad j = 0, 1, 2, \ldots . \tag{2.81}$$

where the operator $\delta/\delta u$ is called the *Euler operator* and acts on functions $F(u, u_x, u_{xx}, \ldots)$ of u and its derivatives:

$$\frac{\delta}{\delta u} := \frac{\partial}{\partial u} - \partial_x \frac{\partial}{\partial u_x} + \partial_x^2 \frac{\partial}{\partial u_{xx}} - + \cdots$$

The important thing to note is that all the equations (2.81) are mutually compatible, i.e. we have $\partial_{t_i}(\partial_{t_j} u) = \partial_{t_j}(\partial_{t_i} u)$ and as a consequence we have simultaneous solutions $u = u(x, t_1, t_2, t_3, \dots)$ of all equations, so that u can be considered as function of all higher time-flow variables.

The existence of this hierarchy is a key characteristic related to the integrability not only of the KdV equation itself, but of any equation in this hierarchy. As such, the notion of a hierarchy of compatible equations is a general phenomenon which can be extended also to discrete equations, as we shall see in Chapter 3, namely that of *multidimensional consistency*.

2.5.1 The Bäcklund transformation

Let us now turn to the derivation of Bäcklund transformations. We use the Miura transformation (2.75), and combine it with the simple observation that the equation (2.76) is *invariant* under the replacement $v \mapsto -v$. The key idea in deriving the Bäcklund transformation is to use both possible signs $\pm v$ in the Miura transformation (2.75), namely one sign in transforming from u to v and another in transforming from v to $\widetilde{u}$; that is, we will have the Miura transformations

$$\widetilde{u} = \lambda + v_x - v^2 , \tag{2.82a}$$

$$u = \lambda - v_x - v^2 . \tag{2.82b}$$

It is surprising that the trivial transformation $v \mapsto -v$ implies a highly nontrivial transformation $u \mapsto \widetilde{u}$ on the solutions of the KdV equation.

Adding and subtracting the two relations above we obtain

$$\widetilde{u} + u = 2(\lambda - v^2) \tag{2.83a}$$

$$\widetilde{u} - u = 2v_x . \tag{2.83b}$$

The latter can be integrated if we introduce the variable w by taking $u = w_x$. For the KdV equation this change of variables leads to

$$w_t = w_{xxx} + 3w_x^2, \tag{2.84}$$

after one integration in x. (Note that we have omitted an irrelevant integration constant.) This equation for w is called the **potential KdV equation** (pKdV).

In terms of the dependent variable w the equation (2.83b) can be integrated to $\widetilde{w} - w = 2v$, and inserting it into the first relation (2.83a) we obtain

$$(\widetilde{w}+w)_x = 2\lambda - \frac{1}{2}(\widetilde{w}-w)^2 \tag{2.85a}$$

written entirely in terms of w.

Equation (2.85a) provides us with the x-dependent part of the BT. To fully characterize the solution $\widetilde{w}$ we need also a t-dependent equation, which can be readily found by using the pKdV equation itself. Adding (2.84) for w and $\widetilde{w}$ and using (2.85a) to reduce $w_{xxx} + \widetilde{w}_{xxx}$ we obtain the relation

$$(\widetilde{w}+w)_t = (\widetilde{w}-w)(w_{xx}-\widetilde{w}_{xx}) + 2(w_x^2 + w_x\widetilde{w}_x + \widetilde{w}_x^2), \tag{2.85b}$$

and the relations (2.85a), (2.85b) together constitute the Bäcklund transformation for the KdV equation. (Note that, in practice, one could use the pKdV equation itself rather than (2.85b) to implement the BT.)

The system of relations (2.85) defines a transformation from a given solution $w(x,t)$ of the pKdV to a new solution $\widetilde{w}(x,t)$ of the pKdV equation. To prove this statement, differentiate (2.85a) with respect to t and (2.85b) with respect to x. Then we get respectively:

$$\begin{aligned}
(\widetilde{w}+w)_{xt} &= -(\widetilde{w}-w)(\widetilde{w}_t - w_t) \\
(\widetilde{w}+w)_{tx} &= (\widetilde{w}_x - w_x)(w_{xx}-\widetilde{w}_{xx}) + (\widetilde{w}-w)(w_{xxx}-\widetilde{w}_{xxx}) \\
&\quad +2\,(2w_x w_{xx} + 2\widetilde{w}_x\widetilde{w}_{xx} + \widetilde{w}_x w_{xx} + w_x\widetilde{w}_{xx}) \\
&= (\widetilde{w}-w)(w_{xxx}-\widetilde{w}_{xxx}) + 3(\widetilde{w}_x + w_x)(\widetilde{w}_{xx}+w_{xx})
\end{aligned}$$

and subtracting the first from the second we get, using also the x-derivative of (2.85a),

$$\begin{aligned}
0 &= (\widetilde{w}-w)\,[(\widetilde{w}_t - \widetilde{w}_{xxx}) - (w_t - w_{xxx})] + 3(\widetilde{w}_x + w_x)\,[-(\widetilde{w}-w)(\widetilde{w}_x - w_x)] \\
&= (\widetilde{w}-w)\left[(\widetilde{w}_t - \widetilde{w}_{xxx} - 3\widetilde{w}_x^2) - (w_t - w_{xxx} - 3w_x^2)\right],
\end{aligned}$$

and hence, if w solves the potential KdV equation (2.84) then either $\widetilde{w} = w$, or $\widetilde{w}$ provides another solution of the potential KdV equation.

2.5.2 Using BTs to generate multi-soliton solutions

Note that the Bäcklund pair (2.85) contains the parameter λ that does not appear in its base equation (2.84). This parameter can be used to generate more complicated solutions from simpler ones.

Suppose we know a given "seed solution" w of the potential KdV, then inserting this into (2.85a) we obtain a first-order nonlinear ODE for $\widetilde{w}$. This ODE is of the form:

$$\widetilde{w}_x = -\frac{1}{2}\widetilde{w}^2 + a(x)\widetilde{w} + b(x),$$

where the right-hand side is a quadratic in $\widetilde{w}$ (with x-dependent coefficients). This is the Riccati equation of Section 1.2. These equations are generally solvable through a linearization procedure. After solving this equation we have some integration constants that may depend on t; they can be determined from (2.85b).

Example 2.5.1 As a specific example, consider the simplest case where the seed solution w of the potential KdV equation (2.84) is the trivial solution $w \equiv 0$. Setting $w \equiv 0$ in (2.85a) yields

$$\widetilde{w}_x = 2\lambda - \frac{1}{2}\widetilde{w}^2,$$

which can be integrated by separation of variables and yields

$$\widetilde{w}(x,t) = 2k \tanh\left(kx + c(t)\right), \quad \lambda = k^2.$$

Substituting this expression into the (2.85b) (with $w = 0$) reveals that $c_t = 4k^3$ and hence we have $c(t) = 4k^3 t + c_0$, where c_0 is a constant. Thus we have obtained for (2.84) the solution:

$$\widetilde{w}(x,t) = 2k \tanh\left(kx + 4k^3 t + c_0\right). \tag{2.86}$$

Note that $\widetilde{w}(x,t) = 2k \coth\left(kx + 4k^3 t + c_0\right)$ is also a solution, albeit singular.

2.5.3 Permutability property of BTs

The solution we obtained above can now be regarded as the starting point for applying the BT once again to obtain yet another solution of the potential KdV equation. Carrying out this procedure we can iteratively obtain an infinite sequence of increasingly complicated solutions of the same nonlinear PDE. Solving the corresponding Riccati equation at each successive step obviously becomes increasingly more cumbersome as we go along. However, there is a powerful new ingredient that can be used to simplify the iteration, namely the *permutability property* of the BTs.

Suppose we want to compose two different BTs: one with a parameter λ, as in (2.85), and one with another parameter, say μ, given by

$$BT_\lambda : w \overset{\lambda}{\mapsto} \widetilde{w} \qquad (\widetilde{w} + w)_x = 2\lambda - \frac{1}{2}(\widetilde{w} - w)^2, \tag{2.87a}$$

$$BT_\mu : w \overset{\mu}{\mapsto} \widehat{w} \qquad (\widehat{w} + w)_x = 2\mu - \frac{1}{2}(\widehat{w} - w)^2, \tag{2.87b}$$

where we have used the notation $\widetilde{w}$, $\widehat{w}$ to denote the solution obtained by applying the BT with parameter λ, μ, respectively.

The two ways to compose these BTs, either start with BT_λ and subsequently apply BT_μ, or the other way round, will yield iterated solutions which we can denote by $\widehat{\widetilde{w}}$ and $\widetilde{\widehat{w}}$ respectively,

$$\widehat{\widetilde{w}} = BT_\mu \circ BT_\lambda\, w, \qquad \widetilde{\widehat{w}} = BT_\lambda \circ BT_\mu\, w.$$

The highly nontrivial outcome is that (under certain circumstances) both ways of composing BTs lead to the same result, $\widehat{\widetilde{w}} = \widetilde{\widehat{w}}$, and hence the two BTs commute. This is the famous *permutability property* of the BTs derived in the seminal 1973 paper by Wahlquist and Estabrook. (In that paper it was also observed that in order to get a regular two-soliton solution $\widetilde{\widehat{w}}$, the ingredient solitons $\widetilde{w}$, $\widehat{w}$ cannot both be regular.)

Theorem 2.5.1 *The BTs given by* (2.87a), (2.87b) *for different parameters* λ *and* μ *generate solutions (with a suitable choice of integration constants) for which we have the following commutation diagram of BTs given in Figure* 2.5.

The proof of the permutability property is quite deep and relies on the spectral properties that play at the background of the equations; see Lamb (1980), p. 247 for a proof based on the inverse scattering scheme.

Proof We outline a direct proof, using the additional relations:

$$BT_\mu : \widetilde{w} \overset{\mu}{\mapsto} \widehat{\widetilde{w}} \qquad \left(\widehat{\widetilde{w}} + \widetilde{w}\right)_x = 2\mu - \frac{1}{2}\left(\widehat{\widetilde{w}} - \widetilde{w}\right)^2, \tag{2.87c}$$

$$BT_\lambda : \widehat{w} \overset{\lambda}{\mapsto} \widetilde{\widehat{w}} \qquad \left(\widetilde{\widehat{w}} + \widehat{w}\right)_x = 2\lambda - \frac{1}{2}\left(\widetilde{\widehat{w}} - \widehat{w}\right)^2. \tag{2.87d}$$

Following the upper trajectory of the diagram we eliminate $\widetilde{w}_x$ using (2.87a), (2.87c) we can solve $\widetilde{w}$ in terms of w and $\widehat{\widetilde{w}}$ as follows:

$$\widetilde{w} = \frac{1}{2}(\widehat{\widetilde{w}} + w) + \frac{(\widehat{\widetilde{w}} - w)_x + 2(\lambda - \mu)}{\widehat{\widetilde{w}} - w}.$$

Reinserting this into (2.87c) we obtain:

$$\lambda + \mu = (\widehat{\widetilde{w}} + w)_x + \partial_x^2 \log(\widehat{\widetilde{w}} - w) + \frac{1}{2}\left(\partial_x \log(\widehat{\widetilde{w}} - w)\right)^2 \\ + \frac{1}{8}(\widehat{\widetilde{w}} - w)^2 + 2\frac{(\lambda - \mu)^2}{(\widehat{\widetilde{w}} - w)^2}, \tag{2.87e}$$

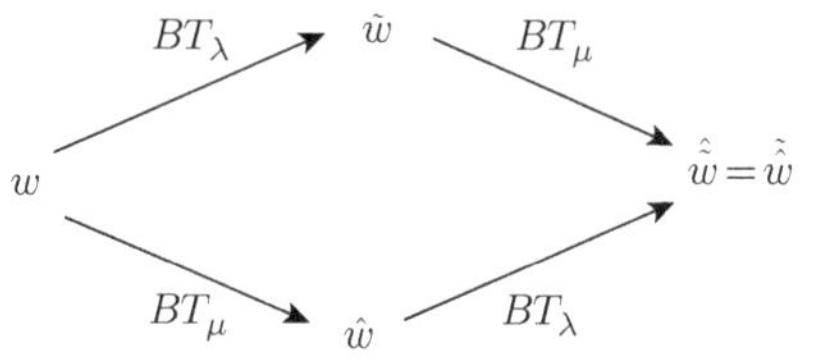

Figure 2.5 The permutability property of Bäcklund transformations.

which is symmetric under the interchange of λ and μ, therefore one gets an identical equation for $\widetilde{\widehat{w}}$. A similar symmetry can be derived for the t-part of the BT. Now the key point is that for a given seed solution w of the original equation, any solution $\widehat{\widetilde{w}}$ of (2.87e) also solves the identical equation for $\widetilde{\widehat{w}}$, therefore we can identify $\widehat{\widetilde{w}} = \widetilde{\widehat{w}}$. This amounts to choosing particular sets of integration constants in constructing $\widetilde{\widehat{w}}$.

The consequences of the permutability property are far reaching, and we will give an explicit realization as follows. In fact, using all four Bäcklund relations (2.87a)–(2.87d), and setting $\widehat{\widetilde{w}} = \widetilde{\widehat{w}}$, we can now eliminate all the derivatives from the four equations (2.87) and obtain a purely algebraic equation of the form:

$$(\widehat{\widetilde{w}} - w)(\widehat{w} - \widetilde{w}) = 4(\mu - \lambda). \tag{2.88}$$

This allows us to obtain directly the iterated BT transformed variable $\widehat{\widetilde{w}}$ from $\widehat{w}$, $\widetilde{w}$, w without having to derive the solution through the Riccati equations of the BT.

2.5.4 Bäcklund transformation for the sine-Gordon equation

BTs exist for other integrable evolution equations as well. In fact the first BT, the one proposed by Bäcklund himself, is associated with the **sine-Gordon** equation,

$$\theta_{xt} = \sin\theta. \tag{2.89}$$

For this equation the Bäcklund transformation $\theta \overset{\lambda}{\mapsto} \widetilde{\theta}$ is given by the following relations:

$$\left(\widetilde{\theta} - \theta\right)_x = 2\lambda \sin\left(\tfrac{1}{2}(\widetilde{\theta} + \theta)\right), \tag{2.90a}$$

$$\left(\widetilde{\theta} + \theta\right)_t = \frac{2}{\lambda} \sin\left(\tfrac{1}{2}(\widetilde{\theta} - \theta)\right), \tag{2.90b}$$

connecting a variable $\theta(x,t)$ to a new variable $\widetilde{\theta}(x,t)$. By calculating the t derivative (2.90a) and the x-derivative (2.90b) and then taking a sum or difference, one can easily derive (2.89) for $\widetilde{\theta}$ or θ, respectively. Thus (2.90) is a one-parameter auto-Bäcklund transformation for the sine-Gordon equation.

As before, we can now introduce a second BT $\theta \overset{\mu}{\mapsto} \widehat{\theta}$ of the form (2.90) with parameter μ, namely

$$\left(\widehat{\theta} - \theta\right)_x = 2\mu \sin\left(\tfrac{1}{2}(\widehat{\theta} + \theta)\right), \tag{2.91a}$$

$$\left(\widehat{\theta} + \theta\right)_t = \frac{2}{\mu} \sin\left(\tfrac{1}{2}(\widehat{\theta} - \theta)\right). \tag{2.91b}$$

We can also apply BT_μ on $\widetilde{\theta}$ and BT_λ on $\widehat{\theta}$, and obtain eight equations; the two additional equations with x-derivatives are

$$\left(\widehat{\widetilde{\theta}} - \widehat{\theta}\right)_x = 2\lambda \sin\left(\tfrac{1}{2}(\widehat{\widetilde{\theta}} + \widehat{\theta})\right), \tag{2.92a}$$

$$\left(\widetilde{\widehat{\theta}} - \widetilde{\theta}\right)_x = 2\mu \sin\left(\tfrac{1}{2}(\widetilde{\widehat{\theta}} + \widetilde{\theta})\right). \tag{2.92b}$$

(2.92c)

Eliminating the x derivatives between (2.90a, 2.91a, 2.92) we obtain (under the assumptions as explained above on the permutability of the BTs, i.e. $\widehat{\widetilde{\theta}} = \widetilde{\widehat{\theta}}$ (Lamb, 1980)), the following equation

$$\lambda \sin\left(\tfrac{1}{4}(\widehat{\widetilde{\theta}} + \widehat{\theta} - \widetilde{\theta} - \theta)\right) = \mu \sin\left(\tfrac{1}{4}(\widehat{\widetilde{\theta}} + \widetilde{\theta} - \widehat{\theta} - \theta)\right). \tag{2.93}$$

If, for simplification, we denote $e^{\frac{i}{2}\theta} = w$, we can write (2.93) as

$$\lambda(\widehat{\widetilde{w}}\widehat{w} - \widetilde{w}w) = \mu(\widehat{\widetilde{w}}\widetilde{w} - \widehat{w}w). \tag{2.94}$$

Thus again we can get $\widehat{\widetilde{w}}$ by a simple algebraic equation from $\widehat{w},\ \widetilde{w},\ w$. (It should be noted that this equation (2.94) arises also as the lattice potential mKdV equation, cf. (3.11).)

We have now discussed two integrable PDEs and their BTs, but in fact BTs have been constructed for almost all integrable (1+1)-dimensional PDEs. For (2+1)-dimensional PDEs, such as the KP equation (3.101), some results also exists, but then it would be necessary to consider more general transformations, like the Moutard transformation (Nimmo and Schief, 1997).

2.5.5 Transition to lattice equations

It is obvious from the discussion in Section 2.5.3 that by iterating the BTs with two different parameters we obtain from one seed solution w an entire *lattice of solutions*; see Figure 2.6. In practice this is not so interesting, because the construction of $\widetilde{\widetilde{w}}$ by applying the same BT twice usually produces a solution with singularities. Indeed, each new soliton should have its own parameter and therefore the picture for a three-soliton solution is that of a cube, with the BTs for λ, μ, ν giving the shifts in the three dimensions. (A flattened form of this is the Bianchi lattice; see Lamb (1980), pp. 249–253.) And the miracle of integrable soliton equations (elaborated in the next chapter) is that the cube can be closed to contain the same three-soliton solution no matter which order we compose the BTs.

We have derived the permutability equations (2.88) and (2.94) from the properties of the potential KdV and sG, respectively; thus these equations are *descriptive*, as they describe yet another property of the sequence of functions derived using BTs. Introducing an enumeration of the solutions as $w_{n,m} = BT_\lambda^n \circ BT_\mu^m\, w$, we can write (2.88) and (2.94) as difference equations of the form:

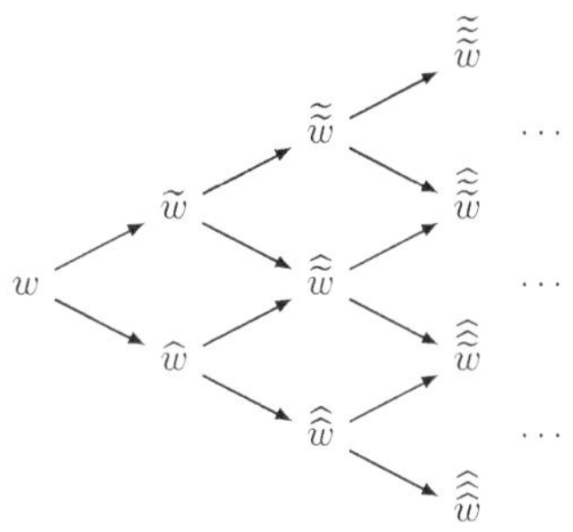

Figure 2.6 A lattice of BTs.

$$(w_{n+1,m+1} - w_{n,m})(w_{n,m+1} - w_{n+1,m}) = 4(\mu - \lambda), \tag{2.95}$$

and

$$\lambda(w_{n+1,m+1}w_{n,m+1} - w_{n+1,m}w_{n,m}) = \mu(w_{n+1,m+1}w_{n+1,m} - w_{n,m+1}w_{n,m}), \tag{2.96}$$

respectively, the shifts along the lattice $w_{n,m} \mapsto w_{n+1,m}$ and $w_{n,m} \mapsto w_{n,m+1}$ corresponding to the application of the Bäcklund transformations BT_λ and BT_μ.

In a sense, having reached this level we can "forget" about the original PDEs from whence the construction originated, and place the permutability equations at the center of our focus. In the next chapter we will *change our point of view*: at each elementary plaquette of the above lattice of solutions we have a relation of the form (2.95) or (2.96) (or something else for other equations), and we will now elevate these equations to becoming the main equations of interest.

2.6 Notes

In this chapter we have seen how a change of perspective between independent variable and parameter allowed us to deduce difference equations from differential equations.

This duality between continuous and discrete equations is common to families of classical special functions, and in particular for classical orthogonal polynomials (Szegö, 1939; Chihara, 2014). There is always some parameter(s), and discrete equations arise as addition formulae for such functions. More generally, the hypergeometric family of functions obey *contiguity* relations, which relate members of the family with different parameters. In the modern theory of special functions, such relations are generalized to the multivariate hypergeometric functions as well as the q- and elliptic deformations (Andrews et al., 1999; Ismail, 2005; Gasper and Rahman, 2004; Olver et al., 2010).

The natural progression of this duality from the linear to the nonlinear case leads on the one hand to elliptic functions, and on the other hand to the Painlevé transcendents. We provided examples in this chapter. Further elaborations can be found in Gromak et al. (2002); Iwasaki et al. (1991).

The idea that BTs can be interpreted as shifts on a chain goes back to Wahlquist and Estabrook (1973); see also Levi (1981); Levi and Benguria (1980) in the context of differential-difference equations. The interpretation of the permutability formula of the BT as a lattice equation goes back to Nijhoff et al. (1983a).

Exercises

2.1 Apply a transformation of variables $w(x) = u(\xi(x))$ to Bessel's equation (2.21) and show that the choice $\xi = \log(x)$ converts it to the form

$$u_{\xi\xi} + (e^{2\xi} - \nu^2)\, u = 0.$$

Apply Darboux's theorem directly to this equation and show that the result leads to the recurrence relations (2.27) found in Section 2.1.3.

2.2 Consider the *Tchebysheff* equation given by the second-order ordinary differential equation

$$(1 - x^2)\frac{d^2T}{dx^2} - x\frac{dT}{dx} + \nu^2 T = 0, \tag{2.97}$$

for a function $T = T(x;\, \nu)$, in which ν is a parameter.

(a) By considering a power series expansion of the form $T(x;\, \nu) = \sum_{k=0}^{\infty} a_k x^k$, and inserting this expansion into the Tchebysheff equation (2.97), derive a recurrence relation for the coefficients a_k. By distinguishing the following two cases of initial values for the recurrence: (i) $a_0 = 1, a_1 = 0$; (ii) $a_0 = 0, a_1 = 1$, show that one obtains an *even* respectively *odd* function as solution of equation (2.97) and give in either case the first three terms in the series expansions.

(b) Verify that (2.97) can be written in factorized form as follows:

$$\left[(\nu + 1)x + (1 - x^2)\frac{d}{dx}\right]\left[\nu x - (1 - x^2)\frac{d}{dx}\right] T = \nu(\nu + 1)T. \tag{2.98}$$

Hence, by introducing the new function $\nu\widetilde{T} := [\nu x - (1 - x^2)\,(d/dx)]T$, write the Tchebysheff equation as a coupled system of linear first-order differential equations.

(c) From the result of part (a) derive a second-order differential equation for the new function $\widetilde{T}$ showing that it obeys the Tchebysheff equation with parameter $\nu + 1$ instead of ν. Identifying $T_\nu =: T(x;\, \nu)$, $T_{\nu+1} =: \widetilde{T}(x;\, \nu)$ and, eliminating the derivatives from the first-order system, derive the following second-order difference equation for the function T_ν:

$$T_{\nu+1} + T_{\nu-1} = 2x T_\nu,$$

where the parameter ν now plays the role of the independent discrete variable.

(d) Setting $\nu = n$, with n a positive integer (or zero), introduce the generating function

$$C(x,z) = \sum_{n=0}^{\infty} T_n(x) z^n,$$

where $T_n(x)$ is the solution of the difference equation of part (d) subject to the initial conditions $T_0(x) = 1$, $T_1(x) = x$. Find the explicit form of the generating function $C(x,z)$.

2.3 Consider Gauss' hypergeometric function ${}_2F_1$.

(a) Verify that ${}_2F_1$ satisfies the differential equation (2.30).

(b) Derive the following relations by using the first parts of equations (2.32), (2.34):

$$\frac{d}{dx}\,{}_2F_1\left(\begin{matrix} a, b \\ c \end{matrix}; x\right) = \frac{ab}{c}\,{}_2F_1\left(\begin{matrix} a+1, b+1 \\ c+1 \end{matrix}; x\right),$$

$$x(x-1)\frac{(a+1)(b+1)}{c(c+1)}\,{}_2F_1\left(\begin{matrix} a+2, b+2 \\ c+2 \end{matrix}; x\right)$$
$$+\frac{c-(a+b+1)x}{c}\,{}_2F_1\left(\begin{matrix} a+1, b+1 \\ c+1 \end{matrix}; x\right) - {}_2F_1\left(\begin{matrix} a, b \\ c \end{matrix}; x\right) = 0,$$

the second of which involves only shifts in the parameters.

(c) Equation (2.33) relates different copies of ${}_2F_1$ involving changes only in the parameter a. Derive the following contiguous relation which relates ${}_2F_1$ involving changes only in the parameter c:

$$(1-x)\,{}_2F_1\left(\begin{matrix} a, b \\ c \end{matrix}; x\right) = \frac{c-(2c-a-b+1)x}{c)}\,{}_2F_1\left(\begin{matrix} a, b \\ c+1 \end{matrix}; x\right)$$
$$+\frac{(c-a+1)(c-b+1)}{c(c+1)}\,{}_2F_1\left(\begin{matrix} a, b \\ c+2 \end{matrix}; x\right)$$

by combining (2.34) and (2.32) or by using the hypergeometric series.

2.4 Use equations (2.44a) and (2.44b) to show the following

$$\mathrm{sn}(x+\delta)\,\mathrm{sn}(x-\delta) = \frac{\mathrm{sn}^2(x) - \mathrm{sn}^2(\delta)}{1 - k^2\mathrm{sn}^2(x)\,\mathrm{sn}^2(\delta)}$$

leading to the "multiplicative" difference equation:

$$y_{n+1}y_{n-1} = \frac{y_n^2 - \mathrm{sn}^2(\delta)}{1 - k^2\mathrm{sn}^2(\delta)\,y_n^2}.$$

Compare this to the additive equation given by the McMillan map (1.19).

2.5 In deriving equations (2.63) and (2.64), we assumed that each mapping of w to $\widehat{w}$ or $\widetilde{w}$ represented one step in the evolution of n. Consider an evolution of step-size two: that is, $w(t,\alpha_n) \mapsto \widehat{w}(t,\alpha_{n+2})$, $w(t,\alpha_n) \mapsto \widetilde{w}(t,\alpha_{n-2})$. Show that then the equations $\alpha_{n+2}=\alpha_n+1, \alpha_{n-2}=\alpha_n-1$, can be solved by $\alpha_n=\alpha_0+\alpha_1(-1)^n+n/2$. Find the resulting difference equation by elimination of $w_n{}'$ and show that taking $w_{2k-1}=u_k$ and $w_{2k}=v_k$ yields the system of decoupled difference equations

$$\frac{(\alpha_0-\alpha_1)+k}{u_{k+1}+u_k}+\frac{(\alpha_0-\alpha_1-1)+k}{u_{k-1}+u_k}=-(2u_k^2+t),$$
$$\frac{(\alpha_0+\alpha_1+1/2)+k}{v_{k+1}+v_k}+\frac{(\alpha_0+\alpha_1-1/2)+k}{v_{k-1}+v_k}=-(2v_k^2+t).$$

Each is a separate copy of alt-dP$_{\mathrm{I}}$ given in (2.59).

2.6 Consider the linear spectral problem:

$$\psi_{xx}+u\psi=\lambda\psi\ , \tag{2.99}$$

with $u=u(x)$ a potential function.

(a) Consider two independent solutions, ψ_1 and ψ_2, of equation (2.99), for the same value λ of the spectral parameter. Introduce now the variable $z=\psi_1/\psi_2$, and derive the *Cole–Hopf relation*:

$$v\equiv\partial_x\ln\psi_2=-\frac{1}{2}\frac{z_{xx}}{z_x}\ . \tag{2.100}$$

(b) Show that by back-substitution of the expression for ψ_2 from (2.100) into the relation (2.99), one can express the potential u in terms of z, giving rise to the relation:

$$u=\lambda+\frac{1}{2}\left(\frac{z_{xxx}}{z_x}-\frac{3}{2}\frac{z_{xx}^2}{z_x^2}\right)\ . \tag{2.101}$$

NB: The expression between brackets on the right-hand side of (2.101) is sometimes denoted as $\{z,x\}$, and is called the *Schwarzian derivative* of z.

(c) Consider next the linear isospectral deformation of (2.99) given by:

$$\psi_t=4\psi_{xxx}+6u\psi_x+3u_x\psi\ , \tag{2.102}$$

the consistency condition of which, with equation (2.99), is given by the KdV equation (2.69). Use (2.102) to derive the following equation for z

$$z_t=6\lambda z_x+z_{xxx}-\frac{3}{2}\frac{z_{xx}^2}{z_x}\ , \tag{2.103}$$

which for $\lambda=0$ is called the *Schwarzian KdV equation* (SKdV).

(d) Derive the following formula for the Schwarzian derivative of the composed function $h(x) = f(g(x))$

$$\{h, x\} = g'^2 \{f, y\} + \{g, x\} ,$$

where $y = g(x)$.

2.7 Consider the following Lax pair for φ:

$$\varphi_{xx} + 2v\varphi_x = -\lambda\varphi ,$$
$$\varphi_t = 2u\varphi_x - 4\lambda v\varphi .$$

Under the assumption of isospectrality (i.e. $\lambda_t = 0$) show that the compatibility conditions by cross-differentiation of the latter linear system for φ give rise to both the derivative of the Miura transform, i.e. $u_x = -v_{xx} - 2vv_x$, and the modified KdV equation:

$$v_t = v_{xxx} - 6v^2 v_x + 6\lambda v_x .$$

2.8 Consider the following Bäcklund transform

$$\widetilde{z}_x z_x = \bar{\lambda}(\widetilde{z} - z)^2 \tag{2.104a}$$
$$z_t \widetilde{z}_x + z_x \widetilde{z}_t = z_{xx}\widetilde{z}_{xx} + 2\lambda(\widetilde{z} - z)(\widetilde{z}_{xx} - z_{xx}) - 4\bar{\lambda}(\widetilde{z}_x - z_x)^2 \tag{2.104b}$$

for the Schwarzian KdV equation (2.103) for $\lambda = 0$.

(a) Show that if z obeys the SKdV equation, then $\tilde{z}$ will also obey the SKdV equation.

(b) By considering a second BT with parameter μ: $\widehat{z}_x z_x = \bar{\mu}(\widehat{z} - z)^2$ and the permutability diagram for the BTs as in Figure 2.5, where we now assume $\widehat{\widetilde{z}} = \widetilde{\widehat{z}}$, derive a permutability condition for $z, \widetilde{z}, \widehat{z}$ and $\widehat{\widetilde{z}}$.

2.9 Consider the potential modified Korteweg–deVries (mKdV) equation:

$$v_t = v_{xxx} - 3\frac{v_x v_{xx}}{v} \tag{2.105}$$

and consider the BT $v \mapsto \widetilde{v}$ given by the following set of equations:

$$(\widetilde{v}v)_x = p\left(\widetilde{v}^2 - v^2\right) , \tag{2.106}$$

$$(\widetilde{v}v)_t = 2p\left(\widetilde{v}\widetilde{v}_{xx} - vv_{xx} - 2\widetilde{v}_x^2 + 2v_x^2\right) . \tag{2.107}$$

(a) By cross-differentiating equations (2.106) and (2.107), show that if v is a solution of (2.105) then $\widetilde{v}$ is also a solution of the same equation.

(b) Let $v \mapsto \widehat{v}$ denote a similar BT with parameter q instead of p, for which we have

$$(\widehat{v}v)_x = q\left(\widehat{v}^2 - v^2\right) . \qquad (2.108)$$

By imposing that the BTs $v \mapsto \widetilde{v}$ and $v \mapsto \widehat{v}$ commute, show that the combination

$$A = pv\widehat{v} + q\widetilde{v}\widehat{\widetilde{v}} - qv\widetilde{v} - p\widehat{v}\widehat{\widetilde{v}}$$

is constant with respect to both x and t. Furthermore, combine (2.106), (2.108) and their counterparts obtained by taking the $\widehat{\ }$ shift of the first and the $\widetilde{\ }$ shift of the second, to eliminate the x-derivatives, and show that either the quantity A vanishes, or $v\widehat{\widetilde{v}} + \widetilde{v}\widehat{v} = 0$. Hint: rewrite (2.106) and (2.108) as

$$\partial_x \ln \widetilde{v} + \partial_x \ln v = p\left(\frac{\widetilde{v}}{v} - \frac{v}{\widetilde{v}}\right), \quad \partial_x \ln \widehat{v} + \partial_x \ln v = q\left(\frac{\widehat{v}}{v} - \frac{v}{\widehat{v}}\right).$$

(c) Verify that $v(x,t) \equiv 1$ is a solution of equation (2.105). Substitute this solution in the BT (2.106) and solve the resulting differential equation for $\widetilde{v}$ by separation of variables to yield:

$$\widetilde{v}(x,t) = -\tanh(px + c(t)),$$

in which $c(t)$ arises as the integration constant (i.e. constant with respect to x). Determine the dependence of $c(t)$ on t either by substituting the solution $\widetilde{v}(x,t)$ back into (2.105), or by using (2.107).

2.10 Construct a two-soliton solution (2SS) by starting from the seed solution $w \equiv 0$, and two one-soliton solutions of the type (2.86) (with different parameters k and l, where $\lambda = k^2$, $\mu = l^2$, respectively) and solving for $\widehat{\widetilde{w}}$ from (2.88). Note that the phases c_0 may be taken to be different for $\widetilde{w}$ and $\widehat{w}$. Verify explicitly that the $\widehat{\widetilde{w}}$ so constructed actually solves (2.84), and that it is singular. Construct also a two-soliton solution from one tanh and one cosh-type soliton and find the condition for it to be regular.

2.11 Starting from the trivial solution $\theta \equiv 0$ of the sine-Gordon equation, use the BT (2.90) to obtain a solution $\widetilde{\theta}(x,t)$ of the same equation containing the parameter λ, namely:

$$\widetilde{\theta}(x,t) = 2\tan^{-1}\left\{\exp\left(\lambda x + \frac{t}{\lambda} + \varphi_0\right)\right\} , \qquad (2.109)$$

where φ_0 is a (constant) phase. Hint: use the integral

$$\int \frac{d\varphi}{\sin\varphi} = \ln\left(\tan\frac{\varphi}{2}\right) + c.$$

3

Integrability of P∆Es

The permutability property of Bäcklund transformations of solutions (of soliton equations), as discussed at the end of the previous chapter, forms one of the main sources of integrable partial difference equations (P∆Es) on the two-dimensional space–time lattice. In this chapter we study these equations in their own right, i.e. ignore their origin, and no longer assume that the dependent variable is a solution of any previously given PDE. In particular, we will consider the "integrability" of these P∆Es.

In general terms the concept of integrability is associated with regularity, solvability and amenability to exact and explicit treatments. These notions will have different mathematical meanings in different contexts, and it is indeed an open problem to give a universal definition of integrability. There are many characteristics associated with integrability, such as the existence of a Lax pair, compatibility structures such as Bäcklund transforms and hierarchies of compatible equations, good behavior of solutions with respect to singularities, regularity with regard to growth properties and complexity of solutions. These aspects of integrability are reflected in various different approaches in constructing and solving P∆Es, such as the τ-function approach of Hirota and the Kyoto school, symmetry methods and inverse methods such as "direct linearization", some of which will be discussed in later chapters. Many of these features can be posed as definitions of integrability, but they are often restricted to specific classes of integrable systems. It is remarkable that if any two different definitions of integrability can be applied to some joint set of equations, they agree on which equations should be called integrable.

The presence of free parameters in the equations (namely the Bäcklund parameters λ and μ, which we will now reinterpret as *lattice parameters*) will play a crucial role in the development of the theory in this chapter. Furthermore, as we will see later, the parameters render the P∆Es very rich: since they can be seen to represent the widths of the underlying lattice grid they allow us to recover, through continuum limits, a great wealth of other equations, semi-continuous (i.e. differential-difference type) as well as fully continuous (i.e. partial differential type) ones. The interplay between the discrete and continuous structures will prove to be one of the emerging features of the integrable systems that we study.

In this chapter we will first discuss "quadrilateral" equations: that is, equations defined on an elementary square of the Cartesian lattice (see Section 1.1.10), and for them the property of multidimensional consistency, which has turned out to be a very effective concept for integrability. However, not all equations that we consider integrable fit into this concept and we will discuss these as well. Finally, it should be stressed that the full classification of integrable lattice equations is still very much an open problem, and only some specific subclasses have been conclusively classified so far.

3.1 Quadrilateral PΔEs

We have already discussed some basic aspects related to PΔEs in Sections 1.1.8 and 1.5. We assume that the *ambient space*, i.e. the underlying lattice, is an infinite uniform and rectangular Cartesian lattice, as in Figures 1.3, 1.6 and 1.7.

3.1.1 Definition

We will investigate here partial difference equations (PΔEs) of the following form, which we will call **quadrilateral PΔEs**:

$$Q(u_{n,m}, u_{n+1,m}, u_{n,m+1}, u_{n+1,m+1}) = 0\,. \tag{3.1}$$

Such an equation gives a relation connecting the values at the vertices of an elementary quadrilateral. Thus, n and m play the role of independent (discrete) variables, very much like x and t being the continuous variables in typical PDEs. It is important to note that the *same* equation is assumed to hold at *every* elementary square of the underlying infinite lattice.

We adopt here the following shorthand notation for vertices surrounding an elementary plaquette[1]

$$u := u_{n,m}, \quad \widetilde{u} := u_{n+1,m}, \quad \widehat{u} := u_{n,m+1}, \quad \widehat{\widetilde{u}} := u_{n+1,m+1},$$

and we will also sometimes use under-accents for shifts in the negative direction

$$\underset{\sim}{u} := u_{n-1,m}, \quad \underset{\wedge}{u} := u_{n,m-1}.$$

Equation (3.1) can then be written as

$$Q(u, \widetilde{u}, \widehat{u}, \widehat{\widetilde{u}}) = 0. \tag{3.2}$$

Schematically, this configuration of points is given in Figure 3.1

[1] In the literature, several other notations are used, e.g. $u_{n+1,m} = u_{10} = u_{[1]}$, $u_{n+1,m+1} = u_{11} = u_{[12]}$.

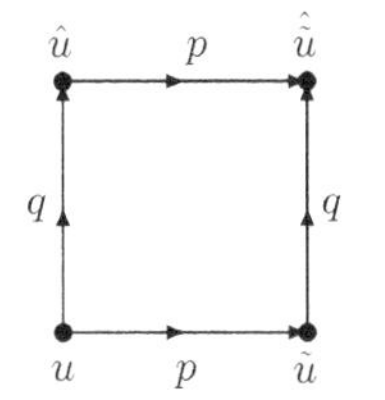

Figure 3.1 The lattice points involved in equation (3.3), along with the lattice parameters associated with the directions.

The notation is inspired by the one for Bäcklund transformations, which (as we have seen in the previous chapter) gives rise to purely algebraic equation as a consequence of their permutability property. Even though the form (3.2) seems very restrictive and appears to be the most elementary form as a model type of equations, it is at the same time remarkably rich.

We want equation (3.2) to define *unique* evolution from any set of initial values defined on a staircase-like configurations as discussed in Section 1.3 (such as in Figure 1.6). For this purpose we should be able to solve uniquely for any corner value; in other words the equation should be *linear* in each of the variables around the quadrilateral, i.e. **affine multilinear**. The most general quadrilateral lattice equation having this property is the following:

$$\begin{aligned}&k\, u\widetilde{u}\widehat{u}\widehat{\widetilde{u}} + l_1\, u\widetilde{u}\widehat{u} + l_2\, u\widetilde{u}\widehat{\widetilde{u}} + l_3\, u\widehat{u}\widehat{\widetilde{u}} + l_4\, \widetilde{u}\widehat{u}\widehat{\widetilde{u}}\\ &+ p_1\, u\widetilde{u} + p_2\, \widetilde{u}\widehat{u} + p_3\, \widehat{u}\widehat{\widetilde{u}} + p_4\, \widehat{\widetilde{u}}u + p_5\, u\widehat{u} + p_6\, \widetilde{u}\widehat{\widetilde{u}}\\ &+ q_1\, u + q_2\, \widetilde{u} + q_3\, \widehat{u} + q_4\, \widehat{\widetilde{u}} + c =: Q(u, \widetilde{u}, \widehat{u}, \widehat{\widetilde{u}};\, p, q) = 0.\end{aligned} \tag{3.3}$$

Here the 16 parameters k, l_i, p_i, q_i, c may depend on the lattice parameters p, q associated with the lattice directions n and m, respectively. (In cases where the P∆E originates from the permutability condition of BTs, these lattice parameters are directly related to the Bäcklund parameters.) If the parameters do not depend on n, m the equation is called **autonomous**, otherwise it is called **nonautonomous**.

3.1.2 Transformations, symmetries and invariances

Möbius transformation

The equation (3.3) is form invariant under the general fractional linear transformation or Möbius transformation

$$u \mapsto U = \frac{\alpha u + \beta}{\gamma u + \delta}, \quad \alpha\delta - \beta\gamma \neq 0,$$

where the *same* transformation is applied to each vertex of the quadrilateral (i.e. the parameters $\alpha, \beta, \gamma, \delta$ are n, m-independent). Special cases that will often be useful are translations $U = u + a$, inversion $U = 1/u$ and scalings $U = c\, u$.

Reflections of the square

Quite often it is assumed that the equation is invariant under the reflections of the square:

- *Diagonal reflection:* $\widetilde{\ } \leftrightarrow \widehat{\ }$
  i.e. $Q(u, \widetilde{u}, \widehat{u}, \widehat{\widetilde{u}}; p, q) = \pm Q(u, \widehat{u}, \widetilde{u}, \widehat{\widetilde{u}}; q, p)$.
- *Horizontal reflection*, i.e. symmetry under the reflection: $\widetilde{\ } \leftrightarrow \underset{\sim}{\ }$ followed by a tilde shift of the whole equation,
  i.e. $Q(u, \widetilde{u}, \widehat{u}, \widehat{\widetilde{u}}; p, q) = \pm Q(\widetilde{u}, u, \widehat{\widetilde{u}}, \widehat{u}; \pm p, q)$.
- *Vertical reflection*, i.e. symmetry under the reflection: $\widehat{\ } \leftrightarrow \underset{\frown}{\ }$ followed by a hat shift of the equation,
  i.e. $Q(u, \widetilde{u}, \widehat{u}, \widehat{\widetilde{u}}; p, q) = \pm Q(\widehat{u}, \widehat{\widetilde{u}}, u, \widetilde{u}; p, \pm q)$.

The last two together constitute the group of so-called *Kleinian symmetries*, and together with diagonal reflection they generate the group of symmetries of the square (i.e. the dihedral group D_4); see Figure 3.2. We note that the equation may sometimes change sign in such reflections and that in the reversal the sign of the associated lattice parameter could also change. For example, (3.11) is invariant (up to an overall sign) under vertical reflection if and only if q changes sign.[2]

If we now demand that the multilinear equation (3.3) also respects the last two reversal symmetries with respect to the shifts $\widetilde{\ }$ and $\widehat{\ }$ on the lattice, we end up with

$$\begin{aligned}&k_0 u\widehat{u}\widetilde{u}\widehat{\widetilde{u}} + k_1(u\widehat{u}\widetilde{u} + u\widehat{u}\widehat{\widetilde{u}} + u\widetilde{u}\widehat{\widetilde{u}} + \widehat{u}\widetilde{u}\widehat{\widetilde{u}}) + k_2(\widehat{u}\widetilde{u} + u\widehat{\widetilde{u}})\\ &+k_3(u\widetilde{u} + \widehat{u}\widehat{\widetilde{u}}) + k_4(u\widehat{u} + \widetilde{u}\widehat{\widetilde{u}}) + k_5(u + \widetilde{u} + \widehat{u} + \widehat{\widetilde{u}}) + k_6 = 0\,,\end{aligned} \tag{3.4}$$

where $k_0, \ldots, k_6$ are coefficients (which may depend on additional parameters). This class of equations contains many (but not all) of the equations discussed below.

n, m-dependent transformations

Transformations that are not global, but depend explicitly on n, m, may also be useful, but we allow only those that keep the equation autonomous. Thus we may consider

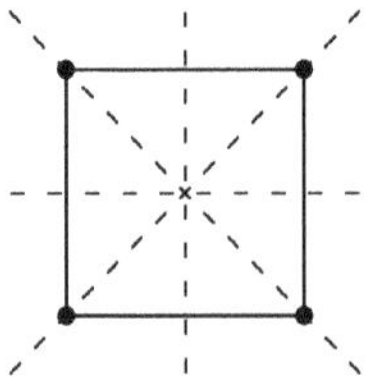

Figure 3.2 Reflections of the square on which the equation is defined.

[2] Often we think of the lattice parameters as being real or complex numbers related to the physical grid. More generally, the parameters of the equation can take values in a commutative group, such as an elliptic curve (see Appendix B, where the group law on the curve relates the parameters).

- translations $u = v + an + bm$,
- scalings $u = a^n b^m \, v$,
- inversion on alternate points $u = v^{(-1)^{n+m}}$,

but only if the transformed equation is again n, m-independent and identical throughout the lattice. For example, using $v = i^{n+m} z$ $(i^2 = -1)$ we can transform (3.11) to $p(\widehat{\widetilde{z}}\,\widehat{z} + \widetilde{z}z) = q(\widehat{\widetilde{z}}\,\widetilde{z} + \widehat{z}z)$.

3.2 Consistency-around-the-cube as integrability

We will now consider a class of quadrilateral PΔEs (3.2) in which, apart from the independent discrete variables n, m on which the variable $u_{n,m}$ depends, there are parameters which we associate with these independent variables. We can think of these parameters as being the parameters which measure the *lattice spacing* in the directions associated with n and m, and we refer to them as *lattice parameters*. Thus, denoting these parameters by p, q, respectively, the equations under question will take the form:

$$Q(u, \widetilde{u}, \widehat{u}, \widehat{\widetilde{u}}; p, q) = 0 \,. \tag{3.5}$$

Now we want to restrict ourselves to quadrilateral PΔEs which we regard to be *integrable*. But what is the proper definition of integrability? We will explore the definition of integrability by means of the following example, namely the example of (2.88), arising from the BTs for the KdV equation, in which we will identify (for later convenience) the parameters $4\lambda = p^2$ and $4\mu = q^2$,

$$(w_{n,m+1} - w_{n+1,m})(w_{n,m} - w_{n+1,m+1}) = p^2 - q^2. \tag{3.6}$$

First, we remark that the presence of the lattice parameters p, q is here crucial. Normally we would like to consider the parameters to be chosen once and for all, i.e. for given $p, q \in \mathbb{C}$ we are to find $w : \mathbb{Z}^2 \to \mathbb{C}$. Here we will argue that we should look at (3.6) as defining a whole *parameter-family* of equations, and that it makes sense to look at them together with the p and q variables, i.e. $w : \mathbb{Z}^2 \times \mathbb{C}^2 \to \mathbb{C}, \quad (n, m; p, q) \mapsto w$. In doing this, however, we must attach each parameter to a specific discrete variable such as p being associated with the variable n, and q with m. This is also natural from the way in which we derived (3.6) from the BT construction: each BT is attached to a parameter $(\lambda, \mu, \dots)$, and with each parameter we can build a new direction in an *infinite-dimensional* lattice of BTs.

These considerations lead to the following definition:

Definition 3.2.1 *If we can consistently extend a quadrilateral lattice equation of the form* (3.5) *to an infinite parameter-family of PΔEs with infinite number of independent variables, then the equation is said to be* **multidimensionally consistent**.

In other words, for integrable systems we can *consistently* impose a copy of the PΔE (in terms of the relevant discrete variables and associated lattice parameters) in *each* quadrilateral sublattice of the infinite-dimensional lattice. That is, we can extend w to a map $w : \mathbb{Z}^N \times \mathbb{C}^N \to \mathbb{C}$, $(n_1, n_2, \dots, ; p_1, p_2, \dots) \mapsto w$, for arbitrary N. Still another way of saying this is that we must be able to construct a solution w in the infinite-dimensional lattice such that the projection to any 2D sublattice is a solution of the basic 2D equation. The existence of an infinite parameter-family of consistent equations is the discrete analogue of the hierarchy of compatible higher-order equations in the continuous case, as discussed in Section 2.5.

Let us illustrate this by means of the example of (3.6). What the statement in definition (3.2.1) suggests is that we should be able to "embed" the equation in a multidimensional lattice by considering the dependent variable w, not only to depend on n and m (with associated lattice parameters p and q, respectively), but on an infinity of lattice variables each associated with its parameter, as follows:

$$w = w_{n,m,h,\dots} = w(n, m, h, \dots; p, q, r, \dots)$$

and with each of these variables we have a corresponding elementary shift on the lattice:

$$\widetilde{w} := w_{n+1,m,h,\dots}, \quad \widehat{w} := w_{n,m+1,h,\dots}, \quad \overline{w} := w_{n,m,h+1,\dots} \cdots$$

Let us now see what this means if we extend the system (3.6) from two to three dimensions; see Figure 3.3. Then we should impose a copy of the same equation in all three lattice directions. This leads to the system of equations

$$(\widehat{w} - \widetilde{w})(w - \widehat{\widetilde{w}}) = p^2 - q^2, \tag{3.7a}$$
$$(\overline{w} - \widetilde{w})(w - \overline{\widetilde{w}}) = p^2 - r^2, \tag{3.7b}$$
$$(\widehat{w} - \overline{w})(w - \widehat{\overline{w}}) = r^2 - q^2, \tag{3.7c}$$

obtained from the bottom, left and back faces of the cube, respectively. In addition to this we have the shifted equations

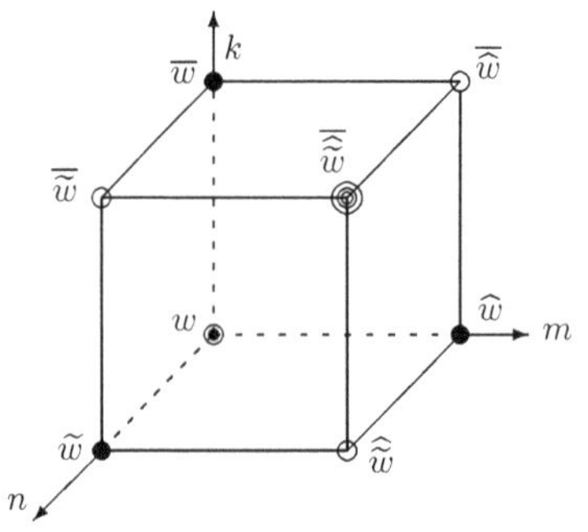

Figure 3.3 Verifying consistency: Values at the black disks are initial values, values at open circles are uniquely determined from them, but there are three different ways to compute $\widehat{\overline{\widetilde{w}}}$.

$$(\widehat{\overline{w}} - \widetilde{\overline{w}})(\overline{w} - \widehat{\widetilde{\overline{w}}}) = p^2 - q^2, \tag{3.7d}$$
$$(\widehat{\overline{w}} - \widehat{\widetilde{w}})(\widehat{w} - \widehat{\widetilde{\overline{w}}}) = p^2 - r^2, \tag{3.7e}$$
$$(\widetilde{\widehat{w}} - \widetilde{\overline{w}})(\widetilde{w} - \widetilde{\widehat{\overline{w}}}) = r^2 - q^2, \tag{3.7f}$$

from the top, right and front sides, respectively.

Consider w and the once-shifted quantities (black disks in Figure 3.3) as given initial values. We can now use (3.7a–c) to solve for the doubly-shifted quantities and then equations (3.7d–f) present three different ways to compute the triply-shifted w. Consistency now manifests itself in that these three a priori different maps all do give the same result, namely (Wahlquist and Estabrook, 1973)

$$\widehat{\widetilde{\overline{w}}} = \frac{(p^2 - q^2)\widetilde{w}\widehat{w} + (q^2 - r^2)\widehat{w}\overline{w} + (r^2 - p^2)\overline{w}\widetilde{w}}{(r^2 - q^2)\widetilde{w} + (q^2 - p^2)\overline{w} + (p^2 - r^2)\widehat{w}}. \tag{3.8}$$

In the generic case we would have the following six equations on the sides of the cube:

$$Q(w, \widetilde{w}, \widehat{w}, \widehat{\widetilde{w}}; p, q) = 0, \tag{3.9a}$$
$$Q(w, \widehat{w}, \overline{w}, \overline{\widehat{w}}; q, r) = 0, \tag{3.9b}$$
$$Q(w, \overline{w}, \widetilde{w}, \overline{\widetilde{w}}; r, p) = 0, \tag{3.9c}$$
$$Q(\overline{w}, \overline{\widetilde{w}}, \overline{\widehat{w}}, \overline{\widehat{\widetilde{w}}}; p, q) = 0, \tag{3.9d}$$
$$Q(\widetilde{w}, \widehat{\widetilde{w}}, \overline{\widetilde{w}}, \overline{\widehat{\widetilde{w}}}; q, r) = 0, \tag{3.9e}$$
$$Q(\widehat{w}, \overline{\widehat{w}}, \widehat{\widetilde{w}}, \overline{\widehat{\widetilde{w}}}; r, p) = 0. \tag{3.9f}$$

With w, $\widetilde{w}$, $\widehat{w}$, $\overline{w}$ given (as suggested by Figure 3.3) we would solve for $\widehat{\widetilde{w}}$, $\overline{\widehat{w}}$, $\overline{\widetilde{w}}$ from the first three equations and then the last three equations should give the same value for $\overline{\widehat{\widetilde{w}}}$. (Alternatively one could give the initial values on the "dog-leg" configuration $\overline{w}$, w, $\widetilde{w}$, $\widehat{\widetilde{w}}$, in which case we would solve for $\widehat{w}$, $\overline{\widehat{w}}$, $\overline{\widetilde{w}}$ from the first three equations; see Exercise 3.1.)

The property that, under the relevant initial value problems on the three-dimensional lattice, the iteration of the solution can be performed in an unambiguous way (namely, independent of the way in which we perform the calculation of the iterates) will be referred to as the **consistency-around-the-cube (CAC) property**. It is this property that we shall consider to be *the definition of integrability* of these types of equations.

Remark: We note also, for later reference, that the formula (3.8) for $\widehat{\widetilde{\overline{w}}}$ is independent of the value w at the opposite end of the main diagonal across the cube. This property we will refer to as the **tetrahedron property**, since the variables $\widehat{w}$, $\widetilde{w}$, $\overline{w}$, $\widehat{\widetilde{\overline{w}}}$ actually appearing in (3.8) form a tetrahedron in Figure 3.3.

Other examples of integrable quadrilateral P∆Es

There are many examples of integrable quadrilateral P∆Es that have been discovered over the years (and we present many of them later). A number of these belong to the *lattice KdV-family* of equations, comprising the following cases:

Lattice potential KdV equation:

$$(p-q+\widehat{u}-\widetilde{u})(p+q+u-\widehat{\widetilde{u}})=p^2-q^2\,, \tag{3.10}$$

which is, in fact, equivalent to the equation (3.6) by the change of (dependent) variable $w_{nm}=u_{nm}-np-mq-c$ (c a constant with respect to n and m). We will show later that this equation reduces to the potential KdV equation after a double continuum limit, so that we rightly regard it as a discretization of the latter equation.

Lattice potential mKdV equation: (cf. (2.94))

$$p(v\widehat{v}-\widetilde{v}\widehat{\widetilde{v}})=q(v\widetilde{v}-\widehat{v}\widehat{\widetilde{v}}). \tag{3.11}$$

Solutions of this equation are related to the solutions of the previous one (3.10) via the relations

$$p-q+\widehat{u}-\widetilde{u}=\frac{p\widetilde{v}-q\widehat{v}}{v},\quad p+q+u-\widehat{\widetilde{u}}=\frac{pv+q\widehat{\widetilde{v}}}{\widetilde{v}}, \tag{3.12}$$

which constitute the analogues of the Miura transformation discussed in Section 2.4.2.

The lattice potential mKdV is also multidimensionally consistent. In fact, embedding the system in a three-dimensional lattice, we obtain

$$\widehat{\widetilde{v}}=\frac{p\widehat{v}-q\widetilde{v}}{p\widetilde{v}-q\widehat{v}}v,\quad \widehat{\overline{v}}=\frac{q\overline{v}-r\widehat{v}}{q\widehat{v}-r\overline{v}}v,\quad \widetilde{\overline{v}}=\frac{p\overline{v}-r\widetilde{v}}{p\widetilde{v}-r\overline{v}}v$$

and subsequently

$$\widehat{\widetilde{\overline{v}}}=\frac{p(r^2-q^2)\widehat{v}\,\overline{v}+q(p^2-r^2)\overline{v}\widetilde{v}+r(q^2-p^2)\widetilde{v}\,\widehat{v}}{p(r^2-q^2)\widetilde{v}+q(p^2-r^2)\widehat{v}+r(q^2-p^2)\overline{v}},$$

which is symmetric under permutations of lattice shifts and corresponding lattice parameters. Note again the tetrahedron property.

Lattice SKdV equation:

$$\frac{(z-\widehat{z})(\widetilde{z}-\widehat{\widetilde{z}})}{(z-\widetilde{z})(\widehat{z}-\widehat{\widetilde{z}})}=\frac{p^2}{q^2} \tag{3.13}$$

which is the lattice version of the "Schwarzian KdV equation" or the "cross-ratio equation" (Nijhoff and Capel, 1995). This equation is invariant under the Möbius transformation. Solutions of equation (3.13) are related to the ones of (3.11) via the relations:

$$p(z - \widetilde{z}) = v\widetilde{v}, \quad q(z - \widehat{z}) = v\widehat{v}, \tag{3.14}$$

which form a discrete analogue of the so-called *Cole–Hopf (CH) transformation*. It is easy to see that eliminating v from (3.14), together with its shifts, leads to the lattice SKdV equation, while eliminating z gives us back the lattice mKdV. The lattice SKdV is again multidimensionally consistent as we get the following expression for the triple shift

$$\widehat{\widetilde{\overline{z}}} = -\frac{p^2(r^2 - q^2)\widehat{z}\overline{z} + q^2(p^2 - r^2)\overline{z}\widetilde{z} + r^2(q^2 - p^2)\widetilde{z}\widehat{z}}{p^2(r^2 - q^2)\widetilde{z} + q^2(p^2 - r^2)\widehat{z} + r^2(q^2 - p^2)\overline{z}},$$

which is manifestly invariant under the permutation of shifts and their associated parameters. It also has the tetrahedron property, i.e. it does not depend on z.

Remark 1: The permutability condition for BTs goes back at least to Bianchi (1899). The recent history seems to start with Nimmo and Schief (1997), which already contains the diagram of the consistency cube. The point of view in that paper is that the concrete constructed sequence of Moutard transformations does satisfy CAC. The next step of abstraction is to consider CAC as a property of PΔEs defined on a lattice. For a generalized cross-ratio equation this was an observation made in Nijhoff et al. (2001) and was explicitly shown to hold for the lattice mKdV in Nijhoff and Walker (2001). Then it was realized that CAC is in fact *a constructive property* in the sense that it directly provides a Lax pair for the system in question (Walker, 2001; Bobenko and Suris, 2002; Nijhoff, 2002). Then CAC was used as a method to search and classify integrable lattice equations in Adler et al. (2003), and in many subsequent papers.

Remark 2: The method employed so far in this book to derive integrable discretizations of integrable PDEs has been based on BTs, which leads directly to the notion of multidimensional consistency. Other methods exist for deriving integrable discretizations based on other principles. For example, Hirota used the bilinear method (discussed in Chapter 8), which provides another way to discretize integrable equations, and indirectly also found quadrilateral equations this way. In contrast, the pioneering work of Ablowitz and Ladik (1976) is based on the discretization of the Lax pair and leads to higher-order lattice equations, which are not of quadrilateral type. The third approach was through the so-called *direct linearization method* based on linear integral equations (Nijhoff et al., 1983a; Quispel et al., 1984).

3.3 Lax pairs and Bäcklund transformation from CAC

If a set of equations is consistent around a cube then this relationship can also be written in the more familiar terms of a Lax pair or as a BT. Examples of this are given in this section.

3.3.1 Lax pair for lattice potential KdV

We shall next give an important application of the CAC property discussed in the previous section: we will show that it guarantees the existence of a Lax pair, i.e. an over-determined linear system of difference equations, the compatibility of which is verified if and only if the nonlinear lattice equation is satisfied (see also Nijhoff, 2002).

The idea is the following: having verified the consistency of the equation around the cube, we know that we can add a new lattice direction to the original lattice and impose simultaneously a version of the old equation on the new two-dimensional quadrilateral sub-lattices. The main idea is now to consider the additional lattice variable $h \in \mathbb{Z}$ associated with lattice parameter k as an auxiliary "virtual" variable and the corresponding bar-shifted variables as completely new variables $\overline{w} := W$.

Proceeding in this way, we rewrite (3.7b) and (3.7c) as follows:

$$(W - \widetilde{w})(\widetilde{W} - w) = k^2 - p^2 \Rightarrow \widetilde{W} = \frac{wW + (k^2 - p^2 - \widetilde{w}w)}{W - \widetilde{w}}, \tag{3.15a}$$

$$(W - \widehat{w})(\widehat{W} - w) = k^2 - q^2 \Rightarrow \widehat{W} = \frac{wW + (k^2 - q^2 - \widehat{w}w)}{W - \widehat{w}}. \tag{3.15b}$$

Noting that equations (3.15) are both fractional linear in W, we can linearize these equations by the substitution:

$$W = \frac{F}{G},$$

leading to

$$\frac{\widetilde{F}}{\widetilde{G}} = \frac{wF + (k^2 - p^2 - w\widetilde{w})G}{F - \widetilde{w}G},$$

$$\frac{\widehat{F}}{\widehat{G}} = \frac{wF + (k^2 - q^2 - w\widehat{w})G}{F - \widehat{w}G},$$

and since at least one of the two functions F or G can be chosen freely, we may split in each of these equations the numerator and denominator to give:[3]

$$\begin{cases} \widetilde{F} = \gamma\left(wF + [k^2 - p^2 - w\widetilde{w}]G\right), \\ \widetilde{G} = \gamma\,(F - \widetilde{w}G), \end{cases} \quad \begin{cases} \widehat{F} = \gamma'\left(wF + [k^2 - q^2 - w\widehat{w}]G\right), \\ \widehat{G} = \gamma'\,(F - \widehat{w}G), \end{cases} \tag{3.16}$$

[3] Since F, G came from a ratio they are only defined up to an overall constant. A natural alternative is then to consider the pair F, G as a point in the projective space $\mathbb{CP}^1$, defined by pairs (x, y) with equivalence relation $(x, y) \simeq (\lambda x, \lambda y)$, where $\lambda \neq 0$. From this point of view, all pairs (F, G) are equivalent modulo common factors and the actual factor is irrelevant.

in which γ and γ' are to be specified later. What happens next is obvious: we introduce the two-component vector

$$\boldsymbol{\phi} = \begin{pmatrix} F \\ G \end{pmatrix} ,$$

and write (3.16) as a system of two 2×2 matrix equations:

$$\widetilde{\boldsymbol{\phi}} = \boldsymbol{L}\boldsymbol{\phi}, \quad \widehat{\boldsymbol{\phi}} = \boldsymbol{M}\boldsymbol{\phi}, \tag{3.17a}$$

with the matrices

$$\boldsymbol{L} = \gamma \begin{pmatrix} w & k^2 - p^2 - w\widetilde{w} \\ 1 & -\widetilde{w} \end{pmatrix}, \quad \boldsymbol{M} = \gamma' \begin{pmatrix} w & k^2 - q^2 - w\widehat{w} \\ 1 & -\widehat{w} \end{pmatrix}. \tag{3.17b}$$

How does this linear system work? The consistency relation of the linear problem (3.17a) is the condition that $\widehat{\widetilde{\boldsymbol{\phi}}} = \widetilde{\widehat{\boldsymbol{\phi}}}$, which simply expresses the fact that $\boldsymbol{\phi}$ must be a proper function of the lattice variables n and m. Calculating the left-hand side of this condition we get

$$\widehat{\widetilde{\boldsymbol{\phi}}} = \widehat{(\boldsymbol{L}\boldsymbol{\phi})} = \widehat{\boldsymbol{L}}\widehat{\boldsymbol{\phi}} = \widehat{\boldsymbol{L}}\boldsymbol{M}\boldsymbol{\phi},$$

whereas the right-hand side yields

$$\widetilde{\widehat{\boldsymbol{\phi}}} = \widetilde{(\boldsymbol{M}\boldsymbol{\phi})} = \widetilde{\boldsymbol{M}}\widetilde{\boldsymbol{\phi}} = \widetilde{\boldsymbol{M}}\boldsymbol{L}\boldsymbol{\phi} \; .$$

Equating both sides we see that a sufficient condition for the consistency is the matrix equation:

$$\widehat{\boldsymbol{L}}\boldsymbol{M} = \widetilde{\boldsymbol{M}}\boldsymbol{L} \; , \tag{3.18}$$

which we will loosely refer to as the *Lax equation*, but it is sometimes also referred to as a *discrete zero-curvature condition* (the reason for this terminology will be explained in the remark below). Pictorially, the Lax equation is illustrated by the following diagram:

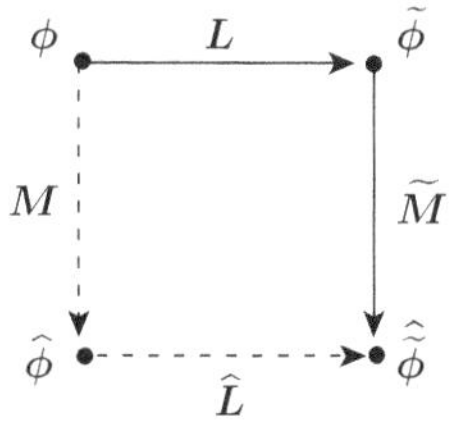

where the vectors $\boldsymbol{\phi}$ are located at the vertices of the quadrilateral and the matrices $\boldsymbol{L}$ and $\boldsymbol{M}$ are attached to the edges linking the vertices.

Using the explicit form (3.17b) of the matrices $\boldsymbol{L}$ and $\boldsymbol{M}$, and working out the condition (3.18) we find

$$\widehat{\gamma}\gamma'\begin{pmatrix} \widehat{w} & k^2-p^2-\widehat{w}\widehat{\widetilde{w}} \\ 1 & -\widehat{\widetilde{w}} \end{pmatrix}\begin{pmatrix} w & k^2-q^2-w\widehat{w} \\ 1 & -\widehat{w} \end{pmatrix}$$
$$=\widetilde{\gamma}'\gamma\begin{pmatrix} \widetilde{w} & k^2-q^2-\widetilde{w}\widehat{\widetilde{w}} \\ 1 & -\widehat{\widetilde{w}} \end{pmatrix}\begin{pmatrix} w & k^2-p^2-w\widetilde{w} \\ 1 & -\widetilde{w} \end{pmatrix}. \tag{3.19}$$

In order to determine γ and γ' (which were unspecified so far) we require that the relation for the determinants of this equation (note the squares):

$$(\widehat{\gamma}\gamma')^2\,\det(\widehat{\boldsymbol{L}})\,\det(\boldsymbol{M}) = (\widetilde{\gamma}'\gamma)^2\,\det(\widetilde{\boldsymbol{M}})\,\det(\boldsymbol{L}) \tag{3.20}$$

is trivially satisfied. Since in this example the determinants of $\boldsymbol{L}$ and $\boldsymbol{M}$ are given by

$$\det(\boldsymbol{L}) = p^2-k^2, \quad \det(\boldsymbol{M}) = q^2-k^2\,,$$

respectively, and are in particular n, m independent, it is clear that the condition (3.20) is satisfied by simply taking $\gamma = \gamma' = 1$. (But we will encounter other examples where a nontrivial choice of γ and γ' is needed in order to satisfy the determinantal condition.)

Working out all the entries on both sides of the matrix products (3.19), it is straightforward to verify that the (1, 1) and (2, 2) entries of the matrix equation both yield the same condition on w, namely equation (3.6). Moreover, the (2, 1) entry is trivially satisfied, and the remaining (1, 2) entry of the equation is satisfied by virtue of the determinantal condition (3.20). In conclusion, we see thus that from the Lax equation (3.18) we recover the lattice equation (3.6). We observe, furthermore, that even though both of the Lax matrices $\boldsymbol{L}$ and $\boldsymbol{M}$ depend on the auxiliary variable k, the final nonlinear equation for w does not depend on k. In fact, k plays the role of a spectral parameter of the Lax pair (3.17).

Example 3.3.1 In the case of the lattice potential mKdV equation

$$p(\widehat{\widetilde{v}}\,\widetilde{v}-\widehat{v}\,v) = q(\widehat{\widetilde{v}}\,\widehat{v}-\widetilde{v}\,v)$$

we obtain the Lax matrices:

$$\boldsymbol{L} = \gamma\begin{pmatrix} -pv & k\widetilde{v}v \\ k & -p\widetilde{v} \end{pmatrix}, \quad \boldsymbol{M} = \gamma'\begin{pmatrix} -qv & k\widehat{v}v \\ k & -q\widehat{v} \end{pmatrix}, \tag{3.21}$$

with determinants $\det(\boldsymbol{L}) = \gamma^2(p^2-k^2)v\widetilde{v}$, $\det(\boldsymbol{M}) = \gamma'^2(q^2-k^2)v\widehat{v}$. Thus, we can here choose the factors $\gamma = \gamma' = 1/v$. In that case the resulting determinantal equation $\det(\widehat{\boldsymbol{L}}\boldsymbol{M}) = \det(\widetilde{\boldsymbol{M}}\boldsymbol{L})$ holds, and $\widehat{\boldsymbol{L}}\boldsymbol{M} = \widetilde{\boldsymbol{M}}\boldsymbol{L}$ yields the lattice potential mKdV.

Example 3.3.2 The lattice SKdV

$$p^2(\widehat{\widetilde{z}} - \widehat{z})(\widetilde{z} - z) = q^2(\widehat{\widetilde{z}} - \widetilde{z})(\widehat{z} - z),$$

leads to the Lax matrices

$$\boldsymbol{L} = \gamma \begin{pmatrix} p^2(\widetilde{z} - z) - k^2\widetilde{z} & k^2 z\widetilde{z} \\ -k^2 & p^2(\widetilde{z} - z) + k^2 z \end{pmatrix}, \tag{3.22}$$

and a similar formula for $\boldsymbol{M}$ (with $p \to q, \widetilde{z} \to \widehat{z}$ and $\gamma \to \gamma'$). The determinants are

$$\det(\boldsymbol{L}) = \gamma^2 p^2(p^2 - k^2)(\widetilde{z} - z)^2, \quad \det(\boldsymbol{M}) = \gamma'^2 q^2(q^2 - k^2)(\widehat{z} - z)^2,$$

and choosing $\gamma = 1/[p^2(\widetilde{z} - z)], \gamma' = 1/[q^2(\widehat{z} - z)]$ the determinant relation $\det(\widehat{\boldsymbol{L}}\boldsymbol{M}) = \det(\widetilde{\boldsymbol{M}}\boldsymbol{L})$ is satisfied, leading to the Lax pair

$$\boldsymbol{L} = \begin{pmatrix} 1 - \dfrac{k^2}{p^2}\dfrac{\widetilde{z}}{\widetilde{z} - z} & \dfrac{k^2}{p^2}\dfrac{\widetilde{z}z}{\widetilde{z} - z} \\ -\dfrac{k^2}{p^2}\dfrac{1}{\widetilde{z} - z} & 1 + \dfrac{k^2}{p^2}\dfrac{z}{\widetilde{z} - z} \end{pmatrix}, \quad \boldsymbol{M} = \begin{pmatrix} 1 - \dfrac{k^2}{q^2}\dfrac{\widehat{z}}{\widehat{z} - z} & \dfrac{k^2}{q^2}\dfrac{\widehat{z}z}{\widehat{z} - z} \\ -\dfrac{k^2}{q^2}\dfrac{1}{\widehat{z} - z} & 1 + \dfrac{k^2}{q^2}\dfrac{z}{\widehat{z} - z} \end{pmatrix}. \tag{3.23}$$

Remark: Suppose that the Lax matrices $\boldsymbol{L}$ and $\boldsymbol{M}$ can be expanded in a power series in a small parameter δ and ϵ respectively as follows:

$$\boldsymbol{L} = \boldsymbol{I} + \delta\boldsymbol{L}_1 + \cdots, \quad \boldsymbol{M} = \boldsymbol{I} + \epsilon\boldsymbol{M}_1 + \cdots,$$

and we expand the shifted variable by Taylor expansion

$$\widehat{\boldsymbol{L}} = \boldsymbol{L} + \epsilon\partial_t\boldsymbol{L} + \cdots, \quad \widetilde{\boldsymbol{M}} = \boldsymbol{L} + \delta\partial_x\boldsymbol{M} + \cdots,$$

then one can show that by collecting the first nontrivial terms in this expansion one obtains to leading order (namely terms proportional to $\epsilon\delta$) the following matrix equation:

$$\partial_t\boldsymbol{L}_1 - \partial_x\boldsymbol{M}_1 + [\boldsymbol{L}_1, \boldsymbol{M}_1] = 0, \tag{3.24}$$

where $[\boldsymbol{L}, \boldsymbol{M}] = \boldsymbol{L}\boldsymbol{M} - \boldsymbol{M}\boldsymbol{L}$ denotes the usual matrix commutator bracket. Equation (3.24) arises also in differential geometry and it is from there that it has the interpretation as a "zero-curvature condition" of differentiable manifolds (with the appropriate interpretation of the matrices $\boldsymbol{L}$ and $\boldsymbol{M}$).

3.3.2 Bäcklund transformation from CAC

In the previous section we have shown how to derive a Lax pair for a PΔE from the consistency around a cube. In the same vein we will now derive Bäcklund transformations (BT) from multidimensional consistency. Recall that the lattice equation was obtained from a BT of a continuous equation, and then the BT was interpreted as a shift in a lattice. Due to the multidimensional consistency, we can single out any shift of the multidimensional lattice and reinterpret it as a BT, but this time as a BT of the lattice equation. The beauty of this idea is that in a sense the equation is its own BT.

We recall that a BT for a nonlinear PDE was given as a pair of differential equations which provides a transformation from one solution of the original PDE to a solution of a different PDE, or to another solution of the same PDE. In the same vein a BT for a PΔE will consist a pair of difference equations providing a transformation from a solution of a PΔE to a solution of a different PΔE or to another solution of the same PΔE. The former case is often called a non-auto-Bäcklund transformation or a Miura transform, and we have seen examples of them in (3.12) and (3.14). The latter is sometimes called an auto-Bäcklund transformation, and it is this type of transformation that we will now discuss.

To be more precise, consider a multidimensionally consistent quad-equation of the form

$$Q(u, \widetilde{u}, \widehat{u}, \widehat{\widetilde{u}};\, p, q) = 0, \tag{3.25}$$

and introduce as before a third direction associated with a distinguished lattice parameter k, which will be re-interpreted as a BT parameter. The associated shift of u in the lattice is denoted by a $\overline{u}$, but in the present context, in order to simplify notation, we will denote the bar-shifted quantities by v and interpret this as the Bäcklund transformed solution associated with u. From multidimensional consistency we can supplement (3.25) with the following pair of equations

$$Q(u, \widetilde{u}, v, \widetilde{v};\, p, r) = 0, \tag{3.26a}$$

$$Q(u, \widehat{u}, v, \widehat{v};\, q, r) = 0, \tag{3.26b}$$

and interpret this as a BT $u \to v$ for (3.25). In fact, again from multidimensional consistency, it follows that v also solves

$$Q(v, \widetilde{v}, \widehat{v}, \widehat{\widetilde{v}};\, p, q) = 0.$$

Note that if Q is affine multilinear then for a given u both equations (3.26) are discrete Riccati equations in the variable v and hence can be solved using the methods of Section 1.2.2.

As an example, let us consider once again the lattice potential KdV equation (3.6), which in the present context we write as

$$(u - \widehat{\widetilde{u}})(\widehat{u} - \widetilde{u}) = p^2 - q^2. \tag{3.27}$$

As in the continuous case (see Section 2.5.2) we need a seed solution in order to start the construction. In the continuous case we could take the solution $w = 0$, but now $u = 0$ is not a solution of (3.27). In fact, the simplest solution is

$$u \equiv u^{(0)}_{n,m} = pn + qm + u^{(0)}_{0,0}, \tag{3.28}$$

as can be easily verified by substitution. Here $u^{(0)}_{0,0}$ is the initial value, which is independent of n, m but could still contain other variables associated with other lattice variables (even if we do not write them out), in accordance with multidimensional consistency. The BT is now given by the following pair of discrete Riccati equations for v:

$$(u - \widetilde{v})(\widetilde{u} - v) = p^2 - k^2, \tag{3.29a}$$

$$(u - \widehat{v})(\widehat{u} - v) = q^2 - k^2, \tag{3.29b}$$

where k is the Bäcklund parameter. In order to solve these equations simultaneously, it is convenient to take

$$v_{n,m} = pn + qm + k + \gamma + v'_{n,m} \tag{3.30}$$

where the first term is just the $u^{(0)}_{n,m}$ of (3.28), and the additional term, containing the Bäcklund parameter k, can be thought of as arising from a shift on the seed solution in the direction of the BT. The v' in (3.30) is the function to be found. After substituting (3.30) into both Riccati equations in (3.29) we find

$$\widetilde{v'} = \frac{(p+k)v'}{p-k-v'}, \quad \widehat{v'} = \frac{(q+k)v'}{q-k-v'}. \tag{3.31}$$

As was explained in Section 1.2.2, we can linearize these by writing $v' = f/g$ and then considering the vector $\boldsymbol{\phi} = (f, g)^T$, for which (3.31) yields the matrix relations

$$\widetilde{\boldsymbol{\phi}} = \boldsymbol{P}\boldsymbol{\phi}\,, \quad \widehat{\boldsymbol{\phi}} = \boldsymbol{Q}\boldsymbol{\phi}, \tag{3.32a}$$

where

$$\boldsymbol{P} = \begin{pmatrix} p+k & 0 \\ -1 & p-k \end{pmatrix}, \quad \boldsymbol{Q} = \begin{pmatrix} q+k & 0 \\ -1 & q-k \end{pmatrix}. \tag{3.32b}$$

In order to find a simultaneous solution for $\boldsymbol{\phi}$ of (3.32) it is essential that the matrices $\boldsymbol{P}$ and $\boldsymbol{Q}$ commute (which can be verified directly). Thus we find the solution

$$\begin{aligned} \boldsymbol{\phi}_{n,m} &= \boldsymbol{P}^n\boldsymbol{Q}^m\boldsymbol{\phi}_{0,0} \\ &= \begin{pmatrix} (p+k)^n(q+k)^m & 0 \\ -\frac{1}{2k}\left[(p+k)^n(q+k)^m - (p-k)^n(q-k)^m\right] & (p-k)^n(q-k)^m \end{pmatrix} \boldsymbol{\phi}_{0,0}, \end{aligned}$$

and this leads to the solution

$$v = v_{n,m} = \gamma + np + mq + k\frac{1+\rho_{n,m}}{1-\rho_{n,m}}\,, \; \rho_{n,m} := \frac{v'_{0,0}}{2k + v'_{0,0}}\left(\frac{p+k}{p-k}\right)^n \left(\frac{q+k}{q-k}\right)^m , \tag{3.33}$$

where $v'_{0,0}$ is an initial value for v'. The solution (3.33) is the discrete analogue of the one-soliton solution (2.86) of the potential KdV equation.

The above method works for other multilinear, multidimensionally consistent quad equations, although sometimes it might be difficult to find an appropriate seed solution. In principle we could take any solution as the seed solution, even complicated ones, but it is not guaranteed that the solution obtained by applying a BT is essentially different from the seed solution. However, if the BT produces a new solution then the process can be iterated to produce multi-soliton solutions.

3.4 Yang–Baxter maps

A different point of view on multidimensional consistency is provided by the notion of Yang–Baxter (YB) maps and the corresponding YB equation. The idea of YB maps is essentially due to Veselov (2003), but the origins go back to certain re-interpretations of the original YB equation. The construction presented here, which makes the connection with lattice equations, follows in broad lines the paper of Papageorgiou et al. (2006).

We will focus on the lpKdV equation (3.6) and construct the YB map for it. First consider the diagram (Figure 3.4a) consisting of three adjacent faces of the cube on each of which we have a copy of one of the three equations (3.7a), (3.7e) and (3.7c). We introduce the variables

$$x := \widetilde{w} - w\,, \quad y := \widehat{\widetilde{w}} - \widetilde{w}\,, \quad z := \widetilde{\widehat{\widetilde{w}}} - \widehat{\widetilde{w}}\,,$$

which can be associated with a succession of three connected links along the cube connecting the vertex w to the vertex $\widetilde{\widehat{\widetilde{w}}}$. From the embedding of (3.6) into the three-dimensional setting given by (3.7), we can derive a sequence of 2–2 mappings, as follows.

On Face I we have the set of relations

$$x + y = \widehat{x} + \underset{\sim}{y}\,, \quad \text{and} \quad (x+y)(x-\underset{\sim}{y}) = p^2 - q^2\,,$$

the first of which is a consequence of the definitions. From these we can obtain the following expressions for $\widehat{x}$ and $\underset{\sim}{y}$

$$\widehat{x} = y + \frac{p^2-q^2}{x+y}\,, \quad \underset{\sim}{y} = x - \frac{p^2-q^2}{x+y}\,.$$

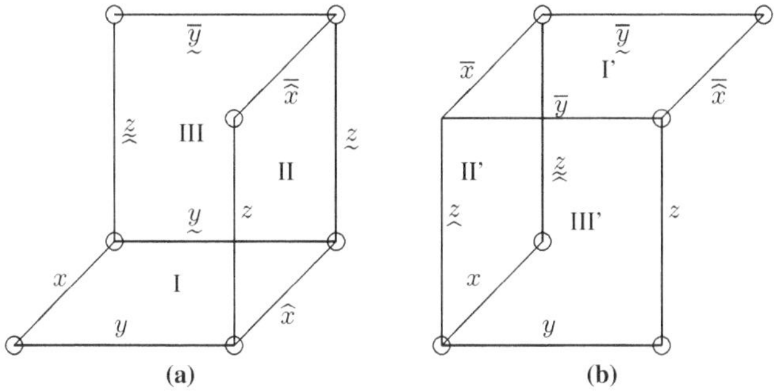

Figure 3.4 Constructing the Yang–Baxter maps and the Yang–Baxter relation.

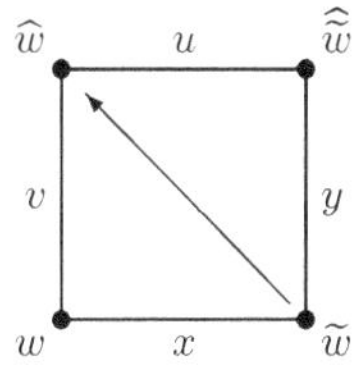

Figure 3.5 Vertex and edge variables in the construction of the Yang–Baxter map.

This map associates with two variables (x, y) two new variables (u, v) according to the diagram in Figure 3.5, in this case $u = \widehat{x}$ and $v = \underset{\sim}{y}$. We denote this map by $R(p, q)$, indicating also dependence on the lattice parameters:

$$(x, y) \overset{R(p,q)}{\longrightarrow} (\widehat{x}, \underset{\sim}{y}) = \left(y + \frac{p^2 - q^2}{x + y}\, , \; x - \frac{p^2 - q^2}{x + y} \right). \tag{3.34a}$$

In the next step we start from the equation (3.7e) on Face II of the cube, and derive a similar map in terms of the variables $\widehat{x}$ and z, which reads:

$$(\widehat{x}, z) \overset{R(p,r)}{\longrightarrow} (\widehat{\widehat{x}}, \underset{\sim}{z}) = \left(z + \frac{p^2 - r^2}{\widehat{x} + z}\, , \; \widehat{x} - \frac{p^2 - r^2}{\widehat{x} + z} \right). \tag{3.34b}$$

Note that the form of the map is exactly the same as in (3.34a). Next we derive a map from the equation (3.7c) which belongs to Face III of the cube, and which acts on the variables $(\underset{\sim}{y}, \underset{\sim}{z})$. This leads to the map:

$$(\underset{\sim}{y}, \underset{\sim}{z}) \overset{R(q,r)}{\longrightarrow} (\underset{\sim}{\overline{y}}, \underset{\approx}{z}) = \left(\underset{\sim}{z} + \frac{q^2 - r^2}{\underset{\sim}{y} + \underset{\sim}{z}}\, , \; \underset{\sim}{y} - \frac{q^2 - r^2}{\underset{\sim}{y} + \underset{\sim}{z}} \right). \tag{3.34c}$$

Thus the form of the map is the same but the parameter dependence is different.

The map $R(p, q)$ can be extended to a map on three variables by indicating which pair of variables it acts. We use the notation $R = R_{ij}$ to mean that R acts on the ith and j^{th}

variable. The three successive maps over Faces I, II and III can now be represented by a sequence of nested maps for the three variables x, y and z

$$(x, y, z) \stackrel{R_{12}(p,q)}{\longrightarrow} (\widehat{x}, \utilde{y}, z) \stackrel{R_{13}(p,r)}{\longrightarrow} (\widehat{\overline{x}}, \utilde{y}, \utilde{z}) \stackrel{R_{23}(q,r)}{\longrightarrow} (\widehat{\overline{x}}, \utilde{\overline{y}}, \underset{\wedge}{\utilde{z}}).$$

An alternative sequence of maps is obtained by considering the remaining three adjacent Faces I', II' and III' of the cube, as indicated in Figure 3.4b, and the 2–2 maps arising from the lattice equation associated with those faces. Starting from Face III' and subsequently II' and I' we obtain in a similar way as before a sequence of maps. From equation (3.7f) on Face III' in Figure 3.4 we get

$$(y, z) \stackrel{R(q,r)}{\longrightarrow} (\overline{y}, \utilde{z}) = \left(z + \frac{q^2 - r^2}{y + z}\,,\ y - \frac{q^2 - r^2}{y + z} \right). \tag{3.35a}$$

Next, from the equation on Face II' (3.7b) we get the map:

$$(x, \utilde{z}) \stackrel{R(p,r)}{\longrightarrow} (\overline{x}, \underset{\wedge}{\utilde{z}}) = \left(\utilde{z} + \frac{p^2 - r^2}{x + \utilde{z}}\,,\ x - \frac{p^2 - r^2}{x + \utilde{z}} \right), \tag{3.35b}$$

and finally from the equation (3.7d) on Face I' we are led to the map:

$$(\overline{x}, \overline{y}) \stackrel{R(p,q)}{\longrightarrow} (\widehat{\overline{x}}, \utilde{\overline{y}}) = \left(\overline{y} + \frac{p^2 - q^2}{\overline{x} + \overline{y}}\,,\ \overline{x} - \frac{p^2 - q^2}{\overline{x} + \overline{y}} \right). \tag{3.35c}$$

Thus, the alternative sequence of successive maps viewed as acting on a triplet of variables (x, y, z) is represented as follows:

$$(x, y, z) \stackrel{R_{23}(q,r)}{\longrightarrow} (x, \overline{y}, \utilde{z}) \stackrel{R_{13}(p,r)}{\longrightarrow} (\overline{x}, \overline{y}, \underset{\wedge}{\utilde{z}}) \stackrel{R_{12}(p,q)}{\longrightarrow} (\widehat{\overline{x}}, \utilde{\overline{y}}, \underset{\wedge}{\utilde{z}}).$$

The two alternative sequences of 2–2 maps transforming the triplet (x, y, z) into the triplet $(\widehat{\overline{x}}, \utilde{\overline{y}}, \underset{\wedge}{\utilde{z}})$ give the same result, relying on the multidimensional consistency which guarantees that these quantities (all of which are expressible in terms of the variables w) are uniquely determined, since $\widehat{\widetilde{\overline{w}}}$ is uniquely determined. Therefore the following relation holds for the composition of these maps:

$$R_{12}(p, q) \circ R_{13}(p, r) \circ R_{23}(q, r) = R_{23}(q, r) \circ R_{13}(p, r) \circ R_{12}(p, q)\,. \tag{3.36}$$

This is the famous *Yang–Baxter (YB) relation*, which in this context we prefer to call **functional Yang–Baxter relation** (in the literature it is often called set-theoretic YB equation). In general, a 2–2 map, like the one constructed above, which obeys (3.36) is called a **Yang–Baxter map**.

The validity of the YB-relation can be checked directly for the map

$$(x, y) \stackrel{R(p,q)}{\longrightarrow} (u, v) = \left(y + \frac{p^2 - q^2}{x + y}, \, x - \frac{p^2 - q^2}{x + y} \right). \tag{3.37}$$

Evaluating the three entries of $R_{12}(p, q) \circ R_{13}(p, r) \circ R_{23}(q, r)(x, y, z)$ in terms of x, y, z and equating these to the three entries of $R_{23}(q, r) \circ R_{13}(p, r) \circ R_{12}(p, q)(x, y, z)$ we get three relations. The first of these is

$$z + \frac{p^2 - r^2}{y + z + \frac{p^2 - q^2}{x + y}} = z + \frac{q^2 - r^2}{z + y} + \frac{p^2 - q^2}{y + z + \frac{p^2 - r^2}{x + y - \frac{q^2 - r^2}{y + z}}},$$

which can be verified by direct computation, for arbitrary x, y and z. The other two entries of the transformed triplet can be verified in a similar way (they follow by using a symmetry between the variables and the parameters).

Note that the Yang–Baxter map (3.37) is involutive, i.e. when the same map is applied twice in a row it produced the identity map

$$R(p, q) \circ R(p, q)(x, y) = (x, y).$$

Thus, by itself the YB-map (for fixed p, q and acting on the same space) does not produce interesting dynamics (it has period one), but when composed with maps operating on other variables their compositions give nontrivial dynamics.

We can view the Yang–Baxter map constructed above as a re-formulation of the lattice potential KdV equation, and then we can also give a reinterpretation of the corresponding Lax pair, in terms of the variables of the YB map. Notice that the Lax matrix $\boldsymbol{L}$ in (3.17b) of Section 3.3.1 involves two variables w and $\widetilde{w}$ (and $\boldsymbol{M}$ involves w and $\widehat{w}$), but this combination can be rewritten as a product of matrices, each of which depends only on one variable. That is, we have

$$\boldsymbol{L} = \boldsymbol{W}\boldsymbol{P}\widetilde{\boldsymbol{W}}^{-1}, \quad \boldsymbol{M} = \boldsymbol{W}\boldsymbol{Q}\widehat{\boldsymbol{W}}^{-1},$$

in which the 2×2 matrices $\boldsymbol{W}$ and $\boldsymbol{P}$ and $\boldsymbol{Q}$ are given by:

$$\boldsymbol{W} = \begin{pmatrix} 1 & w \\ 0 & 1 \end{pmatrix}, \quad \boldsymbol{P} = \begin{pmatrix} 0 & k^2 - p^2 \\ 1 & 0 \end{pmatrix}, \quad \boldsymbol{Q} = \begin{pmatrix} 0 & k^2 - q^2 \\ 1 & 0 \end{pmatrix}.$$

Writing out the Lax compatibility (3.18) we have

$$\widehat{\boldsymbol{W}}\boldsymbol{P}\widehat{\widetilde{\boldsymbol{W}}}^{-1}\boldsymbol{W}\boldsymbol{Q}\widehat{\boldsymbol{W}}^{-1} = \widetilde{\boldsymbol{W}}\boldsymbol{Q}\widehat{\widetilde{\boldsymbol{W}}}^{-1}\boldsymbol{W}\boldsymbol{P}\widetilde{\boldsymbol{W}}^{-1}. \tag{3.38}$$

Multiplying both sides of this matrix relation from the left by $\widehat{\boldsymbol{W}}^{-1}$ and from the right by $\widetilde{\boldsymbol{W}}$, the matrices containing the variable w combine as $\widehat{\widetilde{\boldsymbol{W}}}^{-1}\boldsymbol{W}$ or $\widehat{\boldsymbol{W}}^{-1}\widetilde{\boldsymbol{W}}$ which depend on either the difference $w - \widehat{\widetilde{w}} = -x - y = -\widehat{x} - \underset{\sim}{y}$ or the difference $\widetilde{w} - \widehat{w} = \widehat{x} - y = x - \underset{\sim}{y}$.

We can write the products in (3.38) differently using the matrices

$$X = \begin{pmatrix} 1 & x \\ 0 & 1 \end{pmatrix}, \quad Y = \begin{pmatrix} 1 & y \\ 0 & 1 \end{pmatrix},$$

and their shifted variants. Thus, after inserting these relations in the appropriate places the matrix identity (3.38) it be written in the form

$$YPY^{-1}X^{-1}QX = \widehat{X}Q\widehat{X}^{-1}\underset{\sim}{Y}^{-1}P\underset{\sim}{Y}. \tag{3.39}$$

Introducing now the Lax matrix

$$\mathcal{L}(x;\lambda) := \begin{pmatrix} x & \lambda - x^2 \\ 1 & -x \end{pmatrix} = X\begin{pmatrix} 0 & \lambda \\ 1 & 0 \end{pmatrix}X^{-1}, \tag{3.40}$$

the relation (3.39) can be written as

$$\mathcal{L}(y; k^2 - p^2)\mathcal{L}(-x; k^2 - q^2) = \mathcal{L}(\widehat{x}; k^2 - q^2)\mathcal{L}(-\underset{\sim}{y}; k^2 - p^2). \tag{3.41}$$

This is the Lax formulation for the YB map, in the sense that the above refactorization formula (3.41) reproduces the map (3.34a). The phenomenon of generating dynamical maps through a factorization procedure, like the one implied by (3.41), is observed in various contexts, e.g. in the LR-algorithm of numerical analysis; see the discussion in Section 4.3.2. Matrix factorization has played an important role in the history of DIS; see e.g. Moser and Veselov (1991); Deift et al. (1989).

The construction given here can be applied to other lattice equations as well. In other maps one may use ratios or some other suitable combinations for the Yang–Baxter variables, which typically are invariants of symmetry transformations of the lattice.

3.5 Classification of quadrilateral PΔEs

In an important paper, Adler et al. (2003) posed the problem of classifying scalar (i.e. single-component) quadrilateral lattices that are integrable in the sense of multidimensional consistency (discussed in Section 3.2). Their result is as follows.

Consider quadrilateral PΔEs of the general form:

$$Q(u, \widetilde{u}, \widehat{u}, \widehat{\widetilde{u}}; p, q) = 0,$$

using the notation indicated in Figure 3.1 subject to the following restrictions: (a) Q has the affine multilinear form (3.3), (b) it is invariant under the group D_4 of symmetries of the square (see Section 3.1.2), generated by the interchanges: $Q(u, \widetilde{u}, \widehat{u}, \widehat{\widetilde{u}}; p, q) =$

$\pm Q(u, \widehat{u}, \widetilde{u}, \widehat{\widetilde{u}}; q, p) = \pm Q(\widetilde{u}, u, \widehat{\widetilde{u}}, \widehat{u}; p, q)$, and (c) it has the **tetrahedron property**[4]: in the consistency check, the computed value for $\widehat{\widetilde{\overline{u}}}$ is *independent of* u.

In this class the only equations that satisfy the CAC-condition are the following:

Q-list:

$$\mathrm{Q}_1: \quad p(u-\widehat{u})(\widetilde{u}-\widehat{\widetilde{u}}) - q(u-\widetilde{u})(\widehat{u}-\widehat{\widetilde{u}}) = \delta^2 pq(q-p) \tag{3.42a}$$

$$\begin{aligned}\mathrm{Q}_2: \quad & p(u-\widehat{u})(\widetilde{u}-\widehat{\widetilde{u}}) - q(u-\widetilde{u})(\widehat{u}-\widehat{\widetilde{u}}) + pq(p-q)(u+\widetilde{u}+\widehat{u}+\widehat{\widetilde{u}}) \\ & = pq(p-q)(p^2-pq+q^2)\end{aligned} \tag{3.42b}$$

$$\begin{aligned}\mathrm{Q}_3: \quad & p(1-q^2)(u\widehat{u}+\widetilde{u}\widehat{\widetilde{u}}) - q(1-p^2)(u\widetilde{u}+\widehat{u}\widehat{\widetilde{u}}) \\ & = (p^2-q^2)\left((\widehat{u}\widetilde{u}+u\widehat{\widetilde{u}}) + \delta^2\frac{(1-p^2)(1-q^2)}{4pq}\right)\end{aligned} \tag{3.42c}$$

$$\begin{aligned}& \mathrm{Q}_4 \text{ from Hietarinta (2005)}: \mathrm{sn}(\alpha)(u\widetilde{u}+\widehat{u}\widehat{\widetilde{u}}) - \mathrm{sn}(\beta)(u\widehat{u}+\widetilde{u}\widehat{\widetilde{u}}) \\ & - \mathrm{sn}(\alpha-\beta)(\widetilde{u}\widehat{u}+u\widehat{\widetilde{u}}) + k\,\mathrm{sn}(\alpha)\mathrm{sn}(\beta)\mathrm{sn}(\alpha-\beta)(1+u\widetilde{u}\widehat{u}\widehat{\widetilde{u}}) = 0.\end{aligned} \tag{3.42d}$$

Here sn is the Jacobi elliptic function of modulus k, i.e. $\mathrm{sn}(\alpha) = \mathrm{sn}(\alpha; k)$, etc.

H-list:

$$\mathrm{H}_1: \quad (u-\widehat{\widetilde{u}})(\widehat{u}-\widetilde{u}) = p^2-q^2 \tag{3.43a}$$

$$\mathrm{H}_2: \quad (u-\widehat{\widetilde{u}})(\widetilde{u}-\widehat{u}) = (p-q)(u+\widetilde{u}+\widehat{u}+\widehat{\widetilde{u}}) + p^2 - q^2 \tag{3.43b}$$

$$\mathrm{H}_3: \quad p(u\widetilde{u}+\widehat{u}\widehat{\widetilde{u}}) - q(u\widehat{u}+\widetilde{u}\widehat{\widetilde{u}}) = \delta^2(p^2-q^2) \tag{3.43c}$$

A-list:

$$\mathrm{A}_1: \quad p(u+\widehat{u})(\widetilde{u}+\widehat{\widetilde{u}}) - q(u+\widetilde{u})(\widehat{u}+\widehat{\widetilde{u}}) = \delta^2 pq(p-q) \tag{3.44a}$$

$$\begin{aligned}\mathrm{A}_2: \quad & q(1-p^2)(u\widehat{u}+\widetilde{u}\widehat{\widetilde{u}}) - p(1-q^2)(u\widetilde{u}+\widehat{u}\widehat{\widetilde{u}}) \\ & + (p^2-q^2)(1+u\widetilde{u}\widehat{u}\widehat{\widetilde{u}}) = 0.\end{aligned} \tag{3.44b}$$

Remarks on Q4: It is in itself remarkable that the list of equations found is so short. In fact, all equations in the above lists can be obtained as limits from the last equation, Q4. Equation Q4 was actually discovered earlier by V. Adler (1998), when he constructed the BT for discrete analogue of a famous soliton equation discovered in 1980 by I. Krichever and S. Novikov (1979, 1980), which reads:

$$u_t = u_{xxx} - \frac{3}{2}\frac{u_{xx}^2 - (4u^3 - g_2 u - g_3)}{u_x}, \tag{3.45}$$

and which generalizes the Schwarzian KdV equation.

[4] This property was an important technical assumption in the classification, but there are also cases where this assumption is violated. We will discuss this later.

In Adler (1998) (and in Adler et al., 2003), Q4 was parameterized using Weierstrass elliptic functions as follows:

$$\begin{aligned} &k_0 u\widetilde{u}\widehat{u}\widehat{\widetilde{u}} + k_1(u\widetilde{u}\widehat{u} + \widetilde{u}\widehat{u}\widehat{\widetilde{u}} + \widehat{u}\widehat{\widetilde{u}}u + \widehat{\widetilde{u}}u\widetilde{u}) + k_2(u\widehat{\widetilde{u}} + \widetilde{u}\widehat{u}) + \\ &\quad \bar{k}_2(u\widetilde{u} + \widehat{u}\widehat{\widetilde{u}}) + \widetilde{k}_2(u\widehat{u} + \widetilde{u}\widehat{\widetilde{u}}) + k_3(u + \widetilde{u} + \widehat{u} + \widehat{\widetilde{u}}) + k_4 = 0, \end{aligned} \tag{3.46}$$

where the coefficients k_i are given by

$$k_0 = A + B, \quad k_1 = aB + bA, \quad k_2 = a^2 B + b^2 A$$

$$k_3 = \tfrac{AB(A+B)}{2(b-a)} - b^2 A + B\left(2a^2 - \tfrac{g_2}{4}\right)$$

$$k_4 = \tfrac{AB(A+B)}{2(a-b)} - a^2 B + A\left(2b^2 - \tfrac{g_2}{4}\right)$$

$$k_5 = \tfrac{g_3}{2}k_0 + \tfrac{g_2}{4}k_1, \quad k_6 = \tfrac{g_2^2}{16}k_0 + g_3 k_1$$

where the parameters $(a, A) = (\wp(\alpha), \wp'(\alpha))$ and $(b, B) = (\wp(\beta), \wp'(\beta))$ live on the (Weierstrass) elliptic curve, i.e. they are related by $A^2 = 4a^3 - g_2 a - g_3$, $B^2 = 4b^3 - g_2 b - g_3$ where g_2 and g_3 are fixed constants.

Adler's equation (3.46) can also be expressed as:

$$\begin{aligned} &A\left[(u-b)(\widehat{u}-b) - (a-b)(c-b)\right]\left[(\widetilde{u}-b)(\widehat{\widetilde{u}}-b) - (a-b)(c-b)\right] + \\ &+B\left[(u-a)(\widetilde{u}-a) - (b-a)(c-a)\right]\left[(\widehat{u}-a)(\widehat{\widetilde{u}}-a) - (b-a)(c-a)\right] \\ &= ABC(a-b), \end{aligned} \tag{3.47}$$

cf. Nijhoff (2002), with the previous parameterization of a, A, b, B, while the parameters c, C are related to them by $c = \wp(\beta - \alpha)$, $C = \wp'(\beta - \alpha)$.

A simpler *Jacobi* form of Adler's equation was discovered later by Hietarinta (2005). It reads

$$\begin{aligned} &(1 - p^2q^2)[p(u\widetilde{u} + \widehat{u}\widehat{\widetilde{u}}) - q(u\widehat{u} + \widetilde{u}\widehat{\widetilde{u}})] \\ &\qquad = (pQ - qP)[(\widehat{u}\widetilde{u} + u\widehat{\widetilde{u}}) - pq(1 + u\widetilde{u}\widehat{u}\widehat{\widetilde{u}})] \end{aligned} \tag{3.48}$$

where now the parameter relations are $P^2 = p^4 - \gamma p^2 + 1$, $Q^2 = q^4 - \gamma q^2 + 1$. These relationships can be parameterized by Jacobi elliptic functions of modulus k, if we take $p = \sqrt{k}\,\mathrm{sn}(\alpha)$, $q = \sqrt{k}\,\mathrm{sn}(\beta)$, $P = \mathrm{sn}'(\alpha)$, $Q = \mathrm{sn}'(\beta)$ and $\gamma = (k + 1/k)$. Using further the sn addition formula (see Section 2.2.2), we get from (3.48) the form (3.42d) given for Q_4 in the list (3.42–3.44) above.

Viallet (2009) proposed that equation (3.46) could also be considered integrable for arbitrary values of k_j, $j = 0, \ldots, 4$, based on calculations of algebraic entropy (see

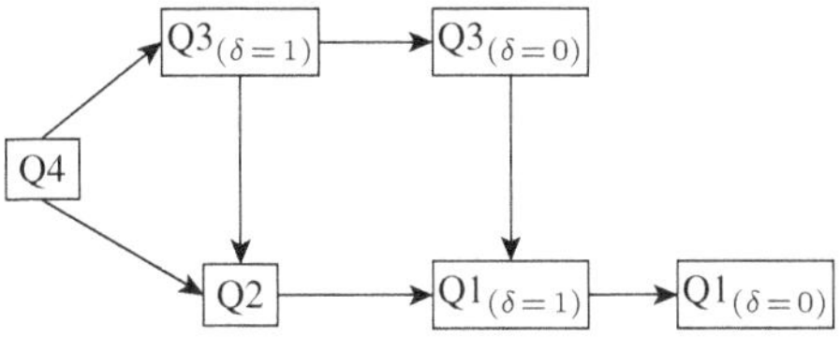

Figure 3.6 Coalescence diagram of the Q-equations (Adler and Suris, 2004).

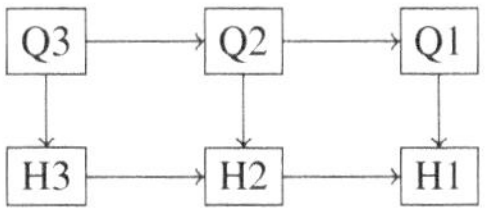

Figure 3.7 The coalescence diagram connecting the Q and H equations (Nijhoff et al., 2009).

Section 7.2). This discovery looks like a generalization but is really a reformulation of Q_4. Viallet's equation also called Q_V can be regarded as the result of multidimensional consistency if we allow another parameterization.

It should be noted that all examples presented in Section 3.2 can be recognized as special cases of the equations in the lists Q, H or A (possibly modulo some simple changes of parameters and variables). Furthermore, as was demonstrated in Adler and Suris (2004) (see its Appendix A), all equations can be obtained as "degenerations" of Adler's equation Q4, in one or another of its presentations. In fact, the degenerations of the Q-type equations in the list follow the **coalescence diagram** of Figure 3.6. As an example, consider the limit of Q4 (3.42d) to Q3 (3.42c): first scale $u \mapsto \delta^{-1}\sqrt{k}u$, divide by k, and then take the $k \to 0$ limit in which $\mathrm{sn}(x) \to \sin(x)$. After this, one just needs to rewrite the sin function using $\exp(i\alpha) = p$, $\exp(i\beta) = q$. Furthermore, the H-type equations can also be obtained from those of Q-type by degenerations of a slightly different form, following the pattern of Figure 3.7 (Nijhoff et al., 2009). The A- and Q-equations, on the other hand, are related by point transformations: A1 to Q1 by $u \mapsto (-1)^{n+m}u$ and A2 to Q3($\delta = 0$) by $u \mapsto u^{(-1)^{n+m}}$.

In Section 3.3.1 we discussed the construction of a Lax pair from the CAC-approach. In Nijhoff (2002) it was shown that Q4 has a Lax pair, and hence by implication all the other equations have a Lax pair as well.

Remark on the tetrahedron condition: The tetrahedron property used in the ABS classification seems an unusual assumption at first sight, but it is in fact related to deep structures underlying the equations listed above. Nevertheless, it is interesting to note that there exist obviously integrable systems that do not have the tetrahedron property, namely the linear equation

$$y - \widehat{y} - \widetilde{y} + \widehat{\widetilde{y}} = 0, \tag{3.49a}$$

and its multiplicative version

$$\widehat{\widetilde{x}}x = \widehat{x}\widetilde{x}. \tag{3.49b}$$

Both of these are consistent around the cube and respectively yield the triply shifted variables as

$$\overline{\widehat{\widetilde{y}}} = \widehat{y} + \widetilde{y} + \overline{y} - 2y,$$

and

$$\overline{\widehat{\widetilde{x}}} = \widehat{x}\,\widetilde{x}\,\overline{x}\,x^{-2}.$$

Since these depend on y and x respectively, they do not satisfy the tetrahedron property.

A nonlinear example is provided by the equation (Hietarinta, 2004)

$$\frac{x + e_2}{x + e_1}\frac{\widehat{\widetilde{x}} + o_2}{\widehat{\widetilde{x}} + o_1} = \frac{\widetilde{x} + e_2}{\widetilde{x} + o_1}\frac{\widehat{x} + o_2}{\widehat{x} + e_1}. \tag{3.50}$$

Note that the four variables and the four parameters are in symmetric position. (For a geometric interpretation, see Adler (2006).) It turns out that this equation is also linearizable (although the linearization is not achieved by a point transformation); see Ramani et al. (2006). Based on our present state of knowledge, all nontrivial one-component CAC equations seem to have the tetrahedron property.

The classification result (Adler et al., 2003) of multidimensionally consistent quad-equations was proven by considering two types of functions related to the polynomial $Q(u, \widetilde{u}, \widehat{u}, \widehat{\widetilde{u}}; \alpha, \beta)$. If we consider the polynomial to be an affine linear function $Q(x, y, z, w; \alpha, \beta)$ of generic variables x, y, z, w and parameters α, β, then the first type of function is the symmetric biquadratic

$$g(x, y; \alpha, \beta) = QQ_{zw} - Q_zQ_w = k(\alpha, \beta)\,h(x, y), \tag{3.51}$$

while the second type is the discriminant $R(x) = h_y^2 - 2hh_{yy}$, where h is defined above. The function $R(x)$ is shown to be a polynomial of degree at most four and the equations are then classified by the factors of $R(x)$.

3.6 Different equations on different faces of the consistency cube

Since the consistency approach led to a classification under certain symmetry assumptions, there have been attempts to see if new multilinear quadrilateral equations could be found that relax some of these assumptions. In the following, we consider two cases: in this section the possibility of allowing equations on the sides of the cube to be different, while in the next section we consider multi-component equations. This area is still under active development.

First of all, it should be noted that a consistent set of equations stays consistent if we change any particular dependent variable by a reversible transformation. For example, if we replace $\widehat{w}$ by $(a\widehat{w}+b)/(c\widehat{w}+d)$ in the set (3.9) without any other changes, then the equations would no longer look similar, but this would not disturb the consistency of the set.

Applying Möbius transformation on each corner of the consistency cube (Figure 3.3) has been used effectively in the classification work (Adler et al., 2009). However, there are severe restrictions in applying this to lattice models under the assumption of a uniform lattice, in most cases the freedom that is left is to make the *same* Möbius transformation on every corner.

3.6.1 Deriving Bäcklund transformations by limits on a consistent set of equations

As noted before, the consistent set can be interpreted as a Bäcklund transformation as was done in Section 3.3.2. If one now makes a suitable (Möbius) transformation and then takes a limit, one can obtain a BT between different integrable equations (Atkinson, 2008a).

Consider for example the BT for H2 as obtained from (3.26a, 3.26b)

$$(u-\widetilde{v})(\widetilde{u}-v)=(p-r)(u+\widetilde{u}+v+\widetilde{v})+p^2-r^2, \tag{3.52a}$$
$$(u-\widehat{v})(\widehat{u}-v)=(q-r)(u+\widehat{u}+v+\widehat{v})+q^2-r^2, \tag{3.52b}$$

together with their hat and tilde shifts, respectively. This is an auto-BT, because if $\widehat{\widetilde{u}},\widetilde{u},\widehat{u}$ are now eliminated using the H2 equation and the above two equations, respectively, then both of the shifted equations yield H2 equation for v, and conversely.

Let us now make the transformation $u\to\epsilon^{-2}+2\epsilon^{-1}u$ together with the parameter choice $r=-\epsilon^{-2}$, then in the limit $\epsilon\to 0$ H2 for u becomes H1 and equations (3.52) become

$$2u\widetilde{u}=v+\widetilde{v}+p,\quad 2u\widehat{u}=v+\widehat{v}+q. \tag{3.53}$$

This is then the BT between H2 for v and H1 for u (albeit without its own a spectral parameter), as can be easily verified (Exercise 3.5). Several such pairs were constructed in Atkinson (2008a). These provide examples of consistent equations where the side equations are similar but the top and bottom equations are different.

3.6.2 Asymmetric deformations

Several examples of quadrilateral equations, without the assumed symmetry of the ABS classification, have been found subsequently; see e.g. Adler et al. (2009), Xenitidis and Papageorgiou (2009), Hydon and Viallet (2010) and Levi and Yamilov (2009b, 2011).

As an example let us consider the following deformation of the H1 equation

$$(u-\widehat{\widetilde{u}})(\widetilde{u}-\widehat{u})+(p-q)(1+\epsilon\widetilde{u}\widehat{u})=0, \tag{3.54}$$

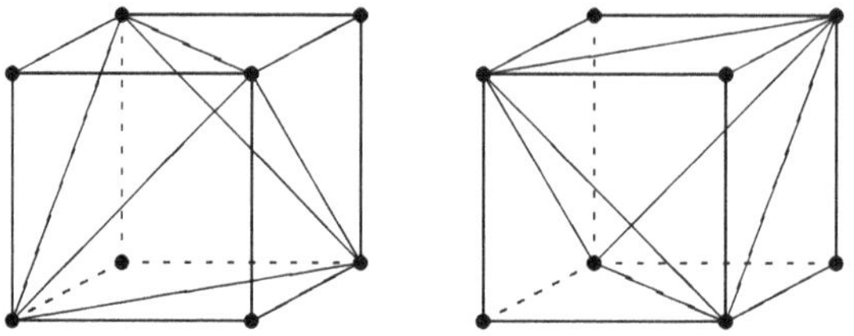

(a) (b)

Figure 3.8 Arranging equations on the consistency cube. The deformation term must connect the corner terms indicated by a thick line.

which includes an additional ϵ term containing a product across one of the diagonals. The equation is asymmetric since only one of the two possible diagonals is used. There is the companion model with the other diagonal product, i.e.

$$(u - \widehat{\widetilde{u}})(\widetilde{u} - \widehat{u}) + (p - q)(1 + \epsilon u \widehat{\widetilde{u}}) = 0. \tag{3.55}$$

Now the question arises as to how such equations can be put on the CAC cube in a consistent way. The answer is given in Figure 3.8: the diagonal terms on the faces must be chosen so that together they connect precisely four of the corner points. There are two ways of doing it, as indicated in the figure: case (a) is for (3.54) and case (b) for (3.55).

The choices given in Figure 3.8 make the models consistent around the cube. But what about putting the equation on a lattice? We must construct the full lattice by gluing the consistent cubes together so that their common sides have the same diagonal deformation term. From the figure it is clear that after translation we can glue cube (b) to any side of cube (a) and vice versa, but we cannot glue a translated cube (a) to any side of cube (a), and the same for (b). If we consider an infinite black-white checker-board lattice, then we can use equation (3.54) on all white squares and (3.55) on all black squares (Adler et al., 2009; Xenitidis and Papageorgiou, 2009).

3.6.3 Sets of equations with a "weak" Lax pair

A somewhat more complicated combination of equations around the cube is given by (Atkinson, 2008a; Adler et al., 2009)

$$p(u - \widehat{u})(\widetilde{u} - \widehat{\widetilde{u}}) + q(u - \widetilde{u})(\widehat{\widetilde{u}} - \widehat{u}) = 0, \tag{3.56a}$$

$$(u - \widetilde{u})(\overline{\widetilde{u}} - \overline{u}) - p(1 + \epsilon u \widetilde{u}) = 0, \tag{3.56b}$$

$$(\widehat{u} - \widehat{\widetilde{u}})(\overline{\widehat{\widetilde{u}}} - \overline{\widehat{u}}) - p(1 + \epsilon \widehat{u} \widehat{\widetilde{u}}) = 0, \tag{3.56c}$$

$$(u - \widehat{u})(\overline{\widehat{u}} - \overline{u}) - q(1 + \epsilon u \widehat{u}) = 0, \tag{3.56d}$$

$$(\widetilde{u} - \widehat{\widetilde{u}})(\overline{\widehat{\widetilde{u}}} - \overline{\widetilde{u}}) - q(1 + \epsilon \widetilde{u}\widehat{\widetilde{u}}) = 0, \tag{3.56e}$$

$$p(\overline{u} - \overline{\widehat{u}})(\overline{\widetilde{u}} - \overline{\widehat{\widetilde{u}}}) + q(\overline{u} - \overline{\widetilde{u}})(\overline{\widehat{\widetilde{u}}} - \overline{\widehat{u}}) + \epsilon(p - q)pq = 0. \tag{3.56f}$$

As can be seen, it looks like a BT between Q1($\delta = 0$) and Q1($\delta = \epsilon$). It is straightforward to verify that when these equations are placed on the appropriate sides of the consistency cube they have the CAC property. It is also easy to construct the Lax pair from (3.56b–3.56e): denoting $\overline{u} = W$ we have

$$\widetilde{W} = W + p\frac{1 + \epsilon u \widetilde{u}}{u - \widetilde{u}}, \qquad \widehat{W} = W + q\frac{1 + \epsilon u \widehat{u}}{u - \widehat{u}},$$

from which it is easy to get the following consistency condition:

$$\left[\epsilon u \widetilde{u}\widehat{u}\widehat{\widetilde{u}}(1/u - 1/\widetilde{u} - 1/\widehat{u} + 1/\widehat{\widetilde{u}}) - u + \widetilde{u} + \widehat{u} - \widehat{\widetilde{u}}\right] \times \\ \left[p(u - \widehat{u})(\widetilde{u} - \widehat{\widetilde{u}}) + q(u - \widetilde{u})(\widehat{\widetilde{u}} - \widehat{u})\right] = 0. \tag{3.57}$$

Thus the side equations and the Lax pair constructed from them do not give a unique equation as their consistency condition: The bottom equation could be either of the factors in (3.57). We already know that the set (3.56) is consistent, but it turns out that replacing (3.56a, 3.56f) with the pair

$$\epsilon u \widetilde{u}\widehat{u}\widehat{\widetilde{u}}(1/u - 1/\widetilde{u} - 1/\widehat{u} + 1/\widehat{\widetilde{u}}) - u + \widetilde{u} + \widehat{u} - \widehat{\widetilde{u}} = 0, \tag{3.58a}$$

$$\overline{u} - \overline{\widetilde{u}} - \overline{\widehat{u}} - \overline{\widehat{\widetilde{u}}} = 0, \tag{3.58b}$$

also yields a consistent set. Thus we have found a set of side equations, i.e. a Lax pair, that allows two different consistent sets. Such a Lax pair may be called "weak" (Hietarinta and Viallet, 2012).

Since the side equations are the same for the two models one can glue them together to make a lattice with an arbitrary arrangement white (3.56a) and black (3.58a) squares, and from the point of view of CAC they are all equally integrable. If one now resorts to a stricter test of integrability, namely algebraic entropy, one finds that only some black–white arrangements are integrable (Hietarinta and Viallet, 2012). This is another example that all Lax pairs are not equally good: a fact well known in the continuum case.

Several further isolated cases have been presented in the literature; see e.g. Atkinson (2008a); Levi and Yamilov (2009b); Hydon and Viallet (2010); Levi and Yamilov (2011). A systematic search program for quadrilateral equations with 3D consistency, while allowing different equations on different sides, has been undertaken in Boll (2011, 2012).

3.7 CAC for multi-component equations

The consistency idea explained in Section 3.2 was developed for a single dependent variable defined at each lattice point. But there is no fundamental reason why this could not be extended to the multicomponent case, and in some cases this is necessary. In this section we discuss such multicomponent discretizations: first the Boussinesq equation and then the nonlinear Schrödinger equation.

3.7.1 Discrete versions of the Boussinesq equation

The Boussinesq (BSQ) equation is a PDE, arising in the theory of water waves (see Boussinesq, 1877) of the form:

$$u_{tt} + \tfrac{1}{3}u_{xxxx} + 4u_x u_{xx} = 0. \tag{3.59}$$

Note in particular the second derivative in time. This means that it cannot be discretized by a one-component quadrilateral equation; rather, we need either a multicomponent quadrilateral system or an equation defined on a larger stencil. Both are discussed below.

The equations and their 3D consistency

Let us first consider the multicomponent quadrilateral system and its CAC-property.

The following lattice Boussinesq equation (lBSQ) was proposed in Tongas and Nijhoff (2005a) (see equation (9)) as a discrete version of BSQ:

$$\widetilde{w} = u\widetilde{u} - v, \quad \widehat{w} = u\widehat{u} - v, \quad \widehat{\widetilde{w}} = \widehat{u}\widehat{\widetilde{u}} - \widehat{v}, \quad \widehat{\widetilde{w}} = \widetilde{u}\widehat{\widetilde{u}} - \widetilde{v}, \tag{3.60a}$$

$$\widehat{\widetilde{v}} = u\widehat{\widetilde{u}} - w + \frac{p^3 - q^3}{\widetilde{u} - \widehat{u}}. \tag{3.60b}$$

This is a set of equations for the three components u, w, v, and the additional new feature here is that four of the equations are defined on the four sides (or links or bonds) of the fundamental quadrilateral.

One of the questions one can ask is about the possible initial value problems that one can impose on such a system. Equation (3.60a) can be used to propagate one corner value to the other corners, for example assuming $\{u, w, v, \widetilde{u}, \widetilde{v}, \widehat{u}, \widehat{v}\}$ as given one can compute $\widehat{w}$ and $\widetilde{w}$ and then have two expressions for $\widehat{\widetilde{w}}$. This is like two-dimensional consistency for one-dimensional equations and yields

$$\widehat{\widetilde{w}} = \frac{\widehat{u}\,\widetilde{v} - \widetilde{u}\,\widehat{v}}{\widetilde{u} - \widehat{u}}, \quad \widehat{\widetilde{u}} = \frac{\widetilde{v} - \widehat{v}}{\widetilde{u} - \widehat{u}}. \tag{3.61}$$

These together with (3.60b) mean that we can take such initial values on the basic square and compute the rest. We can solve the initial value problem for this system, although setting initial values on a staircase is more involved, because (3.61) does impose relations on the staircase itself.

Since the equations are defined on an elementary quadrilateral, one can ask for 3D consistency: as initial values on the cube we take $\{u, w, v, \widetilde{u}, \widetilde{v}, \widehat{u}, \widehat{v}, \overline{u}, \overline{v}\}$, use the edge equations on all twelve edges and (3.60b) on all six faces and proceed to compute the remaining values. The result is that the triply shifted variables are given by manifestly symmetric forms:

$$\overline{\widehat{\widetilde{u}}} = u + \frac{(q^3 - r^3)\widetilde{u} + (r^3 - p^3)\widehat{u} + (p^3 - q^3)\overline{u}}{\widetilde{u}(\widehat{v} - \overline{v}) + \widehat{u}(\overline{v} - \widetilde{v}) + \overline{u}(\widetilde{v} - \widehat{v})}, \tag{3.62a}$$

$$\overline{\widehat{\widetilde{w}}} = w + \frac{(q^3 - r^3)\widetilde{v} + (r^3 - p^3)\widehat{v} + (p^3 - q^3)\overline{v}}{\widetilde{u}(\widehat{v} - \overline{v}) + \widehat{u}(\overline{v} - \widetilde{v}) + \overline{u}(\widetilde{v} - \widehat{v})}, \tag{3.62b}$$

$$\overline{\widehat{\widetilde{v}}} = v + \frac{(q^3 - p^3)\widetilde{u}\widehat{u} + (r^3 - q^3)\widehat{u}\overline{u} + (p^3 - r^3)\overline{u}\widetilde{u}}{\widetilde{u}(\widehat{v} - \overline{v}) + \widehat{u}(\overline{v} - \widetilde{v}) + \overline{u}(\widetilde{v} - \widehat{v})}. \tag{3.62c}$$

Note also that these do not satisfy the tetrahedron condition as there is explicit dependence on u, w, v.

A partial classification

The above can be generalized to a classification problem, restricted to four edge equations and one quadrilateral equation as was done in Hietarinta (2011). Assuming that the edge equations were linear in shifted and unshifted variables separately, and that the coefficients did not depend on the lattice parameters, it was found that the edge equations were transformable into one of the three types

$$\widetilde{\mathsf{x}}\mathsf{z} = \widetilde{\mathsf{y}} + \mathsf{x}, \quad \mathsf{x}\widetilde{\mathsf{x}} = \widetilde{\mathsf{y}} + \mathsf{z}, \quad \mathsf{z}\widetilde{\mathsf{y}} = \widetilde{\mathsf{x}} - \mathsf{x}.$$

(The naming convention was that the unshifted variable was z, the shifted one y, and the one that appears in both forms x.)

Starting with the consistency of edge equations around the quadrilateral and then proceeding to consistency around a cube, leads to a classification. Here we present the results using variable names and parameters that show better the relationships between the results and their connection to the continuous equations. The results contain the following set of three component systems (for the full list and terminology, see Hietarinta, 2011):

- A-2

$$\widetilde{v}u = \widetilde{s} + v, \quad \widehat{v}u = \widehat{s} + v, \quad \widehat{\widetilde{u}} = \frac{s}{v} + \frac{1}{v}\,\frac{p^3\,\widetilde{v} - q^3\,\widehat{v}}{\widetilde{u} - \widehat{u}}, \tag{3.63}$$

- B-2 A generalization of (3.60b):

$$u\widetilde{u} = \widetilde{v} + t, \quad u\widehat{u} = \widehat{v} + t, \quad \widehat{\widetilde{t}} + v = b_0(\widehat{\widetilde{u}} - u) + u\,\widehat{\widetilde{u}} + \frac{p^3 - q^3}{\widetilde{u} - \widehat{u}}, \tag{3.64}$$

- C-3 A generalizations of a model in Nijhoff (1999):

$$w\widetilde{v} = \widetilde{z} - z, \quad w\widehat{v} = \widehat{z} - z, \quad \widehat{\widetilde{w}} = \frac{d_2 z + d_1}{v} + \frac{w}{v}\,\frac{p^3\,\widetilde{v}\widehat{w} - q^3\,\widehat{v}\widetilde{w}}{\widetilde{w} - \widehat{w}}, \tag{3.65}$$

- C-4 or

$$w\widetilde{v} = \widetilde{z} - z, \quad w\widehat{v} = \widehat{z} - z, \quad \widehat{\widetilde{w}} = \frac{z\widehat{\widetilde{z}} + d_0}{v} + \frac{w}{v}\,\frac{p^3\,\widetilde{v}\widehat{w} - q^3\,\widehat{v}\widetilde{w}}{\widetilde{w} - \widehat{w}}. \tag{3.66}$$

Here u, v, w, s, t, z are dependent variables and p, q are lattice parameters and b_0, d_1, d_2 are fixed parameters. The multiple appearances of the same variable in the different systems of equations reflect the fact that there are Miura-type relations between them.[5]

For all these equations one can compute the triply shifted quantities, with results that are similar to (3.62): in particular, the results do not have the tetrahedron property.

Lax pairs

Since these lattice equations fit to the CAC approach, one may ask whether a Lax pair can also be constructed for them, following the procedure given in Section 3.3.1. Since there are now more components we may expect the Lax matrices to be bigger. Starting with (3.60b, 3.61) we consider their tilde-bar versions and write $X = \overline{u}$, $Z = \overline{v}$, i.e.

$$\widetilde{Z} = \frac{p^3 - r^3 + u(\widetilde{v} - Z) - w(\widetilde{u} - X)}{\widetilde{u} - X}, \tag{3.67a}$$

$$\widetilde{X} = \frac{\widetilde{v} - Z}{\widetilde{u} - X}. \tag{3.67b}$$

Next we introduce the projective variables F, G, H by

$$X = G/F, \quad Z = H/F$$

and find that the equations (3.67) can be split as

$$\begin{aligned}
\widetilde{F} &= \gamma(\widetilde{u}F - G),\\
\widetilde{G} &= \gamma(\widetilde{v}F - H),\\
\widetilde{H} &= \gamma[(p^3 - r^3 + u\widetilde{v} - w\widetilde{u})F + wG - uH].
\end{aligned}$$

Next introducing the vector $\Psi = (F, G, H)^t$ we can write the above as $\widetilde{\Psi} = L\Psi$, where (see Tongas and Nijhoff (2005a), Equation (18))

5 Note that C-4 can be transformed into C-3: if $d_0 \neq 0$ a Möbius transformation takes C-4 into C-3 with $d_1 = 0$, while if $d_0 = 0$ a rational transformation takes it into C-3 with $d_2 = 0$.

$$L = \gamma \begin{pmatrix} \widetilde{u} & -1 & 0 \\ \widetilde{v} & 0 & -1 \\ p^3 - r^3 + u\widetilde{v} - w\widetilde{u} & w & -u \end{pmatrix}. \tag{3.68}$$

The determinant of this matrix is $p^3 - r^3$ so we can take $\gamma = 1$. The M matrix is obtained by replacing tilde-shifts by hat-shifts and p by q and then computing the zero curvature condition $\widehat{L}M = \widetilde{M}L$ yields a 3×3 matrix that should vanish. Four of the nine matrix elements vanish automatically, while the remaining yield (3.60b) and

$$\widehat{\widetilde{u}} = \frac{\widetilde{v} - \widehat{v}}{\widetilde{u} - \widehat{u}}, \qquad \widetilde{w} - \widehat{w} = u(\widetilde{u} - \widehat{u}).$$

Thus we cannot quite generate the original equation $\widetilde{w} = u\widehat{u} - v$, but only two of its consequences. (However, on the basis of the second equation we can set $\widetilde{w} = u\widetilde{u} - v + A$, $\widehat{w} = u\widehat{u} - v + A$ for some A and then the first equation implies $\widetilde{A} = \widehat{A}$. If we assume that A only depends on u, w, v and their shifts then A must in fact be a constant.)

The same method works for the other models listed above. For example, from equation (3.63) we get the Lax matrix

$$L = \begin{pmatrix} \widetilde{v} & 0 & -1 \\ \widetilde{v} & -1 & 0 \\ (p^3\widetilde{v} + s\widetilde{v})/v & -r^3/v & -s/v \end{pmatrix} \tag{3.69}$$

which yields the $\widehat{\widetilde{v}}$ equation in (3.63) and

$$\widehat{\widetilde{v}} = \frac{\widetilde{v} - \widehat{v}}{\widetilde{v} - \widehat{v}}, \qquad \widehat{v}\widetilde{s} - \widetilde{v}\widehat{s} = v(\widetilde{v} - \widehat{v}).$$

Similarly for (3.65) the Lax matrix is

$$L = \frac{1}{w} \begin{pmatrix} \widetilde{w} & 0 & -1 \\ 0 & \widetilde{w} & -\widetilde{z} \\ \widetilde{w}(d_1 + r^3 z)/v & \widetilde{w}(d_2 - r^3)/v & (-d_1 - p^3(z - \widetilde{z}) - d_2\, z)/v \end{pmatrix} \tag{3.70}$$

and in addition to $\widehat{\widetilde{w}}$ equation of (3.65) (with $\widehat{v}$ and $\widetilde{v}$ eliminated) it yields

$$\widehat{\widetilde{z}} = \frac{\widetilde{w}\widehat{z} - \widehat{w}\widetilde{z}}{\widetilde{w} - \widehat{w}}, \qquad \widehat{z}\widetilde{v} - \widetilde{z}\widehat{v} = z(\widetilde{v} - \widehat{v}).$$

Remark: The derivation described above was done using the CAC approach, but the results (without the extra parameters and in a different gauge) were actually derived much earlier in Nijhoff et al. (1992); Nijhoff (1999, 1997) using the direct linearization approach (see also Zhang et al., 2012). It is gratifying, and quite common among integrable systems, that the same equations can be derived using different methods. In addition, different methods give different points of view and emphasize different connections.

One component versions on a 3×3 *stencil*

Recall that the continuous BSQ equation has one component and is second-order in time-derivatives while KdV is first-order. Thus we may expect that when the discrete analogues of BSQ are written in a one-component version they would live on a bigger stencil, on which we can give initial values, e.g. on two consecutive staircases. In fact we need to consider a 3×3 stencil as given in Figure 3.9 and extend our equations on it.

Let us consider the B-2 equations (3.64) and try to eliminate v, t. From the left-hand side equations we can derive

$$\widetilde{t} - \widehat{t} = \widehat{\widetilde{u}}(\widetilde{u} - \widehat{u}), \quad \widetilde{v} - \widehat{v} = u(\widetilde{u} - \widehat{u}).$$

Subtracting the tilde- and hat-shifts of the right-hand side equation of (3.64), the v, t dependence can be eliminated using the above with the result

$$\frac{p^3 - q^3}{\widetilde{\widetilde{u}} - \widehat{\widetilde{u}}} - \frac{p^3 - q^3}{\widehat{\widetilde{u}} - \widehat{\widehat{u}}} = (\widehat{\widehat{\widetilde{\widetilde{u}}}} - b_0)(\widehat{\widetilde{\widetilde{u}}} - \widehat{\widehat{\widetilde{u}}}) - \widehat{\widetilde{\widetilde{u}}}\widetilde{u} + \widehat{\widehat{\widetilde{u}}}\widehat{u} + (u + b_0)(\widetilde{u} - \widehat{u}). \tag{3.71}$$

For $b_0 = 0$ this equation was first given in Nijhoff et al. (1992).[6] It is called the regular one-component **lattice (potential) BSQ equation**.

Now it turns out that if we eliminate from the A-2 equation (3.63) all but the u variable we again get equation (3.71), up to some differences in parameters.

If, on the other hand, we eliminate s, u from A-2 we obtain the **lattice modified BSQ equation** (MBSQ) (Nijhoff et al., 1992)

$$\left(\frac{p^3\widehat{\widetilde{v}} - q^3\widehat{\widehat{v}}}{\widehat{\widehat{v}} - \widehat{\widetilde{v}}}\right)\frac{\widehat{\widehat{\widetilde{v}}}}{\widehat{v}} - \left(\frac{p^3\widetilde{\widetilde{v}} - q^3\widehat{\widetilde{v}}}{\widehat{\widetilde{v}} - \widetilde{\widetilde{v}}}\right)\frac{\widehat{\widetilde{\widetilde{v}}}}{\widetilde{v}} = \frac{v}{\widetilde{v}} - \frac{v}{\widehat{v}} - \frac{\widehat{\widehat{\widetilde{v}}}}{\widehat{\widehat{\widetilde{\widetilde{v}}}}} + \frac{\widehat{\widetilde{\widetilde{v}}}}{\widehat{\widehat{\widetilde{\widetilde{v}}}}}. \tag{3.72}$$

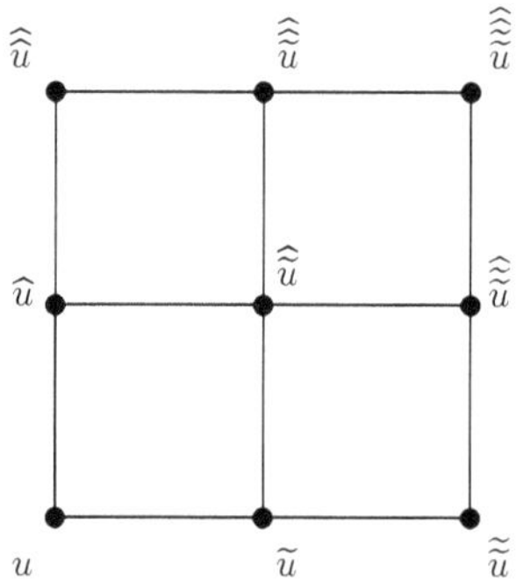

Figure 3.9 The 3×3 stencil needed for the one-component versions of lattice BSQ equations. Here u stands for the one remaining variable.

[6] See Equation (1.3) in Nijhoff et al. (1992), with replacements $u \to u + np^3 + mq^3 + c$; c.f. also equation (5.3) in Nijhoff (1997).

Since both of these one-component models come from the same set of equations (3.63) by different elimination procedures, we may consider this set as a kind of Miura transformation connecting them. The discrete BSQ equation given in Date et al. (1983a) (Example 4, p. 391) is the right-hand side of (3.72).

Finally, if we eliminate v, w variables from (3.65) we obtain the **lattice Schwarzian BSQ equation** (SBSQ)

$$\frac{(\widehat{\widehat{\widetilde{\widetilde{z}}}}-\widehat{\widehat{\widetilde{z}}})(\widehat{\widehat{z}}-\widehat{\widetilde{z}})(\widehat{z}-z)}{(\widehat{\widehat{\widetilde{\widetilde{z}}}}-\widehat{\widetilde{\widetilde{z}}})(\widetilde{\widetilde{z}}-\widehat{\widetilde{z}})(\widetilde{z}-z)}=\frac{(d_1+d_2\widehat{\widetilde{z}})(\widehat{\widetilde{\widetilde{z}}}-\widehat{\widehat{\widetilde{z}}})+p^3(\widehat{\widehat{\widetilde{\widetilde{z}}}}-\widehat{\widehat{\widetilde{z}}})(\widehat{\widetilde{z}}-\widehat{z})-q^3(\widehat{\widehat{\widetilde{\widetilde{z}}}}-\widehat{\widetilde{\widetilde{z}}})(\widehat{\widehat{z}}-\widehat{z})}{(d_1+d_2\widehat{\widetilde{z}})(\widehat{\widehat{\widetilde{z}}}-\widehat{\widetilde{\widetilde{z}}})+q^3(\widehat{\widehat{\widetilde{\widetilde{z}}}}-\widehat{\widetilde{\widetilde{z}}})(\widehat{\widetilde{z}}-\widetilde{z})-p^3(\widehat{\widehat{\widetilde{\widetilde{z}}}}-\widehat{\widehat{\widetilde{z}}})(\widetilde{\widetilde{z}}-\widetilde{z})}. \tag{3.73}$$

For $d_1 = d_2 = 0$ this was given in Nijhoff (1997) (see equation (5.4)). On the other hand, if we instead eliminate z, w from (3.65) then we get again (3.72) and therefore (3.65) can also interpreted as providing a Miura transform between (3.72) and (3.73).

The most general equation in this class can be written in the following form (Zhang et al., 2012)

$$\frac{\mathcal{Q}_{p,q}(\widehat{u},\widehat{\widetilde{u}},\widehat{\widehat{u}},\widehat{\widehat{\widetilde{u}}})}{\mathcal{Q}_{p,q}(\widetilde{u},\widetilde{\widetilde{u}},\widehat{\widetilde{u}},\widehat{\widetilde{\widetilde{u}}})}=\frac{(Q_a u-Q_b\widehat{u})\,(P_a\widehat{\widehat{\widetilde{u}}}-P_b\widehat{\widehat{\widetilde{\widetilde{u}}}})\,(Q_aP_b\widehat{\widetilde{u}}-P_aQ_b\widehat{\widehat{u}})}{(P_a u-P_b\widetilde{u})\,(Q_a\widehat{\widetilde{\widetilde{u}}}-Q_b\widehat{\widehat{\widetilde{\widetilde{u}}}})\,(Q_aP_b\widetilde{\widetilde{u}}-P_aQ_b\widehat{\widetilde{u}})} \tag{3.74}$$

where the quadrilateral multilinear function $\mathcal{Q}$ (reminiscent of Q3) is given by

$$\mathcal{Q}_{p,q}(u,\widetilde{u},\widehat{u},\widehat{\widetilde{u}}):=P_aP_b(u\widehat{u}+\widetilde{u}\widehat{\widetilde{u}})-Q_aQ_b(u\widetilde{u}+\widehat{u}\widehat{\widetilde{u}})-G(p,q)(\widehat{u}\widetilde{u}+u\widehat{\widetilde{u}}) ,$$

and in which

$$P_a=\sqrt{G(p,a)},\quad Q_a=\sqrt{G(q,a)},\quad G(p,q)=p^3-q^3+\alpha_2(p^2-q^2)+\alpha_1(p-q) .$$

From equation (3.74) one can obtain the previous cases (3.71–3.73) by performing point transformations and taking suitable limits on the parameters a and b, such as letting a or b tend to ∞ or by coalescing a and b.

We finish this section by mentioning that the BSQ system forms part of a hierarchy of equations, coined the **lattice Gel'fand–Dikii hierarchy** in Nijhoff et al. (1992). These form a class of discrete equations of increasingly higher order, labeled by an integer N, where $N = 2$ for the KdV class of equations (examined in Section 3.2) while $N = 3$ corresponds to the lattice BSQ class. In the continuous case, similar hierarchies of equations were constructed and studied; see e.g. Drinfel'd and Sokolov (1985). We mention, furthermore, that unlike the ABS class of lattice equations (see Section 3.4) the class of higher-order equations of BSQ type have not yet been completely classified, although partial results exist (Hietarinta, 2011).

3.7.2 Complex and multi-component PΔEs

Lattice nonlinear Schrödinger equation

One of the most important integrable PDEs is the nonlinear Schrödinger (NLS) equation, which reads

$$i\phi_t + \phi_{xx} \pm 2|\phi|^2\phi = 0 \tag{3.75a}$$

where the field $\phi(x,t)$ is complex and $|\phi|^2 = \phi^*\phi$. This equation plays an important role in various areas of applied sciences, notably in plasma physics, Bose–Einstein condensates and in the theory of optical fibers. It is called the *nonlinear* Schrödinger equation (or, in physical contexts, often the Gross–Pitaevskii equation), because in comparison with the standard Schrödinger equation of quantum mechanics, the external potential $V(x)$ has been replaced by nonlinear self-coupling $|\phi|^2$. We could view (3.75a) as one member of a pair of equations for ϕ and its complex conjugate $\phi^*(x,t)$, by supplementing it with its complex conjugate

$$-i\phi_t^* + \phi_{xx}^* - 2|\phi|^2\phi^* = 0. \tag{3.75b}$$

A discrete version of the system consisting of (3.75a) and (3.75b) was given in Date et al. (1983b), and reads

$$(1 + pq\widehat{\widetilde{\phi}}^*\phi)(p\widehat{\phi} - q\widetilde{\phi}) = (p-q)\phi\,, \tag{3.76a}$$

$$(1 + pq\widehat{\widetilde{\phi}}^*\phi)(p\widetilde{\phi}^* - q\widehat{\phi}^*) = (p-q)\widehat{\widetilde{\phi}}^*. \tag{3.76b}$$

Note, however, that the two equations are no longer necessarily complex conjugates of each other, and hence this equation should be read as a system for two different component fields ϕ and ϕ^*. In this case the Lax pair, similarly to those exhibited in Section 3.3, is given by a 2×2 matrix system consisting of

$$\widetilde{\boldsymbol{\psi}} = \begin{pmatrix} 1 & -p\widetilde{\phi}^* \\ -p\phi & 1 - pk + p^2\widetilde{\phi}^*\phi \end{pmatrix}\boldsymbol{\psi}\,,\quad \widehat{\boldsymbol{\psi}} = \begin{pmatrix} 1 & -q\widehat{\phi}^* \\ -q\phi & 1 - qk + q^2\widehat{\phi}^*\phi \end{pmatrix}\boldsymbol{\psi}\,, \tag{3.77}$$

where $\boldsymbol{\psi}$ is a two-component vector function of the discrete variables n and m. Again the condition $(\widetilde{\boldsymbol{\psi}})\widehat{\ } = (\widehat{\boldsymbol{\psi}})\widetilde{\ }$ yields equations (3.76), but one cannot straightforwardly use the vector function $\boldsymbol{\psi}$ to construct a multidimensionally consistent set of equations.

Alternative forms of the discrete NLS exist and were derived starting from different perspectives, cf. e.g. Quispel et al. (1984) and Suris (1997a). For instance, an alternative lattice NLS given in Quispel et al. (1984) reads:

$$\begin{aligned} &|p|^2 + |q|^2 + 2Re\,\left(\theta pq\,\phi_{n,m}\phi_{n,m+1}^*\right) \\ =\; &|p|^2(1 + |\phi_{n,m+1}|^2)\frac{q^*\phi_{n+1,m+1} - \theta q\phi_{n,m}}{\theta p\phi_{n+1,m} - p^*\phi_{n,m+1}} \\ &+ |q|^2(1 + |\phi_{n,m}|^2)\frac{\theta p\phi_{n+1,m} - p^*\phi_{n,m+1}}{q^*\phi_{n+1,m+1} - \theta q\phi_{n,m}} \end{aligned} \tag{3.78}$$

in which the lattice parameters p and q are now complex valued such that $Re(p) = Re(q)$, and where θ is a pure phase, i.e. a fixed parameter such that $|\theta| = 1$.

Lattice derivative nonlinear Schrödinger equation

The derivative NLS is a variant of the PDE (3.75) containing an extra derivative in the nonlinear term:

$$i\phi_t + \phi_{xx} - 2i|\phi|^2\phi_x = 0 \tag{3.79}$$

which can also be supplemented by its complex conjugate. The lattice version of this equation (as given in Date et al., 1983b) reads:

$$(p-q)\phi - p\widehat{\phi} + q\widetilde{\phi} = pq\widehat{\widetilde{\phi}}^{*}\phi\,(\widetilde{\phi} - \widehat{\phi})\,, \tag{3.80a}$$
$$(p-q)\widehat{\widetilde{\phi}}^{*} - p\widetilde{\phi}^{*} + q\widehat{\phi}^{*} = pq\widehat{\widetilde{\phi}}^{*}\phi\,(\widehat{\phi}^{*} - \widetilde{\phi}^{*})\,. \tag{3.80b}$$

(Again, the two equations are not necessarily complex conjugates of each other.) This system of equations derives from the following Lax pair:

$$\begin{aligned}\widetilde{\psi} &= \frac{1}{1-p\widetilde{\phi}^{*}\phi}\begin{pmatrix} 1 & -pk\widetilde{\phi}^{*} \\ p\phi & 1-pk-p\widetilde{\phi}^{*}\phi \end{pmatrix}\psi,\\ \widehat{\psi} &= \frac{1}{1-q\widehat{\phi}^{*}\phi}\begin{pmatrix} 1 & -qk\widehat{\phi}^{*} \\ q\phi & 1-qk-q\widehat{\phi}^{*}\phi \end{pmatrix}\psi,\end{aligned} \tag{3.81}$$

which looks almost similar to the Lax pair (3.77) apart from an extra factor k in the upper right entries of the Lax matrices and the presence of the prefactors.

An alternative, more general lattice analogue of the derivative NLS equation (through a connection to a closely related model, namely the so-called massive Thirring model) was given in Nijhoff et al. (1983b).

The lattice isotropic Heisenberg spin equation

The last example we give here is that of the lattice version of the isotropic Heisenberg spin chain which is the following equation for a three-component spin vector $\boldsymbol{S}(x,t)$ normalized to unity, $|\boldsymbol{S}|^2 = \boldsymbol{S}\cdot\boldsymbol{S} = 1$:

$$\boldsymbol{S}_t = \boldsymbol{S}\times\boldsymbol{S}_{xx}. \tag{3.82}$$

This equation could be seen as a classical equation of motion associated with a corresponding quantum system called the XXX Heisenberg ferromagnet, cf. Faddeev and Takhtajan (2007) for further details. A lattice version of (3.82) was proposed in Date et al. (1982a) and reads:

$$\begin{aligned}&\frac{1}{1+\widehat{\boldsymbol{S}}\cdot\widehat{\widetilde{\boldsymbol{S}}}}\left(\widehat{\boldsymbol{S}} + \widehat{\widetilde{\boldsymbol{S}}} + i\widehat{\boldsymbol{S}}\times\widehat{\widetilde{\boldsymbol{S}}}\right) - \frac{p}{p-q}\widehat{\boldsymbol{S}}\\ &= \frac{1}{1+\boldsymbol{S}\cdot\widetilde{\boldsymbol{S}}}\left(\boldsymbol{S} + \widetilde{\boldsymbol{S}} + i\boldsymbol{S}\times\widetilde{\boldsymbol{S}}\right) - \frac{p}{p-q}\widetilde{\boldsymbol{S}}\,,\end{aligned} \tag{3.83}$$

arising from the Lax representation

$$\widetilde{\boldsymbol{\Psi}} = \left[(1 - pk)\boldsymbol{I} + pk\boldsymbol{S} \cdot \boldsymbol{\sigma}\right] \boldsymbol{\Psi}, \tag{3.84a}$$

$$\widehat{\boldsymbol{\Psi}} = \left[(1 - qk)\boldsymbol{I} + qk\boldsymbol{T} \cdot \boldsymbol{\sigma}\right] \boldsymbol{\Psi}, \tag{3.84b}$$

in which $\boldsymbol{T}$ is a companion spin matrix, which can be eliminated using the compatibility relations of the Lax pair. In (3.84) $\boldsymbol{\sigma}$ denotes the vector of Pauli spin matrices, i.e. $\boldsymbol{\sigma} = (\sigma_x, \sigma_y, \sigma_z)$, where

$$\sigma_x = \begin{pmatrix} 0 & 1 \\ 1 & 0 \end{pmatrix}, \quad \sigma_y = \begin{pmatrix} 0 & -i \\ i & 0 \end{pmatrix}, \quad \sigma_z = \begin{pmatrix} 1 & 0 \\ 0 & -1 \end{pmatrix}.$$

However, equation (3.83) seems to be valid only for complex valued spin vectors and hence would not be a suitable discretization of (3.82) from the point of view of physics. In Quispel et al. (1984) a discrete version of the Heisenberg spin chain equation was constructed which does allow for real-valued spin vector solutions, but this is a more complicated system of equations (containing square roots of inner products of spin vectors) and hence we will not mention it here.

We note that coupled systems of partial difference equations discussed in this subsection are still defined on the quadrilateral, and can therefore be considered as multi-component versions of the systems considered earlier in this chapter. However, this point of view is a bit deceptive, since the equations of NLS type have a distinctly different structure and do not allow the same type of multidimensional consistency.

3.7.3 Elliptic KdV

In addition to Adler's equation Q4, there exists another integrable quadrilateral lattice system which is naturally associated with an elliptic curve, and which is given by the following coupled system of equations in the dependent variables u, s and w:

$$\left(p + q + u - \widehat{\widetilde{u}}\right)(p - q + \widehat{u} - \widetilde{u}) = p^2 - q^2 + f\,(\widetilde{s} - \widehat{s})\left(\widehat{\widetilde{s}} - s\right) \tag{3.85a}$$

$$\left(\widehat{\widetilde{s}} - s\right)(\widetilde{w} - \widehat{w}) = \left[(p + u)\widetilde{s} - (q + u)\widehat{s}\right]\widehat{\widetilde{s}} - \left[(p - \widehat{\widetilde{u}})\widehat{s} - (q - \widehat{\widetilde{u}})\widetilde{s}\right] s \tag{3.85b}$$

$$(\widehat{s} - \widetilde{s})\left(\widehat{\widetilde{w}} - w\right) = \left[(p - \widetilde{u})s + (q + \widetilde{u})\widehat{\widetilde{s}}\right]\widehat{s} - \left[(p + \widehat{u})\widehat{\widetilde{s}} + (q - \widehat{u})s\right]\widetilde{s} \tag{3.85c}$$

$$\left(p + u - \frac{\widetilde{w}}{\widetilde{s}}\right)\left(p - \widetilde{u} + \frac{w}{s}\right) = p^2 - \mathcal{R}(s\widetilde{s}) \tag{3.85d}$$

$$\left(q + u - \frac{\widehat{w}}{\widehat{s}}\right)\left(q - \widehat{u} + \frac{w}{s}\right) = q^2 - \mathcal{R}(s\widehat{s}) \tag{3.85e}$$

in which

$$\mathcal{R}(x) \equiv \frac{1}{x} + 3e + gx,$$

with e and g moduli of a rational form of an elliptic curve $y^2 = \mathcal{R}(1/x)$. The curve for the parameters is

$$p^2 = P + 3e + \frac{g}{P}, \quad q^2 = Q + 3e + \frac{g}{Q}.$$

The system (3.85) is a natural extension of the lattice potential KdV equation (3.10), the latter being obviously recovered when $g = 0$, i.e. when the elliptic curve degenerates. The system is integrable in that it has a Lax pair, is multidimensionally consistent and has explicit N-soliton solutions with parameters on the above elliptic curve; see Nijhoff and Puttock (2003).

3.8 Lattice KdV, SKdV and mKdV equations

There are further integrable PΔEs that do not fit into the CAC approach. For such equations, integrability manifests itself in a different way, for example through a nontrivial Lax pair. In this section we will look at three important examples.

3.8.1 Lattice KdV equation

Here we derive the lattice analogue of the KdV equation, which is different from the lattice *potential* KdV equation (3.6).

Recall that the solution u of the continuous KdV equation (2.69) and w of its potential version (2.84), as discussed in Chapter 2, are connected through the differential relation $u = w_x$. However, on the lattice there are many ways in which we can take the discrete analogue of a partial derivative, for instance we can take a difference involving two neighboring vertices, such as $\Delta_n w_{n,m} \equiv w_{n+1,m} - w_{n,m}$ or $\Delta_m w_{n,m} \equiv w_{n,m+1} - w_{n,m}$. But these are not the only choices. Alternatively, we can take a difference between vertices farther away or across diagonals, such as:

$$\Delta w_{n,m} = w_{n+1,m+1} - w_{n,m} \quad \text{or} \quad \Delta w_{n,m} = w_{n,m+1} - w_{n+1,m},$$

or a host of other choices. As long as a well-chosen continuum limit reduces these to a derivative they can be justifiably considered to be the discrete analogues of the operation of taking a partial derivative. The most sensible way to make a choice is to look at the equation at hand and then decide what would be the most natural choice to apply in that case.

In the case of the lattice potential KdV equation (3.6) the natural choice of the discrete analogue of the derivative seems to be either one of the choices given above as a *difference across the diagonal*. This suggests that as lattice KdV variables we would take either one of the two choices:

$$Q = w - \widehat{\widetilde{w}} \quad \text{or} \quad R = \widehat{w} - \widetilde{w}\,. \tag{3.86}$$

It is then straightforward from (3.6) (using the notation in terms of ~ and ^ shifts) to derive the following equations for Q, R:

$$\widehat{Q} - \widetilde{Q} = \frac{a}{Q} - \frac{a}{\widehat{\widetilde{Q}}} \quad \Leftrightarrow \quad R - \widehat{\widetilde{R}} = \frac{a}{\widehat{R}} - \frac{a}{\widetilde{R}} , \tag{3.87}$$

where $a \equiv p^2 - q^2$. The equations for Q and R are simply related by the fact that $QR = a$.

A Lax pair for either of the equations (3.87) can be derived by starting from the Lax pair for the lattice potential KdV equation (3.17b). In fact, from the relations:

$$\begin{cases} \widetilde{F} = wF + [k^2 - p^2 - w\widetilde{w}]G \\ \widetilde{G} = F - \widetilde{w}G \end{cases} \quad \text{and} \quad \begin{cases} \widehat{F} = wF + [k^2 - q^2 - w\widehat{w}]G \\ \widehat{G} = F - \widehat{w}G \end{cases}$$

setting $\psi := G$ and $\Lambda := k^2 - q^2$, eliminating F, we obtain the *scalar* Lax pair:

$$\widehat{\widehat{\psi}} = Q\,\widehat{\psi} + \Lambda\,\psi, \tag{3.88a}$$

$$\widetilde{\psi} = \widehat{\psi} + (a/Q)\,\psi. \tag{3.88b}$$

This is analogous to the conventional scalar Lax pair of the continuous case, because we can view (3.88a) as a spectral problem and (3.88b) as time evolution (along the diagonal). The Lax pair (3.88) seems to have been first given in Date et al. (1983a) (in a different notation), while the lattice KdV equation (3.87) itself was already obtained by Hirota (1977a).

As usual the Lax pair is a set of two equations on one function and the requirement of consistency leads to some equation relating the coefficients. But how can one derive a condition for a pair such as (3.88)? Previously in Section 3.3.1 the Lax pair was first-order, i.e. each equation contained ψ at two adjacent points on the lattice, and the consistency condition was obtained by going around a square, for which we needed four equations. Now both equations in our Lax pair (3.88) contain ψ at three lattice points, as indicated in Figure 3.10a. In this case we have to consider a bigger diagram to derive a consistency condition, this is described in Figure 3.10b. Thus we need six equations, shifted at the proper places in the lattice.

Another point of view is obtained if one solves $\widehat{\psi}$ from (3.88b) and substitutes into (3.88a), one gets a scalar spectral problem, of the form[7]

$$\widetilde{\widetilde{\psi}} - (Q + a/\widetilde{Q})\widetilde{\psi} = (\Lambda - a)\psi. \tag{3.89}$$

This spectral problem has been investigated by Shabat (2002), Ruijsenaars (2002) and Boiti et al. (2001) in connection with semi-discrete analogues of the KdV equation, from the point of view of the Inverse Scattering Transform Method (ISTM). In the context of

[7] Note that equation (3.89) can be studied by itself, but for the discrete KdV the companion part (3.88b) is essential and it cannot be retrieved from (3.89) alone. See Exercise 3.8.

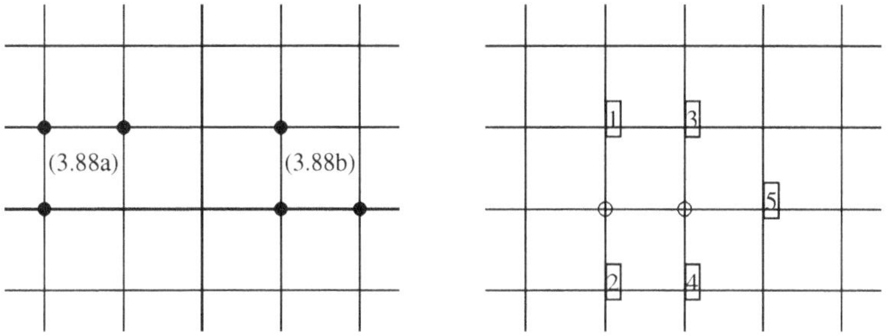

(a) (b)

Figure 3.10 (a) Points involved in equations (3.88). (b) Computational steps leading to a consistency condition at point 5: The values of ψ at open circles are given, then the values at points 1–4 are determined uniquely (in that sequence) and finally at point 5 we can use both equations and thereby obtain a consistency condition.

fully discrete KdV, the ISTM has been recently applied by Butler and Joshi (2010). In general, so far, the ISTM for discrete systems remains to be developed.

3.8.2 The lattice Schwarzian KdV revisited

Consider next the Lax pair for the lattice KdV equation (3.88) from a slightly different perspective. By eliminating Q from this Lax system, one can derive the nonlinear equation for ψ

$$(\widetilde{\psi}-\widehat{\psi})(\widehat{\widetilde{\psi}}-\lambda\psi) = a\widehat{\psi}\psi$$

which is the discrete analogue of the potential mKdV. This derivation is similar to what was done in the continuous case leading to (2.73). In fact, this equation is equivalent to (3.11) up to a transformation of the dependent variable.

Another way to proceed is to consider two independent solutions ψ_1, ψ_2 of the Lax pair (3.88), associated with the same potential Q and the same value of the spectral parameter λ. Introduce now the variable $z = \psi_1/\psi_2$, then from the first equation of the Lax pair we have:

$$\left.\begin{array}{l} (z\psi_2)\widehat{\widetilde{}} = Q(z\psi_2)\widehat{} + \lambda(z\psi_2) \\ \widehat{\widetilde{\psi}}_2 = Q\widehat{\psi}_2 + \lambda\psi_2 \end{array}\right\} \Rightarrow \frac{\widehat{\widetilde{\psi}}_2}{\widehat{\psi}_2} = Q\frac{\widehat{z}-z}{\widehat{\widetilde{z}}-z} \text{ and } \frac{\widehat{\widetilde{\psi}}_2}{\psi_2} = -\lambda\frac{\widehat{z}-z}{\widehat{\widetilde{z}}-\widehat{z}}.$$

Similarly, from the second relation we obtain:

$$\left.\begin{array}{l} (z\psi_2)\widetilde{} = (z\psi_2)\widehat{} + \frac{a}{Q}(z\psi_2) \\ \widetilde{\psi}_2 = \widehat{\psi}_2 + \frac{a}{Q}\psi_2 \end{array}\right\} \Rightarrow \frac{\widetilde{\psi}_2}{\psi_2} = \frac{a}{Q}\frac{\widehat{z}-z}{\widehat{z}-\widetilde{z}} \text{ and } \frac{\widetilde{\psi}_2}{\widehat{\psi}_2} = \frac{\widehat{z}-z}{\widetilde{z}-z}.$$

Using these relations and eliminating ψ_2 we obtain the lattice SKdV in the form (3.13):

$$\frac{(\widehat{\widetilde{z}}-z)(\widetilde{z}-\widehat{z})}{(\widehat{\widetilde{z}}-\widehat{z})(\widetilde{z}-z)}=\frac{a}{\lambda},$$

where we may consider λ to be fixed. We also find the following expression for the potential Q in terms of z:

$$Q\,\underset{\sim}{Q}=a\,\frac{(\widetilde{z}-z)(\widehat{z}-\underset{\sim}{z})}{(\widehat{z}-\widetilde{z})(z-\underset{\sim}{z})}.$$

The above procedure is analogous to what happens in the continuous case, which was discussed in Exercise 2.6.

3.8.3 The lattice modified KdV and sine-Gordon equation

In the same way as in Section 3.8.1 we can derive from the lattice potential mKdV equation (3.11) a "non-potential" equation

$$\frac{\widehat{\widetilde{W}}}{W}=\frac{(\gamma\widehat{W}-1)}{(\gamma-\widehat{W})}\,\frac{(\gamma-\widetilde{W})}{(\gamma\widetilde{W}-1)},\tag{3.90}$$

for the quantity $W\equiv\widetilde{v}/\widehat{v}$. Equation (3.90) is a derived equation in analogue to continuous case, where the mKdV equation (2.76) is derived from potential mKdV equation (2.73).

Equation (3.90) has the following Lax pair for the function φ

$$\widehat{\varphi}=W\widetilde{\varphi}+(1-\gamma W)\varphi,\tag{3.91a}$$

$$\widehat{\widetilde{\varphi}}=(\gamma+U)\widehat{\varphi}+\lambda U\varphi,\qquad U=:\frac{\gamma W-1}{\gamma-W},\tag{3.91b}$$

where $\gamma=p/q$ is a constant (see Exercise 3.10).

There is a close connection between the lattice potential mKdV (3.11) and the lattice sine-Gordon equation which is the following equation first given by Hirota (1977a)

$$p\,\sin\left(\theta-\overline{\theta}-\widetilde{\theta}+\overline{\widetilde{\theta}}\right)=q\,\sin\left(\theta+\overline{\theta}+\widetilde{\theta}+\overline{\widetilde{\theta}}\right).\tag{3.92}$$

This arises from equation (3.11) by considering an additional lattice shift which is associated with lattice parameter $r=0$, leading to the transformation $\mathring{v}=1/v$, which applied to the equation yields

$$p\left(\frac{v}{\underset{\wedge}{\mathring{v}}}-\frac{\widetilde{v}}{\underset{\wedge}{\mathring{\widetilde{v}}}}\right)=q\left(v\widetilde{v}-\frac{1}{\underset{\wedge}{\mathring{v}}}\frac{1}{\underset{\wedge}{\mathring{\widetilde{v}}}}\right).$$

We rewrite this equation in terms of $\widetilde{v}$ and $\widehat{\mathring{v}}$, identifying the combined shift $\widehat{\mathring{v}}$ as a new lattice shift $\overline{v}$ and, at the same time, we set $v=\exp(2\,i\,\theta)$ to obtain (3.92).

Since the equation is no longer symmetric under interchange of p and q and the lattice shifts $\widetilde{\ }$ and $\overline{\ }$, the equation will no longer satisfy the CAC property in the strong sense that the same equation can be uniquely prescribed on all faces of the cube. However, there is a weak form of the CAC property which still applies, in which slightly different equations may be prescribed on different faces. In the present case, one of the accompanying equations would be the original lattice mKdV, but rewritten in terms of the variable θ, see Section 3.6.

It can be shown that the Lax pair for the discrete sine-Gordon equation (3.92) (Quispel et al., 1991), can be derived from the Lax pair of the lattice mKdV (3.21) by introducing the Lax matrix for the circle-shift $\mathring{v}$ and composing it with the Lax matrix for the hat-shift.

Note that in Section 2.5.4, an equation similar to the discrete sine-Gordon equation (2.93) was derived from the permutability condition for the continuous sine-Gordon equation. That equation has different signs inside the sine-functions. Nevertheless, one can be mapped to the other by a change of variables. An alternative approach to mKdV was given in Suris (1997b).

3.8.4 Further equations admitting a scalar Lax pair

In Date et al. (1983a) a list of partial difference equations on the two-dimensional lattice was given, all of which admit scalar Lax pairs, similar to the ones we have found in the previous section for the lattice KdV and lattice mKdV equations. The construction in that paper was different from the considerations we have presented here, and was based on the so-called *Hirota bilinear* approach in combination with the Grasmannian approach developed by the Kyoto school. The details of the latter method will be postponed to Chapter 8. Here we will focus on the Lax pairs and their associated equations, and these can be verified by direct computation. We note, however, that the resulting equations are no longer all of quadrilateral type and may be associated with different graphs or stencils on the lattice.

Discrete-time Toda equation

The time evolution of the Toda lattice equation (1.39) was discretized in Hirota (1977b). His method was based on preserving the multi-soliton solutions using the bilinear formalism (see Chapter 8) and the result was

$$a f_{n,m+1} f_{n,m-1} - b f_{n+1,m} f_{n-1,m} + (b-a) f_{n,m}^2 = 0, \tag{3.93}$$

which is fully symmetric in the old lattice index n and the new discretized time m. Note also that each term is quadratic and the sum of indices is the same in all terms (namely $(2n, 2m)$). Equations of this form are called Hirota bilinear, and we will see that such equations arise time and again in connection with integrable systems (cf. Chapter 8). However, this equation appeared in a slightly different form already in the work by G. Frobenius almost a century earlier; see Section 4.1 equation (4.8).

Following Date et al. (1983a) (see their "Example 3") we will now assert the integrability of the time-discrete Toda equation on the basis of the existence of a Lax pair, which reads (differing slightly from Date et al., 1983a)

$$\widehat{\widetilde{\psi}} = \frac{\widehat{u}}{u}(\widetilde{\psi} + q\widehat{\psi})\,, \tag{3.94a}$$

$$(1-pk)(1-q/k)\psi = \underset{\wedge}{\psi} - p\frac{\underset{\sim}{u}}{\underset{\wedge}{\underset{\sim}{u}}}\underset{\wedge}{\widetilde{\psi}}\,, \tag{3.94b}$$

in which p and q are lattice parameters as usual, and k is a spectral parameter.

In deriving the compatibility condition for this pair note first that the ψ points involved are as in Figure 3.11a, and then following the steps in Figure 3.11b one arrives at the following equation for the potential u

$$\frac{u}{\widehat{u}} - \frac{\underset{\wedge}{\underset{\sim}{u}}}{u} + pq\left(\frac{\widehat{\underset{\sim}{u}}}{u} - \frac{u}{\underset{\wedge}{\widetilde{u}}}\right) = 0\,. \tag{3.95}$$

The lattice points connected by (3.95) are described in Figure 3.12a. Another way of writing (3.95) is obtained if we define (Hirota et al., 1993)

$$I_{n,m} := \frac{u_{n,m-1}}{u_{n,m}}, \quad V_{n,m} := \frac{u_{n,m}}{u_{n+1,m-1}}, \tag{3.96}$$

then we have

$$I_{n,m+1}V_{n,m+1} = I_{n+1,m}V_{n,m}, \tag{3.97a}$$

$$I_{n,m+1} - I_{n,m} = pq(V_{n,m} - V_{n-1,m+1}), \tag{3.97b}$$

where (3.97a) follows immediately from the definition (3.96), and (3.97b) is (3.95) rewritten. (We will meet these equations again in Section 4.1.) Furthermore, if we set $u_{n,m} = f_{n,m}/f_{n+1,m}$ we can write (3.95) as

$$F_{n,m} = F_{n+1,m}, \quad \text{where} \quad F_{n,m} := \frac{f_{n,m-1}f_{n,m+1} + pqf_{n-1,m+1}f_{n+1,m-1}}{f_{n,m}^2}.$$

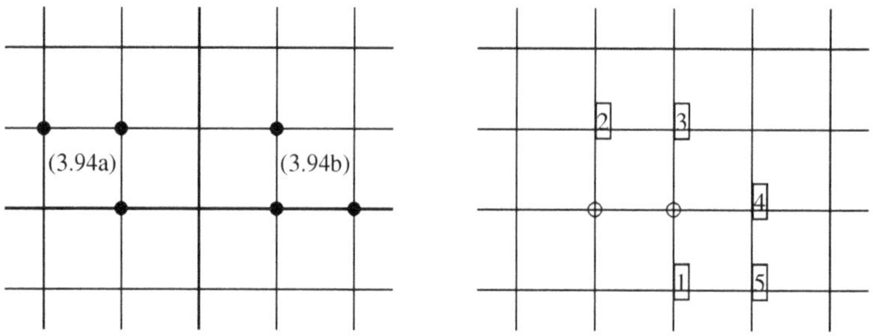

(a) (b)

Figure 3.11 (a) Points involved in equations (3.94). (b) Computational steps leading to a consistency condition at point 5, which can be computed in two ways.

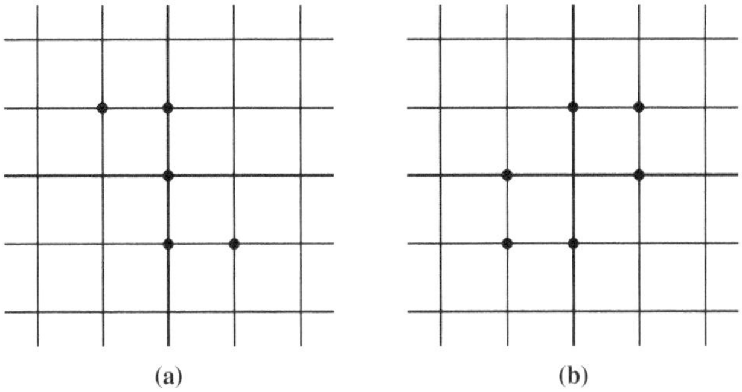

Figure 3.12 Points involved (a) in equation (3.95) and (b) in equation (3.98).

If we now assign a constant value $F_{n,m}$ then the result can be transformed into form (3.93) by simple scaling and transformation on the independent variables.

The discrete-time Toda equation is a universal equation arising in the theory of formal orthogonal polynomials and Padé approximants; see Section 4.1, it has also arisen in physics, for example, as an equation describing the spin–spin correlation functions in the two-dimensional Ising model of ferromagnetism (Perk, 1980). Its integrability follows not only from the existence of a nontrivial Lax pair with spectral parameter (k is notably absent from the equation (3.95)), but also from the existence of multi-soliton solutions (cf. Hirota, 1977b) and other structural features.

Hexagonal lattice Boussinesq equation

A partial difference equation involving a hexagonal configuration on the lattice was proposed in Date et al. (1983a), and reads

$$p(q-r)\left(\frac{\widetilde{u}}{\widehat{\widetilde{u}}}-\frac{\underset{\wedge}{\underset{\sim}{u}}}{\underset{\sim}{u}}\right)=q(p-r)\left(\frac{\widehat{u}}{\widehat{\widetilde{u}}}-\frac{\underset{\wedge}{\underset{\sim}{u}}}{\underset{\wedge}{u}}\right) \tag{3.98}$$

(in which r is an additional fixed parameter). The lattice points of this equation are given in Figure 3.12b. Equation (3.98) arises as the compatibility of the following Lax pair:

$$p(1-pk)(1-qk)(1-rk)\widetilde{\psi}=(p-r)\frac{u}{\widetilde{u}}\underset{\wedge}{\psi}+r\underset{\wedge}{\underset{\sim}{\psi}}\,, \tag{3.99a}$$

$$q(1-pk)(1-qk)(1-rk)\widehat{\psi}=(q-r)\frac{u}{\widehat{u}}\underset{\sim}{\psi}+r\underset{\wedge}{\underset{\sim}{\psi}}\,. \tag{3.99b}$$

The consistency computation is again different for this pair and is explained in Figure 3.13. The Hirota bilinear form following from $u(n,m)=f(n,m)/f(n-1,m-1)$ is

$$p(q-r)\widetilde{f}\underset{\sim}{f}+q(r-p)\widehat{f}\underset{\wedge}{f}+r(p-q)\underset{\wedge}{\underset{\sim}{f}}\widehat{\widetilde{f}}=0. \tag{3.100}$$

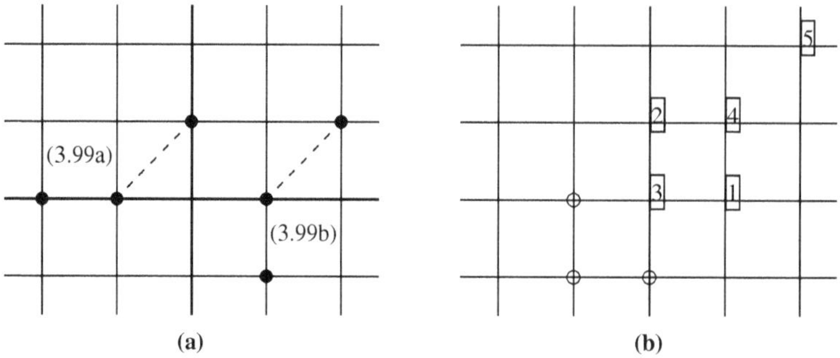

Figure 3.13 (a) Points involved in equations (3.99). (b) Computational steps leading to a consistency condition at point 5. (Note that 3, 4 are determined as a pair.)

The BSQ equation is obtained if we take $p = \omega q = \omega^2 r \to 0$, where ω is a cubic root of unity $\neq 1$, but this equation is integrable for generic parameter values (see e.g. Hietarinta and Zhang, 2013).

This partial difference equation arises as a direct reduction of higher-dimensional equation which we will present later (the lattice modified KP equation), which explains the appearance of the auxiliary parameter r. Recall also that in a previous section we gave alternative lattice equations discretizing the BSQ equation with quite a different structure. This demonstrates that even within the class of "integrable" discretizations of a given PDE different choices are possible.

3.9 Higher-dimensional equations: the KP class

The famous Kadomtsev–Petviashvili (KP) equation, which in its potential version reads

$$\left(u_t - \frac{1}{4}u_{xxx} - \frac{3}{2}u_x^2\right)_x = \frac{3}{4}u_{yy}, \tag{3.101}$$

is a prototype of a three-dimensional integrable equation. In fact, since the function $u = u(x, y, t)$ is dependent on one temporal (t) and two spatial (x and y) independent variables, this equation is often said to be (2+1)-dimensional. As such, it is a natural generalization of the KdV equation (2.69), which is recovered when u becomes independent of the variable y.

What makes the equation (3.101) regarded as integrable is that it possesses a Lax pair, which consists of a pair of linear PDEs of the form:

$$\varphi_y = \varphi_{xx} + 2u_x\varphi, \tag{3.102a}$$

$$\varphi_t = \varphi_{xy} + u_x\varphi_x + \frac{1}{2}(3u_y - u_{xx})\varphi\,. \tag{3.102b}$$

Note that in contrast to the Lax pair for the KdV equation (2.70), (3.102) no longer contains a spectral parameter. Nevertheless, this Lax pair can be seen as the source of many of the integrability characteristics of the KP equation, and in particular, forms the starting point for many of its solution methods, cf. e.g. Kupershmidt (2000); Fokas and Ablowitz (1983); Zakharov and Manakov (1985). Several other related continuous 2+1-dimensional PDEs were studied during the 1980s. In the present monograph, we are interested in its integrable discretizations, which we will discuss next.

An integrable lattice versions of (3.101), i.e. the **lattice (potential) KP equation** reads as follows

$$(p-\widetilde{u})(q-r+\widetilde{\overline{u}}-\widehat{\widetilde{u}})+(q-\widehat{u})(r-p+\widehat{\widetilde{u}}-\widehat{\overline{u}})+(r-\overline{u})(p-q+\widehat{\overline{u}}-\widetilde{\overline{u}})=0\,, \quad (3.103)$$

in which $u=u_{n,m,h}$ is a function depending on three discrete independent variables n,m,h. (The notation for the lattice shifts is as before with $\overline{u}=u_{n,m,h+1}$ as the new shift.) Equation (3.103), which was given in Nijhoff et al. (1984), can be rewritten in various seemingly different but equivalent forms, namely:

$$(p+\widehat{\overline{u}})(q-r+\overline{u}-\widehat{u})+(q+\widetilde{\overline{u}})(r-p+\widetilde{u}-\overline{u})+(r+\widehat{\widetilde{u}})(p-q+\widehat{u}-\widetilde{u})=0\,, \quad (3.104)$$

or

$$\frac{(p-r+\overline{u}-\widetilde{u})\widehat{\ }}{p-r+\overline{u}-\widetilde{u}}=\frac{(q-r+\overline{u}-\widehat{u})\widetilde{\ }}{q-r+\overline{u}-\widehat{u}}=\frac{(p-q+\widehat{u}-\widetilde{u})^{-}}{p-q+\widehat{u}-\widetilde{u}}\,. \quad (3.105)$$

From the last form it is obvious that the p,q,r dependence can be eliminated by the transformation $u=w+np+mq+rh+c$, which was used previously, e.g. to connect (3.6) and (3.10).

It can be shown (see Chapter 5) that by applying a sequence of continuum limits we recover the PDE (3.101) from these equations.

The integrability of (3.103) is a rather subtle matter. The equation is multidimensionally consistent, i.e. it can be embedded in a consistent way in a multidimensional lattice of dimension larger or equal to four with copies of the equation imposed on each three-dimensional sub-lattice. For this three-dimensional equation, this means imposing the equation on three-dimensional faces of a four-dimensional hypercube. However, the consistency check is different, because the equation takes on the general form

$$f(\widetilde{u},\widehat{u},\overline{u},\widehat{\widetilde{u}},\widetilde{\overline{u}},\widehat{\overline{u}};p,q,r)=0\,, \quad (3.106)$$

i.e. they only involve six of the eight vertices on the hypercube as indicated in Figure 3.14, in which the relevant vertices are organized among the hexagonal configuration (indicated by the solid lines).

The paper by Adler et al. (2012) contains a discussion of the multidimensional consistency of equations of the type (3.106) as equations living on an octahedral sub-lattice. This paper also contains a classification of all scalar (i.e. single-field) equations of this type under fairly general conditions. In fact, no special assumptions were made on the linear or nonlinear dependence of f on the variables at the vertices, nor were any specific symmetries imposed. The result was that (up to a group of "admissible" transformations) all

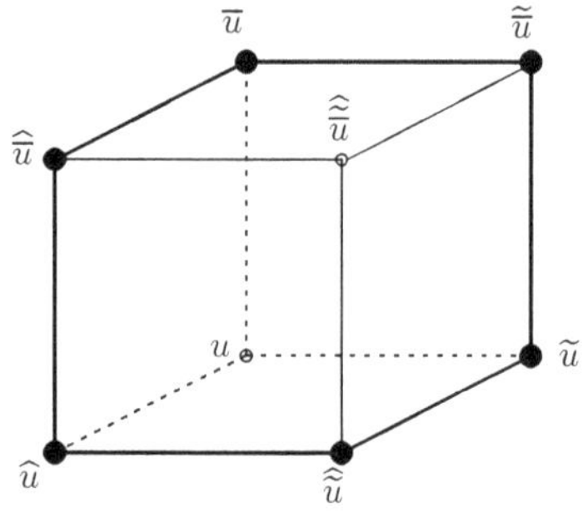

Figure 3.14 Hexagonal configuration of vertices, marked by a black disk, involved in the lattice (potential) KP equation.

octahedral equations of the type (3.106), consistent on the 4D lattice, were among one of five canonical types, namely

$$\textbf{i)}\quad \widehat{\widetilde{\mathsf{u}}}\,\overline{\mathsf{u}} - \widehat{\overline{\mathsf{u}}}\,\widetilde{\mathsf{u}} + \widetilde{\overline{\mathsf{u}}}\,\widehat{\mathsf{u}} = 0, \tag{3.107a}$$

$$\textbf{ii)}\quad \frac{(\widehat{\widetilde{\mathsf{u}}} - \widehat{\mathsf{u}})(\widehat{\overline{\mathsf{u}}} - \overline{\mathsf{u}})(\widetilde{\overline{\mathsf{u}}} - \widetilde{\mathsf{u}})}{(\widehat{\widetilde{\mathsf{u}}} - \widetilde{\mathsf{u}})(\widehat{\overline{\mathsf{u}}} - \widehat{\mathsf{u}})(\widetilde{\overline{\mathsf{u}}} - \overline{\mathsf{u}})} = 1, \tag{3.107b}$$

$$\textbf{iii)}\quad (\widehat{\widetilde{\mathsf{u}}} - \widehat{\overline{\mathsf{u}}})\widehat{\mathsf{u}} + (\widehat{\overline{\mathsf{u}}} - \widetilde{\overline{\mathsf{u}}})\overline{\mathsf{u}} + (\widetilde{\overline{\mathsf{u}}} - \widehat{\widetilde{\mathsf{u}}})\widetilde{\mathsf{u}} = 0, \tag{3.107c}$$

$$\textbf{iv)}\quad \frac{\widehat{\widetilde{\mathsf{u}}} - \widehat{\overline{\mathsf{u}}}}{\widehat{\mathsf{u}}} + \frac{\widehat{\overline{\mathsf{u}}} - \widetilde{\overline{\mathsf{u}}}}{\overline{\mathsf{u}}} + \frac{\widetilde{\overline{\mathsf{u}}} - \widehat{\widetilde{\mathsf{u}}}}{\widetilde{\mathsf{u}}} = 0, \tag{3.107d}$$

$$\textbf{v)}\quad \frac{\widehat{\overline{\mathsf{u}}} - \widetilde{\overline{\mathsf{u}}}}{\overline{\mathsf{u}}} = \widehat{\widetilde{\mathsf{u}}}\left(\frac{1}{\widetilde{\mathsf{u}}} - \frac{1}{\widehat{\mathsf{u}}}\right). \tag{3.107e}$$

All equations of this list were already known prior to the paper Adler et al. (2012), namely (3.107a) had appeared as the bilinear KP equation in Hirota (1981), whereas (3.107b–d) had appeared in Nijhoff et al. (1984) in forms containing additional (removable) parameters (cf. (3.103) and (3.107c)), while (3.107b) in a form similar to this one was given in Dorfman and Nijhoff (1991), and was later rediscovered in Bogdanov and Konopelchenko (1998a). Equation (3.107e) was given in Nijhoff and Capel (1990), where noncommutative (i.e. matrix versions) of the equations were listed. However, the fact that (3.107) forms (up to equivalence) an exhaustive list of such equations defined on an octahedral lattice is a major new insight, which was based on a careful analysis of the multidimensional consistency.

We will now go into some further detail regarding some of the equations in the list (3.107). A Lax representation of (3.103) is given by the following system of (inhomogeneous) linear equations:

$$\left.\begin{aligned} p\widetilde{\varphi} &= (p - \widetilde{u})\varphi + \chi \\ q\widehat{\varphi} &= (q - \widehat{u})\varphi + \chi \\ r\overline{\varphi} &= (r - \overline{u})\varphi + \chi \end{aligned}\right\} \quad \Rightarrow \quad \left\{\begin{aligned} p\widetilde{\varphi} &= r\overline{\varphi} + (p - r + \overline{u} - \widetilde{u})\varphi \\ q\widehat{\varphi} &= r\overline{\varphi} + (q - r + \overline{u} - \widehat{u})\varphi \end{aligned}\right. \tag{3.108}$$

where to obtain the second homogeneous set of linear equations in terms of the function $\varphi = \varphi_{n,m,h}$ the function $\chi = \chi_{n,m,h}$ is eliminated pairwise. The parameters can here be removed using $u = X + np + mq + rh + c$, $\varphi = p^{-n}q^{-m}r^{-h}\psi$, resulting in

$$\begin{cases} \widetilde{\psi} = \overline{\psi} + (\overline{\mathsf{u}} - \widetilde{\mathsf{u}})\psi \\ \widehat{\psi} = \overline{\psi} + (\overline{\mathsf{u}} - \widehat{\mathsf{u}})\psi. \end{cases} \tag{3.109}$$

Next, the compatibility condition $\widehat{\widetilde{\psi}} = \widetilde{\widehat{\psi}}$ of the latter pair gives rise to the lattice KP equation as follows. From the hat-shift of the first relation and the tilde-shift of the second we get:

$$\begin{aligned} \widehat{\widetilde{\psi}} &= \widehat{\overline{\psi}} + (\widehat{\overline{\mathsf{u}}} - \widehat{\widetilde{\mathsf{u}}})\widehat{\psi} \\ &= \overline{\overline{\psi}} + (\overline{\overline{\mathsf{u}}} - \overline{\widehat{\mathsf{u}}})\overline{\psi} + (\widehat{\overline{\mathsf{u}}} - \widehat{\widetilde{\mathsf{u}}})(\overline{\psi} + (\overline{\mathsf{u}} - \widehat{\mathsf{u}})\psi), \\ \widetilde{\widehat{\psi}} &= \widetilde{\overline{\psi}} + (\widetilde{\overline{\mathsf{u}}} - \widetilde{\widehat{\mathsf{u}}})\widetilde{\psi} \\ &= \overline{\overline{\psi}} + (\overline{\overline{\mathsf{u}}} - \overline{\widetilde{\mathsf{u}}})\overline{\psi} + (\widetilde{\overline{\mathsf{u}}} - \widetilde{\widehat{\mathsf{u}}})(\overline{\psi} + (\overline{\mathsf{u}} - \widetilde{\mathsf{u}})\psi). \end{aligned}$$

Now comparing the two results we see that the $\overline{\overline{\psi}}$ and $\overline{\psi}$ terms cancel and from the ψ terms we get $(\widehat{\overline{\mathsf{u}}} - \widehat{\widetilde{\mathsf{u}}})(\overline{\mathsf{u}} - \widehat{\mathsf{u}}) = (\widetilde{\overline{\mathsf{u}}} - \widetilde{\widehat{\mathsf{u}}})(\overline{\mathsf{u}} - \widetilde{\mathsf{u}})$, i.e. (3.107c).

The second example we examine more closely is the so-called **lattice (potential) modified KP equation**, which reads

$$p\left(\frac{\widehat{v}}{\widetilde{\widehat{v}}} - \frac{\overline{v}}{\widetilde{\overline{v}}}\right) + q\left(\frac{\overline{v}}{\widehat{\overline{v}}} - \frac{\widetilde{v}}{\widetilde{\widehat{v}}}\right) + r\left(\frac{\widetilde{v}}{\widetilde{\overline{v}}} - \frac{\widehat{v}}{\widehat{\overline{v}}}\right) = 0, \tag{3.110}$$

for a function $v = v_{n,m,h}$. This has a counterpart for a function $w = w_{n,m,h}$ which is very similar, namely:

$$p\left(\frac{\widehat{\overline{w}}}{\overline{w}} - \frac{\widehat{\widetilde{w}}}{\widetilde{w}}\right) + q\left(\frac{\widetilde{\widehat{w}}}{\widetilde{w}} - \frac{\widetilde{\overline{w}}}{\overline{w}}\right) + r\left(\frac{\widehat{\overline{w}}}{\widehat{w}} - \frac{\widetilde{\overline{w}}}{\widetilde{w}}\right) = 0. \tag{3.111}$$

The equations (3.110) and (3.111) reduce to (3.107d), with $v = p^n q^m r^h/\mathsf{u}$ and $w = \mathsf{u}/(p^n q^m r^h)$, respectively. They are related to the lattice KP equation (3.103) via the following Miura-type relations

$$p - q + \widehat{u} - \widetilde{u} = \frac{p\widehat{v} - q\widetilde{v}}{\widehat{\widetilde{v}}} = \frac{p\widetilde{w} - q\widehat{w}}{w}. \tag{3.112}$$

Furthermore, (3.110) arises from the following inhomogeneous triplet of linear equations

$$\left.\begin{aligned} p\widetilde{\boldsymbol{\psi}} &= p\boldsymbol{\psi} + \widetilde{v}\boldsymbol{\varphi} \\ q\widehat{\boldsymbol{\psi}} &= q\boldsymbol{\psi} + \widehat{v}\boldsymbol{\varphi} \\ r\overline{\boldsymbol{\psi}} &= r\boldsymbol{\psi} + \overline{v}\boldsymbol{\varphi} \end{aligned}\right\} \quad \Rightarrow \quad \begin{cases} \widetilde{\boldsymbol{\psi}} = \frac{r}{p}\frac{\widetilde{v}}{\overline{v}}\overline{\boldsymbol{\psi}} + \left(1 - \frac{r}{p}\frac{\widetilde{v}}{\overline{v}}\right)\boldsymbol{\psi} \\ \widehat{\boldsymbol{\psi}} = \frac{r}{q}\frac{\widehat{v}}{\overline{v}}\overline{\boldsymbol{\psi}} + \left(1 - \frac{r}{q}\frac{\widehat{v}}{\overline{v}}\right)\boldsymbol{\psi} \end{cases} \tag{3.113}$$

where the second set of homogeneous equations for the function $\psi = \psi_{n,m,h}$ is deduced from the first set by eliminating the function φ (which incidentally is the same function as the one appearing in the Lax pair (3.108) for the lattice KP equation). In a very similar way as in the lattice KP case, the compatibility of the latter linear homogeneous system leads to the lattice MKP equation (3.110) for the function v.

The third example is the **lattice Schwarzian KP (SKP) equation**, given by

$$\frac{(\widehat{\widetilde{z}} - \widehat{z})(\widetilde{\overline{z}} - \widetilde{z})(\widehat{\overline{z}} - \overline{z})}{(\widehat{\widetilde{z}} - \widetilde{z})(\widetilde{\overline{z}} - \overline{z})(\widehat{\overline{z}} - \widehat{z})} = 1. \tag{3.114}$$

This equation, which was given in this explicit form in Dorfman and Nijhoff (1991), and was further studied in Bogdanov and Konopelchenko (1998a,b). It is equivalent to an equation with more parameters given in Nijhoff et al. (1984), and it is related to the lattice MKP equations (3.110) and (3.111) via the simple transformations (cf. (3.65)):

$$p(z - \widetilde{z}) = \widetilde{v}w\ , \quad q(z - \widehat{z}) = \widehat{v}w\ , \quad r(z - \overline{z}) = \overline{v}w, \tag{3.115}$$

where by elimination of the functions v and w, in fact by writing out the trivial relation:

$$\frac{\widehat{(\widetilde{v}w)}}{\overline{(\widetilde{v}w)}} = \frac{\widetilde{(\widehat{v}w)}}{\widetilde{(\overline{v}w)}}\,\frac{\widehat{(\overline{v}w)}}{\overline{(\widehat{v}w)}},$$

we obtain (3.114). We observe that equation (3.114) coincides with (3.107b). The equation plays an important role in the connection of these lattice equations with projective and conformal geometry, cf. e.g. Konopelchenko and Schief (2002a); King and Schief (2003).

The first equation in the list (3.107a) is one of the most fundamental discrete equations, because many continuous and discrete integrable equations can be obtained from it in suitable limits, as Hirota (1981) showed. In fact, this was the first fully discrete integrable equation in three dimensions. Hirota himself called it **DAGTE** (discrete analogue of generalized Toda equation) and gave it in the form:

$$\begin{aligned} a_1 f(x+\delta, y, t) f(x-\delta, y, t) + a_2 f(x, y+\varepsilon, t) f(x, y-\varepsilon, t) \\ + a_3 f(x, y, t+\kappa) f(x, y, t-\kappa) = 0, \end{aligned} \tag{3.116}$$

where a_1, a_2, a_3 are three arbitrary constants. This equation is in Hirota bilinear form and we will discuss it further in Chapter 8.

3.10 Notes

The history of integrable partial difference equations goes back to seminal papers by Ablowitz and Ladik (1976) and by Hirota (1977a). Ablowitz and Ladik were motivated by the search for integrable numerical algorithms through finite-difference approximations. This fits well into the general problem of the analysis of finite-difference PΔEs arising from numerical studies of PDEs; see e.g. Garabedian (1964), Chapter 13. On the other hand, the motivation in Hirota's approach was to discretize a given continuum equation while keeping its (soliton) solution structure. We will return to Hirota's method in Chapter 8.

Subsequently, systematic methods for the construction of integrable nonlinear finite-difference PΔEs were proposed, e.g. through the representation theory of infinite-dimensional Lie algebras (Date et al., 1982a,b, 1983a,b,c), and through singular linear integral equations (also known as direct linearization method) and connections with Bäcklund transformations, cf. Nijhoff et al. (1983a); Quispel et al. (1984). Further developments included extensions to higher-order PΔEs (Nijhoff et al., 1992) and multidimensions (Hirota, 1981; Levi et al., 1981; Miwa, 1982; Nijhoff et al., 1984; Nijhoff, 1985; Nijhoff and Capel, 1990).

The idea of multidimensional consistency moved to the foreground of attention in the late 1990s (Doliwa and Santini, 1997; Nimmo and Schief, 1997; Nijhoff et al., 2001; Nijhoff and Walker, 2001; Bobenko and Suris, 2002), although it was already implicit in the constructions of integrable PΔEs both by the bilinear approach as well in the approach through singular integral equations, while consistency-around-the-cube was already formulated from a slightly different perspective in Maillet and Nijhoff (1989).

It is remarkable that in the context of *classical* integrable systems it is the *quantum* Yang–Baxter equation (3.36) that emerges as a structural relation. Originally the YB equation arose in the exactly solvable model in statistical mechanics (Baxter, 1982) and in quantum inverse scattering (Korepin et al., 1997). In the context of classical lattice systems the quantum YB equation has been the basic starting point in the approaches of Maillet and Nijhoff (1989); Zabrodin (1997b, 1998). The idea of a Yang–Baxter *map* discussed in Section 3.4 goes back to the set-theoretical solutions of YB equation (Drinfeld, 1992; Hietarinta, 1997; Etingof et al., 1999) and was made concrete in Veselov (2003); Goncharenko and Veselov (2004). A classification of YB maps was given in Adler et al. (2004) and the construction of Lax representations by Suris and Veselov (2003). The connection to quadrilateral lattice equations and multidimensional consistency was explained by Papageorgiou et al. (2006), cf. also Papageorgiou and Tongas (2007, 2009); Papageorgiou et al. (2010). The study of Yang–Baxter maps is a thriving subject, cf. Kakei et al. (2009, 2010); Kouloukas and Papageorgiou (2009, 2012); Konstantinou-Rizos and Mikhailov (2013); Konstantinou-Rizos (2014).

After the important classification result of Adler et al. (2003, 2009), together with the earlier discovery of Adler's equation (Adler, 1998), a wave of new results followed. This included work on symmetries and conservation laws (Levi and Yamilov, 2009a, 2011, 2009b; Levi and Petrera, 2007; Rasin and Hydon, 2007a,b; Tongas et al., 2007; Xenitidis, 2011; Mikhailov et al., 2011a,b; Zhang et al., 2013), soliton solutions (see Chapters 8, 9). Further generalization and variants of the equations presented here include equations of back-white lattices (Xenitidis and Papageorgiou, 2009; Adler et al., 2009; Boll, 2011, 2012) and multiquadratic equations (Adler and Veselov, 2004; Kassotakis and Nieszporski, 2011, 2012; Atkinson, 2012; Atkinson and Nieszporski, 2014).

The origin of many lattice equations is in the Bianchi permutability diagram (see p. 461 of Bianchi, 1899) of Bäcklund transformations, which were originally devised as transformations between surfaces in differential geometry. From this connection it is clear that there is an intimate connection between discrete differential geometry (difference geometry) and the theory of integrable PΔEs. These play in particular a role in the development of a discrete analogue of differential geometry, which was elaborated in the 1930s as *Differenzengeometrie* by R. Sauer (Sauer, 1970). Furthermore, the connection with BTs, viewed as transformations of discrete surfaces, appeared already in the work by Wunderlich (1951) and to conjugate nets as discrete coordinate systems in the work by G. Koenigs during the late 19th century (Koenigs, 1891, 1892). The modern theory along this line of development was rebaptized *discrete differential geometry*, and this was amply covered in the monograph by Bobenko and Suris (2008). The geometric point of view has given rise to many reinterpretations of lattice equations as equations generating special surfaces, e.g. in Bobenko and Pinkall (1996b); Schief (1996, 2006, 2007); Schief and Rogers (1999); Nimmo and Schief (1997); Konopelchenko and Schief (1998, 2002a); Doliwa et al. (2007, 2004); Doliwa (2009) as well as in algebraic geometry and incidence theory (King and Schief, 2006; Adler, 2006). In particular integrable discrete systems in three dimensions have formed a rich source of geometric connections, (Doliwa, 2010, 2005; Konopelchenko and Schief, 2002b; Doliwa et al., 1999) where they often arise as the compatibility conditions of discrete Laplace and Darboux systems (Doliwa and Santini, 1997, 1999; Doliwa et al., 2000; Doliwa and Santini, 2000).

Exercises

3.1 One can also verify the consistency-around-the-cube when the initial values are w, $\overline{w}$, $\widetilde{w}$, $\widehat{\widetilde{w}}$. Take the equations (3.7) and use the bottom equation to solve for $\widehat{w}$ and left equation to solve for $\overline{\widetilde{w}}$, and then the back and front equations to solve for $\overline{\widehat{w}}$ and $\overline{\widehat{\widetilde{w}}}$, respectively. This leaves top and right equations: verify that they are satisfied.

3.2 Show that equation (3.11) is related to the lattice KdV equation (3.6) via the transformations:

$$\widehat{w} - \widetilde{w} = \frac{p\widetilde{v} - q\widehat{v}}{v}, \quad w - \widehat{\widetilde{w}} = \frac{pv + q\widehat{\widetilde{v}}}{\widetilde{v}} = \frac{qv + p\widehat{\widetilde{v}}}{\widehat{v}}$$

which forms the discrete analogue of the Miura transformation.

3.3 Consider the *discrete Cole–Hopf transformation*, given by the set of relations:

$$p\,(\widetilde{z} - z) = \widetilde{v}v, \quad q\,(\widehat{z} - z) = \widehat{v}v, \tag{3.117}$$

for two different dependent variable v and z defined on the two-dimensional lattice, with the usual notation of the lattice and lattice parameters. By eliminating the variable v, show that the variable z obeys the cross-ratio equation (3.13), and by eliminating z show that v obeys the lattice potential mKdV equation (3.11).

3.4 For the SKdV (3.13) derive the Lax pair given in Example 3.3.2.

3.5 Verify that (3.53) is a Bäcklund transformation between H2 and H1.

3.6 Verify that (3.54) is consistent around the cube if the deformation terms in the equations are as given in Figure 3.8.

3.7 Show by direct computation that the matrix equation $\widehat{L}M = \widetilde{M}L$ for (3.23) gives, entry-by-entry, the lattice SKdV equation (3.13), and hence that the linear system $\widetilde{\psi} = L\psi$, $\widehat{\psi} = M\psi$ is compatible if and only if this equation is satisfied for z.

3.8 (a) Show by combining the two equations (3.88) that we have the second-order difference equation

$$\widehat{\widehat{\psi}} = (Q - a/\widehat{Q})\,\widehat{\psi} + \Lambda\,\psi. \tag{3.118}$$

Use this and (3.88) to compute $\widetilde{\widehat{\widehat{\psi}}}$ in two different ways to show that Q obeys the lattice KdV equation. [Write the result in terms of $\widehat{\psi}$ and $\widehat{\psi}$.]

(b) Now defining

$$\boldsymbol{\phi} = \begin{pmatrix} \widehat{\psi} \\ \psi \end{pmatrix},$$

we can write the Lax pair (3.88, 3.118) in the form $\widetilde{\boldsymbol{\phi}} = \boldsymbol{L}\boldsymbol{\phi}$, $\widehat{\boldsymbol{\phi}} = \boldsymbol{M}\boldsymbol{\phi}$, with

$$\boldsymbol{L} = \begin{pmatrix} Q & \Lambda \\ 1 & a/Q \end{pmatrix}, \quad \boldsymbol{M} = \begin{pmatrix} Q - a/\widehat{Q} & \Lambda \\ 1 & 0 \end{pmatrix} \tag{3.119}$$

and their compatibility implies (3.87). Note that this system separates the mixture of tilde- and hat-shifts in the system (3.88).

(c) Finally, if we let $x \equiv x_{n,m,k} := Q_{n,m}$ and introduce $\overline{x} = x_{n,m,k+1}$ by $\widehat{\psi}/\psi = a\,\overline{x}/x$, with other shifts as usual, we can write these equations on the cube as follows, with $\Lambda = a\lambda$:

$$x(\widehat{x} - \widetilde{x})\widehat{\widetilde{x}} + a(x - \widehat{\widetilde{x}}) = 0, \tag{3.120a}$$

$$x\widetilde{x}(\overline{x} + \lambda) - a\overline{\widetilde{x}}(\overline{x} + 1) = 0, \tag{3.120b}$$

$$x\widehat{x}(\overline{x} + \lambda) - a\overline{x}(\overline{\widehat{x}} + 1) = 0, \tag{3.120c}$$

$$\overline{x}(\overline{\widehat{x}} - \overline{\widetilde{x}})\overline{\widehat{\widetilde{x}}} + \overline{x}\,\overline{\widehat{x}} - \overline{\widetilde{x}}\,\overline{\widehat{\widetilde{x}}} + \lambda(\overline{x} - \overline{\widehat{\widetilde{x}}}) = 0, \tag{3.120d}$$

from the bottom, left, back and top equations, respectively, along with right and front equations by the relevant shifts. Now we observe certain asymmetry in the form, so it does not fit into the assumption of having the same equation on all sides, up to parameter changes. Verify that (3.120) is consistent around the cube.

3.9 Derive the Lax pair (3.88) from the set of equations (3.16) (with $\gamma = \gamma' = 1$) by setting $\varphi = G$ and identifying the spectral variable by $\Lambda = k^2 - q^2$. Verify also that it can be written as (3.119) and that its consistency condition leads to the lattice KdV equation (3.87).

3.10 (a) Show that the equation for W (3.90) when $\gamma = p/q$ can be derived from (3.11) by setting $W \equiv \widetilde{v}/\widehat{v}$. Derive also a quadrilateral lattice equation for $U = v/\widehat{\widetilde{v}}$.
(b) Show that from the compatibility of the Lax pair (3.91) one obtains the equation (3.90).

3.11 This exercise concerns a simple singular solution of the lattice potential KdV equation.
(a) Let $\xi = \xi_{n,m} = \xi_0 + \frac{n}{p} + \frac{m}{q}$, where ξ_0 is constant, and p and q parameters. Consider the function $u = u_{n,m} = 1/\xi$ and prove the following relations:

$$R \equiv p - q + \widehat{u} - \widetilde{u} = (p-q)\frac{\xi\,\widehat{\widetilde{\xi}}}{\widetilde{\xi}\,\widehat{\xi}}\,, \tag{3.121a}$$

$$Q \equiv p + q + u - \widehat{\widetilde{u}} = (p+q)\frac{\widetilde{\xi}\,\widehat{\xi}}{\xi\,\widehat{\widetilde{\xi}}}\,. \tag{3.121b}$$

Hence, conclude that $u = 1/\xi$ gives a solution of the lattice potential KdV equation.
(b) By multidimensional consistency, we can assume $\xi_0 \mapsto \xi_0 + h/k$, where h is a lattice variable associated with lattice parameter k. Indicating the shift in h by a bar, show that

$$\varphi = \varphi_{n,m} = \frac{\overline{\xi}}{\xi}(p-k)^n(q-k)^m\,, \tag{3.122}$$

forms a solution of the linear equations (3.88a) and (3.88b) and find the corresponding value of Λ. Find another independent solution of the Lax pair.

3.12 Consider the linear PΔE

$$(p+q)(\widetilde{w} - \widehat{w}) = (p-q)(\widehat{\widetilde{w}} - w).$$

Following the method of Section 3.4 derive the linear YB-map

$$R_{p,q} : (x, y) \mapsto \left(y - \frac{p-q}{p+q}(x+y),\, x + \frac{p-q}{p+q}(x+y) \right).$$

Show that it is an involution and that it obeys the YB-equation (3.36).

4

Interlude: Lattice equations and numerical algorithms

Discrete integrable systems have some fundamental connections to numerical algorithms; it turns out that some integrable lattice equations first appeared in the development of numerical algorithms. In this chapter we will survey a number of such numerical techniques. We discuss the Padé approximants developed to compute values of a function, the Shanks–Wynn algorithm that accelerates the convergence of a series, and Rutishauer's QD algorithm for matrix diagonalization.

In some applications, standard numerical approaches are not sufficient to obtain an accurate description of the problem under consideration. This is particularly the case when that problem has certain quantities that must be conserved during the calculation, such as energy or the eigenvalues of a matrix. In such problems, it is natural to encounter integrable systems, because by definition the latter possess a sufficient number of conserved quantities.

Discrete equations appear in numerical analysis in two contexts. First, they arise when we approximate a differential equation by a difference scheme. Second, they arise as an iterative method for approximating a function, a limit of a series or eigenvalues of matrices. In this chapter, we focus on effective numerical schemes for which the algorithm by itself is a discrete integrable system.

4.1 Padé approximants

Padé approximants are rational expressions matching the Taylor series of the original function. With such rational Padé approximants one can compute accurate numerical values for $y(x)$ faster than from its Taylor series. The original idea goes back at least to Frobenius, who in 1881 introduced the rational approximations and developed determinantal techniques (Frobenius, 1881). In this seminal paper he refers to an earlier idea of Jacobi (1846), who used Lagrange interpolation for rational approximations. In his thesis, Padé organized the approximations into what is now called the "Padé table" (a lattice of Padé approximants) and studied its properties (Padé, 1892).

Figure 4.1 G. Frobenius

In this section, we will derive the basic formulae and show how this approach gives rise to integrable equations, including the Toda-lattice equation and the cross-ratio equation.

4.1.1 Padé's method of approximation

Given a function $y(x)$ with a known power series expansion up to some order, consider approximating the function by a rational expression:

$$y(x) = \frac{P_L^{(L-M)}(x)}{Q_M^{(L-M)}(x)} + O(x^{L+M+1}), \tag{4.1}$$

where $P_L^{(L-M)}(x)$ and $Q_M^{(L-M)}(x)$ are polynomials of degree L and M, respectively. We will write

$$\begin{aligned} y(x) &= a_0 + a_1\,x + a_2\,x^2 + \cdots, \\ P_L^{(L-M)}(x) &= p_0 + p_1\,x + p_2\,x^2 + \cdots + p_L\,x^L, \\ Q_M^{(L-M)}(x) &= 1 + q_1\,x + q_2\,x^2 + \cdots + q_M\,x^M, \end{aligned}$$

and normalize $q_0 = 1$ without loss of generality (since only the ratio P/Q is important). (We follow here the notation used in the Padé literature; see e.g. Baker, 1975.)

In order to find the unknown coefficients p_j, q_j for given a_j (and L, M), multiply both sides of (4.1) by $Q_M^{(L-M)}(x)$ and take a Taylor expansion of both sides of the resulting expression

$$Q_M^{(L-M)}(x)\,y(x) = P_L^{(L-M)}(x).$$

The first $L+1$ terms of the expansion yield

$$\begin{array}{lllllllll} a_0 & & & & & & & = & p_0 \\ a_1 & + & a_0\,q_1 & & & & & = & p_1 \\ \vdots & & & & & & & \vdots & \\ a_L & + & a_{L-1}\,q_1 & + & \ldots & + & a_0\,q_L & = & p_L \end{array} \tag{4.2a}$$

(where we have to put $q_j = 0$ for $j > M$) and the next M terms lead to

$$\begin{array}{lllllllll} a_{L+1} & + & a_L\,q_1 & + & \ldots & + & a_{L-M+1}\,q_M & = & 0 \\ \vdots & & & & & & & \vdots & \\ a_{L+M} & + & a_{L+M-1}\,q_1 & + & \ldots & + & a_L\,q_M & = & 0 \end{array} \tag{4.2b}$$

where we take $a_j = 0$, for $j < 0$, which is needed if $M > L + 1$. These equations can be solved by using standard methods of linear algebra. With $L + 1$ coefficients $\{p_j\}$ and

M coefficients $\{q_j\}$, we can match $L + M + 1$ coefficients a_j. Note that the solutions for p_i, q_j depend on both L and M so we will use notation that keeps track of them both.

Equation (4.2b) can be written as the following Toeplitz (see Appendix D) matrix equation

$$\begin{pmatrix} a_L & a_{L-1} & \cdots & a_{L-M+1} \\ a_{L+1} & a_L & \cdots & a_{L-M+2} \\ \vdots & \vdots & & \vdots \\ a_{L+M-1} & a_{L+M-2} & \cdots & a_L \end{pmatrix} \begin{pmatrix} q_1 \\ q_2 \\ \vdots \\ q_M \end{pmatrix} = \begin{pmatrix} -a_{L+1} \\ -a_{L+2} \\ \vdots \\ -a_{L+M} \end{pmatrix}.$$

Since the inhomogeneous column is of the same form as the columns of the matrix on the left-hand side, we can conveniently write the solution by using Cramer's rule as

$$q_j = \frac{\begin{vmatrix} a_{L-M+1} & \cdots & a_{L-j+1} & \cdots & a_{L+1} \\ \vdots & & \vdots & & \vdots \\ a_L & \cdots & a_{L-j+M} & \cdots & a_{L+M} \\ 0 & \cdots & 1 & \cdots & 0 \end{vmatrix}}{C(L/M)} \tag{4.3}$$

where $C(L/M)$ is the Hankel determinant (see Appendix D)

$$\Delta_M^{(L)} \equiv C(L/M) := \begin{vmatrix} a_{L-M+1} & a_{L-M+2} & \cdots & a_L \\ \vdots & \vdots & & \vdots \\ a_L & a_{L+1} & \cdots & a_{L+M-1} \end{vmatrix}. \tag{4.4}$$

In this notation, M gives the size of the matrix and L the index of the top right element of the Hankel matrix.

Collecting the various q_i terms to construct $Q_M^{(L-M)}$ we get[1]

$$Q_M^{(L-M)} = \frac{\begin{vmatrix} a_{L-M+1} & \cdots & a_{L-j+1} & \cdots & a_{L+1} \\ \vdots & & \vdots & & \vdots \\ a_L & \cdots & a_{L+M-j} & \cdots & a_{L+M} \\ x^M & \cdots & x^j & \cdots & 1 \end{vmatrix}}{\Delta_M^{(L-M+1)}}. \tag{4.5}$$

Once the coefficients q_j have been found as in (4.3) one can proceed to solve for p_j from the system (4.2a), and insert them into the expansion for P. This leads to the following expression

[1] Here we use the normalization $Q_M^{(L-M)}|_{x=0} = 1$, but sometimes one also uses the normalization in which P is monic, i.e. $P_M^{(L-M)}(x) = x^M + \dots$.

$$P_L^{(L-M)} = \frac{\begin{vmatrix} a_{L-M+1} & \cdots & a_{L+1} \\ \vdots & & \vdots \\ a_L & \cdots & a_{L+M} \\ \sum_{j=M}^{L} a_{j-M}\, x^j & \cdots & \sum_{j=0}^{L} a_j\, x^j \end{vmatrix}}{\Delta_M^{(L-M+1)}}. \tag{4.6}$$

The *Padé approximant* is defined as the ratio $P_L^{(L-M)}(x)/Q_M^{(L-M)}(x)$ (Baker, 1975, p. 9), i.e.

$$[L/M] := \frac{P_L^{(L-M)}}{Q_M^{(L-M)}} = \frac{\begin{vmatrix} a_{L-M+1} & a_{L-M+2} & \cdots & a_{L+1} \\ \vdots & \vdots & & \vdots \\ a_L & a_{L+1} & \cdots & a_{L+M} \\ \sum_{j=M}^{L} a_{j-M}\, x^j & \sum_{j=M-1}^{L} a_{j-M+1}\, x^j & \cdots & \sum_{j=0}^{L} a_j\, x^j \end{vmatrix}}{\begin{vmatrix} a_{L-M+1} & a_{L-M+2} & \cdots & a_{L+1} \\ \vdots & \vdots & & \vdots \\ a_L & a_{L+1} & \cdots & a_{L+M} \\ x^M & x^{M-1} & \cdots & 1 \end{vmatrix}}. \tag{4.7}$$

This formula gives an explicit expression of the Padé approximants in terms of the Taylor coefficients of the original function $y(x)$.

4.1.2 The Padé table and its structure

In practical computations one constructs a *Padé table*, which is a collection of $[L/M]$ as defined in (4.7) for nonnegative values of L, M

$$\begin{matrix} [0/0] & [0/1] & [0/2] & \cdots \\ [1/0] & [1/1] & [1/2] & \cdots \\ [2/0] & [2/1] & [2/2] & \cdots \\ \vdots & \vdots & \vdots & \ddots \end{matrix}$$

See e.g., the Padé table for e^x given in Padé (1892), p. 16, or Baker (1975), p. 11. Note that the entries in the table can be constructed provided that the determinants $C(L/M)$ of (4.4) do not vanish.

One of the issues with the Padé approximants is that the C-table (consisting of the determinants $C(L/M)$) may contain zeros, which would correspond to singularities in the Padé approximation. However, it turns out that the zeros appear in blocks surrounded by nonzero entries (Baker, 1975). In other words, the singularities are isolated. This is reminiscent of singularity confinement, discussed in Chapter 7.

We will now derive a number of identities for the quantities related to the Padé table and show that they are connected to integrable lattice equations. Since the ingredients in Padé approximation theory are given in terms of Hankel determinants, we get interesting equations by using various determinantal identities.

In order to derive the first identity, we will apply Sylvester's identity (D.7) to $\Delta_n^{(m)}$ defined in (4.4). We choose $A = \Delta_n^{(m)}$, $r = 1, s = n+1, p = 1, q = n+1$ and then $A_{1;1} = \Delta_{n-1}^{(m+2)}$, $A_{1;n+1} = \Delta_{n-1}^{(m+1)}$, $A_{n+1;1} = \Delta_{n-1}^{(m+1)}$, $A_{n+1;n+1} = \Delta_{n-1}^{(m)}$, $A_{1,n+1;1,n+1} = \Delta_{n-2}^{(m+2)}$. Thus we get the result

$$\Delta_n^{(m)}\Delta_{n-2}^{(m+2)} = \Delta_{n-1}^{(m+2)}\Delta_{n-1}^{(m)} - (\Delta_{n-1}^{(m+1)})^2, \tag{4.8}$$

which is a variant of the bilinear Toda lattice equation (3.93).

In order to derive further relations in the Padé table, let us consider two approximations differing by one unit in L. By construction we have

$$y(x) - \frac{P_L^{(L-M)}(x)}{Q_M^{(L-M)}(x)} = O(x^{L+M+1}), \tag{4.9}$$

$$y(x) - \frac{P_{L+1}^{(L-M+1)}(x)}{Q_M^{(L-M+1)}(x)} = O(x^{L+M+2}). \tag{4.10}$$

After subtracting both sides we get

$$\frac{P_L^{(L-M)}(x)}{Q_M^{(L-M)}(x)} - \frac{P_{L+1}^{(L-M+1)}(x)}{Q_M^{(L-M+1)}(x)} = O(x^{L+M+1}).$$

In the following derivation we use the normalization of P, Q in which they are defined to be the numerators of the expressions on the right in equations (4.5–4.6). After multiplying the above equation by $Q_M^{(L-M)}(x)\, Q_M^{(L-M+1)}(x)$ we obtain

$$P_L^{(L-M)}(x)Q_M^{(L-M+1)}(x) - P_{L+1}^{(L-M+1)}(x)Q_M^{(L-M)}(x) = O(x^{L+M+1}),$$

where the right side still has the same order. On the other hand, the left side is at most of order x^{L+M+1}, and their common value can be computed by taking the leading power of x on the left-hand side, which is $C(L+1/M)\,C(L+1/M+1)\,x^{L+M+1}$. In this way we obtain

$$[L+1/M] - [L/M] = \frac{C(L+1/M)\,C(L+1/M+1)\,x^{L+M+1}}{Q_M^{(L-M+1)}(x)Q_M^{(L-M)}(x)}. \tag{4.11a}$$

In the same way one can derive

$$[L/M+1]-[L/M]=\frac{C(L/M+1)\,C(L+1/M+1)\,x^{L+M+1}}{Q_{M+1}^{(L-M-1)}(x)Q_M^{(L-M)}(x)}. \tag{4.11b}$$

The left sides of these expressions connect adjacent sites in the Padé table either vertically or horizontally. From (4.11), we can eliminate the x dependence on the right side by taking suitable combinations of these equations and their shifts (see Baker, 1975, p. 29) and obtain

$$\frac{([L/M]-[L/M+1])([L+1/M]-[L+1/M+1])}{([L/M]-[L+1/M])([L/M+1]-[L+1/M+1])}=\text{const.}$$

This is identical to the "cross-ratio" equation, which was given in (3.13) and called the lattice Schwarzian KdV equation.

There are many other identities that can be derived this way. We mention one more identity, called variously "Wynn's identity" (Baker, 1975, p. 34) or the "missing identity of Frobenius":

$$\begin{aligned}\frac{1}{[L+1/M]-[L/M]}&+\frac{1}{[L-1/M]-[L/M]}\\&=\frac{1}{[L/M+1]-[L/M]}+\frac{1}{[L/M-1]-[L/M]}.\end{aligned} \tag{4.12}$$

This equation, connecting points on a cross-shaped stencil, can be derived from the previous relations under the assumption of "normality" (see Gragg, 1972, p. 29, Theorem 5.5). For its relation with other discrete integrable equations in the KdV class, see e.g. Konstantinou-Rizos et al. (2015).

4.1.3 Orthogonal polynomials

We will now show how the polynomial arising from the Padé approach are connected to orthogonal polynomials (OP). We have previously used the normalization $Q_M|_{x=0}=1$, but in the theory of OP, many other normalization are used, usually fixing the coefficient of the highest power to be unity (in which case the polynomials are called "monic"). In the following we will use monic polynomials associated with Q which we will call $\bar{P}$, i.e.

$$\bar{P}_n^{(m)}=\frac{1}{\Delta_{n-1}^{(m)}}\begin{vmatrix} c_m & c_{m+1} & \cdots & c_{m+n}\\ \vdots & \vdots & & \vdots\\ c_{m+n-1} & c_{m+n} & \cdots & c_{m+2n-1}\\ 1 & x & \cdots & x^n\end{vmatrix}. \tag{4.13}$$

The relation to the previous definition is given by $x^n\,Q_m^{(m-1)}(1/x)\,\Delta_M^{(L-M+1)}$ $=\bar{P}_n^{(m)}(x)\,\Delta_{n-1}^{(m)}$, where $L-M+1=m,\ M=n, a=c$.

Let us now apply Sylvester's identity (D.7) to the numerator of $\overline{P}_{n+1}^{(m)}$, with $r = 1$, $s = n+1$, $p = 1$, $q = n+1$. Then the result is

$$\overline{P}_{n+1}^{(m)} = x\,\overline{P}_{n}^{(m+2)} - W_n^{(m)}\,\overline{P}_{n}^{(m+1)}, \qquad W_n^{(m)} := \frac{\Delta_n^{(m+1)}\Delta_{n-1}^{(m+1)}}{\Delta_n^{(m)}\Delta_{n-1}^{(m+2)}}.$$

If we choose the same columns but different rows, i.e. $r = n$, $s = n+1$, we get instead

$$\overline{P}_{n+1}^{(m)} = x\,\overline{P}_{n}^{(m+1)} - V_n^{(m)}\,\overline{P}_{n}^{(m)}, \qquad V_n^{(m)} := \frac{\Delta_n^{(m+1)}\Delta_{n-1}^{(m)}}{\Delta_n^{(m)}\Delta_{n-1}^{(m+1)}}.$$

Using both of these equation we can connect the Ps with the same upper index by an equation of the standard form for orthogonal polynomials (DLMF, 2010, Section 18.9)

$$\overline{P}_{n+1}^{(m)} = (x\,A_n^{(m)} + B_n^{(m)})\,\overline{P}_{n}^{(m)} - C_n^{(m)}\,\overline{P}_{n-1}^{(m)}, \tag{4.14}$$

where now

$$A_n^{(m)} = 1,\ B_n^{(m)} = -(V_{n-1}^{(m+1)} + V_n^{(m)} - W_{n-1}^{(m)}),\ C_n^{(m)} = (V_{n-1}^{(m+1)} - W_{n-1}^{(m)})V_{n-1}^{(m)}.$$

We now explain how the polynomials are understood to form an orthogonal family. We need a linear functional $\mathfrak{L}$ that defines the moments

$$d_k = \mathfrak{L}(x^k)$$

such that $\mathfrak{L}(\overline{P}_r^{(m)}\,\overline{P}_s^{(m)}) = 0$ for all $r \neq s$. In practice this means that we must have $\mathfrak{L}(x^k\,\overline{P}_n^{(m)}) = 0$ for all $k < n$. The so-called "moment problem" in orthogonal polynomial theory is the question of whether, for a given sequence of moments, one can express $\mathfrak{L}$ as an integral

$$\mathfrak{L}(x^k) = \int_\Gamma x^k\,d\mu(x), \tag{4.15}$$

for some curve Γ in the complex plane x and some measure $d\mu$ (Shohat and Tamarkin, 1943). Applying the operator $\mathfrak{L}$ on the determinantal form of the polynomial $\overline{P}_n^{(m)}$ we obtain

$$\mathfrak{L}(x^k\,\bar{P}_n^{(m)}) \propto \begin{vmatrix} c_m & c_{m+1} & \cdots & c_{m+n} \\ \vdots & \vdots & & \vdots \\ c_{m+n-1} & c_{m+n} & \cdots & c_{m+2n-1} \\ d_k & d_{k+1} & \cdots & d_{k+n} \end{vmatrix}$$

which should vanish for all $0 \le k \le n-1$. From this it is clear that if we choose $d_k = c_{m+k}$ then for all $0 \le k \le n-1$, the determinant has the same row twice and therefore it vanishes. Thus, for any choice of numbers c_j for which the relevant Hankel determinants do not vanish, we can define several sequences of orthogonal polynomials and the very meaning of "orthogonality" depends on the choice of c_j.

Returning to the original function $y(x)$ for which the Padé table was built, we see that any function with a Taylor series expression defines orthogonal polynomials. In this construction, a pivotal role is played by the representation of polynomials using Hankel determinants.

4.2 Convergence acceleration algorithm

The general problem is the following: suppose we have some recursively defined convergent sequence of numbers S_n; is it possible to transform the sequence into another sequence, say T_n, such that it converges to the limit faster? In other words, we want the following to hold

$$\lim_{n\to\infty} \frac{T_n - S_\infty}{S_n - S_\infty} = 0.$$

If it is possible, then we say that the method of computing T_n from S_n is an acceleration algorithm. In this section we look at one such method: the ϵ-algorithm of Wynn.

One approach to the convergence problem was given by Schmidt (1941) and Shanks (1955), who found the transformation $S_n \mapsto e_m(S_n)$ for a given sequence S_n, where

$$e_m(S_n) = \frac{\begin{vmatrix} S_n & S_{n+1} & \cdots & S_{n+m} \\ \Delta S_n & \Delta S_{n+1} & \cdots & \Delta S_{n+m} \\ \vdots & \vdots & \vdots & \vdots \\ \Delta S_{n+m-1} & \Delta S_{n+m} & \cdots & \Delta S_{n+2m-1} \end{vmatrix}}{\begin{vmatrix} 1 & 1 & \cdots & 1 \\ \Delta S_n & \Delta S_{n+1} & \cdots & \Delta S_{n+m} \\ \vdots & \vdots & \vdots & \vdots \\ \Delta S_{n+m-1} & \Delta S_{n+m} & \cdots & \Delta S_{n+2m-1} \end{vmatrix}} \tag{4.16}$$

or in terms of Hankel determinants $H_k(\{u_n\}) \equiv \det(u_{n+i+j})_{i,j=0,\ldots,k-1}$:

$$e_m(S_n) = \frac{H_{m+1}(S_n)}{H_m(\Delta^2 S_n)}. \tag{4.17}$$

(In these formulae Δ is the forward difference, i.e. $\Delta S_k = S_{k+1} - S_k$.) Thus we have a series of sequences, labeled by m, and the larger the index m is the faster the sequence converges. It was actually proven that for certain sequences this *Shank's transformation* gives the exact result, for some finite m.

Wynn (1961) noted that this method has the drawback that computation of the Hankel determinants is demanding and furthermore the intermediate results were not reused. He then went on to show that the sequence of transformation e_m can be computed in a more cost-effective way.

Wynn's algorithm is as follows: instead of e_m let us consider ϵ_m defined by

$$\epsilon_{2m}(S_n) = e_m(S_n),$$
$$\epsilon_{2m+1}(S_n) = \frac{1}{e_m(\Delta S_n)};$$

then, by manipulating the Hankel determinants, Wynn showed that the following equation holds (Wynn, 1961, Equation (4)):

$$\epsilon_{s+1}(S_n) = \epsilon_{s-1}(S_{n+1}) + \frac{1}{\epsilon_s(S_{n+1}) - \epsilon_s(S_n)}, \qquad s = 1, 2, \ldots \tag{4.18}$$

provided that none of the quantities $\epsilon_{2m}(S_n)$ becomes infinite. Usually $\epsilon_s(S_n)$ is written as $\epsilon_s^{(n)}$, and the formula for calculating $\epsilon_{s+1}^{(n)}$

$$\epsilon_{s+1}^{(n)} = \epsilon_{s-1}^{(n+1)} + \frac{1}{\epsilon_s^{(n+1)} - \epsilon_s^{(n)}} \tag{4.19}$$

is called the *rhombus rule*. These quantities are displayed in a double entry table, the ϵ-table, where the index s is a column and n a descending diagonal.

$$\begin{array}{ccccccc}
\epsilon_{-1}^{(0)} & & & & & & \\
 & \epsilon_0^{(0)} & & & & & \\
\epsilon_{-1}^{(1)} & & \epsilon_1^{(0)} & & & & \\
 & \epsilon_0^{(1)} & & \epsilon_2^{(0)} & & & \\
\epsilon_{-1}^{(2)} & & \epsilon_1^{(1)} & & \ddots & & \\
 & \epsilon_0^{(2)} & & \epsilon_2^{(1)} & & & \\
\epsilon_{-1}^{(3)} & & \epsilon_1^{(2)} & & \ddots & & \\
 & \vdots & & \vdots & & &
\end{array}$$

The boundary conditions applied are

$$\epsilon_{-1}^{(n)} = 0, \quad \epsilon_0^{(n)} = S_n. \tag{4.20}$$

It is remarkable that (4.19) coincides with the lattice KdV equation (3.6) when we use the connection $\epsilon_s^{(n)} = w_{s+n,n}$.

4.3 Rutishauser's QD algorithm

In this section we consider algorithms for computing the eigenvalues of a matrix.

4.3.1 Derivation of the QD algorithm

The following algorithm was developed by Rutishauser (1954a) and is called the quotient-difference (QD) algorithm for matrix eigenvalues. We follow the description in Gutknecht and Parlett (2011).

Given an $N \times N$ matrix $\boldsymbol{A}$, and N-component vectors y_0, x_0, define

$$f(z) := \langle y_0, (z\boldsymbol{I} - \boldsymbol{A})^{-1} x_0 \rangle,$$

where $\langle \cdot, \cdot \rangle$ is the inner product. Note that the eigenvalues λ_k of $\boldsymbol{A}$ are now the poles of the function $f(z)$.

We expand f around the point $z = \infty$ as

$$f(z) = \sum_{m=0}^{\infty} \frac{s_m}{z^{m+1}}. \tag{4.21}$$

In order to reconstruct the eigenvalues, a key result by Hadamard leads us to construct the following Hankel determinants with s_m as matrix elements:

$$H_n^{(m)} = \begin{vmatrix} s_m & s_{m+1} & \cdots & s_{m+n-1} \\ \vdots & \vdots & & \vdots \\ s_{m+n-1} & s_{m+n} & \cdots & s_{m+2n-2} \end{vmatrix}. \tag{4.22}$$

Here the lower index n gives the size of the matrix. Hadamard proved in 1892 the following result (Gutknecht and Parlett, 2011):

Theorem 4.3.1 *Assume the series* (4.21) *has N poles, counted with multiplicities, and ordered in decreasing order $|\lambda_1| \geq |\lambda_2| \geq \cdots |\lambda_N|$. If $1 \leq k < N$ and $|\lambda_{k+1}| < \Lambda < |\lambda_k|$, or if $k = N$ and $\Lambda < |\lambda_N|$, then*

$$H_k^{(m)} = const.(\lambda_1 \cdots \lambda_k)^m \left[1 + O(\Lambda/|\lambda_k|)^m\right], \quad as \quad m \to \infty.$$

Using this theorem we find for ratios

$$\frac{H_n^{(m+1)}}{H_n^{(m)}} \to \lambda_1 \cdots \lambda_n\,, \qquad \text{as } m \to \infty,$$

and for double ratios

$$q_n^{(m)} := \frac{H_n^{(m+1)}}{H_n^{(m)}} \frac{H_{n-1}^{(m)}}{H_{n-1}^{(m+1)}} \to \lambda_n\,, \qquad \text{as } m \to \infty. \tag{4.23}$$

Thus, in order to get the eigenvalues, one needs to compute bigger and bigger Hankel determinants, which is very time consuming.

In order to make the process more efficient, Rutishauser defined another quantity

$$e_n^{(m)} := \frac{H_{n+1}^{(m)}}{H_n^{(m)}} \frac{H_{n-1}^{(m+1)}}{H_n^{(m+1)}} \tag{4.24}$$

and found the following relations between them

$$e_n^{(m)} + q_n^{(m)} = e_{n-1}^{(m+1)} + q_n^{(m+1)}, \tag{4.25a}$$

$$q_{n+1}^{(m)}\, e_n^{(m)} = q_n^{(m+1)}\, e_n^{(m+1)}. \tag{4.25b}$$

This is called *Rutishauser's QD algorithm* and using it one can construct a QD-table in the increasing m direction, in which only the boundary Hankel determinants need to be computed (Gutknecht and Parlett, 2011). To construct the table we take the initial values $e_0^{(m)} = 0$ and $q_1^{(m)} = s_{m+1}/s_m$ and build the table from the first column using the "rhombus rule" and increasing the lower index.

Note that equations (4.25) are identical to the fully discrete Toda lattice equations (3.97b) with identification $I_{n,m} = q_n^{(m)}$, $pqV_{n,m} = e_n^{(m)}$. In fact, Rutishauser (1954b) also derived the continuum limit of (4.25) and obtained a version of the continuous time Toda lattice. For that purpose first change variables by

$$q_k^{(m)} = 1 + \delta Q_k^{(m)}, \quad e_k^{(m)} = \delta^2 E_k^{(m)},$$

and then take a continuum limit in the m variable by setting in general $f(m) = f(t_0 + \delta m) = f(t)$ so that $f(m+1) = f(t_0 + \delta(m+1)) = f(t+\delta)$. Then in the limit $\delta \to 0$ one obtains at leading order

$$Q_n'(t) = E_n(t) - E_{n-1}(t), \tag{4.26a}$$

$$E_n'(t) = E_n(t)[Q_{n+1}(t) - Q_n(t)]. \tag{4.26b}$$

Furthermore we set $Q_n(t) = -\varphi_n'(t)$, after which we can integrate the second equation as $E_n = e^{-(\varphi_{n+1}-\varphi_n)}$, and then the first equation yields (1.39) for φ.

4.3.2 From a QD algorithm to an LR algorithm

There is another interpretation of the QD algorithm, also made by Rutishauser, which makes the integrability of equations (4.25) very clear.

Given the quantities defined by equations (4.23, 4.24) consider the tridiagonal matrix

$$\boldsymbol{T}^{(m)} = \begin{pmatrix} q_1^{(m)} & 1 & 0 & 0 & \cdots & 0 \\ e_1^{(m)}q_1^{(m)} & e_1^{(m)}+q_2^{(m)} & 1 & 0 & \ddots & \vdots \\ 0 & e_2^{(m)}q_2^{(m)} & e_2^{(m)}+q_3^{(m)} & 1 & \ddots & \vdots \\ 0 & 0 & \ddots & \ddots & \ddots & 0 \\ \vdots & \vdots & \vdots & \vdots & \ddots & 1 \\ 0 & 0 & \cdots & 0 & e_{N-1}^{(m)}q_{N-1}^{(m)} & e_{N-1}^{(m)}+q_N^{(m)} \end{pmatrix}. \tag{4.27}$$

This matrix can be factorized as $\boldsymbol{T}^{(m)} = \boldsymbol{L}^{(m)}\boldsymbol{R}^{(m)}$, where

$$\boldsymbol{L}^{(m)} = \begin{pmatrix} 1 & 0 & 0 & \cdots & 0 \\ e_1^{(m)} & 1 & 0 & \ddots & \vdots \\ 0 & e_2^{(m)} & \ddots & \ddots & 0 \\ \vdots & \ddots & \ddots & 1 & 0 \\ 0 & \cdots & 0 & e_{N-1}^{(m)} & 1 \end{pmatrix}, \quad \boldsymbol{R}^{(m)} = \begin{pmatrix} q_1^{(m)} & 1 & 0 & \cdots & 0 \\ 0 & q_2^{(m)} & 1 & \ddots & \vdots \\ 0 & \ddots & \ddots & \ddots & 0 \\ \vdots & \ddots & \ddots & \ddots & 1 \\ 0 & \cdots & 0 & 0 & q_N^{(m)} \end{pmatrix}.$$

It was the observation of Rutishauser (1954a) that his rules (4.25) can be used to compute the matrix elements of the matrix when the order of $\boldsymbol{L}$ and $\boldsymbol{R}$ is reversed:

$$\boldsymbol{R}^{(m)}\boldsymbol{L}^{(m)} = \begin{pmatrix} e_1^{(m)}+q_1^{(m)} & 1 & 0 & 0 & \cdots & 0 \\ e_1^{(m)}q_2^{(m)} & e_2^{(m)}+q_2^{(m)} & 1 & 0 & \ddots & \vdots \\ 0 & e_2^{(m)}q_3^{(m)} & e_3^{(m)}+q_3^{(m)} & 1 & \ddots & \vdots \\ 0 & 0 & \ddots & \ddots & \ddots & 0 \\ \vdots & \vdots & \vdots & \vdots & \ddots & 1 \\ 0 & 0 & \cdots & 0 & e_{N-1}^{(m)}q_N^{(m)} & q_N^{(m)} \end{pmatrix}.$$

$$
\begin{array}{ccccccccc}
 & q_1^{(0)} & & & & & & & \\
0 & + & e_1^{(0)} & & & & & & \\
 & q_1^{(1)} & \times & q_2^{(0)} & & & & & \\
0 & + & e_1^{(1)} & + & e_2^{(0)} & & & & \\
 & q_1^{(2)} & \times & q_2^{(1)} & \times & q_3^{(0)} & & & \\
0 & + & e_1^{(2)} & + & e_2^{(1)} & + & e_3^{(0)} & & \\
 & \vdots & & \vdots & & \vdots & & & \\
0 & & \vdots & & \vdots & & \vdots & & \\
 & q_1^{(N-1)} & \times & q_2^{(N-2)} & \times & \cdots & \times & q_N^{(0)} & \\
0 & + & e_1^{(N-1)} & + & e_2^{(N-2)} & \cdots & e_{N-1}^{(1)} & + & 0 \\
 & \vdots & & \vdots & & \vdots & & \vdots & \\
0 & & \vdots & & \vdots & & \vdots & & 0
\end{array}
$$

Figure 4.2 Iterating in the QR algorithm. The $q_j^{(0)}, e_j^{(0)}$ can be extracted from the given matrix, after which new values can be computed going down by diagonal row, increasing the upper index. The symbol inside the rhombus of four terms tell which equation to use: with $+$ use (4.25a) and for $\times$ use (4.25b).

That is, with the boundary conditions $e_N^{(m)} = 0, e_0^{(m)} = 0$ one finds from (4.25) and the definition (4.27) that $\boldsymbol{R}^{(m)}\boldsymbol{L}^{(m)} = \boldsymbol{T}^{(m+1)}$.

Consider the following linear equations (Hirota et al., 1993; Papageorgiou et al., 1995)

$$\boldsymbol{T}^{(m)}\boldsymbol{\phi}^{(m)} = \lambda\boldsymbol{\phi}^{(m)}, \quad \boldsymbol{L}^{(m)}\boldsymbol{\phi}^{(m+1)} = \boldsymbol{\phi}^{(m)} \tag{4.28}$$

for an n-component vector $\boldsymbol{\phi}$.This linear system consists of an eigenvalue problem for the matrix $\boldsymbol{T}$ along with a shift in the variable m given by $\boldsymbol{L}$. Imposing the property of *isospectrality*, i.e. that λ does not depend on m, we can derive the condition

$$\boldsymbol{T}^{(m+1)} = (\boldsymbol{L}^{(m)})^{-1}\boldsymbol{T}^{(m)}\boldsymbol{L}^{(m)}. \tag{4.29}$$

This is the same equation that is obtained by eliminating $R^{(m)}$ from $\boldsymbol{T}^{(m)} = \boldsymbol{L}^{(m)}\boldsymbol{R}^{(m)}$ and $\boldsymbol{T}^{(m+1)} = \boldsymbol{R}^{(m)}\boldsymbol{L}^{(m)}$ derived above. Thus Rutishauser's QD algorithm, rewritten in the above matrix form, is an isospectral deformation of the eigenvalue problem for $\boldsymbol{T}$. That is, the discrete flow in the variable m preserves the eigenvalues of the matrix $\boldsymbol{T}$. In particular, from (4.29) we find that the quantities

$$I_k := \mathrm{Tr}(\boldsymbol{T}^k)$$

are preserved under this flow (due to the cyclicity of the trace) and therefore are conserved quantities or "integrals" of that flow.

It follows from Hadamard's Theorem 4.3.1 that asymptotically as $m \to \infty$, the matrix $\boldsymbol{L}^{(m)} \to \boldsymbol{1}$ (the unit matrix) while $\boldsymbol{R}$ (and hence $\boldsymbol{T}$, according to (4.29)) becomes a matrix with eigenvalues of $\boldsymbol{A}$ on the diagonal.

4.3.3 From the LR algorithm to the Toda lattice

Historically, another approach leading to (4.25) started with the (continuous time) Toda lattice equation (1.39). If we change variables according to

$$a_n = \tfrac{1}{2}e^{-\frac{1}{2}(q_{n+1}-q_n)}, \quad b_n = -\tfrac{1}{2}\tfrac{d}{dt}q_n,$$

then the molecule version of (1.39) (with $a_0 = a_N = 0$) becomes (cf. (4.26))

$$\tfrac{d}{dt}a_n = a_n(b_{n+1} - b_n), \quad \tfrac{d}{dt}b_n = 2(a_n^2 - a_{n-1}^2), n = 1, \ldots, N. \tag{4.30}$$

These equations are the consistency conditions for the Lax pair (Flaschka, 1974) (modified for the molecule version)

$$\left\{ \begin{array}{lcl} A\psi & = & \lambda\psi, \\ B\psi & = & \partial_t\psi \end{array} \right. \quad \Rightarrow \quad \tfrac{d}{dt}A = [B, A] \tag{4.31}$$

where

$$A = \begin{pmatrix} b_1 & a_1 & 0 & \ldots & 0 \\ a_1 & b_2 & a_2 & \ddots & \vdots \\ 0 & \ddots & \ddots & \ddots & 0 \\ \vdots & \ddots & \ddots & b_{N-1} & a_{N-1} \\ 0 & \ldots & 0 & a_{N-1} & b_N \end{pmatrix} \quad B = \begin{pmatrix} 0 & a_1 & 0 & \ldots & 0 \\ -a_1 & 0 & a_2 & \ddots & \vdots \\ 0 & \ddots & \ddots & \ddots & 0 \\ \vdots & \ddots & \ddots & \ddots & a_{N-1} \\ 0 & \ldots & 0 & -a_{N-1} & 0 \end{pmatrix}$$

If we define $I_k := Tr(A^k)$ then due to (4.31) we have

$$\tfrac{d}{dt}I_k = kTr(A^{k-1}[B, A]) = 0.$$

The tridiagonal matrix A may look rather restrictive if we want to find the eigenvalues of a generic symmetric matrix. However, there is a relatively simple Hausholder's algorithm that takes a symmetric $N \times N$ matrix to a tridiagonal form in $N - 2$ steps (Bauer, 1959).

An essential property of the Toda lattice (1.39) is that it is a scattering system (Moser, 1975), i.e. the long time behavior is given by

$$q_{k+1} - q_k \to \infty \quad \text{as} \quad t \to \infty, \quad \text{hence} \quad a_n \to 0.$$

From this it follows that A becomes diagonal as $t \to \infty$. Thus given a symmetric tridiagonal matrix with positive off-diagonal elements, we use the matrix entries to define the starting positions and velocities of the particles in the Toda lattice (matrix A above), then its time evolution (1.39) diagonalizes the matrix. That the Toda lattice is related to the LR and QD algorithms was then noted in Symes (1982).

For a numerical algorithm one just has to discretize the time of the Toda molecule equation. This was done in Hirota et al. (1993) on the basis of the bilinear method and solution structure; the results was nothing but (4.25). The corresponding bilinear discrete Toda lattice was also given in Hirota et al. (1993), the transformation

$$e_{n,m} = \frac{\tau_{n+1,m}\tau_{n-1,m+1}}{\tau_{n,m}\tau_{n,m+1}}, \qquad q_{n,m} = \frac{\tau_{n-1,m}\tau_{n,m+1}}{\tau_{n-1,m+1}\tau_{n,m}},$$

satisfies (4.25b) by construction, while (4.25a) yields

$$F_{n,m+1} = F_{n-1,m+1}, \quad \text{where} \quad F_{n,m} := \frac{\tau_{n-1,m+1}\tau_{n+1,m-1} - \tau_{n,m-1}\tau_{n,m+1}}{\tau_{n,m}^2},$$

and if we choose $F_{n,m} = -1$ we get (4.8).

For further discussion on the Toda lattice and integrable algorithms, see Sogo (1993); Papageorgiou et al. (1993); Hirota et al. (1993); Nagai and Satsuma (1995); Nagai et al. (1998).

4.4 Notes

In numerical analysis there exist still more algorithms, connected to integrable lattice equations, than are covered in this chapter. We mention four approaches here as directions for further exploration.

As is well known, there is an intimate connection between the structure of the Padé table to integrable lattice equations. In the literature there exist many generalizations of the Padé approach, using vector or matrix Padé approximants (Graves-Morris and Roberts, 1997; Brezinski, 2013). Connections of the latter with integrable PΔE remain largely to be explored.

There are also many connections to integrable lattice equations in the theory of acceleration algorithms (Brezinski, 1977; Papageorgiou et al., 1993; Nagai and Satsuma, 1995; Papageorgiou et al., 1996; He et al., 2011; Brezinski et al., 2012).

The problem of computing matrix eigenvalues has been studied further by Y. Nakamura, based on various extensions of the discrete hungry Lotka–Volterra system (Fukuda et al., 2009, 2013); these produce algorithms that have high relative accuracy and guaranteed convergence for certain types of matrices. Rutishauser's QD algorithm has also been generalized to the "QQD" algorithm (Spicer et al., 2011) (see also Exercise 4.6).

One of the questions that remain of perennial interest is: how do we find the best numerical approximation of a given differential equation? There are many answers according to which property we wish to preserve. For example, if we wish to preserve the structure of soliton solutions of an integrable PDE, then Hirota provided an efficient method for doing so. (See Chapter 8.) Alternatively, Ablowitz and Ladik (1976) discretized the Lax pair associated with the differential equation of interest, at the same time as providing a numerical approximation of the equation. This line of research has been pursued further, in particular the subtle difference between integrable and nonintegrable numerical approximations has been studied in e.g. Taha and Ablowitz (1984a,b); Herbst and Ablowitz (1989); Ablowitz and Herbst (1990); Herbst et al. (1994).

Exercises

4.1 Compute the Padé approximant $[L/M]$ given in (4.7) for the function e^x in the cases 1) $L = 2, M = 1$ and 2) $L = 1, M = 2$. Plot the function and its approximants in the interval $x \in [-2, 2]$.

4.2 Consider the equation for $C[L/M] \equiv \Delta_M^{(L)}$ given in (4.8). Construct the entries that are needed for $n = 3, m = 1$ using (4.4) but use undetermined a_j. Show by explicit computation that (4.8) is satisfied.

4.3 Find a gauge transformation of the form

$$f_{n,m} = \alpha^{n^2} \beta^{nm} \gamma^{m^2} \Delta_n^{(m)},$$

which takes the bilinear Toda lattice (3.93), where the sum of parameters is zero, into the Frobenius form (4.8) (after some index shifts).

4.4 Consider the sequence $S_{n+1} = \frac{1}{4}(S_n^2 + 2)$ with initial value $S_0 = 1$.

1 Show that $S_\infty = 2 - \sqrt{2} = 0.585786437626904951\ldots$.

2 Show that $S_{10} - S_\infty = 0.00000317\ldots$.

3 Using (4.16) derive the formula

$$e_1(S_n) = \frac{S_n S_{n+2} - S_{n-1}^2}{S_{n+2} - 2S_{n+1} + S_n}.$$

4 Show that $e_1(S_{10}) - S_\infty = -1.04239\ldots\ 10^{12}$.

4.5 (a) Write a computer algebra program that iterates as described in Figure 4.2 for $N = 4$ (i.e. every iteration step increases the upper index).

(b) Consider the matrix

$$\begin{pmatrix} 2 & 1 & 0 & 0 \\ 4 & 3 & 1 & 0 \\ 0 & 3 & 11 & 1 \\ 0 & 0 & 8 & 2 \end{pmatrix}.$$

Extract from it the initial values $q_i^{(0)}, e_i^{(0)}$, as defined in (4.27). Then iterate them using the program constructed in (a) and observe how the $e_i^{(M)}$ approach zero as M increases, while $q_i^{(M)}$ approach the eigenvalues.

4.6 Consider the Lax pair:

$$\Psi_{l+1,m} = L_{l,m}\Psi_{l,m}\,, \quad \Psi_{l,m+1} = M_{l,m}\Psi_{l,m}\,, \tag{4.32}$$

where the matrices L and M are given by

$$L_{l,m} = \begin{pmatrix} -u_{l,m+2} & 1 & 0 & 0 \\ 0 & -u_{l+1,m+1} & 1 & 0 \\ 0 & 0 & -u_{l+2,m} & 1 \\ 0 & 0 & -w_{l+2,m}u_{l+2,m} & -v_{l+2,m}+w_{l+2,m} \end{pmatrix}$$
$$+\lambda \begin{pmatrix} 0 & 0 & 0 & 0 \\ 0 & 0 & 0 & 0 \\ 0 & 0 & 0 & 0 \\ u_{l+1,m+2}u_{l,m+2} & -u_{l+1,m+1}-u_{l+1,m+2} & 1 & 0 \end{pmatrix},$$

and

$$M_{l,m} = \begin{pmatrix} -\frac{u_{l,m+2}}{u_{l,m+3}} & \frac{1}{u_{l,m+3}} & 0 & 0 \\ -u_{l,m+2} & 1 & 0 & 0 \\ 0 & -u_{l+1,m+1} & 1 & 0 \\ 0 & 0 & -u_{l+2,m} & 1 \end{pmatrix}$$
$$+\lambda^{-1} \begin{pmatrix} 0 & -w_{l,m+1} & 1+\frac{w_{l+1,m}-v_{l+1,m}}{u_{l,m+3}} & -\frac{1}{u_{l,m+3}} \\ 0 & 0 & 0 & 0 \\ 0 & 0 & 0 & 0 \\ 0 & 0 & 0 & 0 \end{pmatrix}.$$

Show that it gives rise to the following coupled set of equations:

$$u_{l+2,m} + v_{l+1,m} + w_{l+1,m+1} = u_{l,m+3} + v_{l+1,m+1} + w_{l+1,m}\,, \tag{4.33a}$$
$$u_{l,m+3}v_{l,m+1} = v_{l+1,m}u_{l+1,m}\,, \tag{4.33b}$$
$$u_{l,m+2}w_{l,m} = w_{l+1,m}u_{l+1,m}\,. \tag{4.33c}$$

This constitutes a generalization of the QD algorithm, found in Spicer et al. (2011), and which was called the QQD algorithm.

5

Continuum limits of lattice PΔE

We consider in this chapter the *continuum limits* of the lattice equations, i.e. the limiting equations (typically differential equations) that we retrieve from the discrete equations by shrinking the lattice grid to a continuous set of values corresponding to spatial and temporal coordinates. Having discussed in Chapter 3 the integrability aspects of quadrilateral PΔEs it is now in order to see what these equations are, and whether they can be identified as discrete analogues or discretizations of PDEs.

In numerical analysis the conventional picture is the one where discrete equations, namely finite-difference schemes, are used as numerical approximations to differential equations. The choice of difference schemes is sought on the basis of a variety of criteria such as numerical stability and speed of convergence. In most cases the continuum limit itself is a *reductive* procedure: we lose the lattice grid parameters by performing the limit. One major philosophical question in this picture is whether the differential equation is an approximation of a given discrete equation, or conversely, whether it is the finite-difference equation that plays the role of an approximation.

It is our perspective that continuum limits of discrete equations are the degenerations and approximations of the latter. One would expect, therefore, that the continuous equations are less rich in parameters than the discrete ones. In studying continuum limits of *integrable* discrete systems, however, it turns out that the latter is only partially true. We will observe that in some respects we do not really lose the "richness" of the relevant equations, as long as we keep intact the structure of *parameter families of compatible equations*. In their most explicit form the discrete equations can be shown to generate entire infinite families of continuous differential equations, which constitute the hierarchies associated with the famous soliton systems.

In practice we should perform continuum limits in such a way that integrability is retained at all levels. We will find that continuum limits are not necessarily unique, because starting from a two-dimensional PΔE, we can separately perform a continuum limit in each direction, leading in first instance to semi-continuous (semi-discrete) equations.

5.1 How to take a continuum limit

In a continuum limit involving a step-size parameter, say h, the difference operator $\Delta_h y$, defined in (1.9), will tend to a derivative, namely by (1.8), as was explained in Section

1.1.2. We obtain the transition from difference to differential equation by carrying out Taylor expansions on shift operators $T_h y(x)$, defined in (A.7), namely by using expansions of the form

$$y(x+h) = y(x) + \frac{h}{1!}\frac{dy}{dx} + \frac{h^2}{2!}\frac{d^2y}{dx^2} + \cdots \tag{5.1}$$

in the discrete equation and then by expanding power-by-power in the lattice parameter h. Typically the equation that emerges as "the continuum limit" of the discrete equation is the coefficient of the dominant term in this expansion as $h \to 0$.

In PΔEs of the form we have seen, we have independent variables $u_{n,m}$ and two shifts

$$\widetilde{u} = u_{n+1,m}, \quad \widehat{u} = u_{n,m+1}.$$

By interpreting these variables as

$$u_{n,m} = u(x_0 + n\delta, t_0 + m\varepsilon) = u(x,t), \tag{5.2a}$$

$$\widetilde{u} \equiv u_{n+1,m} = u(x_0 + (n+1)\delta, t_0 + m\varepsilon) = u(x+\delta, t), \tag{5.2b}$$

$$\widehat{u} \equiv u_{n,m+1} = u(x_0 + n\delta, t_0 + (m+1)\varepsilon) = u(x, t+\varepsilon), \tag{5.2c}$$

we can do Taylor expansions with respect to δ and ε:

$$u_{n+1,m} = T_\delta u = u + \delta\partial_x u + \ldots, \quad u_{n,m+1} = T_\varepsilon u = u + \varepsilon\partial_t u + \ldots$$

For this to work we may need delicate relations between the expansion parameters, depending on the form of the PΔE.

Performing this sequence of steps on a given lattice equation there are two questions to answer, namely

- among the various parameters present in the lattice equation, how do we identify the parameter (or combination of parameters) to take as the one tending to zero in order to shrink the lattice;
- how do we determine the behavior of the independent and dependent variables under the limit on this chosen parameter?

5.2 Plane-wave factors and linearization

As an example for our ideas we will study the continuum limits of (3.6), i.e. the PΔE associated with the Bäcklund transformations of the KdV equation, which we constructed in Chapter 2. In Chapter 3 we pointed out that the Bäcklund parameters λ and μ, which can be associated with the directions of the lattice, could be interpreted as parameters measuring somehow the grid size in each direction. It is, thus, these parameters that can be used as "tuning devices" by which the lattice can be shrunk or expanded in a certain direction, eventually allowing us to shrink the lattice points together to create a continuum of points.

However, there may be many different ways in which this can happen and, thus, in principle there may be various limits that we could perform on a given lattice equation. In general a *brute force* or naive continuum limit may easily lead to a total collapse of the equation, where the Taylor expansions applied to the equation lead to mismatch of orders, and hence to a situation where the limit results in no equation at all (due to conflicting constraints emerging from the expansion) or to a trivial equation in leading order in the lattice parameter. To avoid this problem and to answer the questions above, it turns out it is useful to first study the continuum limits of the *linearized equation* before we attack the full nonlinear equation. We will, thus, first derive the linearized form of the lattice equations under consideration and derive a special class of solutions of these linear equations to use these as a guidance on how to take nontrivial and consistent limits.

By a *linearization* of a nonlinear lattice equation such as (3.6) we mean the linear equation obtained by expanding the dependent variable around a specific known solution of the nonlinear equation and taking the dominant term. The simplest linearizations are obtained by taking a trivial solution, such as the *zero solution* (if it exists). In the case of the example (3.6), taking $w \equiv 0$ is not allowed since it does not lead to a solution of that equation, but we can modify the equation slightly by changing it into

$$(p - q + u_{n,m+1} - u_{n+1,m})(p + q + u_{n,m} - u_{n+1,m+1}) = p^2 - q^2 \,, \tag{5.3}$$

which is (3.10) in explicit form, by setting as before $4\lambda = p^2$, $4\mu = q^2$, and $w_{n,m} = u_{n,m} - np - mq$. It is easy to see that (5.3) admits the solution $u_{n,m} = 0$, $\forall n, m$. By next setting

$$u_{n,m} = \epsilon \rho_{n,m} \,,$$

and expanding up to linear terms in the small parameter ϵ, we obtain the following linear equation for ρ:

$$(p + q)(\rho_{n,m+1} - \rho_{n+1,m}) = (p - q)(\rho_{n+1,m+1} - \rho_{n,m}) \,. \tag{5.4}$$

It is easily verified that the linear lattice equation (5.4) obeys the consistency-around-the-cube property of Section 3.2 in the same way as the full nonlinear equation, and hence we can consistently embed this equation in a higher-dimensional lattice by writing the compatible system:

$$(p + q)(\widehat{\rho} - \widetilde{\rho}) = (p - q)(\widehat{\widetilde{\rho}} - \rho), \tag{5.5a}$$

$$(p + k)(\overline{\rho} - \widetilde{\rho}) = (p - k)(\overline{\widetilde{\rho}} - \rho), \tag{5.5b}$$

$$(q + k)(\overline{\rho} - \widehat{\rho}) = (q - k)(\overline{\widehat{\rho}} - \rho), \tag{5.5c}$$

where $\bar{\ }$ denotes the shift in the third direction associated with lattice parameter k. The linear system (5.5) has many solutions, but we will fix a specific class of solution by *demanding that the specific variable k is associated with a shift* $\rho \mapsto \overline{\rho}$ *such that* $\overline{\rho} = 0$. (One way of understanding this is that the solution ρ is obtained from an (inverse) Bäcklund transformation $\overline{\rho} \overset{k}{\mapsto} \rho$ with seed solution $\overline{\rho} = 0$.)

Solving the two relations (5.5b) and (5.5c) with $\overline{\rho} = 0$:

$$\widetilde{\rho} = \frac{p-k}{p+k}\rho, \quad \widehat{\rho} = \frac{q-k}{q+k}\rho,$$

we obtain the solution:

$$\rho_{n,m} = \left(\frac{p-k}{p+k}\right)^n \left(\frac{q-k}{q+k}\right)^m \rho_{0,0}, \tag{5.6}$$

with $\rho_{0,0}$ some arbitrary initial value. It is straightforward to show by direct computation that (5.6) is a solution of (5.4).

We will refer to the solution (5.6) as a *lattice plane-wave factor* and we will see later that they play not only a role as approximate solutions (namely as solutions of the linearized version of the nonlinear lattice equation (5.3)) but also an important role in the *exact* solution of the full nonlinear equation. We will exploit these solutions in the next section to formulate the precise limits on parameters and discrete variables to get nontrivial limiting equations.

5.3 The semi-continuous limits

We will now study the various limits that we can perform on the plane-wave factors ρ of the specific form found in (5.6). The guiding principle is to seek ways in which this solution will approach exponential factors with continuous variables in the exponents. The fundamental formula that we will use is the following well-known limit

$$\lim_{n\to\infty}\left(1+\frac{\alpha}{n}\right)^n = e^{\alpha}\,. \tag{5.7}$$

5.3.1 Straight continuum limit

Our objective is first to rewrite the second factor in (5.6) in the form (5.7). Note that

$$\left(\frac{q-k}{q+k}\right)^m = \left(1+\frac{-2k}{q+k}\right)^m .$$

Now since k is a soliton parameter (that does not appear in the equation) we leave it intact and concentrate on the lattice parameter q. Clearly we should let

$$q, m \to \infty \quad \text{while keeping} \quad \xi := m/q \quad \text{finite}. \tag{5.8}$$

Then we obtain:

$$\lim_{\substack{m\to\infty\\ q\to\infty\\ m=\xi q}} \left(\frac{q-k}{q+k}\right)^m = \lim_{m\to\infty}\left(1+\frac{\xi}{m}\,\frac{(-2k)}{(1+k\xi/m)}\right)^m = e^{-2k\xi}\,, \tag{5.9}$$

where ξ is the new continuous coordinate. (Note that the extra term in the denominator within the brackets becomes negligible as $m\to\infty$.)

Let us now investigate the effect of this limit on the equations, first on the linear equation (5.5a). The idea is to reinterpret (cf. (5.2)) the dependent variable as

$$\rho = \rho_{n,m} =: \rho_n(\xi)\,, \quad \xi = \xi_0 + \frac{m}{q}\,, \tag{5.10}$$

where ξ_0 is some initial value. Note that $m \mapsto m+1$ implies $\xi \mapsto \xi + 1/q$. Now since $q\to\infty$, the parameter $1/q$ is small and we may expand

$$\rho_{n,m+1} = \rho_n(\xi+\tfrac{1}{q}) = \rho_n(\xi) + \tfrac{1}{q}\partial_\xi\rho_n(\xi) + \tfrac{1}{2}\tfrac{1}{q^2}\partial_\xi^2\rho_n(\xi) + \dots\,.$$

Inserting this into the equation (5.4) we get:

$$\begin{aligned}&\left(1+\frac{p}{q}\right)\left[-\rho_{n+1}+\rho_n+\tfrac{1}{q}\partial_\xi\rho_n+\tfrac{1}{2}\tfrac{1}{q^2}\partial^2\rho_n+\cdots\right]\\ &=\left(-1+\frac{p}{q}\right)\left[-\rho_n+\rho_{n+1}+\tfrac{1}{q}\partial_\xi\rho_{n+1}+\tfrac{1}{2}\tfrac{1}{q^2}\partial^2\rho_{n+1}+\cdots\right],\end{aligned}$$

and then to leading order (in terms of $1/q$) we obtain the *differential-difference* equation (DΔE):

$$\partial_\xi\,(\rho_{n+1}+\rho_n) = 2p(\rho_{n+1}-\rho_n)\,. \tag{5.11}$$

This linear equation is a mixed form of differential equation (with respect to variable ξ) and discrete (with respect to variable n), and hence it is called a *semi-discrete* equation. It is clear (and can be directly verified) that

$$\rho_n(\xi) = \left(\frac{p-k}{p+k}\right)^n e^{-2k\xi}\,\rho_0(0)\,, \tag{5.12}$$

is a solution of (5.11).

Inspired by this result, let us now turn to the nonlinear equation for u, (5.3), and perform exactly the same limit there. Introducing in a similar way as for the linear equation the reinterpretation of the discrete variables as in (5.10), we derive

$$\begin{aligned}p^2-q^2 = &\left[p-q+\left(u_n+\frac{1}{q}\partial_\xi u_n+\frac{1}{2q^2}\partial_\xi^2u_n+\cdots\right)-u_{n+1}\right]\\ &\times\left[p+q+u_n-\left(u_{n+1}+\frac{1}{q}\partial_\xi u_{n+1}+\frac{1}{2q^2}\partial_\xi^2u_{n+1}+\cdots\right)\right].\end{aligned}$$

Expanding in powers of $1/q$, and noting that the terms of order $\mathcal{O}(q^2)$ and $\mathcal{O}(q)$ cancel identically, we obtain as coefficient of the leading term of order $\mathcal{O}(1)$ the following equation

$$\partial_\xi \left(u_n + u_{n+1}\right) = 2p(u_{n+1} - u_n) - (u_{n+1} - u_n)^2 \,, \tag{5.13}$$

which is the continuum limit of the lattice equation (5.3). It is nonlinear, first order in the derivative with respect to ξ and first order in the difference in n.

Remark: Note that the change of variables $x = 2\xi$, $p^2 = 4\lambda^2$, $\widetilde{w} - w = u_{n+1} - u_n - p$, $\partial_\xi u_n = 2w_x$, allows us to recover the spatial part of the BT of the KdV equation (2.85a). Furthermore, (5.13) is a truncated form of the dressing chain relation (Shabat, 1992; Veselov and Shabat, 1993).

Other lattice equations: We can perform similar limits on the other members of the KdV family of lattice equations, namely on (3.9) and (3.11), leading to

$$\partial_\xi (v_{n+1} v_n) = p(v_{n+1}^2 - v_n^2) \tag{5.14}$$

and

$$(\partial_\xi z_n)(\partial_\xi z_{n+1}) = p^2(z_n - z_{n+1})^2, \tag{5.15}$$

respectively. We shall show later that all these DΔEs are *integrable* by virtue of the existence of semi-discrete analogues of the Lax pairs in all cases.

5.3.2 Skew continuum limit

The limit described above is not the only continuum limit that we can perform on the lattice equation. Instead of taking a limit in one of the variables n and m separately, one could also mix them up, by means of a change of independent variables on the lattice, before taking a limit. As we shall see this will lead to quite different semi-continuous equations.

To describe the "skew" limit, let us first consider the linearized equation (5.4) under the following change of variables:

$$\rho_{n+k,m+l} = R_{n+m+k+l,m+l} \equiv R_{N+k+l,m+l}, \ \forall k,l \text{ where } N = n+m. \tag{5.16}$$

This leads to

$$\begin{aligned}
\rho &= R, \\
\widetilde{\rho} &= \rho_{n+1,m} = R_{N+1,m} = \widetilde{R}, \\
\widehat{\rho} &= \rho_{n,m+1} = R_{N+1,m+1} = \widehat{\widetilde{R}}, \\
\widehat{\widetilde{\rho}} &= \rho_{n+1,m+1} = R_{N+2,m+1} = \widehat{\widetilde{\widetilde{R}}},
\end{aligned}$$

which can be visualized in the diagram of Figure 5.1.

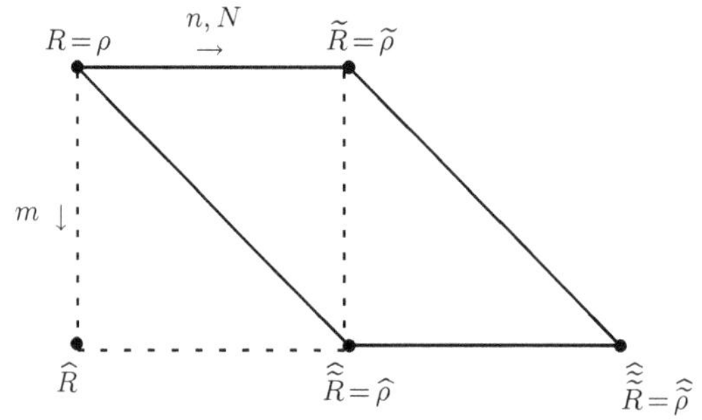

Figure 5.1 The change of variables in the skew limit.

The change of independent variables $(n, m) \mapsto (N = n + m, m)$ brings the linear equation (5.5a) to the form

$$(p+q)(R_{N+1,m+1} - R_{N+1,m}) = (p-q)(R_{N+2,m+1} - R_{N,m}). \tag{5.17}$$

Let us investigate what happens to the plane-wave solution as $m \to \infty$. Rearranging factors we get:

$$\begin{aligned} R_{N,m} &= \left(\frac{p-k}{p+k}\right)^{n+m} \left(\frac{q-k}{q+k}\,\frac{p+k}{p-k}\right)^{m} R_{0,0} \\ &= \left(\frac{p-k}{p+k}\right)^{N} \left(1 + \frac{2(q-p)k}{(q+k)(p-k)}\right)^{m} R_{0,0}. \end{aligned}$$

Keeping N fixed and setting $\delta = q - p$, we can now perform the limit

$$n \to -\infty, \quad m \to \infty, \quad \delta \to 0 \quad \text{while} \quad N \text{ and } \delta m =: \tau' \quad \text{are fixed.} \tag{5.18}$$

Focusing on what happens with the second factor in this limit we observe:

$$\begin{aligned} &\lim_{\substack{m\to\infty\\ \delta\to 0\\ \delta m=\tau'}} \left(1 + \frac{2\delta k}{(p+\delta+k)(p-k)}\right)^{m} \\ &\quad = \lim_{\substack{m\to\infty\\ \delta\to 0\\ \delta m=\tau'}} \left(1 + \frac{2\delta k}{(p^2-k^2)+(p-k)\delta}\right)^{m} \\ &\quad = \lim_{m\to\infty} \left(1 + \frac{\tau'}{m}\,\frac{2k}{(p^2-k^2)+(p-k)\tau'/m}\right)^{m} \\ &\quad = \exp\left(\frac{2k\tau'}{p^2-k^2}\right). \end{aligned} \tag{5.19}$$

Thus, the limit (5.18) makes good sense on the level of the plane-wave factors.

To investigate what happens with the linear lattice equation (5.17) under the limit (5.18), we set

$$R_{N,m} =: R_N(\tau), \qquad \tau = \tau_0 + m\delta, \tag{5.20}$$

allowing for some constant background value τ_0 of the continuous variable, and apply the Taylor expansion:

$$R_{N,m+1} = R_N(\tau+\delta) = R_N(\tau) + \delta\,\partial_\tau R_N(\tau) + \frac{1}{2}\delta^2\,\partial_\tau^2 R_N(\tau) + \dots .$$

Inserting this into equation (5.17) we get (here and in the following we use dot for τ-derivatives):

$$\begin{aligned}
&(2p+\delta)\left[\left(R_{N+1} + \delta\dot{R}_{N+1} + \frac{1}{2}\delta^2\ddot{R}_{N+1} + \cdots\right) - R_{N+1}\right] \\
&= -\delta\left[\left(R_{N+2} + \delta\dot{R}_{N+2} + \frac{1}{2}\delta^2\ddot{R}_{N+2} + \cdots\right) - R_N\right],
\end{aligned}$$

which to leading order $\mathcal{O}(\delta)$ yields:

$$2p\dot{R}_N = R_{N-1} - R_{N+1}\,. \tag{5.21}$$

It is straightforward to check that

$$R_N(\tau) = \left(\frac{p-k}{p+k}\right)^N \exp\left(\frac{2k\tau}{p^2-k^2}\right) R_0(0), \tag{5.22}$$

provides a solution of equation (5.21) with initial value $R_0(0)$.

Let us now move to the nonlinear equation and perform a similar limit there. In fact, applying the change of variables

$$(n,m) \mapsto (N = n+m, m), \qquad u_{n,m} =: U_{n+m,m} \tag{5.23}$$

in equation (5.3) we obtain

$$(p - q + U_{N+1,m+1} - U_{N+1,m})(p + q + U_{N,m} - U_{N+2,m+1}) = p^2 - q^2, \tag{5.24}$$

and then reinterpreting the variable U as

$$U = U_{N,m} =: U_N(\tau), \qquad \tau = \tau_0 + m\delta\,, \tag{5.25}$$

we can perform a similar Taylor expansion on $U_N(\tau+\delta)$ as we did for R. Inserting this into (5.24) we get

$$\left[-\delta + \left(U_{N+1} + \delta \dot{U}_{N+1} + \tfrac{1}{2}\delta^2 \ddot{U}_{N+1} + \cdots\right) - U_{N+1}\right]$$
$$\times \left[2p + \delta + U_N - \left(U_{N+2} + \delta \dot{U}_{N+2} + \tfrac{1}{2}\delta^2 \ddot{U}_{N+2} + \cdots\right)\right] = -(2p+\delta)\delta,$$

which to leading order yields

$$(\dot{U}_{N+1} - 1)(2p + U_N - U_{N+2}) = -2p,$$

or equivalently

$$\dot{U}_N = 1 - \frac{2p}{2p + U_{N-1} - U_{N+1}}. \tag{5.26}$$

In spite of the fact that equation (5.26) was derived starting from the same lattice equation, it is quite a different nonlinear DΔE from (5.13), as is evident by inspecting the orders. We show later that both equations (5.13) and (5.26) are *integrable* in the sense that there exists an associated Lax pair in both cases.

Other examples: We refer to the process described above as the *skew continuum limit* of the lattice equation. It can be applied also to other quadrilateral equations. For the lattice KdV family, namely equations (3.11) and (3.13), we obtain

$$p\partial_\tau \log V_N = \frac{V_{N-1} - V_{N+1}}{V_{N-1} + V_{N+1}}, \tag{5.27}$$

and

$$\dot{Z}_N = \frac{2}{p} \frac{(Z_{N-1} - Z_N)(Z_N - Z_{N+1})}{Z_{N-1} - Z_{N+1}} \tag{5.28}$$

respectively.

5.4 Semi-discrete Lax pairs

In this section we shall show that the integrability characteristics of the equations are preserved when performing the continuum limits presented above. We will do this by applying the limits to the relevant Lax pair on the lattice.

We will illustrate this for the Lax pair (3.17) of lattice potential KdV equation (5.3). As was noted before, the limiting process is delicate. For example, in order to extract a derivative for ψ, the leading term of the Lax matrix should be a unit matrix, because then we would have $\widetilde{\psi} - \psi = L'\psi$. We will therefore first prepare the Lax system as follows: (i) for the Lax matrices we choose the prefactors as $\gamma = 1/(p-k)$, $\gamma' = 1/(q-k)$, (ii) we change to the u variable by $w_{nm} = u_{nm} - (pn + qm + c)$ and (iii) we change the eigenfunctions by the following "gauge" transformation:

$$\boldsymbol{\phi} = G\boldsymbol{\psi}, \quad \text{where} \quad G = \begin{pmatrix} -(np + mq + c) & 1 \\ 1 & 0 \end{pmatrix}.$$

After this the Lax pair becomes

$$(p-k)\widetilde{\boldsymbol{\psi}} = \mathcal{L}\,\boldsymbol{\psi}, \quad (q-k)\widehat{\boldsymbol{\psi}} = \mathcal{M}\,\boldsymbol{\psi}, \tag{5.29a}$$

in which the matrices $\mathcal{L}$ and $\mathcal{M}$ are given by

$$\mathcal{L} := (p-k)\widetilde{G}^{-1}\boldsymbol{L}G = \begin{pmatrix} p-\widetilde{u} & 1 \\ k^2 - p^2 + (p-\widetilde{u})(p+u) & p+u \end{pmatrix}, \tag{5.29b}$$

$$\mathcal{M} := (q-k)\widehat{G}^{-1}\boldsymbol{M}G = \begin{pmatrix} q-\widehat{u} & 1 \\ k^2 - q^2 + (q-\widehat{u})(q+u) & q+u \end{pmatrix}. \tag{5.29c}$$

The prefactors $(p-k)$ and $(q-k)$ are obviously irrelevant for the nonlinear equation that arises from the Lax pair as the consistency condition (since these factors can be easily absorbed into a redefinition of the vector $\boldsymbol{\psi}$), but will turn out to be useful in performing the limits.

5.4.1 Straight continuum limit

We will now perform the limit (5.8) on the Lax pair itself. Obviously this will not change the first member of the Lax pair, i.e. the first equation in (5.29), since it only affects the dependence on the variable m and the parameter q. Setting once again

$$u_{n,m} = u_n(\xi), \quad \boldsymbol{\psi}_{n,m} = \boldsymbol{\psi}_n(\xi), \quad \xi = \xi_0 + \frac{m}{q},$$

and expanding both $\boldsymbol{\psi}_{n,m+1} = \boldsymbol{\psi}_n(\xi + \frac{1}{q})$ and $u_{n,m+1} = u_n(\xi + \frac{1}{q})$ in the entries of the matrix $\mathcal{M}$ by Taylor series in powers of $1/q$, we get from the second member of the Lax pair:

$$\begin{aligned} (q-k)\left(\boldsymbol{\psi}_n + \tfrac{1}{q}\partial_\xi \boldsymbol{\psi}_n + \cdots\right) &= q\boldsymbol{\psi}_n \\ &+ \begin{pmatrix} -u_n - \frac{1}{q}\partial_\xi u_n - \cdots & 1 \\ k^2 + q(-\frac{1}{q}\partial_\xi u_n - \cdots) - u_n(u_n + \frac{1}{q}\partial_\xi u_n + \cdots) & u_n \end{pmatrix} \boldsymbol{\psi}_n, \end{aligned}$$

and retaining only terms of order $\mathcal{O}(1)$ we obtain:

$$\partial_\xi \boldsymbol{\psi}_n = \boldsymbol{N}_n\,\boldsymbol{\psi}_n, \quad \boldsymbol{N}_n = \begin{pmatrix} k-u_n & 1 \\ k^2 - \partial_\xi u_n - u_n^2 & k+u_n \end{pmatrix}. \tag{5.30}$$

Thus we have obtained the semi-discrete Lax pair, consisting of (5.30) and the linear equation which we obtain directly from (5.29b), namely:

$$(p-k)\boldsymbol{\psi}_{n+1} = \boldsymbol{L}_n\,\boldsymbol{\psi}_n, \quad \boldsymbol{L}_n = \begin{pmatrix} p-u_{n+1} & 1 \\ k^2 - p^2 + (p-u_{n+1})(p+u_n) & p+u_n \end{pmatrix}. \tag{5.31}$$

The consistency of

$$
\begin{aligned}
(p-k)\partial_\xi(\boldsymbol{\psi}_{n+1}) &= \partial_\xi(\boldsymbol{L}_n\,\boldsymbol{\psi}_n) = (\partial_\xi \boldsymbol{L}_n)\,\boldsymbol{\psi}_n + \boldsymbol{L}_n\,\boldsymbol{N}_n\,\boldsymbol{\psi}_n \\
(p-k)(\partial_\xi \boldsymbol{\psi})_{n+1} &= (p-k)\boldsymbol{N}_{n+1}\,\boldsymbol{\psi}_{n+1} = \boldsymbol{N}_{n+1}\,\boldsymbol{L}_n\,\boldsymbol{\psi}_n,
\end{aligned}
$$

leads to the semi-discrete equation

$$
\partial_\xi \boldsymbol{L}_n = \boldsymbol{N}_{n+1}\,\boldsymbol{L}_n - \boldsymbol{L}_n\,\boldsymbol{N}_n\,. \tag{5.32}
$$

It is a straightforward exercise to check that the compatibility condition (5.32) indeed gives rise to the nonlinear DΔE (5.13). In a similar way one can derive the semi-discrete limits of other Lax pairs.

5.4.2 Skew continuum limit

We now turn to the derivation of the Lax pair for the skew continuum limit of the lattice equation (5.3) from (5.29). In this case the derivation is slightly more complicated, because the two members of the Lax pair get mixed together in the limit due to the change of variables

$$
\boldsymbol{\psi}_{n,m} := \boldsymbol{\Phi}_N(\tau),\; u_{n,m} = U_N(\tau) \quad \text{with} \quad N = n+m, \quad \tau = \tau_0 + m\delta, \quad \delta = q - p.
$$

The first part of the Lax pair (5.29a) with (5.29b) is similar, we just need to replace the variable n by the new spatial variable N:

$$
\begin{aligned}
(p-k)\boldsymbol{\Phi}_{N+1} &= \mathfrak{L}_N\,\boldsymbol{\Phi}_N, \\
\mathfrak{L}_N &= \begin{pmatrix} p - U_{N+1} & 1 \\ k^2 - p^2 + (p - U_{N+1})(p + U_N) & p + U_N \end{pmatrix}.
\end{aligned} \tag{5.33}
$$

In the second part of (5.29a) the matrix $\mathcal{M}$ of (5.29c) can be expanded as follows:

$$
\mathcal{M}_N = \mathfrak{L}_N + \delta \begin{pmatrix} 1 - \partial_\tau U_{N+1} & 0 \\ U_N - U_{N+1} - p\partial_\tau U_{N+1} - U_N \partial_\tau U_{N+1} & 1 \end{pmatrix}
$$

while the left-hand side can be expanded as

$$
\begin{aligned}
(q-k)\boldsymbol{\psi}_{n,m+1} &= (q-k)\boldsymbol{\Phi}_{N+1}(\tau+\delta) \\
&= (p-k)\boldsymbol{\Phi}_{N+1}(\tau) + \delta\,[\boldsymbol{\Phi}_{N+1} + (p-k)\partial_\tau \boldsymbol{\Phi}_{N+1}(\tau)] + \cdots
\end{aligned}
$$

Using (5.33) in the above expansions we see that the leading order terms cancel and at order $\mathcal{O}(\delta)$ we get the equation

$$\boldsymbol{\Phi}_{N+1} + (p-k)\partial_\tau \boldsymbol{\Phi}_{N+1} = \begin{pmatrix} 1-\partial_\tau U_{N+1} & 0 \\ U_N - U_{N+1} - p\partial_\tau U_{N+1} - U_N \partial_\tau U_{N+1} & 1 \end{pmatrix} \boldsymbol{\Phi}_N .$$

Using again (5.33) on the left-hand side to express $\boldsymbol{\Phi}_{N+1}$ in terms of $\boldsymbol{\Phi}_N$ we arrive at the following temporal part of the Lax pair:

$$\begin{aligned} \dot{\boldsymbol{\Phi}}_N &= \frac{1}{k^2-p^2} \mathcal{N}_N \boldsymbol{\Phi}_N , \\ \mathcal{N}_N &= \begin{pmatrix} k - U_{N+1} - (p-U_{N+1})\dot{U}_N & 1-\dot{U}_N \\ k^2 - U_{N+1}^2 + (p-U_{N+1})^2 \dot{U}_N & k + U_{N+1} + (p-U_{N+1})\dot{U}_N \end{pmatrix} . \end{aligned} \tag{5.34}$$

Thus, equations (5.33) and (5.34) together form the Lax pair for the DΔE (5.26). In fact, by a similar argument as given above in the derivation of (5.32) we now get

$$(k^2-p^2)\,\dot{\mathcal{L}}_N = \mathcal{N}_{N+1}\,\mathcal{L}_N - \mathcal{L}_N\,\mathcal{N}_N \,, \tag{5.35}$$

and one can show that (5.35) is verified if and only if U_N obeys the semi-discrete KdV equation (5.26).

5.5 Full continuum limit

The derivation of the full continuum limit of lattice equations such as equation (5.3) is given here as a two-stage process: we have already established two different semi-discrete (semi-continuous) limits in the previous section, and now we will find the full continuum limit. We start from one of the semi-discrete limits and do the second limit on the remaining discrete variable and turn it into a continuous one. In fact, in order to obtain a nontrivial limit, we shall see that the process is slightly more involved, and that we need to mix spatial and temporal variables in order to obtain a nonlinear PDE, while retaining the integrability.

5.5.1 From the straight continuum limit

Let us first consider the linearized straight continuum limit (cf. Section 5.3.1). It is clear that for the remaining discrete variable n we should use a limit similar to (5.8), i.e.

$$n \to \infty, \quad p \to \infty, \quad p = n/\tau, \quad \tau \quad \text{fixed and finite} \tag{5.36}$$

along with

$$\rho_n(\xi) = \rho(\xi,\tau), \quad \tau = \tau_0 + \frac{n}{p}. \tag{5.37}$$

Starting with (5.11) a simple calculation with Taylor expansions gives

$$2(\rho_\xi - \rho_\tau) + p^{-1}\partial_\tau(\rho_\xi - \rho_\tau) + p^{-2}(\tfrac{1}{2}\rho_{\tau\tau\xi} - \tfrac{1}{3}\rho_{\tau\tau\tau}) + \cdots = 0 \qquad (5.38)$$

so that the leading terms at $\mathcal{O}(1)$ give the very simple equation

$$\rho_\xi = \rho_\tau,$$

which is solved with the corresponding limit of the plane-wave factor

$$\rho(\xi, \tau) = e^{-2k(\xi+\tau)}\rho(0, 0).$$

The above result is not satisfactory as it is too simple. Thus we look at the possibility of getting a more interesting result by combining the terms at $\mathcal{O}(1)$ and $\mathcal{O}(p^{-1})$ by going to new coordinates defined by their derivatives

$$\partial_\tau = \partial_x + \tfrac{1}{12p^2}\partial_t, \quad \partial_\xi = \partial_x, \qquad (5.39a)$$

which imply

$$\xi = x - 12p^2 t, \quad \tau = 12p^2 t, \quad \text{or} \quad t = \tfrac{1}{12p^2}\tau, \quad x = \xi + \tau. \qquad (5.39b)$$

After this change of variables we get at $\mathcal{O}(p^{-2})$ the dispersive equation

$$\rho_{xxx} = \rho_t,$$

which is the linear part of the (p)KdV equation.

What about the plane-wave factor? We must expand (5.12) to higher order in $1/p$:

$$\begin{aligned}\rho_n(\xi) &= \left(\frac{p-k}{p+k}\right)^n e^{-2k\xi}\rho_0(0)\\ &= \exp\left\{-2k\xi + \tau p \log\left[1 - \tfrac{2k}{p+k}\right]\right\}\rho_0(0)\\ &= \exp\left\{-2k(\xi+\tau) - \tfrac{2}{3}p^{-2}k^3\tau + \tau O(p^{-4}\right\}\rho_0(0)\\ &= \exp\left\{-2kx - 8k^3 t + O(p^{-2})\right\}\rho_0(0),\end{aligned}$$

which solves the above equation.

One can treat the nonlinear semi-continuous equation (5.13) the same way. Using (5.37) we find that the nonlinear term contributes at $\mathcal{O}(p^{-2})$ and therefore if we add the term $-\frac{1}{p^2}\partial_\tau u^2$ to (5.38) and use the redefinition (5.39) we get

$$u_t = u_{xxx} + 6u_x^2, \qquad (5.40)$$

i.e. the continuous potential KdV equation.

5.5.2 From the skew continuum limit

The situation is basically the same as for the straight continuum limit. The limit itself would now be

$$N \to \infty, \quad p \to \infty, \quad p = N/\xi, \quad \xi \quad \text{fixed and finite}, \tag{5.41}$$

but in a naive limit of (5.21) with

$$R_N(\tau) = R(\xi, \tau), \quad \xi = \xi_0 + \frac{n}{p}, \tag{5.42}$$

the right-hand side of (5.21) would just disappear.

Let us therefore use the plane-wave factors to guide us in finding the limit we need to impose in order to get nontrivial equations from the semi-discrete one. Starting with the skew limit of the plane-wave factor, i.e. the form of the variable $R_N(\tau)$ as given in (5.22), we can expand as follows

$$\begin{aligned}
&\left(\tfrac{p-k}{p+k}\right)^N \exp\left(\tfrac{2k\tau}{p^2-k^2}\right) \\
&= \exp\left\{\tfrac{2k\tau}{p^2-k^2} + N\left[\ln\left(1-\tfrac{k}{p}\right) - \ln\left(1+\tfrac{k}{p}\right)\right]\right\} \\
&= \exp\left\{\tfrac{2k\tau}{p^2}\left[1+\tfrac{k^2}{p^2}+\tfrac{k^4}{p^4}+\cdots\right] - 2N\left[\tfrac{k}{p}+\tfrac{1}{3}\tfrac{k^3}{p^3}+\tfrac{1}{5}\tfrac{k^5}{p^5}+\cdots\right]\right\} \\
&= \exp\left\{2k\left(\tfrac{\tau}{p^2}-\tfrac{N}{p}\right) + 2k^3\left(\tfrac{\tau}{p^4}-\tfrac{1}{3}\tfrac{N}{p^3}\right) + 2k^5\left(\tfrac{\tau}{p^6}-\tfrac{1}{5}\tfrac{N}{p^5}\right)+\cdots\right\}.
\end{aligned}$$

Since we want this to approach e^{kx+k^3t} when $p \to \infty$, the above leads to the identification of the variables of the full continuum limit. Thus from the first and second terms in the exponent we infer that the good variables should be defined as follows:

$$x = 2\left(\tfrac{\tau}{p^2} - \xi\right), \; t = 2\left(\tfrac{\tau}{p^4} - \tfrac{1}{3}\tfrac{\xi}{p^2}\right), \tag{5.43a}$$

where we have used $N/p =: \xi$. With this choice we get the desired result since the higher-order terms vanish in the limit. For the derivatives the transformation is

$$\partial_\tau = \tfrac{2}{p^2}\partial_x + \tfrac{2}{p^4}\partial_t, \quad \partial_\xi = -2\partial_x - \tfrac{2}{3p^2}\partial_t. \tag{5.43b}$$

From this we can immediately see why the naive limit failed completely: τ should have been scaled $\tau \mapsto p^{-2}\tau$, which would have given at least a simple first-order equation. Furthermore, since the two steps (i.e. Taylor expansion of the shift in N and change of variables $(\xi, \tau) \mapsto (x, t)$) both involve the parameter p, the selection of the dominant term in series expansion in powers of $1/p$ can only be done after having done both of these steps, which we will now do.

First, we identify the solution $R_N(\tau)$ of this DΔE as follows:

$$R_N(\tau) := S(\xi,\tau), \tag{5.44}$$

$$R_{N\pm1}(\tau) = S(\xi \pm \frac{1}{p},\tau) = S \pm \frac{1}{p}S_\xi + \frac{1}{2p^2}S_{\xi\xi} \pm \frac{1}{6p^3}S_{\xi\xi\xi} + \dots . \tag{5.45}$$

Next, we perform the change of variables given in (5.43), and since the two steps of Taylor expansion of the shift in N and the change of variables $(\xi,\tau) \mapsto (x,t)$ both involve the parameter p, the expansion in powers of $1/p$ and the selection of the dominant term can only be done after having done both of these steps. Thus, starting from (5.21) for $R_N(\tau)$ we get

$$2p\partial_\tau S = -\tfrac{2}{p}\partial_\xi S - \tfrac{1}{3p^3}\partial_\xi^3 S - \cdots$$

Next, using (5.43b) and considering S as a function of x,t we get

$$2p\left(\tfrac{2}{p^2}\partial_x + \tfrac{2}{p^4}\partial_t\right)S = -\tfrac{2}{p}\left(-2\partial_x - \tfrac{2}{3p^2}\partial_t\right)S - \tfrac{1}{3p^3}\left(-2\partial_x - \tfrac{2}{3p^2}\partial_t\right)^3 S - \cdots ,$$

and we observe that while the leading term of order $\mathcal{O}(1/p)$ cancels identically, the next term of order $\mathcal{O}(1/p^3)$ yields precisely the equation:

$$S_t = S_{xxx} ,$$

as expected. The limit of the plane-wave factor is e^{kx+k^3t}, as shown above, and it clearly solves this equation.

Turning now to the nonlinear equation (5.26) and performing the same continuum limit (5.41) there, we get first

$$\partial_\tau U = 1 - \left[1 - \tfrac{1}{2p}\left(\tfrac{2}{p}U_\xi + \tfrac{1}{3p^3}U_{\xi\xi\xi} + \cdots\right)\right]^{-1} .$$

Next we change variables according to (5.43a) which for the left-hand side means

$$U_\tau = \left(\tfrac{2}{p^2}\partial_x + \tfrac{2}{p^4}\partial_t\right)U,$$

and expanding the right-hand side yields

$$\begin{aligned} 1 &- \left[1 - \tfrac{1}{p^2}\left(-2\partial_x - \tfrac{2}{3p^2}\partial_t\right)U - \tfrac{1}{6p^4}\left(-2\partial_x - \tfrac{2}{3p^2}\partial_t\right)^3 U - \cdots\right]^{-1} \\ &= \left[\tfrac{1}{p^2}\left(2\partial_x + \tfrac{2}{3p^2}\partial_t\right)U + \tfrac{1}{6p^4}\left(2\partial_x + \tfrac{2}{3p^2}\partial_t\right)^3 U + \cdots\right] \\ &\quad - \left[\tfrac{1}{p^2}\left(2\partial_x + \tfrac{2}{3p^2}\partial_t\right)U + \cdots\right]^2 + \cdots \\ &= \tfrac{2}{p^2}\partial_x U + \tfrac{1}{p^4}\left[\tfrac{2}{3}\partial_t U + \tfrac{4}{3}\partial_x^3 U - (2\partial_x U)^2\right] + \dots \end{aligned}$$

Again we observe that the terms of order $\mathcal{O}(1/p^2)$ cancel identically, and that at the next order $\mathcal{O}(1/p^4)$ we get the equation:

$$U_t = U_{xxx} - 3U_x^2\,, \tag{5.46}$$

which is the potential KdV equation (coinciding with (5.40)) in terms of the variable w, up to a change of sign and scaling in the independent variables.

We have now completed a full circle. We started out with the continuous KdV equation, derived its Bäcklund transformations, which using the permutability theorem led to the construction of a lattice of solutions, the relations between these solutions being reinterpreted as a partial difference equation on the two-dimensional lattice. As a dynamical equation the latter was seen as a discretization of some continuum equations, both semi-discrete as well as fully discrete, and the full continuum limit now turns out to be the equation from which we started, namely the KdV itself.

$$\boxed{\text{KdV}} \xrightarrow{\text{Lax pair}} \boxed{\text{BT}} \xrightarrow{\text{permut. cond.}} \boxed{\text{P}\Delta\text{E}} \xrightarrow{\text{cont. lim.}} \boxed{\text{D}\Delta\text{E}} \xrightarrow{\text{cont. lim.}} \boxed{\text{KdV}}$$

Perhaps this is not so surprising, but what is remarkable is that the equations at all levels are compatible with each other: the continuous and discrete equations can be imposed simultaneously on one and the same variable u.

5.6 All at once, or the double continuum limit

With the experience from the previous limits we may attempt to take both limits at once. The limits (5.39, 5.10, 5.37) suggest that we could try

$$u_{n+k,m+l} = u(x + (a_1\,k + a_2\,l)\epsilon, t + (b_1\,k + b_2\,l)\epsilon^3)$$

for some fixed constant a_i, b_i to be determined, and that for the parameters we should have

$$p = d_1/\epsilon, \quad q = d_2/\epsilon.$$

Substitution all these into (5.3) and expanding we obtain pKdV, i.e. equation (5.40), at order ϵ^2, provided that

$$a_1 d_1 = a_2 d_2, \quad d_1 \neq d_2, \quad a_1 a_2 (a_1^2 - a_2^2) + 12(a_1 b_2 - a_2 b_1) = 0. \tag{5.47}$$

The above works equally well for the other equations in the KdV class, i.e. one obtains the potential modified KdV (2.73) from (3.11), and the Schwarzian KdV (2.103) from (3.13).

5.7 Continuum limits of the 9-point BSQ

In this section, we consider continuum limits of scalar Boussinesq-type lattice equations, which are defined on a 9-point stencil; see Figure 3.9. We will consider in detail only the lattice BSQ equation (3.71), for $b_0 = 0$ and $u_{n,m} \mapsto u_{n,m} - np - mq$

$$\frac{p^3-q^3}{p-q+\hat{\hat{\tilde{u}}}-\hat{\tilde{\tilde{u}}}} - \frac{p^3-q^3}{p-q+\hat{\tilde{u}}-\tilde{\tilde{u}}}$$
$$= (p-q+\hat{u}-\tilde{u})(2p+q+u-\hat{\tilde{\tilde{u}}}) - (p-q+\hat{\hat{\tilde{u}}}-\hat{\tilde{\tilde{u}}})(2p+q+\hat{u}-\hat{\hat{\tilde{\tilde{u}}}}) \, . \tag{5.48}$$

The plane-wave factors are determined, as before, by the linearized lattice BSQ equation, which is the following

$$\frac{p^2+pq+q^2}{p-q}\left(2\hat{\tilde{\rho}}-\tilde{\tilde{\rho}}-\hat{\hat{\rho}}\right) = (p+2q)\left(\hat{\rho}-\hat{\tilde{\tilde{\rho}}}\right) - (q+2p)\left(\tilde{\rho}-\hat{\hat{\tilde{\rho}}}\right) + (p-q)\left(\rho+\hat{\hat{\tilde{\tilde{\rho}}}}\right) . \tag{5.49}$$

which has plane-wave-factor solutions of the form

$$\rho_k(n,m) = \left(\frac{p+k}{p+\omega k}\right)^n \left(\frac{q+k}{q+\omega k}\right)^m , \tag{5.50}$$

where ω is a cube root of unity, with $\omega \neq 1$, and k is a spectral parameter.

In order to determine possible semicontinuous limits of the equation, we consider continuum limits of the plane-wave factors (5.50). We consider two possibilities:

1 The straight limit

$$q \mapsto \infty \ , \quad m \mapsto \infty \ , \quad m/q \mapsto x \ , \tag{5.51}$$

which results in the following limiting behavior of the discrete plane-wave factor

$$\left(\frac{p+k}{p+\omega k}\right)^n \left(\frac{q+k}{q+\omega k}\right)^m \mapsto \left(\frac{p+k}{p+\omega k}\right)^n e^{(1-\omega)kx} \, . \tag{5.52}$$

2 The skew limit

$$\delta \equiv p - q \mapsto 0 \ , \quad m \mapsto \infty \ , \quad \delta m \mapsto \tau \ , \tag{5.53}$$

which leads to

$$\left(\frac{p+k}{p+\omega k}\right)^n \left(\frac{q+k}{q+\omega k}\right)^m \mapsto \left(\frac{p+k}{p+\omega k}\right)^{n'} \exp\left[\frac{(1-\omega)k\tau}{(p+k)(p+\omega k)}\right] \tag{5.54}$$

where $n' = n + m$.

In the first case, we expand the dependent variable as

$$u_n(m) \quad \mapsto \quad u_n(\tau) + \delta \dot{u}_n(\tau) + \frac{1}{2}\ddot{u}_n(\tau) + \ldots,$$

and obtain

$$\begin{aligned} 3p^2\partial_\tau \log(1+\dot{u}_n) &= (3p + u_{n-1} - u_{n+2})(1+\dot{u}_n)(1+\dot{u}_{n+1}) \\ &\quad - (3p + u_{n-2} - u_{n+1})(1+\dot{u}_{n-1})(1+\dot{u}_n)\,, \end{aligned} \tag{5.55}$$

where $\dot{u}_n = \partial_\tau u_n$.

In the second case, we expand

$$u_{n,m} \mapsto u_n(x) + \frac{1}{q}u'_n(x) + \frac{1}{2q^2}u''_n(x) + \ldots\,,$$

where $u'_n \equiv \partial_x u_n$, which leads to

$$\begin{aligned} \partial_x^2(u_{n+1} + u_n + u_{n-1}) &= 3u'_{n+1}(p + u_n - u_{n+1}) - 3u'_{n-1}(p + u_{n-1} - u_n) \\ &\quad + (p + u_n - u_{n+1})^3 - (p + u_{n-1} - u_n)^3. \end{aligned} \tag{5.56}$$

This equation has also appeared as the dressing chain of the Boussinesq equation (Adler et al., 2001).

We can also take a double continuum limit of the one-component lattice BSQ equations. However, when doing that we must make sure the structure of soliton solutions is preserved and this may require some reparametrization. In the case of (3.71) we must reparametrize $(p^3, q^3) \mapsto (p^3 + b_0 p^2, q^3 + b_0 q^2)$, and after this, the double-continuum limit of (3.71) is obtained by taking

$$u_{n+\nu,m+\mu} = (n+\nu)\,p + (m+\mu)\,q - u\big(x + 2(\nu/p + \mu/q),\, y + 2(\nu+\mu)/(pq)\big), \tag{5.57a}$$

setting $p = p'/\epsilon$, $q = q'/\epsilon$ and then letting $\epsilon \to 0$ while keeping p', q' fixed. The result is the BSQ equation

$$u_{xxxx} + 6u_x u_{xx} + 2b_0 u_{xy} + 3u_{yy} = 0. \tag{5.57b}$$

The meaning of b_0 for discrete solitons is discussed in Hietarinta and Zhang (2011).

Next, for equation (3.72) we must similarly reparametrize $(p^3, q^3) \mapsto (p^3 + d_2 p, q^3 + d_2 q)$, and then take

$$v_{n+\nu,m+\mu} = e^{v(x+2(\nu/p+\mu/q),\,y+2(\nu/p^2+\mu/q^2))}/(p^{n+\nu}q^{m+\mu}), \tag{5.58a}$$

set $p = p'/\epsilon$, $q = q'/\epsilon$ and take the $\epsilon \to 0$ limit. This results with the modified BSQ equation

$$v_{xxxx} - 6v_x^2 v_{xx} + 12 v_{xx} v_y + 4 d_2 v_{xx} + 12\, v_{yy} = 0. \tag{5.58b}$$

Finally, the double continuum limit of (3.73) is obtained using

$$z_{n+\nu,m+\mu} = z(x + 2(\nu/p + \mu/q), y + 2(\nu/p^2 + \mu/q^2)) \tag{5.59a}$$

with $p, q \to \infty$ while keeping p/q fixed. The result is

$$3\partial_y \left(\frac{z_y}{z_x}\right) + \partial_x \left(\frac{z_{xxx}}{z_x} + \frac{3}{2}\frac{z_y^2 - z_{xx}^2}{z_x^2} - \frac{1}{2}\frac{d_1 + d_2 z}{z_x}\right) = 0. \tag{5.59b}$$

This is a generalization of the usual Schwarzian Boussinesq equation.

5.8 Notes

The technique used here for obtaining continuum limits, namely the analysis of plane-wave factor solutions of the linearized equation, was introduced in Nijhoff et al. (1983a); Quispel et al. (1984). These techniques can be used to find semi-discrete limits to all equations in the ABS list. However it is best to use a parameterization suitable for this purpose (see e.g. Section 9.5 and Atkinson et al., 2007). Since the method relies on taking limits of plane-wave factors, without altering the way in which these plane-waves enter in the solutions (this relationship is discussed further in Chapters 8 and 9), the limiting equations will have the same structure of solutions and are therefore integrable.

In Wiersma and Capel (1987a) hierarchies of semi-discrete equations were derived by systematic expansions. Such hierarchies can also be interpreted as generalized symmetries of the PΔE, since the corresponding higher continuous flows commute not only among themselves but also with original difference equations.

The subject of continuous symmetries is therefore related to the subject of semi-continuous limits. For PΔEs, this is a broad subject (Levi and Winternitz, 2006). The investigation of such symmetries of integrable PΔEs has received a great deal of recent interest. In particular in Mikhailov et al. (2011b) some of the results of Wiersma and Capel (1987a) were recovered from a more general perspective.

Continuum limits of the BSQ class of difference equations were studied in Nijhoff (1997) and in Walker (2001).

When one starts from a higher-dimensional equation, such as lattice KP, there is a larger number of intermediate steps in taking continuum limits: after the first step we have one continuous and two discrete variables, then two continuous and one discrete, and in the full limit three continuous ones. At each step, one has the possibility of straight or skew limits, in various directions, and the limits provide different types of integrable semi-discrete equations. For the lattice KP these limits were analyzed in Wiersma and Capel (1987b,

1988a,b). These results contain many interesting equations, including the two-dimensional Toda equation, which depends on two continuous and one discrete variable.

Exercises

5.1 Linearize the lattice modified KdV equation (3.11) around the solution $v_{n,m} = 1$, and the Schwarzian KdV equation (3.13) around the solution $z_{n,m} = z_{0,0}+(n/p)+(m/q)$ How do the linearized equations compare to (5.4)?

5.2 Perform similar continuum limits on the semi-discrete equations (5.27) and (5.28) to recover the fully corresponding continuous equations (2.73) and (2.103).

5.3 Consider the following cross-ratio of four variables enumerated by a lattice variable n

$$\mathrm{CR} = \frac{(z_{n+2} - z_n)(z_{n+1} - z_{n-1})}{(z_{n+2} - z_{n+1})(z_n - z_{n-1})} \tag{5.60}$$

and suppose that z_n depends on the discrete variable n by increments of a lattice parameter ϵ, such that $z_n = Z(x) = Z(x_0 + n\epsilon)$. By Taylor expansions in the increments by ϵ show that the cross-ratio (5.60) behaves as:

$$\mathrm{CR} = 4 - 2\epsilon^2\{Z, x\} + \mathcal{O}(\epsilon^4)\,,$$

in which (as before) the bracket $\{Z, x\}$ denotes the Schwarzian derivative of Z given in (2.101). Thus, the cross-ratio (5.60) can be considered to be a discrete version of the Schwarzian derivative.

5.4 Consider the semi-discrete limit of the lattice potential KdV equation (1.38):

$$\partial_\tau u_n = 1 - \frac{2p}{2p + u_{n-1} - u_{n+1}}\,. \tag{5.61}$$

i) Set $a_n = (2p + u_{n-1} - u_{n+1})^{-1}$, and derive from (5.61) the following equation:

$$\partial_\tau a_n = 2pa_n^2(a_{n-1} - a_{n+1})\,. \tag{5.62}$$

This equation is called the *modified Volterra equation*.

ii) Set $b_n = a_n a_{n-1}$, with a_n obeying the equation (5.62). Show that b_n obeys the *Volterra* equation

$$\partial_\tau b_n = 2pb_n(b_{n-1} - b_{n+1})\,. \tag{5.63}$$

This equation is sometimes also referred to as *Kac–van Moerbeke equation* (Kac and van Moerbeke, 1975).

iii) By taking a τ-derivative of equation (5.63), show that the variable $\psi_n \equiv \ln(b_{n+1}b_n)$ obeys the equation:

$$\partial_\tau^2 \psi_n = 4p^2\left(e^{\psi_{n+2}} - 2e^{\psi_n} + e^{\psi_{n-2}}\right)\,, \tag{5.64}$$

which we could call the *modified Toda equation*. **iv)** Show that by setting $\psi_n = \varphi_n - \varphi_{n-2}$, we can derive equation (5.64) from the Toda equation for φ_n which reads (cf. (1.39)):

$$\partial_\tau^2 \varphi_n = 4p^2 \left(e^{\varphi_{n+2}-\varphi_n} - e^{\varphi_n - \varphi_{n-2}} \right) . \tag{5.65}$$

NB: Note that both equations (5.64) and (5.65) effectively "live" on a sublattice of the original lattice labeled by the discrete variable n, namely a sublattice of either even or odd sites.

5.5 Consider the partial difference equation Q_1, i.e.

$$p(u-\widehat{u})(\widetilde{u}-\widehat{\widetilde{u}}) - q(u-\widetilde{u})(\widehat{u}-\widehat{\widetilde{u}}) + \gamma pq(p-q) = 0 \tag{5.66}$$

where p and q are lattice parameters associated with the lattice variables n and m, respectively, and γ is a fixed parameter.

i) Show that the equation (5.66) admits the solution $u = u^0$, where

$$u^0 = u^0_{n,m} = u_{0,0} + nP + mQ \quad \text{where} \quad P^2 = \gamma p^2 + \delta p, \quad Q^2 = \gamma q^2 + \delta q, \tag{5.67}$$

in which δ is an arbitrary constant, $u_{0,0}$ denoting an initial value.

ii) By expanding up to first order in a small parameter ε, derive a linearized equation (i.e. a linear approximation) to equation (5.66) by setting $u = u^0 + \varepsilon\rho$, where u^0 is the solution (5.67), and show that the function $\rho = \rho_{n,m}$ obeys the equation

$$(pQ - qP)(\widehat{\widetilde{\rho}} - \rho) = (pQ + qP)(\widetilde{\rho} - \widehat{\rho}) , \tag{5.68a}$$

with P and Q as given in (5.67).

iii) Assuming that ρ, in addition to equation (5.68a), solves two more linear equations, namely

$$(pK - kP)(\overline{\widetilde{\rho}} - \rho) = (pK + kP)(\widetilde{\rho} - \overline{\rho}) , \tag{5.68b}$$
$$(qK - kQ)(\overline{\widehat{\rho}} - \rho) = (qK + kQ)(\widehat{\rho} - \overline{\rho}) , \tag{5.68c}$$

in terms of a third shift $\overline{\widetilde{\rho}}$ and $\overline{\widehat{\rho}}$ associated with a lattice parameter k, where $K^2 = \gamma k^2 + \delta k$. Applying the shift $\rho \to \overline{\rho}$ to equation (5.68a) and expressing $\widehat{\widetilde{\rho}}$, $\overline{\widetilde{\rho}}$ and $\overline{\widehat{\rho}}$ in terms of $\widetilde{\rho}$, $\widehat{\rho}$ and $\overline{\rho}$, show that the resulting expression for $\overline{\widehat{\widetilde{\rho}}}$ is invariant under any interchange of the parameters p, q and k, and conclude from this that equation (5.68a) is multidimensionally consistent.

iv) Find a nontrivial solution to the equation (5.68a) by considering equations (5.68b) and (5.68c), setting $\overline{\rho} \equiv 0$ for all $n, m \in \mathbb{Z}^2$, and finding a simultaneous solution for both.

5.6 Consider the lattice equation H3, i.e.

$$p(u\widetilde{u} + \widehat{u}\widehat{\widetilde{u}}) - q(u\widehat{u} + \widetilde{u}\widehat{\widetilde{u}}) = \gamma(p^2 - q^2), \tag{5.69}$$

where γ is a fixed parameter, and where p, q are lattice parameters. Equation (5.69) is known to be multidimensionally consistent.
i) If $\gamma = 0$, show that (5.69) allows an elementary solution of the form $u = u_{n,m} = \tau^n \sigma^m$, where τ and σ must be solutions of

$$\frac{p}{\tau + \tau^{-1}} = \frac{q}{\sigma + \sigma^{-1}} = \frac{1}{2}\theta = \text{constant}\ ,$$

where the constant θ is the same for all lattice directions. Show that we can parameterize the corresponding τ and σ as follows:

$$\tau = e^{\alpha}, \quad \sigma = e^{\beta} \quad \text{where} \quad \cosh(\alpha) = \frac{p}{\theta}, \quad \cosh(\beta) = \frac{q}{\theta}\ .$$

ii) Show that if $\gamma \neq 0$ the lattice equation (5.69) has a solution of the form:

$$u = u^{(0)}_{n,m} = \sqrt{\gamma\theta}\,\cosh(\xi_0 + n\alpha + m\beta), \tag{5.70}$$

where ξ_0 is constant with respect to n,m.
Hint: Note that the solution for $\gamma = 0$ allows for the interchange $\tau \leftrightarrow \tau^{-1}$, $\sigma \leftrightarrow \sigma^{-1}$, and consequently show that (5.69) allows for a solution of the form $u = c_+\tau^n\sigma^m + c_-\tau^{-n}\sigma^{-m}$ with coefficients $c_\pm$ obeying $c_+c_- = \gamma\theta/4$.
iii) Linearize (5.69) around the solution $u^{(0)}_{n,m}$, i.e. set $u_{n,m} = u^{(0)}_{n,m} + \epsilon\rho_{n,m}$ and inserting this into the equation obtain the equation for ρ from the first-order terms in the expansion in powers of ϵ:

$$p\left(\widetilde{u}^{(0)}\rho + u^{(0)}\widetilde{\rho} + \widehat{u}^{(0)}\widehat{\widetilde{\rho}} + \widehat{\widetilde{u}}^{(0)}\widehat{\rho}\right) = q\left(\widehat{u}^{(0)}\rho + u^{(0)}\widehat{\rho} + \widetilde{u}^{(0)}\widehat{\widetilde{\rho}} + \widehat{\widetilde{u}}^{(0)}\widetilde{\rho}\right).$$

Based on the multidimensional consistency, considering the "background solution" obtained by setting $\overline{\rho} = 0$ for the shift of the solution ρ in a direction associated with lattice parameter $k = \theta\cosh(\kappa)$, show that a solution of the linear equation takes the form

$$\rho_{n,m} = \left(\frac{\sinh(\alpha-\kappa)}{\sinh(\alpha+\kappa)}\right)^n \left(\frac{\sinh(\beta-\kappa)}{\sinh(\beta+\kappa)}\right)^m \sinh(\xi_0 + \kappa + n\alpha + m\beta). \tag{5.71}$$

iv) Find the straight and continuum limits for the lattice equation (5.69), and compare them to those of the lattice potential mKdV equation.

5.7 Consider here continuum limits of the lattice sine-Gordon equation

$$\begin{aligned}\sin(\theta_{n,m} + \theta_{n+1,m} + \theta_{n,m+1} + \theta_{n+1,m+1}) &= \\ = \frac{p}{q}\sin(\theta_{n,m} - \theta_{n+1,m} - \theta_{n,m+1} + \theta_{n+1,m+1}).\end{aligned} \tag{5.72}$$

i) Approximate equation (5.72) by considering θ to be small, and derive, thus, the linearized equation:

$$(p-q)(\theta_{n,m}+\theta_{n+1,m+1})=(p+q)(\theta_{n+1,m}+\theta_{n,m+1})\,. \tag{5.73}$$

Show that this has a solution of the form:

$$\theta_{n,m}=\left(\frac{p+k}{p-k}\right)^n\left(\frac{k+q}{k-q}\right)^m\theta_{0,0}, \tag{5.74}$$

where k is an arbitrary parameter.
ii) For the solution (5.74) of the linearized equation consider separately the two limits:
a) $n\to\infty$, $p\to\infty$, such that $n/p\to x$,
b) $m\to\infty$, $q\to 0$, such that $mq\to t$.
Consider the limit **a)** applied to the nonlinear equation (5.72) and derive the following differential-difference equation

$$(\Theta_{m+1}-\Theta_m)_x=q\sin 2\,(\Theta_{m+1}+\Theta_m)\,. \tag{5.75}$$

iii) Consider the *full* continuum limit of the lattice sine-Gordon equation by applying the limit **b)** of **ii)** to equation (5.75). Thus, setting:

$$\Theta_m(x)=\vartheta(x,t)\quad\text{with}\quad t=t_0+mq$$

and performing a Taylor expansion in powers of q, show that this leads to the sine-Gordon equation in the form

$$\vartheta_{xt}=\sin(4\vartheta).$$

6

One-dimensional lattices and maps

In this chapter we discuss difference equations with one (discrete) independent variable. Their continuous analogues describe, for example, the motion of point particles. In that context the Hamiltonian formulation and the Liouville–Arnold integrability play a central role. In the discrete-time case the situation is more complicated, because we do not have the analogue of the chain rule of time derivative, and furthermore the motion is given by finite time steps rather than by a continuous time flow.

The difference between continuous and discrete systems is underscored by the observation that whereas a one-degree-of-freedom (autonomous continuous time) Hamiltonian system always has one conserved quantity, namely the Hamiltonian (= energy) itself, it turns out that *generically* one-degree-of-freedom discrete-time systems do not posses a conserved quantity. Surprisingly, there does exist a rather large family of (autonomous) one-degree-of-freedom maps called Quispel–Roberts–Thompson (QRT) maps, which possess a conserved quantity or *invariant*. They are examples of what we will call integrable maps. There are also extensions to more than one-degree-of-freedom maps (variously called higher-dimensional or multicomponent maps) and to nonautonomous maps (i.e. maps depending explicitly to the discrete independent variable, usually denoted by n). A classification of integrable nonautonomous maps corresponding to Painlevé equations is discussed in Chapter 11.

Integrable systems are rare and it is difficult to find such systems. One efficient way of deriving integrable maps is to start with a 2D integrable lattice equation (such as the ones discussed in Chapter 3) and apply a dimensional reduction to it in order to derive a 1D lattice equation (i.e. an equation in one discrete independent variable), which we interpret as a dynamical map. We will show that this can be done so that the integrability properties of the original 2D lattice, such as the existence of a Lax pair, carry over to the map, and can be used to derive their invariants.

6.1 Integrability of maps

There are several properties associated with integrability of maps, for example, the existence of a sufficient number of conserved quantities, symmetries, Lax pair and the behavior around singularities. In this section we will discuss the notion of Liouville integrability and the structures associated with it.

6.1.1 Hamiltonian mechanics and Liouville integrability

Let us first recall the definition of Hamiltonian evolution in the continuous case. In this setting we work with a $2N$-dimensional phase space on which we choose as coordinates N position variables q_i and their N conjugate momenta p_i. In this phase space we can introduce a **Poisson bracket**, $\{\cdot,\cdot\}$, which operates on two functions of the phase space variables and produces a new function, and which has the following properties:

1 Linearity: $\{F, aG + bH\} = a\{F, G\} + b\{F, H\}$, where a, b do not depend on the coordinates of the phase-space;
2 Antisymmetry: $\{F, G\} = -\{G, F\}$;
3 Derivative-like product rule: $\{F, GH\} = \{F, G\}H + G\{F, H\}$;
4 Jacobi identity: $\{F, \{G, H\}\} + \{H, \{F, G\}\} + \{G, \{H, F\}\} = 0$.

When we use the canonically conjugate variables the Poisson bracket has the simple form:

$$\{F, G\}_{q,p} = \sum_{j=1}^{N} \left(\frac{\partial F}{\partial q_j}\frac{\partial G}{\partial p_j} - \frac{\partial F}{\partial p_j}\frac{\partial G}{\partial q_j} \right), \tag{6.1}$$

in particular the elementary brackets are then

$$\{q_i, q_j\} = 0, \quad \{p_i, p_j\} = 0, \quad \{q_i, p_j\} = \delta_{i,j}. \tag{6.2}$$

Corresponding to this we have a **symplectic structure** given by the 2-form[1]

$$\omega(q, p) = \sum_{j=1}^{N} dp_j \wedge dq_j.$$

Now given a Hamiltonian function $H(q, p)$ the Hamiltonian evolution is given by

$$\dot{q}_j = \{q_j, H(q, p)\} = \frac{\partial H(q, p)}{\partial p_j}, \quad \dot{p}_j = \{p_j, H(q, p)\} = -\frac{\partial H(q, p)}{\partial q_j}. \tag{6.3}$$

From this it immediately follows that $dH/dt = 0$; that is, the Hamiltonian itself is a conserved quantity.

Following Arnold (1997) Chapter 10, a system with N degrees of freedom (i.e. with $2N$-dimensional phase space) is called **Liouville integrable** if there are N functions $I_k(p, q)$ (H one of them) such that the I_k

[1] A $2N$-dimensional manifold is said to be symplectic if it admits a globally defined 2-form, which by definition needs to be closed under d-differentiation, i.e. $d\omega = 0$, and nondegenerate, cf. Arnold (1997).

1 are functionally independent,
2 are in involution, i.e. $\{I_n, I_m\} = 0,\ \forall n, m$,
3 are globally defined regular functions.

Liouville's theorem (Arnold, 1997) states that a system with these properties can be "integrated by quadratures" (i.e. given in terms of integrals). Since the I_k are conserved, the motion stays on the intersection of the level sets $I_k(p, q) = \text{const}$, $k = 1, \ldots, N$. Furthermore, *if* the motion is restricted into a compact region of the phase space, then the phase space is "foliated" into tori. Veselov adapted the Liouville–Arnold theorem to discrete maps in Veselov (1991a).

Note also that autonomous Hamiltonian 1D systems could be considered trivially integrable in Liouville sense, since the Hamiltonian is the single required conserved quantity.

6.1.2 Generating functions and canonical coordinates

In the discrete case the analogue of Hamiltonian evolution is taken by a **symplectic map**, that is, a map of the form

$$p_j(n+1) = f_j(q(n), p(n)), \quad q_j(n+1) = g_j(q(n), p(n)), \tag{6.4}$$

which also preserves the symplectic form or equivalently the Poisson bracket:

$$\{p_j(n+1), p_i(n+1)\}_{q(n),p(n)} = 0, \{q_j(n+1), q_i(n+1)\}_{q(n),p(n)} = 0, \tag{6.5}$$

$$\{q_j(n+1), p_i(n+1)\}_{q(n),p(n)} = \delta_{j,i}. \tag{6.6}$$

Here the variable n serves as the discrete time.

In this context a function $I_j = I_j(q(n), p(n))$ is a **conserved quantity** if

$$I_j(q(n+1), p(n+1)) = I_j(q(n), p(n)),$$

and they are in involution if $\{I_j(q(n), p(n)), I_k(q(n), p(n))\}_{q(n),p(n)} = 0$.

The connection of (6.4) to (6.3) can be seen in the continuum limit. Thus if the map (6.4) is near identity, and hence we can expand $f_j(q(n), p(n)) = p_j(n) + h\phi_j(q, p) + O(h^2)$, $g_j(q(n), p(n)) = q_j(n) + h\psi_j(q, p) + O(h^2)$, then in the limit $h \to 0$ we get from (6.6)

$$\frac{\partial \phi_j(q, p)}{\partial p_i} = -\frac{\partial \psi_i(q, p)}{\partial q_j} \quad \Longrightarrow \quad \phi_j = \frac{\partial H(q, p)}{\partial q_j}, \quad \psi_j = -\frac{\partial H(q, p)}{\partial p_j}$$

for some function $H(q, p)$ and with these the continuum limit of (6.4) will be (6.3).

The symplectic map (6.4) can be interpreted as a **canonical transformation** from variables $p(n), q(n)$ to $p(n+1), q(n+1)$, and as such it can be given using generating functions. Thus the map (6.4) can also be given as

$$q_j(n+1) - q_j(n) = \frac{\partial G(q(n), p(n+1))}{\partial p_j(n+1)}, \tag{6.7a}$$

$$p_j(n+1) - p_j(n) = -\frac{\partial G(q(n), p(n+1))}{\partial q_j(n)}, \tag{6.7b}$$

where the generating function G depends on the "old" $q_j(n)$ and the "new" $p_j(n+1)$. If the map is given in the form (6.4), where f, g preserve the Poisson structure, we can find the corresponding generating function by comparing (6.7) and (6.4) and integrating the resulting system of PDEs. (See Exercise 6.1.) Conversely, we could start by giving G, and then the form of (6.7) guarantees that we have a symplectic map.

The above formulation is particularly useful in demonstrating (in the discrete case) the validity of Liouville's theorem that an integrable system can actually be integrated. We will give a brief outline of the procedure (Bruschi et al., 1991). Thus let us assume that the map (6.4) has N conserved quantities $I_j(q, p)$ in involution. Let us make another canonical transformation to new variables defined by $P_j := I_j(q, p)$. Since the I_j are in involution we have $\{P_i, P_j\} = 0$. In order to find the Q_i that have canonical commutation rules with P_j, we construct a generating function $S(P, q)$ for this transformation; it is defined by

$$p_j = \frac{\partial S(P, q)}{\partial q_j}, \qquad Q_j = \frac{\partial S(P, q)}{\partial P_j}. \tag{6.8a}$$

From the first set of equations we can integrate

$$S(P, q) = \sum_{k=1}^{N} \int_{q_k(0)}^{q_k(n)} p_k(P, q')dq', \tag{6.8b}$$

and then we get

$$Q_i(n) = \sum_{k=1}^{N} \int_{q_k(0)}^{q_k(n)} \frac{p_k(P, q')}{\partial P_j} dq'. \tag{6.8c}$$

If we now write equations (6.7) for the new system Q, P, with corresponding generating function H, we find from (6.7b) that H is a function of P only. Then from (6.7a) one finds that the Q_j are linear in n, and thus the system has been integrated. The situation is described with the commutative diagram of canonical transformations

$$\begin{array}{ccc}
(q(n), p(n)) & \xrightarrow{G(q(n), p(n+1))} & (q(n+1), p(n+1)) \\
{\scriptstyle S(q(n), P)}\big\downarrow & & \big\downarrow{\scriptstyle S(q(n+1), P)} \\
(Q(n), P) & \xrightarrow{H(P)} & (Q(n+1), P)
\end{array}$$

where we have denoted $P = P(n) = P(n+1)$. In the diagram the top horizontal arrow is the map (6.7), the vertical arrows are canonical transformations (6.8) to the *action angle variables* P, Q. For the bottom horizontal arrow we note that since $0 = P(n+1) - P(n) = -\partial H/\partial Q$ we have $H = H(P)$. Since P is a constant we have also that $Q(n+1) - Q(n) = \partial H/\partial P$ is a constant and therefore that the map is linear in n:

$$Q_j(n) = \nu_j n + Q_j(0), \quad \nu_j = \sum_{k=1}^{N} \int_{q_k(n)}^{q_k(n)} \frac{p_k(P, q')}{\partial P_j} dq'. \tag{6.9}$$

Thus in this picture a symplectic map appears as a sequence of canonical transformations; in fact, in this context, compositions of canonical transformations play a role not only in the iteration of the map but also in the transformation to action angle variables. This is unlike the conventional use of canonical transformations (see e.g. Goldstein, 1950) where they are only applied once in order to bring the system into a convenient form.

6.1.3 Measure preservation

Some aspects of integrability, e.g. the existence of conserved quantities, still hold for a slightly wider class of maps than the symplectic maps: for example, measure preserving maps.

Measure preservation is a slightly weaker property than symplecticity. If we have a map of the plane $\Phi : (p, q) \to (p', q') = (f(q, p), g(q, p))$ where f, g are differentiable, then we can associate with it the Jacobian matrix and its determinant:

$$J(\Phi) := \begin{vmatrix} \frac{\partial f}{\partial p} & \frac{\partial f}{\partial q} \\ \frac{\partial g}{\partial p} & \frac{\partial g}{\partial q} \end{vmatrix}. \tag{6.10}$$

If $J = 1$ the map is called **conservative**, if $J < 1$ **dissipative** and if $J > 1$ **expansive**. If the map is symplectic then we clearly have $J = 1$ by (6.6) and the map is also area preserving since the area form $\omega = dq \wedge dp$ is preserved. If J can be written as a ratio of expressions in old and new variables,

$$J = \frac{m(p, q)}{m(p', q')}, \tag{6.11}$$

then the map Φ is called **measure preserving**. It means that the measure is preserved in the sense that

$$\int_D m(p, q) dp dq = \int_{D'} m(p', q') dp' dq', \tag{6.12}$$

for any (small) region; here D' is the region D as transformed by Φ.

Measure conservation is necessary but not sufficient for integrability. For example, any map of the form $p' = q$, $q' = p + f(q)$ has Jacobian equal to 1 but only very few of them are integrable.

6.1.4 Lagrangian formulation

An alternative to the Hamiltonian formulation is given by the Lagrangian approach. We discuss this here in the simplest setting, namely that of mappings, with function depending on one independent discrete variable. A more extensive discussion of Lagrangian structures is given in Chapter 12. The approach is based on the **least-action principle** or Hamilton's principle (Goldstein, 1950; Gelfand and Fomin, 2000). This is formulated in terms of an object called the **action**, which physically represents the work done by the system during the motion (i.e. time evolution) of the system.

The action is given by

$$S := \sum_{k\in\mathbb{Z}} \mathcal{L}(q(k), q(k+1), \ldots, q(k+K)), \tag{6.13}$$

where $\mathcal{L}$ is called the **Lagrangian**, which is a function of K consecutive coordinates. In the context of the least-action principle the $q(k)$ are not considered as a sequence of given numbers but as a yet unknown function of k, where the points $\ldots, q(k), q(k+1), \ldots$ describe the trajectory of the system in discrete time. In this point of view S should be considered as a functional, i.e. a function of functions which is often denoted by $S[q]$.

The least-action principle states that the trajectory of the system is the one for which the action is minimized. In the same way as one would minimize a continuous function, we need to find the critical or stationary points of the functional $S[q]$. Analogous to finding the critical point of a function $f(x)$, for which one needs to solve $f'(x) = 0$, finding the critical trajectories of a functional requires solving the set of equation $\delta S/\delta q(n) = 0$, where the latter involves the **functional derivative** defined by

$$\lim_{\epsilon\to 0} \frac{1}{\epsilon}(S[q+\epsilon q'] - S[q]) = \sum_{k\in\mathbb{Z}} \frac{\delta S}{\delta q(k)} q'(k), \tag{6.14}$$

where q' is an arbitrary function of k. Now choosing $q'(k) = \delta_{n,k}$ allows us to single out $\delta S/\delta q(n)$, while the left-hand side can be evaluated using (6.13) and expanding each Lagrangian in the sum with shifted variables in Taylor series up to first order. This yields

$$\frac{\delta S}{\delta q(n)} = \frac{\partial}{\partial q_n} \sum_{s=0}^{K} \mathcal{L}(q_{n-s}, q_{n-s+1}, \ldots, q_{n-s+K}), \forall n \in \mathbb{Z} \tag{6.15}$$

where for notational convenience we have reverted to the subscript notation: $q(n) \equiv q_n$. Now setting the right-hand side of (6.15) to zero we get the **Euler–Lagrange (EL) equations**

$$\partial_{q_n} \sum_{s=0}^{K} \mathcal{L}(q_{n-s}, q_{n-s+1}, \ldots, q_{n-s+K}) = 0, \quad \forall n \in \mathbb{Z}. \tag{6.16}$$

For example, if the Lagrangian only depends on two consecutive points of the trajectory; that is, $\mathcal{L} = \mathcal{L}(q_k, q_{k+1})$, we have $S := \sum_{k\in\mathbb{Z}} \mathcal{L}(q_k, q_{k+1})$ and we get the EL equation

$$\partial_{q_n}\left[\mathcal{L}(q_n, q_{n+1}) + \mathcal{L}(q_{n-1}, q_n)\right] = 0. \tag{6.17}$$

Usually the resulting equation gives a correspondence (a multivalued map) because the left-hand side can be a nonlinear function of q_{n+1}.

Let us now discuss under which conditions the equations coming from a Lagrangian can be presented in the form (6.7). For this purpose let us introduce a function G of two arguments q_n, p_{n+1} defined by

$$G(q_n, p_{n+1}) := (q_{n+1} - q_n)p_{n+1} - L(q_n, q_{n+1}). \tag{6.18}$$

Since here the left-hand side does not depend on q_{n+1} whereas the right-hand side does, the above is not just a definition but also a relation between the three variables. In order to analyze the precise relation between the variables we apply the technique of implicit differentiation on (6.18):

$$\begin{aligned}\frac{\partial G}{\partial q_n}\delta q_n + \frac{\partial G}{\partial p_{n+1}}\delta p_{n+1} = {}& (\delta q_{n+1} - \delta q_n)p_{n+1} + (q_{n+1} - q_n)\delta p_{n+1}\\ & - \frac{\partial L}{\partial q_n}\delta q_n - \frac{\partial L}{\partial q_{n+1}}\delta q_{n+1}.\end{aligned}$$

Now, comparing the coefficients of δq_{n+1}, δp_{n+1}, δq_n, respectively, we get the relations

$$\frac{\partial L(q_n, q_{n+1})}{\partial q_{n+1}} = p_{n+1}, \tag{6.19a}$$

$$\frac{\partial G(q_n, p_{n+1})}{\partial p_{n+1}} = q_{n+1} - q_n, \tag{6.19b}$$

$$\begin{aligned}\frac{\partial G(q_n, p_{n+1})}{\partial q_n} &= -p_{n+1} - \frac{\partial L(q_n, q_{n+1})}{\partial q_n}\\ &= -p_{n+1} + \frac{\partial L(q_{n-1}, q_n)}{\partial q_n}\\ &= -p_{n+1} + p_n.\end{aligned} \tag{6.19c}$$

Equation (6.19a) establishes a relation between p_{n+1}, q_n, q_{n+1}, from which we assume that we can solve for q_{n+1} in terms of q_n, p_{n+1}. Substituting this q_{n+1} into the right-hand side of (6.18) will give us G as a function of q_n, p_{n+1}. Equation (6.19b) coincides with (6.7a) and (6.19c) coincides with (6.7b). In deriving the last relation we have used the EL equation (6.17), as well as (6.19a) after the shift $n \to n-1$.

The transformation defined by (6.18), (6.19) is the discrete analogue of the **Legendre transform** of Hamiltonian mechanics (Goldstein, 1950). This method only works if we can solve for q_{n+1} from (6.19a). This can be done locally, if the Hessian

$$\frac{\partial^2 \mathcal{L}(q_n, q_{n+1})}{\partial q_n \partial q_{n+1}} \neq 0,$$

in which case there is the associated symplectic structure given by

$$\omega = \frac{\partial^2 \mathcal{L}(q_n, q_{n+1})}{\partial q_n \partial q_{n+1}} dq_n \wedge dq_{n+1}.$$

It can be easily shown that this is indeed a symplectic form and invariant under the map in question.

In particular, the *standard map*, which is a three-point map of the form

$$q_{n+1} - 2q_n + q_{n-1} = -V'(q_n), \tag{6.20}$$

is obtained this way from the Lagrangian

$$L := \tfrac{1}{2}(q_{n+1} - q_n)^2 - V(q_n). \tag{6.21}$$

This is the discrete analogue of Newtonian mechanics and the function G will then be

$$G(q_n, p_{n+1}) = \tfrac{1}{2} p_{n+1}^2 + V(q_n). \tag{6.22}$$

In particular, (6.19b) then implies $p_{n+1} = q_{n+1} - q_n$, which is the discrete analogue of the Newtonian relationship between momentum and velocity.

6.1.5 Nonpolynomial maps of the plane

It seems that polynomial standard maps are not integrable, except for the linear case (Veselov, 1991b), and it should therefore be necessary to go beyond polynomial maps to find examples. It is only in recent years that systematic searches for such systems have been undertaken.

Let us next assume that we have a reversible map

$$x_{n+1} = R(x_n, x_{n-1}), \tag{6.23}$$

where R is rational-linear in its last variable (so that we can solve for x_{n-1} which is needed for reversibility). A prospective constant of motion is then a sufficiently regular function of two variables $I(x, y)$ with the property

$$I(x_{n+1}, x_n) = I(x_n, x_{n-1}), \tag{6.24}$$

subject to (6.23); in other words

$$I(R(x, y), x) - I(x, y) = 0, \ \forall x, y.$$

The systematic search for discrete maps of the plane possessing a first integral is, curiously enough, something that has been studied only in recent years. The first integrable map of the plane to itself (case (a) below) was probably found by McMillan (1971); he searched for area-preserving rational maps of the type

$$x_{n+1} + x_{n-1} = f(x_n) \tag{6.25}$$

carrying an invariant and found the case (a) below. McMillan's results were expanded in Suris (1989) leading to the following list, for which the standard map $x_{n+1} - 2x_n + x_{n-1} = f(x_n)$ is used:

(a) Rational maps:

$$f(x) = \epsilon \frac{A + Bx + Cx^2 + Dx^3}{1 - \epsilon(E + \frac{1}{3}Cx + \frac{1}{2}Dx^2)};$$

$$I(x, y) = \tfrac{1}{2}(x - y)^2 + \epsilon \Big[-\tfrac{1}{2}A(x + y) - \tfrac{1}{2}Bxy - \tfrac{1}{6}Cxy(x + y) \\ -\tfrac{1}{4}Dx^2y^2 - \tfrac{1}{2}E(x - y)^2 \Big].$$

(b) Trigonometric maps:

$$f(x) = \frac{2}{\omega} \arctan \left[\frac{\frac{\omega\epsilon}{2}(A \sin \omega x + B \cos \omega x + C \sin 2\omega x + D \cos 2\omega x)}{1 - \frac{\omega\epsilon}{2}(A \cos \omega x - B \sin \omega x + C \cos 2\omega x - D \sin 2\omega x + E)} \right];$$

$$I(x, y) = \frac{1}{\omega^2}(1 - \cos \omega(x - y)) + \frac{\epsilon}{2\omega} \big[A(\cos \omega x + \cos \omega y) - B(\sin \omega x + \sin \omega y) \\ C \cos \omega(x + y) - D \sin \omega(x + y) + E \cos \omega(x - y) \big].$$

(c) Hyperbolic (exponential) maps:

$$f(x) = \frac{1}{\alpha} \ln \left[\frac{1 + \alpha\epsilon(Be^{-\alpha x} + De^{-2\alpha x} - E)}{1 - \alpha\epsilon(Ae^{\alpha x} + Ce^{2\alpha x} + E)} \right];$$

$$I(x, y) = \frac{1}{\alpha^2} \cosh(\alpha(x - y)) + \frac{\epsilon}{2\alpha} \Big[-A\left(e^{\alpha x} + e^{\alpha y}\right) + B\left(e^{-\alpha x} + e^{-\alpha y}\right) \\ -Ce^{\alpha(x+y)} + De^{-\alpha(x+y)} - 2E \cosh \alpha(x - y) \Big].$$

In the above, A, B, C, D, E, as well as ω and α, are constants. Note that the cases (b) and (c) can also be written as rational maps, but in multiplicative form $x_{n-1}x_{n+1} = g(x_n)$. All these cases are included in the QRT family of maps, which we will discuss in Section 6.3.

6.2 The Kahan–Hirota–Kimura discretization

Evidence suggests that integrable discretization is in general difficult to accomplish. However, there is one method attributed to Kahan (1993) and discovered independently by Hirota and Kimura (see Hirota and Kimura, 2000 and Kimura and Hirota, 2000), that seems to be surprisingly successful.

Given a set of equations that are linear in derivatives and at most quadratic in the other terms, the discretization rule is given as follows:

1 The derivative is discretized by a forward difference

$$\frac{dx_j}{dt} \mapsto [x_j(n+1) - x_j(n)]/h$$

where h is the lattice parameter;

2 a first-order monomial is discretized by taking average:

$$x_j \mapsto \tfrac{1}{2}[x_j(n+1) + x_j(n)];$$

3 a quadratic term is discretized by

$$x_j x_k \mapsto \tfrac{1}{2}[x_j(n+1)x_k(n) + x_j(n)x_k(n+1)].$$

After applying these rules one solves for the $x_j(n+1)$ and obtains a rational map, which often is integrable.

As the first example, recall the discretization of the Verhulst equation $\dot{x} = ax(1-x)$ using linearization (Exercise 1.1) with the result

$$x(n+1) - x(n) = ah\left[\tfrac{1}{2}(x(n+1) + x(n)) - x(n+1)x(n)\right]. \tag{6.26}$$

This is exactly what we get from the above prescription, and it yields the Riccati-type map

$$x(n+1) = \frac{(ah+2)x(n)}{2 - ah + 2ahx(n)}. \tag{6.27}$$

As a more complicated and interesting example, consider the equations of the Euler top, given by

$$\begin{cases} \dot{x}_1 = \alpha_1\, x_2\, x_3, \\ \dot{x}_2 = \alpha_2\, x_3\, x_1, \\ \dot{x}_3 = \alpha_3\, x_1\, x_2. \end{cases} \tag{6.28}$$

This was discretized by Hirota and Kimura (2000) (see also Petrera et al., 2011), using the above recipe, with the following elegant final result

$$x(n+1) = A(x(n), \epsilon)^{-1} x(n), \quad A(x, \epsilon) = \begin{pmatrix} 1 & -\epsilon\,\alpha_1\,x_3 & -\epsilon\,\alpha_1\,x_2 \\ -\epsilon\,\alpha_2\,x_3 & 1 & -\epsilon\,\alpha_2\,x_1 \\ -\epsilon\,\alpha_3\,x_2 & -\epsilon\,\alpha_3\,x_1 & 1 \end{pmatrix} \tag{6.29}$$

in which $x = (x_1, x_2, x_3)^T$, with the two independent constants of motion

$$I_1 = \frac{1 - \epsilon^2 \alpha_2 \alpha_3\, x_1^2}{1 - \epsilon^2 \alpha_3 \alpha_1\, x_2^2}, \qquad I_2 = \frac{1 - \epsilon^2 \alpha_3 \alpha_1\, x_2^2}{1 - \epsilon^2 \alpha_1 \alpha_2\, x_3^2}. \tag{6.30}$$

For other successes of this approach see Dragović and Gajić (2008); Hone and Petrera (2009); Petrera et al. (2009); Petrera and Suris (2010); Petrera et al. (2011); Grammaticos et al. (2012). However, it is not clear whether this method always produces an integrable discretization even when the starting ODE is integrable.

6.3 The QRT maps

A large family of integrable mappings of the plane was constructed by Quispel et al. (1988, 1989), and we will refer to this family as the QRT family of mappings. In Duistermaat (2010) the QRT family of maps was studied in detail in the context of algebraic geometry of elliptic surfaces; see also Tsuda (2004).

6.3.1 *Definition of the QRT map*

The QRT family of dynamical maps is given by the following implicit form

$$x \mapsto \widetilde{x} = \frac{f_1(y) - x\, f_2(y)}{f_2(y) - x\, f_3(y)}, \qquad y \mapsto \widetilde{y} = \frac{g_1(\widetilde{x}) - y\, g_2(\widetilde{x})}{g_2(\widetilde{x}) - y\, g_3(\widetilde{x})}, \tag{6.31}$$

in which $f_1, f_2, f_3, g_1, g_2, g_3$ are fourth-order polynomials given as follows. We can view this map as a sequence of steps $(x, y) \mapsto (\widetilde{x}, y) \mapsto (\widetilde{x}, \widetilde{y})$, where in order to get an explicit map we have to substitute expression for $\widetilde{x}$ into the functions g_i.

In order to define the functions f_i, g_i let us first define the following vectors:

$$\mathbf{F} = (f_1, f_2, f_3)^T, \quad \mathbf{G} = (g_1, g_2, g_3)^T, \quad \mathbf{Y} = (y^2, y, 1)^T, \quad \mathbf{X} = (x^2, x, 1)^T,$$

and similarly for tilded quantities, and the matrices

$$\mathbf{A}_i = \begin{pmatrix} \alpha_i & \beta_i & \gamma_i \\ \delta_i & \epsilon_i & \zeta_i \\ \kappa_i & \lambda_i & \mu_i \end{pmatrix},$$

then the polynomials f_i, g_i are given by

$$\mathbf{F} := (\mathbf{A}_0 \cdot \mathbf{Y}) \times (\mathbf{A}_1 \cdot \mathbf{Y}), \quad \mathbf{G} := (\mathbf{A}_0^T \cdot \mathbf{X}) \times (\mathbf{A}_1^T \cdot \mathbf{X}), \tag{6.32}$$

or more explicitly

$$\begin{pmatrix} f_1(y) \\ f_2(y) \\ f_3(y) \end{pmatrix} := \begin{pmatrix} \alpha_0 y^2 + \beta_0 y + \gamma_0 \\ \delta_0 y^2 + \epsilon_0 y + \zeta_0 \\ \kappa_0 y^2 + \lambda_0 y + \mu_0 \end{pmatrix} \times \begin{pmatrix} \alpha_1 y^2 + \beta_1 y + \gamma_1 \\ \delta_1 y^2 + \epsilon_1 y + \zeta_1 \\ \kappa_1 y^2 + \lambda_1 y + \mu_1 \end{pmatrix}$$
$$\begin{pmatrix} g_1(x) \\ g_2(x) \\ g_3(x) \end{pmatrix} := \begin{pmatrix} \alpha_0 x^2 + \delta_0 x + \kappa_0 \\ \beta_0 x^2 + \epsilon_0 x + \lambda_0 \\ \gamma_0 x^2 + \zeta_0 x + \mu_0 \end{pmatrix} \times \begin{pmatrix} \alpha_1 x^2 + \delta_1 x + \kappa_1 \\ \beta_1 x^2 + \epsilon_1 x + \lambda_1 \\ \gamma_1 x^2 + \zeta_1 x + \mu_1 \end{pmatrix}.$$

Thus equation (6.31) contains 18 parameters but only 8 of them are essential, as can be shown by symmetry arguments (Ramani et al., 2002).

Associated with this map is a biquadratic curve:

$$\begin{aligned}
&\Phi(x, y) := \mathbf{X}^T \cdot \mathbf{A}_0 \cdot \mathbf{Y} - K\,(\mathbf{X}^T \cdot \mathbf{A}_1 \cdot \mathbf{Y}) \equiv \\
&(\alpha_0 - K\alpha_1)x^2y^2 + (\beta_0 - K\beta_1)x^2y + (\gamma_0 - K\gamma_1)x^2 + (\delta_0 - K\delta_1)xy^2 + (\epsilon_0 - K\epsilon_1)xy \\
&+ (\zeta_0 - K\zeta_1)x + (\kappa_0 - K\kappa_1)y^2 + (\lambda_0 - K\lambda_1)y + (\mu_0 - K\mu_1) = 0.
\end{aligned} \tag{6.33}$$

In other words, there is a constant of motion which is a ratio of two biquadratics:

$$K = \frac{\alpha_0 x^2y^2 + \beta_0 x^2 y + \gamma_0 x^2 + \delta_0 xy^2 + \epsilon_0 xy + \zeta_0 x + \kappa_0 y^2 + \lambda_0 y + \mu_0}{\alpha_1 x^2y^2 + \beta_1 x^2 y + \gamma_1 x^2 + \delta_1 xy^2 + \epsilon_1 xy + \zeta_1 x + \kappa_1 y^2 + \lambda_1 y + \mu_1}. \tag{6.34}$$

The mapping (6.31) is the most general one that preserves this K-family of invariant biquadratic curves foliating the plane.

Let us give a brief proof of the invariance of K. It is easily noted that the map (6.31) can be written in the form

$$\mathbf{X} \cdot (\mathbf{F} \times \widetilde{\mathbf{X}}) = 0, \quad \mathbf{Y} \cdot (\mathbf{G} \times \widetilde{\mathbf{Y}}) = 0 \tag{6.35}$$

using the definition (6.32). The argument is now very simple: using (6.32) and the rules for expanding vector products we can rewrite

$$\begin{aligned}
0 &= \mathbf{X} \cdot (\mathbf{F} \times \widetilde{\mathbf{X}}) = \mathbf{X} \cdot \left[((\mathbf{A}_0 \cdot \mathbf{Y}) \times (\mathbf{A}_1 \cdot \mathbf{Y})) \times \widetilde{\mathbf{X}}\right] \\
&= \left(\mathbf{X}^T \cdot \mathbf{A}_1 \cdot \mathbf{Y}\right)\left(\widetilde{\mathbf{X}}^T \cdot \mathbf{A}_0 \cdot \mathbf{Y}\right) - \left(\mathbf{X}^T \cdot \mathbf{A}_0 \cdot \mathbf{Y}\right)\left(\widetilde{\mathbf{X}}^T \cdot \mathbf{A}_1 \cdot \mathbf{Y}\right), \\
0 &= \mathbf{Y} \cdot \mathbf{G} \times \widetilde{\mathbf{Y}} = \mathbf{Y} \cdot \left[((\mathbf{A}_0^T \cdot \widetilde{\mathbf{X}}) \times (\mathbf{A}_1^T \cdot \widetilde{\mathbf{X}})) \times \widetilde{\mathbf{Y}}\right] \\
&= \left(\widetilde{\mathbf{X}}^T \cdot \mathbf{A}_0 \cdot \widetilde{\mathbf{Y}}\right)\left(\widetilde{\mathbf{X}}^T \cdot \mathbf{A}_1 \cdot \mathbf{Y}\right) - \left(\widetilde{\mathbf{X}}^T \cdot \mathbf{A}_0 \cdot \mathbf{Y}\right)\left(\widetilde{\mathbf{X}}^T \cdot \mathbf{A}_1 \cdot \widetilde{\mathbf{Y}}\right),
\end{aligned}$$

and from these we get

$$K := \frac{\mathbf{X}^T \cdot \mathbf{A}_0 \cdot \mathbf{Y}}{\mathbf{X}^T \cdot \mathbf{A}_1 \cdot \mathbf{Y}} = \frac{\widetilde{\mathbf{X}}^T \cdot \mathbf{A}_0 \cdot \mathbf{Y}}{\widetilde{\mathbf{X}}^T \cdot \mathbf{A}_1 \cdot \mathbf{Y}} = \frac{\widetilde{\mathbf{X}}^T \cdot \mathbf{A}_0 \cdot \widetilde{\mathbf{Y}}}{\widetilde{\mathbf{X}}^T \cdot \mathbf{A}_1 \cdot \widetilde{\mathbf{Y}}}, \tag{6.36}$$

and thus K is indeed invariant under the evolution (6.31).

6.3.2 Derivation of the QRT map

The QRT map given above is rather complicated and therefore it will be instructive to see how it can be derived. The construction proceeds as follows.

Let us start with one variable and construct a map $x \to \widetilde{x}$ that has the rational invariant

$$I(x) := \frac{ax^2 + bx + c}{dx^2 + ex + f},$$

that is, $I(x) = I(\widetilde{x})$ or

$$\frac{ax^2 + bx + c}{dx^2 + ex + f} = \frac{a\widetilde{x}^2 + b\widetilde{x} + c}{d\widetilde{x}^2 + e\widetilde{x} + f}.$$

After cross-multiplication the numerator factorizes as

$$(x - \widetilde{x})[(ae - db)x\widetilde{x} + (cd - af)(x + \widetilde{x}) + (bf - ec)] = 0,$$

from which, after discarding the identity map, we get a fractional linear transformation of the form

$$\widetilde{x} = \frac{f_1 - f_2 x}{f_2 - f_3 x}, \tag{6.37}$$

where

$$f = \begin{pmatrix} f_1 \\ f_2 \\ f_3 \end{pmatrix} = \begin{pmatrix} a \\ b \\ c \end{pmatrix} \times \begin{pmatrix} d \\ e \\ f \end{pmatrix}.$$

We denote the map (6.37) by ι and note that it is an **involution**; that is, its square is the identity map, $\iota^2 = \iota \circ \iota = \text{id}$. Thus, as a one-variable map (6.37) is almost trivial. It only becomes nontrivial if we "interlace" the map $\iota = \iota_1$ with another map ι_2 of the same form. In order to introduce the second involution we need another variable y on which it acts nontrivially. The second map ι_2 is constructed the same way as ι_1 but with another set of parameters. Thus we take

$$K(y) := \frac{\alpha y^2 + \beta y + \gamma}{\delta y^2 + \epsilon y + \zeta}$$

and the condition $K(y) = K(\widehat{y})$ yields in the same way as before a map $y \xrightarrow{\iota_2} \widehat{y}$, with

$$\widehat{y} = \frac{g_1 - g_2 y}{g_2 - g_3 y} \tag{6.38}$$

where

$$\boldsymbol{g} = \begin{pmatrix} g_1 \\ g_2 \\ g_3 \end{pmatrix} = \begin{pmatrix} \alpha \\ \beta \\ \gamma \end{pmatrix} \times \begin{pmatrix} \delta \\ \epsilon \\ \zeta \end{pmatrix}.$$

The "interlacing" follows when we take the parameters of ι_1 to be functions of y and the parameters of ι_2 to be functions of x (note that ι_1 acts trivially on y and ι_2 acts trivially on x):

$$\iota_1 : \begin{cases} x \mapsto \overline{x} = \dfrac{f_1(y) - x f_2(y)}{f_2(y) - x f_3(y)} \\ y \mapsto \overline{y} = y \end{cases} \tag{6.39a}$$

and

$$\iota_2 : \begin{cases} x \mapsto \widehat{x} = x \\ y \mapsto \widehat{y} = \dfrac{g_1(x) - y g_2(x)}{g_2(x) - y g_3(x)} \end{cases} \tag{6.39b}$$

where f_1, f_2, f_3 are still some undetermined functions of y and g_1, g_2, g_3 functions of x. Since the ι_i maps do not commute we can create a nontrivial map by composing the two involutions and introduce the map $\varphi = \iota_2 \circ \iota_1$ which is invertible since $\varphi^{-1} = \iota_1 \circ \iota_2$. Then writing $(\widetilde{x}, \widetilde{y}) = (\widehat{\overline{x}}, \widehat{\overline{y}}) = \phi(x, y) = \iota_2 \circ \iota_1(x, y)$ we obtain the QRT map (6.31) (with bars rather than tildes denoting the combined shifts).

The final step in the derivation is in fixing the functions: namely, *if both involutions carry the same invariant $I = K$, then the composed map φ also carries this invariant and will be an integrable map!* Since I and K are both rational quadratic, we can have $I = K$ iff the coefficients $a, b, \ldots, e, f$ are quadratic in y, and $\alpha, \beta, \ldots \epsilon, \zeta$ are quadratic in x. When comparing the various terms in $K = I$ we see that the functions f_i, g_i $(i = 1, 2, 3)$ in the map must be of the form (6.32).

6.3.3 Measure preservation

The measure-preservation property was defined in Section 6.1.3 and we will now establish it for the QRT map. In order to do this we have to compute the Jacobian (6.10), which is best done by calculating it for each involution (6.39) separately (Roberts, 1990; Roberts and Quispel, 1992). From the involution ι_1 (6.39a) we find the Jacobian

$$J_1 := \frac{\partial(\overline{x}, \overline{y})}{\partial(x, y)} = \frac{f_1 f_3 - f_2^2}{(f_2 - x f_3)^2} = -\frac{f_2 - \overline{x} f_3}{f_2 - x f_3}$$

so that $(f_2(y) - f_3(y)x)^{-1}$ is the measure $m_1(x, y)$ for this step. Let us rewrite the above using (6.32, 6.36):

$$\begin{aligned}
-(f_2 - xf_3) &= [\mathbf{X}^T \times \mathbf{F}]_1 \\
&= [\mathbf{X}^T \times ((\mathbf{A}_0 \cdot \mathbf{Y}) \times (\mathbf{A}_1 \cdot \mathbf{Y}))]_1 \\
&= (\mathbf{X}^T \cdot \mathbf{A}_1 \cdot \mathbf{Y})[\mathbf{A}_0 \cdot \mathbf{Y}]_1 - (\mathbf{X}^T \cdot \mathbf{A}_0 \cdot \mathbf{Y})[\mathbf{A}_1 \cdot \mathbf{Y}]_1 \\
&= (\mathbf{X}^T \cdot \mathbf{A}_1 \cdot \mathbf{Y})\,([\mathbf{A}_0 \cdot \mathbf{Y}]_1 - K[\mathbf{A}_1 \cdot \mathbf{Y}]_1)
\end{aligned}$$

where $[\,\cdot\,]_1$ stands for the first component of a vector. Since the involution ι_1 does not change y or K we get a similar formula for $f_2 - \overline{x} f_3$, with the same factor and therefore we find

$$J_1 = -\frac{\overline{\mathbf{X}}^T \cdot \mathbf{A}_1 \cdot \mathbf{Y}}{\mathbf{X}^T \cdot \mathbf{A}_1 \cdot \mathbf{Y}}.$$

In order to get the conservation property of the full map we have to consider next the second step (6.39b). For it we get a similar result:

$$J_2 = -\frac{\widehat{\mathbf{Y}}^T \cdot \mathbf{A}_1^T \cdot \mathbf{X}}{\mathbf{Y}^T \cdot \mathbf{A}_1^T \cdot \mathbf{X}}.$$

The Jacobian J for the composed map φ is the product

$$J = \overline{J_2} J_1 = \frac{(\overline{\mathbf{X}}^T \cdot \mathbf{A}_1 \cdot \mathbf{Y})\ (\widehat{\overline{\mathbf{Y}}}^T \cdot \mathbf{A}_1^T \cdot \overline{\mathbf{X}})}{(\mathbf{X}^T \cdot \mathbf{A}_1 \cdot \mathbf{Y})\ (\mathbf{Y}^T \cdot \mathbf{A}_1^T \cdot \overline{\mathbf{X}})}$$

where we have used the fact that $\overline{\mathbf{Y}} = \mathbf{Y}$. Canceling the common factor $\mathbf{Y}^T \cdot \mathbf{A}_1^T \cdot \overline{\mathbf{X}} = \overline{\mathbf{X}}^T \cdot \mathbf{A}_1 \cdot \mathbf{Y}$ and using the fact that the hat shift acts trivially on $\mathbf{X}$ and that the tilde shift acts trivially on $\mathbf{Y}$ we can write this as

$$J = \overline{J_2} J_1 = \frac{\widehat{\overline{\mathbf{X}}}^T \cdot \mathbf{A}_1 \cdot \widehat{\overline{\mathbf{Y}}}}{\mathbf{X}^T \cdot \mathbf{A}_1 \cdot \mathbf{Y}} = \frac{m(x, y)}{m(\widetilde{x}, \widetilde{y})}$$

and therefore the QRT map is measure preserving, with measure

$$m(x, y) = \left(\mathbf{X}^T \cdot \mathbf{A}_1 \cdot \mathbf{Y}\right)^{-1}.$$

In the terminology of Roberts and Quispel (1992) (cf. Appendix A of that paper, where it is called a “working definition”), integrability of a two-component map is equivalent to the existence of a conserved quantity and measure preservation.

6.3.4 The symmetric case

One interesting special case of the QRT map is when the matrices $\boldsymbol{A}_i$ are symmetric, i.e. $\boldsymbol{A}_i^T = \boldsymbol{A}_i$, in which case the set of functions $\boldsymbol{F}, \boldsymbol{G}$ have the same form, as can be seen from (6.32). Furthermore $\Phi(x, y) = \Phi(y, x)$; that is, the biquadratic curve (6.33) gets a symmetric form:

$$\begin{aligned}\Phi(x, y) = (\alpha_0 - K\alpha_1)x^2y^2 + (\beta_0 - K\beta_1)xy(x+y) + (\gamma_0 - K\gamma_1)(x^2 + y^2)\\ + (\epsilon_0 - K\epsilon_1)xy + (\xi_0 - K\xi_1)(x+y) + (\mu_0 - K\mu_1) = 0.\end{aligned} \tag{6.40}$$

Since now $f_i = g_i$ in (6.31), the map reduces to the following three-point map[2]

$$x_{n+1} = \frac{f_1(x_n) - x_{n-1}f_2(x_n)}{f_2(x_n) - x_{n-1}f_3(x_n)} \tag{6.41}$$

where

$$\begin{aligned}
f_1 &= x_n^4(\beta_0\gamma_1 - \beta_1\gamma_0) + x_n^3(\beta_0\xi_1 - \beta_1\xi_0 + \epsilon_0\gamma_1 - \epsilon_1\gamma_0) + x_n^2(\beta_0\mu_1 - \beta_1\mu_0\\
&\quad + \epsilon_0\xi_1 - \epsilon_1\xi_0 - \gamma_0\xi_1 + \gamma_1\xi_0) + x_n(\epsilon_0\mu_1 - \epsilon_1\mu_0) - \mu_0\xi_1 + \mu_1\xi_0,\\
f_2 &= x_n^4(-\alpha_0\gamma_1 + \alpha_1\gamma_0) + x_n^3(-\alpha_0\xi_1 + \alpha_1\xi_0 - \beta_0\gamma_1 + \beta_1\gamma_0) + x_n^2(-\alpha_0\mu_1\\
&\quad + \alpha_1\mu_0 - \beta_0\xi_1 + \beta_1\xi_0) + x_n(-\beta_0\mu_1 + \beta_1\mu_0 - \gamma_0\xi_1 + \gamma_1\xi_0) - \gamma_0\mu_1 + \gamma_1\mu_0\\
f_3 &= x_n^4(\alpha_0\beta_1 - \alpha_1\beta_0) + x_n^3(\alpha_0\epsilon_1 - \alpha_1\epsilon_0) + x_n^2(\alpha_0\xi_1 - \alpha_1\xi_0 + \beta_0\epsilon_1\\
&\quad - \beta_0\gamma_1 - \beta_1\epsilon_0 + \beta_1\gamma_0) + x_n(\beta_0\xi_1 - \beta_1\xi_0 - \epsilon_0\gamma_1 + \epsilon_1\gamma_0) + \gamma_0\xi_1 - \gamma_1\xi_0.
\end{aligned}$$

If we solve $K(x, y)$ from (6.40) and compute $K(x_{n+1}, x_n) - K(x_n, x_{n-1})$, we find that it vanishes by virtue of (6.41).

If we further set

$$\alpha_0 = b\alpha_1,\ \epsilon_0 = b\epsilon_1,\ \gamma_0 = b\gamma_1,\ \xi_0 = b\xi_1,\ \text{where } b = \beta_0/\beta_1, \tag{6.42}$$

then $f_3 = 0$ and f_1, f_2 become quadratic; that is, we get the map

$$x_{n+1} + x_{n-1} = -\frac{\beta_1 x_n^2 + \epsilon_1 x_n + \xi_1}{\alpha_1 x_n^2 + \beta_1 x_n + \gamma_1}. \tag{6.43}$$

This is the same map as case (a) of Section 6.1.5. The QRT map is invariant under the Möbius transformation and if we apply it to (6.43) we get a map of the form (6.41) but with quadratic rather than quartic f_i.

[2] Note that the reduction is not direct but involves working on a staggered chain; see Exercise 6.6.

The symmetric case of QRT is simpler than the general case, and some results have only been established for this case; for example, the Lax pair is only known for (6.43) (Iatrou and Roberts, 2001).

We have seen Lax pairs for discrete 2D equations already since Section 3.3, but with dynamical maps the situation is different because they only involve shifts in one discrete direction rather than in two directions, as is the case with (3.18). For maps such as (6.41) we need first the eigenvalue (spectral) problem

$$L_n(h)\varphi_n(h) = \lambda\varphi_n(h)\,, \tag{6.44}$$

where λ is the eigenvalue and h an additional internal spectral parameter.[3] We also impose on the eigenvector $\varphi_n(h)$ the discrete evolution in terms of the variable n of the form

$$\varphi_{n+1}(h) = M_n(h)\varphi_n(h)\,, \tag{6.45}$$

and then the compatibility of these two equations yields the discrete Lax equation

$$L_{n+1}(h)M_n(h) = M_n(h)L_n(h)\,. \tag{6.46}$$

For (6.43) we choose (Iatrou and Roberts, 2001)

$$L_n(h) = \begin{pmatrix} a-e & b-c+x_{n+1} & 1 & 0 \\ 0 & 0 & b+c-x_n & 1 \\ h & 0 & a+e & b+c-x_{n+1} \\ h(b-c+x_n) & h & 0 & 0 \end{pmatrix}, \tag{6.47a}$$

$$M_n(h) = \begin{pmatrix} \frac{a-e}{b-c+x_{n+1}} & 1 & 0 & 0 \\ 0 & 0 & 1 & 0 \\ 0 & 0 & \frac{a+e}{b+c-x_{n+1}} & 1 \\ h & 0 & 0 & 0 \end{pmatrix}, \tag{6.47b}$$

and obtain from (6.46) the equation

$$x_{n+1} + x_{n-1} = \frac{2c\,x_n^2 + 2(a-2c^2)\,x_n + 2(c^3 + be - b^2c - ac)}{x_n^2 - 2c\,x_n - b^2 + c^2}. \tag{6.48}$$

Comparison with (6.43) gives the relation between the different parameters: $\alpha_1 = 1,\ \beta_1 = -2c,\ \gamma_1 = c^2 - b^2,\ \epsilon_1 = 4c^2 - 2a,\ \xi_1 = 2(ac - eb + cb^2 - c^3)$.

[3] As we will see later, when we derive maps from periodic reductions, the parameter h can be identified with the spectral parameter of the 2D lattice Lax pair, while λ arises as a "Floquet" parameter associated with periodicity along directions in the lattice.

The QRT invariant (6.33) gives

$$\Phi(x_{n-1}, x_n) = \alpha_1 x_{n-1}^2 x_n^2 + \beta_1 (x_{n-1}^2 x_n + x_{n-1} x_n^2) + \gamma_1 (x_{n-1}^2 + x_n^2) \\ +\epsilon_1 x_{n-1} x_n + \xi_1 (x_{n-1} + x_n). \tag{6.49}$$

This invariant follows from (6.40) with parameters given by (6.42). The trajectory of the map (6.48) stays on a level set of the curve given by (6.49), which form a family of elliptic curves in the space of the dependent variables.

On the other hand, we also have another type of invariant, namely the characteristic equation (or spectral curve)

$$P(h, \lambda) = \det(\lambda - L_n(h)) = 0\,, \tag{6.50}$$

which is an elliptic curve in the space of the spectral variables (h, λ). Equation (6.46) basically tells us that the evolution in discrete-time n is obtained from a similarity transformation on the matrix $L_n(h)$ at each step of the iteration. This means that $\det(L_n)$ is an invariant. These two elliptic curves are equivalent and either one can be used to find the solution of the map.

More generally, the symmetric QRT-mapping (6.41) can be solved in terms of elliptic functions parameterizing the symmetric biquadratic curve (cf. e.g. the final section of Baxter, 1982). However, a Lax pair is, to our knowledge, not known for the general symmetric case.

6.3.5 HKY generalization

The QRT map has turned out to be quite important in the study of OΔEs and therefore there have been various attempts to generalize it. One method is to add n-dependence in its coefficients; another is to consider multipoint generalizations. We will return to these later.

The HKY-generalization (Hirota, Kimura and Yahagi 2001) is to reconsider the conservation law $K(x_{n+1}, x_n) = K(x_n, x_{n-1})$: what if we require conservation only up to an involution; that is, if we have

$$K(x_{n+1}, x_n) - \sigma(K(x_n, x_{n-1})) = 0,$$

where σ is an involution, i.e. $\sigma(\sigma(x)) = x$, for example $\sigma(A) = A, -A$, or c/A. Recall that factorizing the condition of conservation yields the map if $\sigma(A) = A$. The same may work on some specific biquadratics with nontrivial involutions σ.

As an example consider

$$K(x, y) = \frac{2xy}{x^2 + y^2 + \beta^2}. \tag{6.51}$$

Then we have

$$K(u_{n+1},u_n)-K(u_n,u_{n-1}) = \frac{-2u_n(u_{n+1}-u_{n-1})[u_{n+1}u_{n-1}-(u_n^2+b^2)]}{(u_{n+1}^2+u_n^2+b^2)(u_n^2+u_{n-1}^2+b^2)},$$

and thus K of (6.51) is the constant of motion for the three-point map

$$u_{n+1}u_{n-1} = u_n^2 + b^2.$$

But another computation yields

$$K(u_{n+1},u_n)+K(u_n,u_{n-1}) = \frac{2u_n(u_{n+1}+u_{n-1})[u_{n+1}u_{n-1}+(u_n^2+b^2)]}{(u_{n+1}^2+u_n^2+b^2)(u_n^2+u_{n-1}^2+b^2)},$$

in which we can identify the equation

$$u_{n+1}u_{n-1} = -(u_n^2 + b^2).$$

It seems that in the second case K is conserved "up to sign," and then $K(x,y)^2$, which is biquartic, should be a genuine invariant. Indeed, a direct computation shows that

$$K(u_{n+1},u_n)^2 - K(u_n,u_{n-1})^2 = \frac{-4u_n^2(u_{n+1}+u_{n-1})(u_{n+1}-u_{n-1})}{(u_{n+1}^2+u_n^2+b^2)^2(u_n^2+u_{n-1}^2+b^2)^2} \times [u_{n+1}u_{n-1}+(u_n^2+b^2)][u_{n+1}u_{n-1}-(u_n^2+b^2)],$$

having both equations as factors.

Other examples of mappings with biquadratic invariants have been given in Hirota et al. (2001); Kimura et al. (2002); Quispel (2003).

6.4 Periodic reductions

We will now show how one may derive maps by applying reductions on 2D lattice equations. For lattice equations that can be thought of as discrete analogues of hyperbolic PDEs (as was discussed in Section 1.3 for quadrilateral lattice equations), initial value problems can be posed by giving initial values on a staircase (or sawtooth) type configurations of lattice points, which can be iterated throughout the lattice.

If we impose periodic initial values in the lattice (in the spirit of Papageorgiou et al., 1990) and if this periodicity is preserved under evolution, then the problem of solving the equation is reduced to a problem involving only a finite number of components. In other words, since we start with a PΔE with two independent (discrete) variables, in the resulting reduction, one of the variables is used to enumerate the components while the other independent variable is interpreted as a time variable. Typically, what happens is that we get a system of OΔEs in terms of this new time variable. If the original 2D lattice equation was integrable (in the sense of possessing a Lax pair, or multidimensional consistency) then the new OΔE is also integrable (in the sense of also possessing a Lax pair and admitting conserved quantities).

We illustrate the method through a number of explicit examples, from which it should become clear how to do it in general.

By a period N reduction we mean that the values of the dependent variable, located at the vertices of the lattice, are repeated according to a fixed configuration involving N elementary steps in the lattice.

In the following we will consider the problem of finding periodic solutions to the lattice potential KdV equation

$$(p - q + \widehat{u} - \widetilde{u})(p + q + u - \widehat{\widetilde{u}}) = p^2 - q^2 \,. \tag{6.52}$$

The simplest case $N = 1$ (if it is compatible with the equation) is trivial because then there is no evolution. The next simplest case is $N = 2$, when we have only two different initial values u_0, u_1 on the staircase, as illustrated in Figure 6.1a. We take the discrete-time evolution along the vertical downward direction and indicate it by the hat-shift. Then we have

$$\widehat{u}_1 = u_0, \quad (p - q + \widehat{u}_0 - u_1)(p + q + u_0 - \widehat{u}_1) = p^2 - q^2 \,.$$

Now since by periodicity we have $\widehat{u}_1 = u_0$ the pair of equations collapses to the linear set $\widehat{u}_1 = u_0, \widehat{u}_0 = u_1$.

In many other lattice equations this simple reduction yields a linear or linearizable equations, however for Q_4 one gets a genuine map of the HKY-type (Joshi et al., 2006).

6.4.1 Period 3 reduction

Next consider the (2, 1) reduction illustrated in Figure 6.1b. We can take u_0, u_1, u_2 as initial data supplied for the configuration in the figure. This configuration is assumed to be periodically repeated through the lattice to form a connected periodic staircase. From the lattice equation (6.52) applied on the two quadrilaterals in Figure 6.1b we

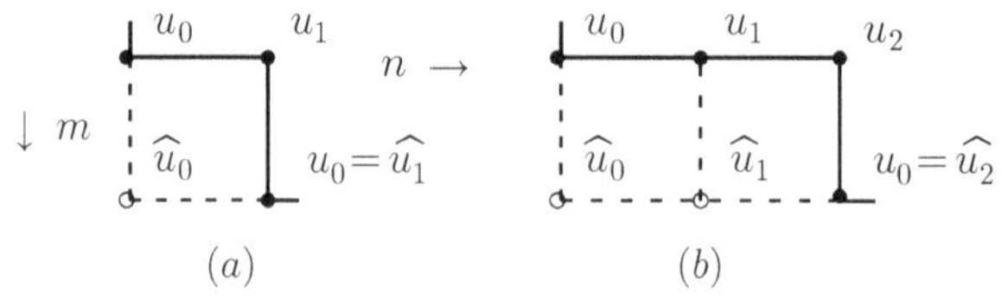

Figure 6.1 Periodic initial values on the lattice. The case (a) is a (1, 1) reduction and case (b) is a (2, 1) reduction.

get the dynamical mapping $(u_0, u_1, u_2) \mapsto (\widehat{u}_0, \widehat{u}_1, \widehat{u}_2)$, associated with this periodic reduction:

$$\widehat{u}_0 = q - p + u_1 + \frac{p^2 - q^2}{p + q + \widehat{u}_2 - \widehat{u}_1}, \tag{6.53a}$$

$$\widehat{u}_1 = q - p + u_2 + \frac{p^2 - q^2}{p + q + u_1 - u_0}, \tag{6.53b}$$

$$\widehat{u}_2 = u_0 . \tag{6.53c}$$

The above three-component map (6.53) can be simplified to a two-component map by introducing the variables

$$x := p + q + u_1 - u_0 , \quad y := p + q + u_2 - u_1 \tag{6.54}$$

after which (6.53) can be written as a two-variable dynamical map $(x, y) \to (\widehat{x}, \widehat{y})$

$$\begin{cases} \widehat{x} &= y + \dfrac{p^2 - q^2}{x} - \dfrac{p^2 - q^2}{\widehat{y}}, \\ \widehat{y} &= 4p + 2q - x - y - \dfrac{p^2 - q^2}{x} . \end{cases} \tag{6.55}$$

In contrast to the previous reduction, this is a nontrivial rational map. This map is integrable because: (1) it possesses an exact invariant (conserved quantity)

$$I(x, y) = (xy + p^2 - q^2)(4p + 2q - x - y), \tag{6.56}$$

i.e. $I(x, y) = I(\widehat{x}, \widehat{y})$ and (2) it is area-preserving, i.e. the Jacobian of the map (6.55)

$$J = \frac{\partial(\widehat{x}, \widehat{y})}{\partial(x, y)} = 1,$$

(3) it has a Lax pair as discussed later, (4) it can be solved explicitly using elliptic functions.

An alternative formulation of this reduction is obtained by posing the periodicity condition

$$u_{n,m} = u_{n+2,m+1}, \quad \forall n, m$$

and introducing a change of variables

$$u_{n,m} = U_{n-2m,m} = U_{N,m}, \quad N := n - 2m. \tag{6.57}$$

The periodicity now implies that the new dependent variable U is constant in the second independent variable m. From (6.52) it then follows that

$$(p - q + U_{N-2} - U_{N+1})(p + q + U_N - U_{N-1}) = p^2 - q^2, \tag{6.58}$$

which is a four-point map, i.e. third-order difference equation.

In order to simplify (6.58) we introduce the new variable

$$Y_N := p + q + U_N - U_{N-1}. \tag{6.59}$$

Now noting that (6.59) implies

$$Y_{N-1} + Y_N + Y_{N+1} = 3(p+q) + U_{N+1} - U_{N-2},$$

and we can write (6.58) as

$$(4p + 2q - Y_{N-1} - Y_N - Y_{N+1})Y_N = p^2 - q^2,$$

which can also be written as

$$Y_{N+1} + Y_N + Y_{N-1} = a + \frac{b}{Y_N}, \tag{6.60}$$

where $a = 4p + 2q$, $b = -p^2 + q^2$. This is the autonomous form of the discrete Painlevé I equation (1.20). It can also be written as a two-component first-order map

$$\begin{cases} Y_{N+1} = -X_N - Y_N + a + \dfrac{b}{Y_N}, \\ X_{N+1} = Y_N. \end{cases} \tag{6.61}$$

This is equivalent to (6.55) and shares the same integrability properties and possesses the same invariant.

6.4.2 Period 4 reduction

As the next example, let us consider (2, 2) reduction illustrated in Figure 6.2. This configuration is assumed to be periodically repeated horizontally (along the staircase) through the lattice. We apply it on the same equation (6.52) as before.

From (6.52) one can derive the four-component mapping $(u_0, u_1, u_2, u_3) \mapsto (\widehat{u}_0, \widehat{u}_1, \widehat{u}_2, \widehat{u}_3)$

$$\begin{cases} \widehat{u}_0 = q - p + u_1 + \dfrac{p^2 - q^2}{p + q + u_0 - u_2}, \\ \widehat{u}_1 = u_2\,, \\ \widehat{u}_2 = q - p + u_3 + \dfrac{p^2 - q^2}{p + q + u_2 - u_0}, \\ \widehat{u}_3 = u_0\,. \end{cases} \tag{6.62}$$

The above three-component map (6.53) can be simplified to a two-component map by introducing the variables

$$x := u_0 - u_2\,, \quad y := u_1 - u_3$$

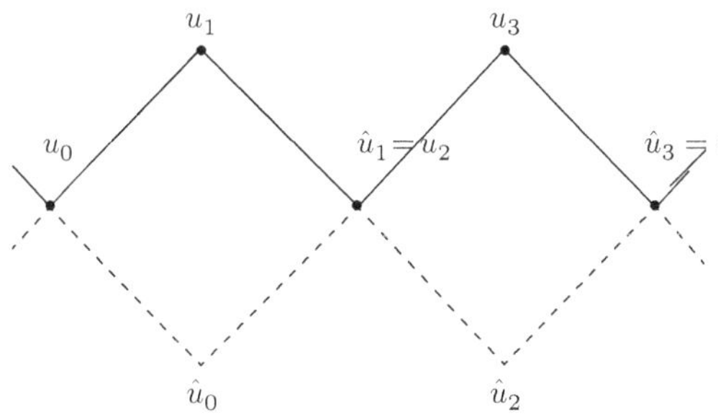

Figure 6.2 Configuration of point for the (2, 2) reduction.

and then from (6.62) we obtain a two-variable dynamical map $(x, y) \to (\widehat{x}, \widehat{y})$

$$\begin{cases} \widehat{x} = y - 2\dfrac{(p^2 - q^2)x}{(p+q)^2 - x^2}, \\ \widehat{y} = -x. \end{cases} \tag{6.63}$$

Once again we get a nontrivial rational map, but this looks even simpler than the map (6.55) resulting from the previous (2, 1) reduction.

As expected, this map is also integrable because: (1) it possesses an exact invariant

$$I(x, y) = [x^2 - (p+q)^2][y^2 - (p+q)^2] + 2(p^2 - q^2)xy, \tag{6.64}$$

(2) it is area-preserving, (3) it has a Lax pair, as discussed later and (4) it can be solved explicitly using elliptic functions.

In this case the periodicity condition $\widehat{u}_3 = u_0$ actually means

$$u_{n+2,m+2} = u_{n,m}.$$

Following the previous recipe, we define

$$u_{n,m} = U_{n-m,m} = U_{N,m}, \Longrightarrow U_{N,m+2} = U_{N,m}.$$

Thus in this case we need two sets of dependent variables $U_{N,0} = U_N$ and $U_{N,1} = V_N$, and in terms of these variables the original equation and its shift read

$$\begin{aligned} (p - q + V_{N-1} - U_{N+1})(p + q + U_N - V_N) &= p^2 - q^2, \\ (p - q + U_{N-1} - V_{N+1})(p + q + V_N - U_N) &= p^2 - q^2. \end{aligned}$$

From this we can solve

$$\begin{cases} U_{N+1} = V_{N-1} + \dfrac{(p-q)(U_N - V_N)}{p+q+U_N - V_N}, \\ V_{N+1} = U_{N-1} - \dfrac{(p-q)(U_N - V_N)}{p+q-U_N + V_N}. \end{cases} \tag{6.65}$$

Due to the obvious symmetry we introduce the new variable

$$X_N = U_N - V_N,$$

and then by subtracting the equations in (6.65) we obtain

$$X_{N+1} + X_{N-1} = \frac{2(p^2-q^2)X_N}{(p+q)^2 - X_N^2}, \tag{6.66}$$

or as the first-order system

$$\begin{cases} X_{N+1} = -Y_N + \dfrac{2(p^2-q^2)X_N}{(p+q)^2 - X_N^2}, \\ Y_{N+1} = X_N \end{cases}. \tag{6.67}$$

Equation (6.66) is a special case of the well-known McMillan map and an autonomous limit of the discrete Painlevé II equation (11.2). Note that by adding the equation in (6.65) and introducing $W_N = U_N + V_N$ we obtain

$$W_{N+1} = -W_{N-1} - \frac{2(p-q)X_N^2}{(p+q)^2 - X_N^2},$$

which is linear in W with a source in term X.

6.4.3 Higher periodic reductions

We have previously looked at the simplest reductions of the lattice KdV, namely the period 1, 2, 3 and 4 reductions. In this section we consider higher period reductions, which illustrate the general mechanism. This is achieved by extending the staircase on which the initial values are posed. For convenience we will consider for now the following two initial data patterns in Figure 6.3, corresponding to even period $2P$, i.e. $u_{2P+j} = u_j$ and odd period $u_{2P+1+j} = u_j$ (Capel et al., 1991).[4] For both diagrams we proceed as before and construct a corresponding mapping, and the evolution is in the downward direction and denoted by the hat-shift.

[4] Note that even though we can connect any two fixed points on the lattice by different staircases, the resulting mappings will be equivalent, up to changes of variables.

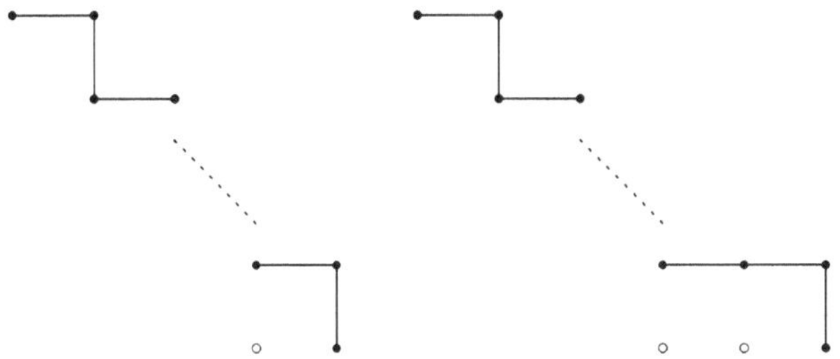

Figure 6.3 Higher-order staircase.

Let us consider the generic multilinear quad equation $Q(u, \widetilde{u}, \widehat{u}, \widehat{\widetilde{u}}) = 0$ from which one of the values, say $\widehat{u}$ can be solved as $\widehat{u} = F(u, \widetilde{u}, \widehat{\widetilde{u}})$, where F is rational. For the new odd sites we get, by virtue of the labeling,

$$\widehat{u}_{2k+1} = u_{2k+2}, \text{ for } k = 0, \ldots, P - 2, \tag{6.68a}$$

and for the even sites we have to use the equation

$$\widehat{u}_{2k} = F(u_{2k}, u_{2k+1}, u_{2k+2}), \text{ for } k = 0, \ldots, P - 2. \tag{6.68b}$$

This takes care of the updates up to the last step in the staircase. Only at the last step of the staircase do we have to distinguish between the even and odd periods by supplementing (6.68) with the following: in the even case

$$\widehat{u}_{2P-1} = u_0, \quad \widehat{u}_{2P-2} = F(u_{2P-2}, u_{2P-1}, u_0),$$

and in the odd case

$$\widehat{u}_{2P-2} = F(u_{2P-2}, u_{2P-1}, \widehat{u}_{2P-1}), \quad \widehat{u}_{2P-1} = F(u_{2P-1}, u_{2P}, u_0), \quad \widehat{u}_{2P} = u_0.$$

This kind of reduction can be performed to all quad-equations that are affine linear in one of the variables (above it was $\widehat{u}$), integrable and nonintegrable. All these reductions take an infinite-dimensional system (the original lattice equation) into a system given in terms of a finite number of variables (finite number of degrees of freedom). However, for reductions of integrable quad equations there is in principle a Lax pair for the reduced system, which can be derived from the Lax pair of the quad-equation. The former can then be used to construct invariants for the finite map and in this higher-dimensional case we get more invariants. The existence of a Lax pair and of conserved quantities is an indicator for the integrability of the map. However in the context of finite-dimensional systems one has to be more specific on the meaning of the term "integrability" and we may need additional structure (e.g. symplecticity).

In the following we show that in some of the integrable equations we can go further and introduce reduced variables in terms of which the map simplifies further to a lower-dimensional map, but that procedure only works for a subclass of affine linear quad-equation. To demonstrate that procedure we now specialize to the lattice potential KdV equation (6.52) for which case F is given by

$$F(x, y, z) := q - p + y + \frac{(p^2 - q^2)}{p + q + x - z}.$$

We now specialize to the even case. In order to derive an integrable map in this case we define the reduced variables by

$$Y_j := p + q + u_{2j} - u_{2j+2}, \quad X_j = p + q + u_{2j-1} - u_{2j+1}, \quad j = 1, \ldots, P$$

with periodicity $Y_P = Y_0$, $X_P = X_0$ in the even periodic case. Furthermore by the construction we have

$$C_1 := \sum_{j=1}^{P} X_j\,, \quad C_2 := \sum_{j=1}^{P} Y_j\,, \quad C_1 = C_2 = (p+q)P. \tag{6.69a}$$

In terms of the reduced variable we can write the equation in the form

$$\begin{aligned} \widehat{X}_j &= Y_j \\ \widehat{Y}_j &= X_{j+1} - \frac{a}{Y_{j+1}} + \frac{a}{Y_j}, \quad (j = 1, \cdots, P), \end{aligned} \tag{6.69b}$$

where $a = p^2 - q^2$. Equations (6.69b) and (6.69a) together define the map $(X_i, Y_i) \mapsto (\widehat{X}_i, \widehat{Y}_i)$ for $i = 1, \ldots, P$, but this is actually a $2(P-1)$-dimensional map after using the constraints (6.69a) to eliminate one pair of variables. Alternatively we can take C_1, C_2 to be arbitrary and treat the system as a $2P$-dimensional map with C_1, C_2 as constants of motion.

Since the equation we started with (6.52) was integrable and has a Lax pair in terms of 2×2 matrices, one can use this inherited Lax pair to construct a 2×2 Lax pair for the periodic reduction; this is done in Section 6.5. Instead of this we will here present a $2P \times 2P$ Lax pair (Capel et al., 1991) of the form

$$\boldsymbol{L}\phi = k^2\phi, \quad \widehat{\phi} = \boldsymbol{M}\phi,$$

where

$$\boldsymbol{L} := \begin{pmatrix} a & -Y_0 & 1 & 0 & \cdots & \cdots & 0 \\ 0 & 0 & -X_1 & 1 & \ddots & & 0 \\ 0 & 0 & a & -Y_1 & 1 & \ddots & \vdots \\ \vdots & \vdots & & \ddots & \ddots & \ddots & 0 \\ 0 & & & & \ddots & \ddots & 1 \\ h & 0 & & & & a & -Y_{P-1} \\ -hX_0 & h & 0 & \cdots & & 0 & 0 \end{pmatrix}.$$

$$
\boldsymbol{M} := \begin{pmatrix}
-aY_0 & 1 & 0 & \dots & \dots & 0 \\
0 & 0 & 1 & \ddots & & 0 \\
0 & 0 & -aY_1 & 1 & \ddots & \vdots \\
\vdots & & \ddots & \ddots & \ddots & 0 \\
 & & & \ddots & \ddots & 1 \\
0 & & & & 0 & -Y_{P-1} \\
h & 0 & \dots & & 0 & 0
\end{pmatrix}.
$$

The isospectral consistency condition is

$$
\widehat{\boldsymbol{L}}\boldsymbol{M} = \boldsymbol{M}\boldsymbol{L}
$$

and it yields the equations (6.69b).

Now since $\widehat{\boldsymbol{L}} = \boldsymbol{M}\boldsymbol{L}\boldsymbol{M}^{-1}$ we find that $I_k := \operatorname{trace}(\boldsymbol{L}^k)$ is an invariant for all k. But for many values of k, the invariant so obtained is trivial and does not depend on X, Y. For example if $P = 4$ we have an 8×8 matrix and I_k is trivially constant for $k = 1, \dots, 5$ and only I_6, I_7, I_8 are functions of $X_j, Y_j,\ j = 0, 1, 2, 3$. Having three constants of motion, as we do in this case, would be sufficient for Liouville integrability, subject to the conditions of being independent and in involution with respect to some Poisson bracket, this was proven in Capel et al. (1991).

6.5 Lax pair for the periodic reductions and construction of invariants

We are now going to construct Lax pairs for the maps derived by reduction from the Lax pair of original lattice equation. Once we have the Lax pairs we can construct the invariant. As an example we take the lattice KdV equation.

6.5.1 *Reduced Lax system*

The Lax matrices were given in Section 3.3.1 for the variable w, but here we would like to use the variable u and the Lax pair

$$
\boldsymbol{L} = \begin{pmatrix} p - \widetilde{u} & 1 \\ k^2 - p^2 + \\ +(p-\widetilde{u})(p+u) & p+u \end{pmatrix}, \quad \boldsymbol{M} = \begin{pmatrix} q - \widehat{u} & 1 \\ k^2 - q^2 + \\ +(q-\widehat{u})(q+u) & q+u \end{pmatrix}. \tag{6.70}
$$

we have $\widehat{\boldsymbol{L}}\boldsymbol{M} = \widetilde{\boldsymbol{M}}\boldsymbol{L}$. These Lax matrices $\boldsymbol{L}$ and $\boldsymbol{M}$ factorize as follows:

$$
\boldsymbol{L} = \boldsymbol{U}\boldsymbol{P}\widetilde{\boldsymbol{U}}^{-1}, \quad \boldsymbol{M} = \boldsymbol{U}\boldsymbol{Q}\widehat{\boldsymbol{U}}^{-1}, \tag{6.71}
$$

where

$$
\boldsymbol{U} = \begin{pmatrix} 1 & 0 \\ u & 1 \end{pmatrix}, \quad \boldsymbol{P} = \begin{pmatrix} p & 1 \\ k^2 & p \end{pmatrix}, \quad \boldsymbol{Q} = \begin{pmatrix} q & 1 \\ k^2 & q \end{pmatrix}. \tag{6.72}
$$

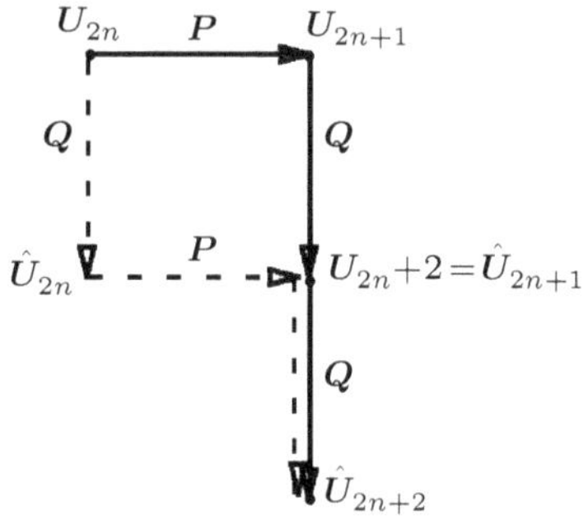

Figure 6.4 The basic configuration for constructing the Lax pair for the reduction.

We now consider the configuration of Lax matrices described in Figure 6.4. The matrices $\boldsymbol{U}_i$ (containing the relevant value u_i of the dependent variable) are associated with the vertices, while the matrices Lax matrices $\boldsymbol{L}$ (horizontally) or $\boldsymbol{M}$ (vertically) interpolate between the vertices, implying that $\boldsymbol{P}$ and $\boldsymbol{Q}$ are associated with the links.

The two (solid and dashed) trajectories along the lattice must lead to the same result, as a consequence of the Lax consistency. We then get the following two chains of Lax matrices involved:

$$\text{(solid arrows)}: \quad \boldsymbol{U}_{2n+2}\,\boldsymbol{Q}\,\widehat{\boldsymbol{U}}_{2n+2}^{-1}\,\boldsymbol{U}_{2n+1}\,\boldsymbol{Q}\,\boldsymbol{U}_{2n+2}^{-1}\,\boldsymbol{U}_{2n}\,\boldsymbol{P}\,\boldsymbol{U}_{2n+1}^{-1}\,,$$

by following the solid trajectory , and by pursuing the dashed arrows we get:

$$\text{(dashed arrows)}: \quad \widehat{\boldsymbol{U}}_{2n+1}\,\boldsymbol{Q}\,\widehat{\boldsymbol{U}}_{2n+2}^{-1}\,\widehat{\boldsymbol{U}}_{2n}\,\boldsymbol{P}\,\widehat{\boldsymbol{U}}_{2n+1}^{-1}\,\boldsymbol{U}_{2n}\,\boldsymbol{Q}\,\widehat{\boldsymbol{U}}_{2n}^{-1}\,.$$

We cancel the leftmost factor by dividing both chains from the left by the matrix $\widehat{\boldsymbol{U}}_{2n+1} = \boldsymbol{U}_{2n+2}$, and multiply both chains from the right by $\boldsymbol{U}_{2n-1}$. Furthermore in the second chain we replace $\widehat{\boldsymbol{U}}_{2n-1}$ with $\boldsymbol{U}_{2n}$, and then the equality of these paths yield

$$\begin{aligned}&\left(\boldsymbol{Q}\,\widehat{\boldsymbol{U}}_{2n+2}^{-1}\,\boldsymbol{U}_{2n+1}\right)\left(\boldsymbol{Q}\,\boldsymbol{U}_{2n+2}^{-1}\,\boldsymbol{U}_{2n}\,\boldsymbol{P}\,\boldsymbol{U}_{2n+1}^{-1}\,\boldsymbol{U}_{2n-1}\right) = \\ &= \left(\boldsymbol{Q}\,\widehat{\boldsymbol{U}}_{2n+2}^{-1}\,\widehat{\boldsymbol{U}}_{2n}\,\boldsymbol{P}\,\widehat{\boldsymbol{U}}_{2n+1}^{-1}\,\widehat{\boldsymbol{U}}_{2n-1}\right)\left(\boldsymbol{Q}\,\widehat{\boldsymbol{U}}_{2n}^{-1}\,\boldsymbol{U}_{2n-1}\right)\,.\end{aligned}$$

The factors in this equation will be called

$$\mathcal{L}_n = \boldsymbol{Q}\,\boldsymbol{U}_{2n+2}^{-1}\,\boldsymbol{U}_{2n}\,\boldsymbol{P}\,\boldsymbol{U}_{2n+1}^{-1}\,\boldsymbol{U}_{2n-1}, \quad \mathcal{M}_n = \boldsymbol{Q}\,\widehat{\boldsymbol{U}}_{2n}^{-1}\,\boldsymbol{U}_{2n-1}, \tag{6.73}$$

leading to the matrix relation

$$\mathcal{M}_{n+1}\,\mathcal{L}_n = \widehat{\mathcal{L}}_n\,\mathcal{M}_n\,. \tag{6.74}$$

This is what happens in an elementary quadrilateral of the lattice. If the reduction staircase contains edges of several quadrilateral, we have to follow the full path using the above procedure. For example in the three-periodic reduction we get

$$U_{2n+2}\, Q\, \widehat{U}_{2n+2}^{-1}\, U_{2n+1}\, Q\, U_{2n+2}^{-1}\, U_{2n}\, P\, U_{2n+1}^{-1}\, U_{2n-1}\, P\, U_{2n}^{-1} = \\ \widehat{U}_{2n+1}\, Q\, \widehat{U}_{2n+2}^{-1}\, \widehat{U}_{2n}\, P\, \widehat{U}_{2n+1}^{-1}\, \widehat{U}_{2n-1}\, P\, \widehat{U}_{2n}^{-1}\, U_{2n-1}\, Q\, \widehat{U}_{2n-1}^{-1}\, .$$

As before, we cancel from the left by $U_{2n+2} = \widehat{U}_{2n+1}$ and multiply from the right by U_{2n-2} and change U_{2n-1} to $\widehat{U}_{2n-2}$. Then after defining

$$\mathcal{L}_{2n+2} = Q\, U_{2n+2}^{-1}\, U_{2n}\, P\, U_{2n+1}^{-1}\, U_{2n-1}\, P\, U_{2n}^{-1} U_{2n-2}, \quad \mathcal{M}_{\nu} = Q\, \widehat{U}_{\nu}^{-1}\, U_{\nu-1},$$

and applying the $(2, 1)$ reduction we find that $\mathcal{M}_{2n+2} = \mathcal{M}_{2n-1}$ and then the equation becomes

$$\mathcal{M}_{2n+2}\, \mathcal{L}_{2n+2} = \widehat{\mathcal{L}}_{2n+2}\, \mathcal{M}_{2n+2}\, . \tag{6.75}$$

From this we find that $\mathcal{L}$ evolves by a similarity transformation and therefore its trace and determinant are conserved quantities.

Note that the Lax matrices are ordered products composed of elementary blocks of the form

$$\mathcal{P}_i = P U_i^{-1} U_{i-2} \quad \text{or} \quad \mathcal{Q}_i = Q U_i^{-1} U_{i-2} \quad \text{and} \quad \mathcal{M}_i = Q \widehat{U}_i^{-1} U_{i-1}$$

where the label i follows the chain. Thus, over one full period of the chain, we build a matrix $\mathcal{T}$, called the *monodromy matrix*. The mapping $\mathcal{T} \to \widehat{\mathcal{T}}$ for the monodromy matrix involves the same matrix $\mathcal{M}$ at the beginning and end of the chain, and thus the map implies

$$\widehat{\mathcal{T}} = \mathcal{M}\, \mathcal{T}\, \mathcal{M}^{-1} \quad \text{with} \quad \mathcal{M} \equiv \mathcal{M}_0 = \mathcal{M}_P \quad \Rightarrow \quad \mathrm{tr}(\widehat{\mathcal{T}}) = \mathrm{tr}(\mathcal{T}) = \text{invariant}\, .$$

The above method applies immediately to any lattice equation for which the Lax matrices allow factorization (6.71).

To illustrate this construction we consider the lattice potential KdV equation, To facilitate the subsequent computation we note that the elementary matrices take the form

$$P U_i^{-1} U_{i-2} = \begin{pmatrix} 1 & 0 \\ p & 1 \end{pmatrix} \begin{pmatrix} p + q + u_{i-2} - u_i & 1 \\ k^2 - p^2 & 0 \end{pmatrix} \begin{pmatrix} 1 & 0 \\ q & 1 \end{pmatrix}^{-1},$$

from which we see that the elementary matrices $\mathcal{P}_i$ depend only on the combination $u_{i-2} - u_i$ which are the reduced variables of the map. The same holds true for the matrices $\mathcal{Q}_i$.

6.5.2 Computation of the invariant for period 3 and 4 reductions

Let us now look at the two cases considered before, namely the period 3 and 4 reductions.

(i) Period $P = 3$ map. In this case we get for the trace of the monodromy matrix:

$$\begin{aligned}
\mathrm{tr}(\mathcal{T}) &= \mathrm{tr}\left(\boldsymbol{Q}\boldsymbol{U}_0^{-1}\boldsymbol{U}_1\boldsymbol{P}\boldsymbol{U}_2^{-1}\boldsymbol{U}_0\boldsymbol{P}\boldsymbol{U}_1^{-1}\boldsymbol{U}_2\right)\\
&= \mathrm{tr}\left\{\begin{pmatrix} X & 1\\ k^2-q^2 & 0\end{pmatrix}\begin{pmatrix} Z & 1\\ k^2-p^2 & 0\end{pmatrix}\begin{pmatrix} Y & 1\\ k^2-p^2 & 0\end{pmatrix}\right\}\\
&= XYZ + (k^2-p^2)(X+Y) + (k^2-q^2)Z\\
&= [XYZ + (p^2-q^2)Z] + (k^2-p^2)(X+Y+Z)\,,
\end{aligned}$$

where we have used shorthand notation $X = p + q + u_1 - u_0$, $Z = 2p + u_0 - u_2$, $Y = p + q + u_2 - u_1$. Since the latter is invariant and the (spectral) parameter k^2 is arbitrary, we obtain the following invariants:

$$I = (XY + p^2 - q^2)Z, \quad \mathcal{C} = X + Y + Z = 4p + 2q.$$

The first invariant is the same as (6.56), since $Z = 4p + 2q - X - Y$. The latter invariant $\mathcal{C}$ is sometimes called a "Casimir", because one can show that it has zero Poisson brackets with other variables.

(ii) Period $P = 4$ map. In this case we get for the trace of the monodromy matrix:

$$\begin{aligned}
\mathrm{tr}(\mathcal{T}) &= \mathrm{tr}\left(\boldsymbol{Q}\boldsymbol{U}_0^{-1}\boldsymbol{U}_2\boldsymbol{P}\boldsymbol{U}_3^{-1}\boldsymbol{U}_1\boldsymbol{Q}\boldsymbol{U}_2^{-1}\boldsymbol{U}_0\boldsymbol{P}\boldsymbol{U}_1^{-1}\boldsymbol{U}_3\right)\\
&= \mathrm{tr}\left\{\begin{pmatrix} p+q-X & 1\\ k^2-q^2 & 0\end{pmatrix}\begin{pmatrix} p+q+Y & 1\\ k^2-p^2 & 0\end{pmatrix}\begin{pmatrix} p+q+X & 1\\ k^2-q^2 & 0\end{pmatrix}\begin{pmatrix} p+q-Y & 1\\ k^2-p^2 & 0\end{pmatrix}\right\}\\
&= [(p+q)^2 - X^2][(p+q)^2 - Y^2] + 2(k^2-p^2)[(p+q)^2 - XY]\\
&\qquad +2(k^2-q^2)[(p+q)^2 + XY] + (k^2-p^2)^2 + (k^2-q^2)^2\\
&= \mathcal{I} + 2(p+q)^2(2k^2 - p^2 - q^2) + (k^2-p^2)^2 + (k^2-q^2)^2\,,
\end{aligned}$$

where $X = u_0 - u_2$ and $Y = u_1 - u_3$.

Since $\mathcal{T}$ changes under the map by a similarity transformation, the eigenvalues of $\mathcal{T}$ are invariant as well. This is referred to as *isospectrality* and the corresponding eigenvalue problem $\mathcal{T}\phi = \lambda\phi$ is called **isospectral**. Note that $\mathcal{T}$ depends on the arbitrary parameter k^2 (which also plays the role of a spectral parameter in the context of the KdV theory) which is distinct from the eigenvalue λ of the eigenvalue problem for $\mathcal{T}$. However, these two spectral variables are related through the characteristic equation

$$\det(\lambda I - \mathcal{T}) = \lambda^2 - \mathrm{tr}(\mathcal{T}(k^2))\lambda + \det(\mathcal{T}(k^2)) = 0\,,$$

which is a polynomial quadratic in λ and quartic in k^2, and which is called the *spectral curve*. This plays a role in constructing solutions of the mapping, which in the case of the above two reductions can be expressed in terms of elliptic functions.

The trajectory of the map will be restricted to a level set $I(x, y) = I_0$ given by some particular value I_0 of the invariant determined by initial conditions. Since the expression for I is a biquadratic in terms of the variables x, y, $I(x, y) = I_0$ also defines an elliptic curve on which the motion stays. The essence of the solution method is in establishing a connection between the two algebraic curves, namely the spectral curve and the curve given by the invariant.

6.6 Pole reduction of the semi-discrete KP equation

In addition to integrable maps arising from periodic (or other) initial value problems for integrable lattice equations, there exist other solutions of those equations that give rise to finite-dimensional integrable systems. One of these is the class of rational solutions arising from pole reduction (Choodnovsky and Choodnovsky, 1977; Airault et al., 1977; Krichever, 1978, 1980; Wojciechowski, 1982). Interestingly, this pole reduction provides a link from soliton equations to integrable many-body systems of Calogero–Moser (CM) type (1.42).

The pole reduction method also works in the discrete case, providing integrable correspondences from lattice equations (Nijhoff and Pang, 1994, 1996). To illustrate the method we use the semi-discrete KP equation (Nijhoff et al., 1984, 1985)

$$\partial_\xi(\widehat{u} - \widetilde{u}) = (p - q + \widehat{u} - \widetilde{u})(u + \widehat{\widetilde{u}} - \widehat{u} - \widetilde{u})\,. \tag{6.76}$$

In the continuous case the KP equation provides the best starting point for this purpose (Krichever, 1978, 1980). We follow this example by analogy.

First note that equation (6.76) arises as the compatibility condition of the Lax pair for a scalar function $\phi(n, m; \xi)$ depending on two discrete and one continuous variable:

$$\widetilde{\phi} = \partial_\xi\phi + (p + u - \widetilde{u})\phi\,, \tag{6.77a}$$

$$\widehat{\phi} = \partial_\xi\phi + (q + u - \widehat{u})\phi\,. \tag{6.77b}$$

The main assumption of the pole reduction is that the function u depends rationally on ξ as follows:

$$u = \sum_{i=1}^{N} \frac{1}{\xi - x_i(n, m)}\,, \tag{6.78}$$

where the poles $x_i(n, m)$ only depend on the discrete variables n, m, and we will derive equations on x_i which imply that u solves equation (6.76). This ansatz is motivated by the observation that (6.76) admits solutions of the form

$$u = \frac{1}{\xi + n/p + m/q}\,. \tag{6.79}$$

If we take this solution then the linear problem (6.77) has a solution of the form

$$\phi = \left(1 - \frac{1/k}{\xi + n/p + m/q}\right)(p + k)^n\,(q + k)^m\,e^{k\xi}\,. \tag{6.80}$$

Note the extra parameter k that plays the role of a spectral variable.

The rational solution (6.78) is a generalization of (6.79) by allowing multiple poles and the solution of the Lax pair generalizing (6.80) is given by

$$\phi = \left(1 - \frac{1}{k}\sum_{i=1}^{N} \frac{b_i(n,m)}{\xi - x_i(n,m)}\right)(p+k)^n\,(q+k)^m\,e^{k\xi}\,, \tag{6.81}$$

where the coefficients b_i are yet to be determined. Substituting (6.78) and (6.81) into the first Lax equation (6.77a) and using partial fraction expansion to separate terms of the type $(\xi - x_i(n,m))^{-1}$ and $(\xi - x_i(n+1,m))^{-1}$ we obtain

$$(p+k)b_i = k + \left(\sum_{l=1}^{N} \frac{1}{x_i - \widetilde{x}_l} - \sum_{\substack{l=1\\ l\neq i}}^{N} \frac{1}{x_i - x_l}\right) b_i - \sum_{\substack{j=1\\ j\neq i}}^{N} \frac{b_j}{x_i - x_j}\,, \tag{6.82a}$$

$$(p+k)\widetilde{b}_i = k - \sum_{j=1}^{N} \frac{b_j}{\widetilde{x}_i - x_j}, \qquad i = 1, 2, \ldots, N. \tag{6.82b}$$

A similar set of equations with tilde replaced with hat and p with q is obtained from the other Lax equation (6.77b). We write these results in matrix form by introducing the vectors $\boldsymbol{b} = (b_1, b_2, \ldots, b_N)^T$, and $\boldsymbol{e} = (1, 1, \ldots, 1)^T$, and the Lax matrices

$$\boldsymbol{L} = \sum_{i,j=1}^{N} \frac{E_{ii}}{x_i - \widetilde{x}_j} - \sum_{\substack{i,j=1\\ j\neq i}}^{N} \frac{E_{ii} + E_{ij}}{x_i - x_j}, \tag{6.83}$$

$$\boldsymbol{M} = -\sum_{i,j=1}^{N} \frac{E_{ij}}{\widetilde{x}_i - x_j}, \tag{6.84}$$

where E_{ij} are matrices with entries $(E_{ij})_{kl} = \delta_{ik}\delta_{jl}$. We can then write (6.82a) and (6.82b) as

$$(p+k)\boldsymbol{b} = k\boldsymbol{e} + \boldsymbol{L}\boldsymbol{b}, \tag{6.85a}$$

$$(p+k)\widetilde{\boldsymbol{b}} = k\boldsymbol{e} + \boldsymbol{M}\boldsymbol{b}, \tag{6.85b}$$

which forms an $N \times N$ matrix Lax pair. Remarkably, the tilde part of the Lax pair of the semidiscrete KP equation alone already yields a Lax pair of the reduced system; thus we have reduced from $(2+1)$D directly to 1D.

Despite the fact that the Lax pair (6.85) is not in the standard form due to the inhomogeneous term $k\boldsymbol{e}$ we can nevertheless compute the compatibility condition and obtain

$$\left(\widetilde{\boldsymbol{L}}\boldsymbol{M} - \boldsymbol{M}\boldsymbol{L}\right)\boldsymbol{b} + k\left(\widetilde{\boldsymbol{L}} - \boldsymbol{M}\right)\boldsymbol{e} = 0. \tag{6.86}$$

It is not obvious that we can split the two terms in (6.86), as the functions b_i may still depend on k, but one can show that the latter relation contains a common factor (see Exercise 6.12) which leads to the system of equations:

$$\sum_{j=1}^{N}\left(\frac{1}{x_i-\widetilde{x}_j}+\frac{1}{x_i-\underset{\sim}{x}_j}\right)-2\sum_{\substack{j=1\\ j\neq i}}^{N}\frac{1}{x_i-x_j}=0\,,\quad i=1,\ldots,N. \tag{6.87}$$

in which $\underset{\sim}{x}_i(n,m)=x_i(n-1,m)$. Conversely, both relations

$$\widetilde{\boldsymbol{L}}\boldsymbol{M}=\boldsymbol{M}\boldsymbol{L}\quad\text{and}\quad(\widetilde{\boldsymbol{L}}-\boldsymbol{M})\boldsymbol{e}=0 \tag{6.88}$$

hold simultaneously if (6.87) are satisfied. The latter is a coupled system of N second-order ordinary difference equations for the quantities $x_i(n,m)$ as functions of n, and as is shown below, they give rise to a multi-valued map or *correspondence*. This system of equations can be considered as a discrete-time analogue of the rational Calogero–Moser system, because the continuum limit of (6.87) yields (Nijhoff and Pang, 1994, 1996)

$$\ddot{x}_i=-\sum_{\substack{j=1\\ j\neq i}}^{N}\frac{\alpha_0}{(x_i-x_j)^3}, \tag{6.89}$$

in which the coupling constant α_0 arises from the continuum limit (see Exercise 6.13). Note that the sign corresponds to the Calogero–Moser system with an attractive inverse squared potential, while the original model had repulsive potential.

In order to make the connection between the *discrete-time Calogero–Moser system* (6.87) and the semi-discrete KP equation (6.76), we need also to consider simultaneously the other member of the Lax pair (6.77). A similar analysis as the one given above shows easily that we obtain once again another copy of the equations (6.87), but now in the hat direction:

$$\sum_{j=1}^{N}\left(\frac{1}{x_i-\widehat{x}_j}+\frac{1}{x_i-\underset{\frown}{x}_j}\right)-2\sum_{\substack{j=1\\ j\neq i}}^{N}\frac{1}{x_i-x_j}=0\,,\quad(i=1,\ldots,N), \tag{6.90}$$

in which the under-shift $\underset{\frown}{x}_i$ is shorthand for a shift in the variable m over one unit in the negative direction, i.e. $\underset{\frown}{x}_i(n,m)=x_i(n,m-1)$. The distinction between the two discrete-time variables, which is not evident from the two identical systems of equations, arises from two additional equations which involve the lattice parameters p and q of (6.76). These are the relations:

$$p-q=\sum_{l=1}^{N}\left(\frac{1}{x_i-\widetilde{x}_l}-\frac{1}{x_i-\widehat{x}_l}\right),\quad i=1,\ldots,N, \tag{6.91a}$$

$$=\sum_{l=1}^{N}\left(\frac{1}{x_i-\underset{\frown}{x}_l}-\frac{1}{x_i-\underset{\sim}{x}_l}\right),\quad i=1,\ldots,N, \tag{6.91b}$$

which not only ensure (if $p \neq q$) that the two flows given by the tilde and hat operation are distinct, but also guarantee that they are compatible (in other words, that the shifts along these flow-directions commute: $\widehat{\widetilde{x}}_i = \widetilde{\widehat{x}}_i$) (Yoo-Kong et al., 2011).

The compatibility of the flows given by (6.87, 6.90, 6.91) is not easy to check by direct computation, since the equations are implicit, but the statement must be true because we can construct a common solution to all these equations. The solution is given by the following statement:

Theorem 6.6.1 *The eigenvalues $x_1(n, m), \ldots, x_N(n, m)$ of the $N \times N$ matrix*

$$\boldsymbol{Y}(n, m) = \boldsymbol{Y}(0, 0) - n(p\boldsymbol{I} + \boldsymbol{\Lambda})^{-1} - m(q\boldsymbol{I} + \boldsymbol{\Lambda})^{-1} \tag{6.92a}$$

where $\boldsymbol{\Lambda}$ is an arbitrary diagonal matrix, and in which the initial value matrix $\boldsymbol{Y}(0, 0)$ is subject to the condition

$$[\,\boldsymbol{Y}(0, 0)\,,\ \boldsymbol{\Lambda}\,] = \boldsymbol{I} + \textit{rank}\ 1\ \textit{matrix} \tag{6.92b}$$

obey both the discrete-time Calogero–Moser systems given by equations (6.87) *and* (6.90) *as well as the systems of constraint equations given by* (6.91).

The proof of this statement is based on the Lax pair and can be found in the Appendix of Yoo-Kong et al. (2011). In the case of $N = 2$ it can be verified by direct computation; see Exercise 6.14. As a consequence of this statement we obtain the simultaneous solutions $x_i(n, m)$ of the system of equations constituting the discrete-time CM system, from a secular problem for a matrix depending. It is evident from the commutativity of the terms containing the variables n, m in (6.92a) that these discrete flows commute, and hence this provides us with a nontrivial system with *commuting discrete flows*.[5] Integrable commuting maps are discussed further in Veselov (1987, 1991a,b).

The multi-valuedness of the solution is a consequence of the fact that the eigenvalue problem provides us only with an *unordered* set of roots of the corresponding algebraic equations, and hence these different roots can be permuted.

Here we see an important new feature of the discrete system contrasting it from the continuous-time systems: In the continuous-time analogue of the CM system the solution is composed of individual particles whose positions can be identified on the basis of the continuity of the time-flow, while in the discrete-time case, at each time step we can have permutation of particles and we cannot identify individual particles by their trajectories. Thus, in the discrete-time case we can have indistinguishability of particles on the classical level while this is often seen as a pure quantum phenomenon.

Finally we note that in Nijhoff et al. (1996) the discrete-time version of the relativistic variant of the CM system, the Ruijsenaars–Schneider system (Ruijsenaars and Schneider, 1986), was constructed.

[5] This means in this context that there is an OΔE in m and an OΔE in n, which are compatible with each other.

6.7 Notes

Mappings and difference equations of the type we have considered in this chapter fit into two established areas of mathematics, namely dynamical systems theory and finite-difference approximations in numerical analysis. However, the systems that we consider stand out by the property of integrability. This property already appeared in the classical problems of geometry (going back to Jacobi 1823), e.g. the elliptic billiard and the Poncelet's porism (Veselov, 1991a,b; Dragović and Radnović, 2011).

In the modern era, the McMillan map (McMillan, 1971) was one of the first examples of a family of maps possessing an exact invariant. Such nonlinear integrable maps have been used to investigate near integrable dynamics (Glasser et al., 1989). Since the late 1980s the reductions of soliton equations have provided a major source for integrable maps (Moser and Veselov, 1991; Quispel et al., 1988, 1989; Papageorgiou et al., 1990; Quispel et al., 1991; Bruschi et al., 1991; Grammaticos and Ramani, 1996). Since the early 1990s there has been a wealth of activity in construction integrable discretizations of integrable ODEs (Suris, 2003; Grammaticos et al., 2004).

The general problem of obtaining finite-dimensional reductions of two-dimensional lattices through initial value problems has also been studied in recent years. A "standard staircase" reduction was introduced in Quispel et al. (1991), generalizing the discussion in Section 6.4.3. Initial value problems defined on more general stencils were discussed in van der Kamp and Quispel (2010). Some general results have been established, e.g. on the well-posedness of initial value problems; see Adler and Veselov (2004); van der Kamp (2009, 2015).

Higher-dimensional[6] analogues of QRT were investigated by several researchers (Capel and Sahadevan, 2001; Roberts and Quispel, 2006; Fordy and Kassotakis, 2006). Special class of higher-dimensional maps is also obtained from higher periodic reductions of lattice equations; see e.g. Section 6.4.3.

Geometric integration is another area where discretizations appear in the form of numerical methods for the integration of (usually ordinary) differential equations, which preserve certain geometric properties such as conservation laws or a symplectic structure. This is a relatively new area of numerical analysis, started by the seminal paper of Yoshida (1990), and which has become a very active field of research (Sanz-Serna and Calvo, 1994; Hairer et al., 2006). A connection with integrability of maps is elaborated in Field and Nijhoff (2003).

Still another interesting direction of research is establishing the connection between integrable recurrences and cluster algebras through the Laurent phenomenon. The Laurent phenomenon is the property that certain recurrences defined by rational functions, where one would expect generically rational functions to appear in the iteration process,

[6] The reader should be aware that the usage of the word "dimensional" here, following the literature on this topic, may lead to confusion. In the present context *higher-dimensional* actually means *multicomponent*, i.e. higher-dimensional with regard to the space of the dependent variables, while elsewhere by "higher-dimensional" we refer to the space of the independent variables.

nevertheless only yield upon iteration Laurent polynomials (polynomial in positive and negative powers) in terms of the initial data (Fomin and Zelevinsky, 2002b). On the basis of this property Fomin and Zelevinsky introduced the new mathematical objects called *cluster algebras* (Fomin and Zelevinsky 2002a, 2003; Marsh 2013) which have since then attracted interest in representation theory and integrable systems (Fordy and Marsh, 2011; Fordy and Hone, 2014). The general connection between cluster algebras and DIS is still being explored. A particular spinoff of this branch of research is the study of integer sequences (the simplest example of which is the Fibonacci sequence) defined by rational recurrences, in particular the famous Somos sequences[7] (Hone, 2007; Hone and Swart, 2008).

Exercises

6.1 Show that when $N = 1$ one can solve for G from (6.7) when (6.4) is given and satisfies (6.6). In the one-dimensional case we can write the equations as

$$p' = f(q, p), \quad q' = g(q, p), \quad q' - q = \partial_{p'} G(q, p'), \quad p' - p = -\partial_q G(q, p').$$

(1) Since G depends only on q, p' invert $p' = f(q, p)$ to obtain $p = \phi(q, p')$ and $q' = g(q, \phi(q, p'))$.

(2) Integrate both G equations to obtain

$$G_1 = -qp' + \int^{p'} g(q, \phi(q, s))\, ds + a(q),$$
$$G_2 = -qp' + \int^{q} \phi(t, p')\, dt + b(p'),$$

where $a(q)$ and $b(p')$ are the respective integration "constants".

(3) Due to the integration constants it is enough to show that $\partial_q \partial_{p'}(G_1 - G_2) = 0$. Show that this yields

$$\partial_q g(q, \phi(q, p')) - \partial_{p'} \phi(q, p') = 0.$$

(4) Now it is necessary to return to q, p variables. Show that

$$\partial_{p'} \phi(q, p')|_{p'=f(p,q)} = \frac{1}{\partial_p f(p, q)}, \quad \partial_q \phi(q, p')|_{p'=f(p,q)} = -\frac{\partial_q f(q, p)}{\partial_p f(p, q)},$$

and using these write the condition as $\partial_q g\, \partial_p f - \partial_p g\, \partial_q f = 1$.

6.2 (a) Show that the map (6.61) is area-preserving, and that

$$\mathfrak{I} := (XY - b)(a - X - Y)$$

[7] http://faculty.uml.edu/jpropp/somos.html

is invariant.

(b) Show that the map (6.67) is area preserving, and possesses the following invariant:

$$\mathfrak{I}_N = (b^2 - X_N^2)(b^2 - X_{N-1}^2) + 2aX_N X_{N-1}, \tag{6.93}$$

where $a = p^2 - q^2$ and $b = p + q$, i.e. show that $\mathfrak{I}_{N+1} = \mathfrak{I}_N$.

6.3 In order to obtain the results described in Section 6.1.5, let us introduce the variable

$$u_n \equiv x_n + \frac{1}{2}f(x_n) - x_{n-1} = x_{n+1} - x_n - \frac{1}{2}f(x_n)$$

and derive the functional equation

$$K(x, x + \frac{1}{2}f(x) - u) = K(x, x + \frac{1}{2}f(x) + u).$$

Introducing also $\varphi(x - y) = K_0(x, y)$ and expanding in orders of $\epsilon, \epsilon^2, \epsilon^3$, show by elimination of the coefficients f_1, f_2, f_3 that φ obeys the equation

$$\varphi'''(u) = c\varphi'(u),$$

and deduce the various possibilities **(a)**–**(c)**.

6.4 In result **(c)**, of Section 6.1.5 make a substitution $x = \frac{1}{\alpha}\log(y)$ and deduce a rational map

$$y_{n+1}y_{n-1} = \frac{y^2 + \epsilon\alpha(D + By - Ey^2)}{1 - \epsilon\alpha(E + Ay + Cy^2)}.$$

What form does the invariant take?

Use a similar method to reduce result **(b)** into a rational map.

6.5 Show that in the symmetric case ($f_i = g_i$), starting with (6.39) and using labeling $x = x_n$, $y = x_{n+1/2}$, $\bar{x} = x_{n+1}$ etc., the involutions are same except for an index shift by $1/2$. Therefore the symmetric QRT map (6.41) can be viewed as the product of two identical maps.

6.6 Show that the map (6.48) arises from (6.46) with the matrices $L_n(h)$ and $M_n(h)$ given by (6.47a) and (6.47b).

6.7 Perform the (1, 1) reduction for Q4 to obtain a map and show that it is HKY-type.

6.8 Solve (6.55) in terms of elliptic functions.

6.9 Establish a direct relation between (6.61) and (6.55).

6.10 Work out the Lagrangians for the autonomous version of discrete $P_{\rm I}$ and $P_{\rm II}$.

6.11 (a) Given a Lax representation $\boldsymbol{L\phi} = \eta\boldsymbol{\phi}$, $\boldsymbol{M\phi} = \widetilde{\boldsymbol{\phi}}$ with Lax matrices

$$\boldsymbol{L} = \begin{pmatrix} x & 1 \\ \lambda + a & 0 \end{pmatrix}\begin{pmatrix} z & 1 \\ \lambda & 0 \end{pmatrix}\begin{pmatrix} y & 1 \\ \lambda & 0 \end{pmatrix}, \quad \boldsymbol{M} = \begin{pmatrix} w & 1 \\ \lambda + a & 0 \end{pmatrix},$$

where λ is a spectral parameter, a a constant, and η an eigenvalue of the matrix $\boldsymbol{L}$, derive from the Lax compatibility the map $(x, y, z) \mapsto (\widetilde{x}, \widetilde{y}, \widetilde{z})$ for the dynamical variables x, y, z. Determine also the dynamical quantity w in terms of x, y, z.
(b) By considering the invariant curve $\det(\eta \boldsymbol{I} - \boldsymbol{L}) = 0$ derive the two invariants $C = x + y + z$ and $I = xyz + az$ of this map. Furthermore, by setting C equal to a constant and eliminating z, derive a reduced map $(x, y) \mapsto (\widetilde{x}, \widetilde{y})$. Show that this map is area-preserving.

6.12 Using the definitions given in this section, show that left-hand side of (6.86) can be written as

$$\sum_{i,j=1}^{N} \left[\sum_{l=1}^{N} \left(\frac{1}{\widetilde{x}_i - \widetilde{\widetilde{x}}_l} + \frac{1}{\widetilde{x}_i - x_l} \right) - 2 \sum_{\substack{l=1 \\ l \neq i}}^{N} \frac{1}{\widetilde{x}_i - \widetilde{x}_l} \right] \left(\frac{k}{N} E_{ij} \boldsymbol{e} - \frac{E_{ij}}{\widetilde{x}_i - x_j} \boldsymbol{b} \right)$$

and hence conclude that this vanishes if (6.87) is satisfied. Furthermore, show that both equations (6.88) are implied by (6.87).

6.13 Show that setting $x_i(n) = X_i(t) + n\alpha$ where $t = t_0 + n\delta$ and taking the limit $n \to \infty$, $\delta \to 0$, such that $n\delta \to \tau$ fixed and $\alpha = \alpha_0 \delta^2$ with α_0 fixed, the equations (6.87) become (6.89).

6.14 Compute the eigenvalues of (6.92a) in the case $N = 2$ and show that they solve the equations (6.87), (6.90), (6.91). (Use arbitrary diagonal $\boldsymbol{\Lambda} = \mathrm{diag}(\lambda_1, \lambda_2)$ and $\boldsymbol{Y}(0, 0) = \begin{pmatrix} a & b \\ c & d \end{pmatrix}$, where $bc(\lambda_1 - \lambda_2)^2 = -1$.)

7

Identifying integrable difference equations

Suppose that we are *given* some difference equation arising in some practical problem: can we say something about the integrability of the equation without actually solving it? Furthermore, can this evaluation be done algorithmically? In the previous chapters we have *derived* equations from some specific mathematical properties, and from this it followed that the equations had some nice regularity properties that allowed us to call them integrable. In this chapter we approach the problem from the opposite point of view.

For differential equations the predictive method that has often been used is local analysis (in complex variables) to check whether solutions have movable singularities (Painlevé method) (Painlevé, 1902, 1906; Ablowitz et al., 1980a,b).

For continuous equations the Painlevé method has turned out to be very effective. Below we will discuss how an analogue can be implemented for difference equations. However, it turns out that this type of local analysis is not enough and it is also necessary to develop global methods. Among these, the most useful are based on (i) geometric description of the initial value space (Okamoto space), (ii) growth of complexity of solutions or algebraic entropy and (iii) application of Nevanlinna theory. In this chapter we will describe some of these through examples. For a general review and introduction, see also Grammaticos et al. (2009).

7.1 Singularity analysis of differential and difference equations

In 1988 Kowalevski was awarded a prestigious Bordin prize for her work on identifying a new spinning top and integrating its equations of motion (Kowalevski, 1889, 1890). Kowalevski's crucial idea was to demand that the solutions be meromorphic: that is, only have poles as singularities. This was known to be true for previously known integrable cases of the top analyzed by Euler and Lagrange, who solved their respective problems in terms of elliptic functions. The positions of the poles of solutions depended on initial values; that is, they are movable. Kowalevski imposed this requirement of meromorphicity, discovered a new integrable case and provided explicit solutions in terms of hyperelliptic functions.

Figure 7.1 S. Kowalevski

This insight of Kowalevski was used by many mathematicians including L. Fuchs and Picard as a starting point in the search for new globally defined functions that arise as solutions of differential equations (see Chapters 13 and 14 of Ince, 1956). This led eventually to the six equations that are now called the Painlevé equations. (A brief history of these developments and a list of the Painlevé equations can be found in Appendix C.) The property that all movable singularities of all solutions be poles is now called the Painlevé property and the method for checking necessary conditions following from this property is called the Painlevé test. This algorithmic test relies on expanding solutions in a power series around movable singularities and has been widely used, in various forms, for both ODEs and PDEs.

It is natural to ask whether there is an analogue of these ideas for difference equations. In the differential case, if a meromorphic solution exists, it can be expanded in a near neighborhood of a singularity. However, in the discrete case, there is no natural concept of a near neighborhood in terms of the independent variable. The situation is still problematic even if we were to consider *analytic* difference equations (see Section 1.1.3). Instead, the essential idea for singularity analysis turns out to be to consider singularities in the initial values.

7.1.1 Singularity analysis in the continuum case

The analysis proceeds by constructing a power series expansion of a solution in the neighborhood of a movable singularity. The first step is to find the possible leading powers of the expansion, by doing a careful analysis of leading terms of the equation and how they balance in the limit as we approach the singularity.

Consider the equation

$$y' = -y^2 + x.$$

Assuming that a singularity is located at an arbitrary point x_0 (which means that it is movable) and substituting an expansion in powers of $x - x_0$, starting with $a_p(x - x_0)^p$ $(p < 0)$, we find that the balance of leading powers gives us $p - 1 = 2p$, and therefore $p = -1$, and $a_{-1} = 1$. Substituting the full expansion starting therefore with $(x - x_0)^{-1} + a_0 + a_1(x - x_0)\ldots$ we find that every successive coefficient is well defined for arbitrary x_0. Since the general solution of a first-order ODE can only have one integration constant (here x_0), this expansion serves to represent the general solution.

To describe general solutions of second-order ODEs, such an expansion should contain a second integration constant in addition to x_0, the position of the pole. For example, for $y'' = 6y^2 + x$ (the first Painlevé equation), one can construct a power series expansion that starts with $(x - x_0)^p$, where, due to the second derivative y'' and the quadratic term y^2, we find that $(x - x_0)^{p-2}$ must balance with $(x - x_0)^{2p}$, and so $p = -2$. We also find that the coefficient a_{-2} of the first terms satisfies $6a_{-2} = 6(a_{-2})^2$, that is, $a_{-2} = 0, 1$. The first case, along with the next term, implies we no longer have a Laurent expansion and so we choose the second case. Consequently, substituting the power series expansion of the solution $y(x) = (x-x_0)^{-2} + a_{-1}(x-x_0)^{-1} + a_0 + a_1(x-x_0)\ldots$ into the equation, we find that each successive coefficient is defined until we get to a_4, the coefficient of $(x-x_0)^4$, which turns out to be undefined; that is, it is the second free constant in the power series. This step arises at the place where the recursion relation for the coefficients a_n becomes an identity, and the index where it occurs is often called a *resonance* in the expansion. A proof of convergence of this series can be found in Joshi and Kruskal (1994).

In contrast, for nonintegrable equations, the recursion relation becomes contradictory at the place of a resonance. To resolve the contradiction, it is usual to introduce logarithms at that point in the expansion in order to construct a general solution containing sufficiently many integration constants. An example is $y'' = 6y^2 + x^2$, for which the Laurent expansion has to be extended to include a logarithmic term $\log(x-x_0)$ multiplying a sub-series starting with the term $(x-x_0)^4 \log(x-x_0)$. In other nonintegrable cases, such as $y'' = 6y^4 + x$, non-integer powers also need to be introduced in the expansion (Ramani et al., 1982; Kruskal and Clarkson, 1992; Babelon et al., 2010).

7.1.2 Singularity analysis for O∆Es

What could be a singularity for a difference equation? First consider the equation

$$y_{n+1} = \frac{y_n + n}{y_n}.$$

We can analyze this equation near the apparent singularity $y = 0$ by using the initial value $y_{n_0} = \epsilon$, $(\epsilon \ll 1)$. The next two iterates are $y_{n_0+1} = n_0/\epsilon + 1$, $y_{n_0+2} = 1 + \mathcal{O}(\epsilon)$. We see that the solution is well defined after two steps and analytic in ϵ. This procedure is the analogue of the Laurent expansion method for ODEs. The initial point n_0 is arbitrary and corresponds to the location of the movable pole. If we were to change the equation to $y_{n+1} = (y_n^2 + n)/y_n$ then after the first two steps the iterates would continue to have the leading term n_0/ϵ forever. We say that the singularity is confined in the first case but not in the second case.

Sometimes the map might not have singularities when iterated forward, but is singular when iterated backward. For example, reversing $y_{n+1} = ay_n(1-y_n)$ would lead to multi-valuedness in the form of square roots, and this multivaluedness would grow exponentially as we go further toward decreasing n-values. In the following we only consider maps that are rational in both directions: they are called *birational.*

For rational maps an infinite value of the iterate y is not, by itself, problematic because we can always transform to projective coordinates where it would be a finite value. However, when the definition of the next iterate is ambiguous such as $0/0$ or $\infty - \infty$, we are at a true singularity as it prevents us from defining further iterates.

dP_I: As an example consider the discrete Painlevé I equation ($\mathrm{dP_I}$) already given in (1.20)

$$x_{n+1} + x_n + x_{n-1} = \frac{\alpha + \beta n}{x_n} + b. \tag{7.1}$$

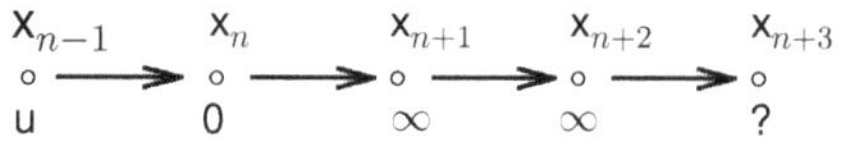

Figure 7.2 The sequence leading to a singularity.

A potential singularity is reached if some previous values of x_n are such that (7.1) yields $x_0 = 0$. How does the sequence continue?

Let us first consider the autonomous case

$$x_{n+1} = -x_n - x_{n-1} + \frac{a}{x_n} + b. \tag{7.2}$$

The sequence starting with $x_{-1} = u,\ x_0 = 0$ continues as:

$$\begin{aligned} x_1 &= -0 - u + a/0 + b = \infty, \\ x_2 &= -\infty - 0 + a/\infty + b = -\infty, \\ x_3 &= +\infty - \infty + a/\infty + b = \ ? \end{aligned}$$

where we have denoted unbounded values by the symbol ∞; see also Figure 7.2. Infinite values are not a problem, as they can be handled with projective coordinates (see Section 7.2.1). The true singularity is the *ambiguity* in the third iterate x_3, which we have denoted by "$\infty - \infty$" and to resolve it we expand around the troublesome initial value $x_0 = 0$.

Thus for a more detailed analysis of what happens near the singularity we assume that $x_0 = \epsilon$ (small) and obtain the following sequence:

$$\begin{aligned} x_{-1} &= u, \\ x_0 &= \epsilon, \\ x_1 &= \tfrac{a}{\epsilon} + b - u - \epsilon, \\ x_2 &= -\left[\tfrac{a}{\epsilon} + b - u - \epsilon\right] - \epsilon + \tfrac{a}{\left[\frac{a}{\epsilon} + b - u - \epsilon\right]} + b \\ &= -\tfrac{a}{\epsilon} + u + \epsilon + \tfrac{u-b}{a}\,\epsilon^2 + O(\epsilon^3), \\ x_3 &= -\left[-\tfrac{a}{\epsilon} + u + \epsilon + \tfrac{u-b}{a}\,\epsilon^2 + O(\epsilon^3)\right] - \left[\tfrac{a}{\epsilon} + b - u - \epsilon\right] \\ &\quad + \tfrac{a}{\left[-\frac{a}{\epsilon} + u + O(\epsilon)\right]} + b \\ &= -\epsilon + \tfrac{b-2u}{a}\,\epsilon^2 + O(\epsilon^3), \\ x_4 &= -[O(\epsilon)] - \left[-\tfrac{a}{\epsilon} + u + O(\epsilon)\right] + \tfrac{a}{\left[-\epsilon + (b-2u)/a\,\epsilon^2 + O(\epsilon^3)\right]} + b \\ &= u + O(\epsilon). \end{aligned}$$

We have found that the previously ambiguous x_3 is now well defined and has the value 0 as $\epsilon \to 0$. We say that in this case the *singularity pattern* is $\ldots, 0, \infty, -\infty, 0, \ldots$.

Furthermore, we found that the initial information u is recovered in x_4; it is as if the initial value was "shaded" by the singularity but emerged finally. Note that the x_4 and later terms are analytic in the initial values u, ϵ, because the map is rational. (Note that if $u \propto \epsilon$ then the singularity is unconfined in both directions. However, we cannot enter into this case from generic regular initial values.)

The **singularity confinement test** proposed in Grammaticos et al. (1991) is based on the above idea: if the step-by-step iteration leads to an ambiguity then after a few more steps one should be able to get out of the singularity, and this should take place *without essential loss of information.*[1]

qP_{III}: As a different example, consider the following equation related to the discrete version of $P_{\rm III}$,

$$x_{n+1}x_{n-1} = \frac{g(x_n - a)(x_n - b)}{(x_n - c)(x_n - d)}. \tag{7.3}$$

Let us see what the singularity confinement test says about this equation. If we solve the equation for forward iteration, potential problem arises when the denominator vanishes, for example when $x_{n-1} = c$. Let us therefore expand around this point:

$$\begin{aligned}
x_{n-2} &= u,\\
x_{n-1} &= c + \epsilon,\\
x_n &= \frac{g(c-a)(c-b)}{(c-d)u\epsilon} + O(1),\\
x_{n+1} &= \frac{g}{c} - \epsilon\frac{(a-c)(b-c)g + (a-c+b-d)(c-d)cu}{(a-c)(b-c)c^2} + O(\epsilon^2),\\
x_{n+2} &= \epsilon\frac{(g-ac)(g-bc)(c-d)u}{(a-c)(b-c)(g-c^2)(g-cd)} + O(\epsilon^2).
\end{aligned}$$

This would continue to give us further singularities, unless the denominator actually vanishes in this expression. In order to confine the singularity here and get a finite and nonzero x_{n+2} we take $g = cd$. Note that we cannot just substitute this into the above expression of x_{n+2}; rather, one has to do the analysis once more with this condition (and in doing that it is necessary to keep the $O(\epsilon)$ term in x_{n+1}). This yields a finite nonzero value for x_{n+2}:

$$x_{n+2} = \frac{(a-d)(b-d)cu}{(a-c)(b-c)d + (a-c+b-d)(c-d)u} + O(\epsilon).$$

Thus we see that (7.3) satisfies a necessary condition of integrability only when $g = cd$. Note that the denominator of x_{n+2} vanishes also if $g = c^2$ but this does not confine, leading instead to a periodic situation.

[1] In fact it has been suggested by M. D. Kruskal that passing the singularity confinement test only means that the system is well posed, even at singular points.

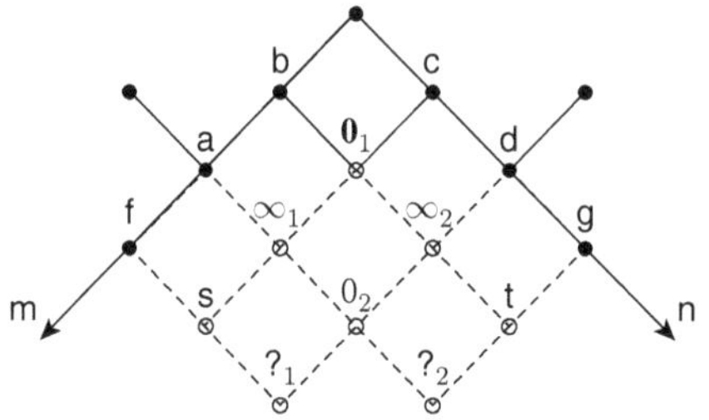

Figure 7.3 Propagation of singularities in a two-dimensional map. Here n grows in the SE direction and m in the SW direction. We have indicated the problematic points with symbols that suggest the size of the iterates, which can be analyzed in more detail by expanding in the initial value $\varepsilon = 0_1$.

7.1.3 Singularity analysis for PΔEs

In the seminal paper of Grammaticos et al. (1991), singularity confinement was also discussed in the two-dimensional setting, with the same idea: *singularities induced by initial data should not propagate.*

As an example, let us consider the Hirota's discretization of the KdV equation (3.87)

$$w_{n+1,m+1} - w_{n,m} = \frac{1}{w_{n+1,m}} - \frac{1}{w_{n,m+1}}. \tag{7.4}$$

We will now study how the initial data progresses, when one hits a singularity. The initial data is given at the solid line on Figure 7.3, $w_{0,2} = a$, $w_{0,1} = b$, $w_{1,1} = 0$, $w_{1,0} = c$, $w_{2,0} = d$. Proceeding from these initial values, we obtain infinite values for $w_{1,2}$ and $w_{2,1}$, then $w_{2,2} = 0$, and finally two ambiguities of the type $\infty - \infty$ for $w_{2,3}$ and $w_{3,2}$.

A more detailed analysis with the initial value $w_{1,1} = \varepsilon$ (small) yields the following values at the subsequent iterations

$$w_{1,2}(= \infty_1) = b + \frac{1}{\varepsilon} - \frac{1}{a}, \quad w_{2,1}(= \infty_2) = c + \frac{1}{d} - \frac{1}{\varepsilon},$$

and then

$$\begin{aligned} s = w_{1,3} &= a + \frac{1}{\infty_1} - \frac{1}{f} = a - \frac{1}{f} + \varepsilon + \mathcal{O}(\varepsilon^2), \\ t = w_{3,1} &= d + \frac{1}{g} - \frac{1}{\infty_2} = d + \frac{1}{g} + \varepsilon + \mathcal{O}(\varepsilon^2), \end{aligned}$$

and subsequently

$$w_{2,2}(= 0_2) = \varepsilon + \frac{1}{\infty_2} - \frac{1}{\infty_1} = -\varepsilon + \left(b - c - \frac{1}{a} - \frac{1}{d}\right)\varepsilon^2 + \mathcal{O}(\varepsilon^3).$$

Then at the next step we can resolve the ambiguities:

$$w_{2,3}(=?_1) = \infty_1 + \frac{1}{0_2} - \frac{1}{s} = c + \frac{1}{d} - \frac{1}{a - 1/f} + \mathcal{O}(\varepsilon), \tag{7.5a}$$

$$w_{3,2}(=?_2) = \infty_2 + \frac{1}{t} - \frac{1}{0_2} = b - \frac{1}{a} + \frac{1}{d + 1/g} + \mathcal{O}(\varepsilon). \tag{7.5b}$$

Thus the singularity is confined for generic initial values a, d, f, g. (Note that for some specific initial values, for example if $af = 1$, more complicated singularity patterns would arise.)

Note that if the lattice equation is deformed, e.g. by taking:

$$w_{n+1,m+1} - w_{n,m} = \frac{1}{w_{n+1,m}} - \frac{\lambda}{w_{n,m+1}}$$

with $\lambda \neq 1$, the above fine cancellations in (7.5) would no longer happen, and singularities would again occur at $w_{2,3}$, $w_{3,2}$, and would persist throughout! In that case the singularities are no longer confined to a finite number of iteration steps, and we conclude that the corresponding map is not integrable.

In the above example we gave the initial values on a (tilted) staircase. In the next example we will use initial values given on a corner, in a variation of the singularity confinement approach called "ultra-local" (Sahadevan and Capel, 2003). In this approach, we choose the initial values so that there is only one place where the iterate is ambiguous and can be resolved by ε-analysis.

As an example, consider the lattice potential KdV equation (3.6)

$$u_{n+1,m+1} = u_{n,m} - \frac{1}{u_{n,m+1} - u_{n+1,m}}.$$

Now the idea is to consider a 3×3 square of points as in Figure 7.4 with initial values given at the black disks, and the values at open circles determined from them. The initial

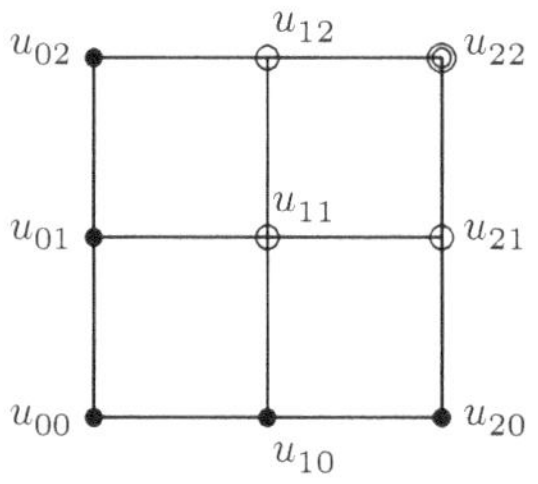

Figure 7.4 Initial value configuration for "ultra-local" singularity confinement, where the values at the black disks are given. These initial values can be chosen so that there is an ambiguity at $u_{2,2}$, which can be resolved using ϵ-analysis.

values can be chosen so that $u_{1,1} = \infty$, and the requirement is that $u_{1,2}$, $u_{2,1}$ are finite and ambiguity at $u_{2,2}$ can be resolved using ϵ-analysis.

Starting with the initial value $u_{0,1} = u_{1,0} = a$ we get

$$
\begin{aligned}
u_{1,1} &= \infty, \\
u_{1,2} &= a, \\
u_{2,1} &= a, \\
u_{2,2} &= \infty - \infty,
\end{aligned}
$$

where the last iterate is ambiguous. In order to resolve this ambiguity we do a detailed analysis with $u_{0,1} = a$, $u_{1,0} = a + \epsilon$ and then

$$
\begin{aligned}
u_{1,1} &= \tfrac{1}{\epsilon} + u_{0,0}, \\
u_{1,2} &= a + \epsilon + \epsilon^2(u_{0,2} - u_{0,0}) + O(\epsilon^3), \\
u_{2,1} &= a + \epsilon^2(-u_{2,0} + u_{0,0}) + O(\epsilon^3), \\
u_{2,2} &= u_{1,1} - \frac{1}{u_{1,2} - u_{2,1}} \\
&= \frac{1}{\epsilon} + u_{0,0} - \frac{1}{\epsilon + \epsilon^2(u_{0,2} + u_{0,2} - 2u_{0,0}) + O(\epsilon^3)} \\
&= u_{0,2} + u_{2,0} - u_{0,0} + O(\epsilon).
\end{aligned}
$$

This resolves the $\infty - \infty$ ambiguity, and recovers the initial value $u_{0,0}$ which was temporarily submerged in earlier iterates.

7.1.4 Singularity confinement for nonautonomous equations

Singularity confinement has been found to be very useful in deriving nonautonomous equations from autonomous singularity confining equations. The central idea is to introduce functions of n in place of constant parameters and then insist that the nonautonomous equations should have the *same* singularity pattern as the autonomous one (Ramani et al., 1991).

As an example, let us consider the deautonomization of (7.2); that is, let us assume that the coefficient a is n-dependent:

$$
x_{n+1} = -x_n - x_{n-1} + \frac{a_n}{x_n} + b. \tag{7.6}
$$

Starting with the same initial values as before, $x_{n-1} = u$, $x_n = \epsilon$, the sequence now continues as follows:

$$\begin{aligned}
x_{n+1} &= \tfrac{a_n}{\epsilon} + b - u - \epsilon, \\
x_{n+2} &= -\tfrac{a_n}{\epsilon} + u + \tfrac{a_{n+1}}{a_n}\epsilon + \tfrac{a_{n+1}}{a_n}\tfrac{u-b}{a_n}\epsilon^2 + O(\epsilon^3), \\
x_{n+3} &= -\tfrac{a_{n+2}+a_{n+1}-a_n}{a_n}\epsilon + (\tfrac{a_{n+1}}{a_n^2}b - \tfrac{a_{n+1}+a_{n+2}}{a_n^2}u)\epsilon^2 + O(\epsilon^3), \\
x_{n+4} &= -\tfrac{a_{n+3}-a_{n+2}-a_{n+1}+a_n}{a_{n+2}+a_{n+1}-a_n}\tfrac{a_n}{\epsilon} + \dots
\end{aligned}$$

If we now insist on the same singularity pattern as before, namely $\dots, 0, \infty, -\infty, 0, \dots$, and that this should hold independently on where we start from, then we get the condition

$$a_{n+3} - a_{n+2} - a_{n+1} + a_n = 0, \ \forall n.$$

This equation has the solution

$$a_n = \alpha + \beta n + \gamma\,(-1)^n.$$

Here the otherwise free parameters α, β, γ should be chosen so that $a_n \neq 0$, $\forall n$ (which also implies that $a_2 + a_1 - a_0 \neq 0$).

With the above choice for a_n the singularity is indeed confined, and we find

$$x_{n+4} := \frac{u(\alpha+\gamma) + 2b\beta}{\alpha + 3\beta - \gamma} + O(\epsilon).$$

The autonomous case is recovered, if $\beta = \gamma = 0$, which implies $x_{n+4} = u + \dots$. Note also that (7.1) corresponds to the special case $\gamma = 0$.

qP_{III}: Let us next see how (7.3) could be deautonomized. We assume that all parameters may depend on n. Starting again with $x_{n-2} = u$ and $x_{n-1} = c_{n-1} + \epsilon$, the same argument as in the autonomous case now implies $x_{n+1} = g_n/c_{n-1} + O(\epsilon)$, and for this to be a root of the denominator we find

$$g_n = c_{n-1}d_{n+1}.$$

Similarly starting with $x_{n-1} = d_{n-1} + \epsilon$ we arrive at $g_n = c_{n+1}d_{n-1}$. Combining the results we find that

$$c_{n-1}/d_{n-1} = c_{n+1}/d_{n+1}$$

which means that the ratio c_n/d_n must be periodic with period two.

For the study of the zeroes of the numerator we can use the above results, if we make the substitution $x_n = 1/y_n$, which leads to

$$y_{n+1}y_{n-1} = \frac{c_n d_n}{g_n a_n b_n}\frac{(y_n - 1/c_n)(y_n - 1/d_n)}{(y_n - 1/a_n)(y_n - 1/b_n)},$$

i.e. (7.3) with replacements $(g, a, b, c, d) \to (cd/(gab), 1/c, 1/d, 1/a, 1/b)$. Thus we obtain also the conditions

$$g_n a_n b_n/(c_n d_n) = a_{n-1} b_{n+1} = a_{n+1} b_{n-1}. \tag{7.7}$$

Again, one of the consequences is that $a_{n-1}/b_{n-1} = a_{n+1}/b_{n+1}$; that is, a/b must be periodic with period two. We will not pursue the full generality but rather assume that c_n/d_n and a_n/b_n are just constants. Furthermore, by making the change of variables $x_n \to x_n c_n/c$ and renaming the parameter functions appropriately, we find that c_n can be replaced by a constant c in the equation without loss of generality. Then d_n is also a constant, denoted by d, and $g_n = cd$. Finally letting $a_n = a\kappa_n$, $b_n = b\kappa_n$ we find from (7.7) that $\kappa_n \kappa_n = \kappa_{n+1}\kappa_{n-1}$, which is solved by $\kappa_n \propto q^n$. Therefore the equation

$$x_{n+1} x_{n-1} = \frac{cd(x_n - aq^n)(x_n - bq^n)}{(x_n - c)(x_n - d)} \tag{7.8}$$

has the singularity confinement property. This equations is called discrete $\mathrm{qP_{III}}$ (Ramani et al., 1991).

Note that the n-dependence in (7.1) and in (7.8) is different in an essential way: in (7.1) it is linear, in (7.8) exponential. Such equations are called q-difference equations and (7.8) is usually denoted by $\mathrm{qP_{III}}$.

7.1.5 Singularity confinement is not sufficient

Although singularity confinement has turned out to be a useful tool as a first test of integrability and in generating integrable nonautonomous equations, it is not sufficient for integrability. A counterexample is provided by the following map (Hietarinta and Viallet, 1998):

$$x_{n+1} + x_{n-1} = x_n + \frac{1}{x_n^2}. \tag{7.9}$$

A detailed analysis of singularity confinement (with $x_{-1} = u$, $x_0 = \epsilon$) now yields

$$\begin{aligned}
x_1 &= \epsilon^{-2} - u + \epsilon,\\
x_2 &= \epsilon^{-2} - u + \epsilon^4 + O(\epsilon^6),\\
x_3 &= -\epsilon + 2\epsilon^4 + O(\epsilon^6),\\
x_4 &= u + 3\epsilon + O(\epsilon^3).
\end{aligned}$$

Thus the singularity is confined with pattern $\ldots, 0, \infty, \infty, 0, \ldots$, and furthermore, the initial information u is recovered in x_4.

From the singularity confinement point of view everything looks fine. The problem is that (7.9) shows numerical chaos (Hietarinta and Viallet, 1998). Thus singularity confinement can only be a necessary condition and we need to develop more precise techniques. This will be done in the next section.

7.2 Algebraic entropy

When we iterate a rational map from some given initial conditions the resulting expression gets more and more complicated, and the degrees of numerator and denominator often grow fast. However, we will show in this section that for integrable systems there are cancellations that reduce this "growth of complexity", and these cancellations are associated with singularity confinement.

In order to get further information it is useful to write the equation in complex projective space, where the singularities reveal their nature best.

7.2.1 Singularity confinement in projective space

Let $\mathbb{P}^2$ be $\mathbb{C}^3\backslash\{0\}$ together with an equivalence relation $\sim$ defined by $(x_0, x_1, x_2) = (\lambda x_0, \lambda x_1, \lambda x_2)$ for all $\lambda \in \mathbb{C}\backslash\{0\}$. Any point $(\xi, \eta) \in \mathbb{C}^2$ can be identified with a point in $\mathbb{P}^2$ by taking

$$\xi = x_0/x_2, \quad \eta = x_1/x_2, \quad (\xi, \eta, 1) = (x_0, x_1, x_2), \quad x_2 \neq 0.$$

$(\xi, \eta, 1)$ are called *affine* coordinates, while (x_0, x_1, x_2) are called *homogeneous* coordinates. The line $x_2 = 0$ is called the *line at infinity* and is denoted here by $\mathcal{L}_0$.

The system (7.6) is a one-dimensional second-order rational map, but can also be written as a first-order polynomial map in $\mathbb{P}^2$ as follows: we start by writing it as a first-order system by introducing $y_n = x_{n-1}$, i.e.

$$\begin{cases} x_{n+1} = -x_n - y_n + \dfrac{a_n}{x_n} + b \\ y_{n+1} = x_n. \end{cases}$$

Next the equation is homogenized by substituting $x_n = u_n/f_n$, $y_n = v_n/f_n$:

$$\begin{cases} \dfrac{u_{n+1}}{f_{n+1}} = -\dfrac{u_n}{f_n} - \dfrac{v_n}{f_n} + a_n\dfrac{f_n}{u_n} + b, \\ \dfrac{v_{n+1}}{f_{n+1}} = \dfrac{u_n}{f_n}, \end{cases}$$

and finally the denominators are cleared by defining f_{n+1} to be the least common multiplier of the denominators of the right-hand sides. As a result we get a *polynomial* map in $\mathbb{P}^2$

$$\begin{cases} u_{n+1} = -u_n(u_n + v_n) + f_n(a_n f_n + b\, u_n), \\ v_{n+1} = u_n^2, \\ f_{n+1} = f_n u_n. \end{cases} \tag{7.10}$$

Since the right-hand side is a homogeneous polynomial of degree 2, we find by iterating this equation (with initial values u_0, v_0, f_0) that the degrees of u_n, v_n or f_n as polynomials of the initial values u_0, v_0, f_0 will be 2^n, unless some cancellations can be done using the equivalence property of the projective space. The actual degree describes the *complexity* of the iterate and the *growth of complexity* characterizes the equation, as we will soon see.

The sequence that led to a singularity in our previous analysis in Section 7.1.2, namely $x_{-1} = u,\ x_0 = 0,\ x_1 = \infty,\ x_2 = \infty,\ x_3 = \infty - \infty = ?$, becomes in the projective space

$$\begin{pmatrix} 0 \\ u \\ 1 \end{pmatrix} \to \begin{pmatrix} 1 \\ 0 \\ 0 \end{pmatrix} \to \begin{pmatrix} -1 \\ 1 \\ 0 \end{pmatrix} \to \begin{pmatrix} 0 \\ 1 \\ 0 \end{pmatrix} \to \begin{pmatrix} 0 \\ 0 \\ 0 \end{pmatrix} \notin \mathbb{P}^2.$$

Thus the infinities appear as regular points and the true singularity is in the last step, which takes us out of the space $\mathbb{P}^2$.

For the detailed ϵ study with $x_{-1} = u,\ x_0 = \epsilon$, the starting point is

$$\begin{pmatrix} x_0 \\ x_{-1} \\ 1 \end{pmatrix} = \begin{pmatrix} \epsilon \\ u \\ 1 \end{pmatrix} =: \begin{pmatrix} u_0 \\ v_0 \\ f_0 \end{pmatrix}$$

and the sequence continues as follows:[2]

$$\begin{pmatrix} u_1 \\ v_1 \\ f_1 \end{pmatrix} = \begin{pmatrix} a_0 + (-u+b)\epsilon + \dots \\ \epsilon^2 \\ \epsilon \end{pmatrix},$$
$$\begin{pmatrix} u_2 \\ v_2 \\ f_2 \end{pmatrix} = \begin{pmatrix} -a_0^2 + \epsilon a_0(2u - b) + \dots \\ a_0^2 + 2\epsilon a_0(-u+b) + \dots \\ \epsilon a_0 + \epsilon^2(-u+b) + \dots \end{pmatrix},$$
$$\begin{pmatrix} u_3 \\ v_3 \\ f_3 \end{pmatrix} = \begin{pmatrix} \epsilon^2 a_0^2(-a_0 + a_1 + a_2) + \dots \\ a_0^4 + 2\epsilon a_0^3(-2u+b) \dots \\ -\epsilon a_0^3 + \epsilon^2 a_0^2(3u - 2b) + \dots \end{pmatrix},$$
$$\begin{pmatrix} u_4 \\ v_4 \\ f_4 \end{pmatrix} = \begin{pmatrix} \epsilon^2 a_0^6 A_3 + \epsilon^3 a_0^5[b(4A_3 + a_0 - a_2) - u(6A_3 + a_0)] + \dots \\ \epsilon^4 a_0^4 A_2^2 + \dots \\ -\epsilon^3 a_0^5 A_2 + \dots \end{pmatrix},$$

where we have used the shorthand notation $A_2 = a_2 + a_1 - a_0$ and $A_3 = a_0 - a_1 - a_2 + a_3$. This last point is potentially singular, because if $\epsilon = 0$ we get $(0, 0, 0) \notin \mathbb{P}^2$.

This is the crucial point of singularity confinement. *If* $A_3 = 0$, then $A_2 = a_3 \neq 0$ and ϵ^3 is a common factor and can be divided out and then the $\epsilon \to 0$ limit yields

[2] During computations it is actually necessary to keep terms up to order ϵ^6, although only one or two leading terms are displayed.

$$\begin{pmatrix} u_4 \\ v_4 \\ f_4 \end{pmatrix} = \begin{pmatrix} (a_0(u-b)+a_2b) \\ 0 \\ a_3 \end{pmatrix}.$$

Thus we have emerged from the singularity and in particular recovered the initial data u. At the same time the growth of complexity was reduced by the cancellation.

Let us consider in the same way the system (7.9). In projective coordinates it reads

$$\begin{cases} u_{n+1} = -v_n u_n^2 + u_n^3 + f_n^3, \\ v_{n+1} = u_n^3, \\ f_{n+1} = f_n u_n^2, \end{cases} \tag{7.11}$$

so the default growth of the degree is 3^n. The ϵ analysis yields

$$\begin{pmatrix} \epsilon \\ u \\ 1 \end{pmatrix} \to \begin{pmatrix} 1-u\epsilon^2+\dots \\ \epsilon^3 \\ \epsilon^2 \end{pmatrix} \to \begin{pmatrix} 1-3u\epsilon^2+\dots \\ 1-3u\epsilon^2+\dots \\ \epsilon^2+\dots \end{pmatrix}$$
$$\to \begin{pmatrix} -\epsilon^3+\dots \\ 1-9u\epsilon^2+\dots \\ \epsilon^2+\dots \end{pmatrix} \to \begin{pmatrix} u\epsilon^8+\dots \\ -\epsilon^9+\dots \\ \epsilon^8+\dots \end{pmatrix} = \begin{pmatrix} u+\dots \\ -\epsilon+\dots \\ 1+\dots \end{pmatrix}.$$

Note that in order to exit from the singularity we canceled ϵ^8. But even though the map is singularity confining (with a rather large cancellation) it is not integrable.

We observe that in each case two things happen at the same time:

- The cancellation of the common power of ϵ *removes the singularity*.
- The cancellation also *reduces growth of complexity*, as defined by the degree of the iterate.

This can be viewed from two different perspectives: singularity confinement implies reduction in growth by cancellation, but on the other hand, cancellation is only possible because the limit $\epsilon \to 0$ would otherwise lead us to the singular point $(0, 0, 0)$.

7.2.2 Complexity of iterates

In this section we show that the amount of cancellation must be sufficient for the system to be integrable. We have already observed that without cancellations the successive iterates of any map would be polynomials of degree d^n, where d is the degree of the projective map. Since the growth of the degrees of the polynomials is an indication of complexity, we expect that for integrable systems the growth should be lower than d^n. All calculations carried out so far show that for integrable systems the growth should be at most polynomial in n rather than exponential (Veselov, 1992; Falqui and Viallet, 1993; Bellon and Viallet, 1999).

Let us consider two maps that are of degree three as polynomial maps in $\mathbb{P}^2$, namely the chaotic map (7.11) and the following integrable map

$$x_{n+1} + x_{n-1} = \frac{a}{x_n} + \frac{b}{x_n^2}, \tag{7.12}$$

which in projective coordinates reads

$$\begin{cases} u_{n+1} = -v_n u_n^2 + a f_n^2 u_n + b f_n^3, \\ v_{n+1} = u_n^3, \\ f_{n+1} = f_n u_n^2. \end{cases} \tag{7.13}$$

When the map (7.11) or (7.13) is iterated the degrees d_n of the resulting homogeneous polynomials in the initial values u_0, v_0, f_0 grow nominally as $d_n = 3^n$ but the actual growth is different due to cancellations. In the chaotic case (7.11) the degrees d_n actually grow as[3]

$$1,\ 3,\ 9,\ 27,\ 73,\ 195,\ 513,\ 1347,\ 3529,\ 9243,\ 24201\ldots \tag{7.14}$$

while for (7.13) the growth is

$$1,\ 3,\ 9,\ 19,\ 33,\ 51,\ 73,\ 99,\ 129,\ 163,\ 201\ldots \tag{7.15}$$

Due to stronger cancellations the growth in the integrable case is much lower and appears to follow the polynomial rule (Hietarinta and Viallet, 2000)

$$d_n = 2n^2 + 1.$$

In the chaotic case we can also see cancellations: the fifth number is reduced from 81 to 73, due to the ϵ^8 cancellation noted in Section 7.1.5.

An interesting observation about the degree growths of polynomial maps is that they can almost always be described by an associated generating function

$$g(x) := \sum_{n=0}^{\infty} d_n x^n.$$

In the integrable case (7.13) the sequence of numbers (7.15) follows from the generating function

$$g(x) = \frac{1 + 3x^2}{(1 - x)^3},$$

[3] The first few iterations can be computed explicitly, but for higher iterates one has to use special techniques: for example, choosing simpler initial values; see Viallet (2006).

while for the chaotic case (7.11) the degree sequence (7.14) follows from

$$g(x) = \frac{1 + 3x^2}{(1 - x)(1 + x)(x^2 - 3x + 1)}. \tag{7.16}$$

Let us now define the **algebraic entropy** by

$$\mathrm{E} := \lim_{n \to \infty} \frac{1}{n} \log(d_n), \quad \varepsilon := e^{\mathrm{E}}, \tag{7.17}$$

and if $E > 0$ the asymptotic growth is given by

$$d_n \propto \varepsilon^n, \quad n \gg 1.$$

The asymptotic growth can be determined from the generating function (Hietarinta and Viallet, 1998). We assume that the generating function is rational, i.e. $g(x) = R(x)/S(x)$ where R and S are polynomials of degree r, s, respectively, and write

$$S(x) = \sum_{k=0}^{s} \beta_k x^k.$$

Then looking at the large N powers of x in $g(x)S(x) = R(x)$ we obtain the recursion formula

$$\sum_{k=0}^{s} \beta_k d_{N-k} = 0, \quad N \gg 1.$$

In the chaotic case (7.16) the denominator is $x^4 - 3x^3 + 3x - 1$ and therefore the recursion rule is

$$d_{N-4} - 3d_{N-3} + 3d_{N-1} - d_N = 0.$$

This is solved by $d_n = \lambda^n$, where $\lambda = 1, -1, \frac{1}{2}(3 \pm \sqrt{5})$. Then the growth of degrees is given by λ with the largest modulus. In the present case $\varepsilon = (3 + \sqrt{5})/2 \approx 2.61803$. For the integrable case we find by similar process that $E = 0$ and $\varepsilon = 1$.

7.2.3 Algebraic entropy analysis for lattices

The algebraic entropy analysis can also be applied to PΔEs (Tremblay et al., 2001; Viallet, 2006). We restrict our attention to quadratic maps in a quadrilateral lattice, as given in equation (3.3) with $k = l_i = 0$. Solving for the doubly-shifted term $\widehat{\widetilde{u}} = u_{[12]}$ and using projective coordinates $u = v/f$ and then separating numerator and denominator we get:

$$\begin{cases} v_{[12]} = p_1\, v\, v_{[1]} f_{[2]} + p_2\, v_{[1]} v_{[2]} f + p_5\, v\, v_{[2]} f_{[1]} \\ \qquad\quad + q_1\, v\, f_{[1]} f_{[2]} + q_2\, v_{[1]} f_{[2]} f + q_3\, v_{[2]} f_{[1]} f + c\, f\, f_{[1]} f_{[2]}, \\ f_{[12]} = p_3\, v_{[2]} f_{[1]} f + p_4\, v\, f_{[1]} f_{[2]} + p_6\, v_{[1]} f_{[2]} f + q_4 f\, f_{[1]} f_{[2]}. \end{cases}$$

We noted earlier that a well-defined evolution can start from a staircase-like initial configuration. We now consider two types of initial value problems: a regular staircase or the edges of a quadrant. Since we are using projective coordinates we may choose the initial values of v freely while fixing all initial values of f to be the same. Note that for both v and f every term on the right-hand side of the map is a product of one unshifted function, one function shifted in the tilde direction and one shifted in the hat direction. Furthermore, since we chose the initial values of f to be the same, one factor of f can be canceled in all iterates since we use projective coordinates. Thus the default degree growth of v is given by

$$\deg(v_{[12]}) = \deg(v) + \deg(v_{[1]}) + \deg(v_{[2]}) - 1. \tag{7.18}$$

The same holds for the degrees of f. The default degree growth for the respective initial value configurations is then

$$\begin{array}{ccccccc} \vdots & \vdots & \vdots & \vdots & \vdots & \vdots & \\ 1 & 4 & 13 & 32 & 65 & 116 & \dots \\ 1 & 3 & 7 & 13 & 21 & 31 & \dots \\ 1 & 2 & 3 & 4 & 5 & 6 & \dots \\ 1 & 1 & 1 & 1 & 1 & 1 & \dots \end{array} \qquad \begin{array}{ccccccccc} \ddots & \ddots & \ddots & \ddots & \ddots & \ddots & \ddots & \ddots & \\ & 1 & 1 & 2 & 4 & 9 & 21 & 50 & \dots \\ & & 1 & 1 & 2 & 4 & 9 & 21 & \dots \\ & & & 1 & 1 & 2 & 4 & 9 & \dots \\ & & & & 1 & 1 & 2 & 4 & \dots \\ & & & & & \ddots & \ddots & \ddots & \end{array}$$

For the staircase initial value problem the degrees are the same along the lines where $n+m$ is constant. We solve the linear problem (7.18) along the perpendicular direction and find that asymptotically ($n + m \gg 1$) the degree growth is given by $(1 + \sqrt{2})^{n+m}$ (Viallet, 2006).

Some well-known models were studied from this point of view in Tremblay et al. (2001). For the lattice KdV the following degrees were obtained

$$\begin{array}{ccccccc} \vdots & \vdots & \vdots & \vdots & \vdots & \vdots & \\ 1 & 4 & 7 & 10 & 13 & 16 & \dots \\ 1 & 3 & 5 & 7 & 9 & 11 & \dots \\ 1 & 2 & 3 & 4 & 5 & 6 & \dots \\ 1 & 1 & 1 & 1 & 1 & 1 & \dots \end{array} \qquad \begin{array}{ccccccccc} \ddots & \ddots & \ddots & \ddots & \ddots & \ddots & \ddots & \ddots & \\ & 1 & 1 & 2 & 4 & 7 & 11 & 16 & \dots \\ & & 1 & 1 & 2 & 4 & 7 & 11 & \dots \\ & & & 1 & 1 & 2 & 4 & 7 & \dots \\ & & & & 1 & 1 & 2 & 4 & \dots \\ & & & & & \ddots & \ddots & \ddots & \end{array}$$

In the quadrant initial value case the degrees follow the rule $d_{n,m} = nm + 1$, while in the staircase initial value problem $d_{n,m} = 1 + (n + m)(n + m - 1)/2$, so that in each case the growth is polynomial and therefore the algebraic entropy vanishes. Thus the lattice KdV equation also passes this test of integrability.

In conclusion, the results of this section, supported by many studies of computing algebraic entropy, have led to the following conjecture:

- If the degree growth is *linear* in the independent variable(s), the equation is linearizable;
- If the degree growth is *polynomial* with degree > 1 the equation is integrable;
- If the degree growth is *exponential* the equation is nonintegrable.

The above idea is related to the relationship between proliferation of images and complexity studied by Veselov (1992), who proved that Liouville integrability implies at most polynomial growth in the number of images.

7.3 Singularities from a geometric point of view

In this section we extend the study of singularities, started in Section 7.1.2, using geometric methods. Recall that a rational map may become singular in the sense that it fails to define the next iterate in a unique analytic way. There is a geometric way to make sense of such behaviors, based on the well-known technique called "resolution of singularities" or "blow-up" in algebraic geometry.

In order to resolve the singularities we will work in the complex projective space $\mathbb{P}^1 \times \mathbb{P}^1$ where $\mathbb{P}^1 = (\mathbb{C}^2 \backslash \{(0,0)\})/\sim$, with $\sim$ being the equivalence relation $[f_1, g_1] = [\lambda f_1, \lambda g_1]$ where λ is a nonzero complex number.

We will see that singularities arise at places called "base points" where the iterate is ambiguous. In order to remove the ambiguity at these points we use the simplest possible coordinate transformation. Since these points can occur anywhere in $\mathbb{P}^1 \times \mathbb{P}^1$ it is necessary to use coordinate charts that include all of $\mathbb{P}^1 \times \mathbb{P}^1$ and are smoothly compatible on their intersection. We will use four affine coordinate charts to cover $\mathbb{P}^1 \times \mathbb{P}^1$. These are

(i) (u_{00}, v_{00}) where $u_{00} = f_1/g_1$, $v_{00} = f_2/g_2$;
(ii) (u_{01}, v_{01}) where $u_{01} = 1/u_{00}$, $v_{01} = v_{00}$;
(iii) (u_{02}, v_{02}) where $u_{02} = u_{00}$, $v_{02} = 1/v_{00}$; and
(iv) (u_{03}, v_{03}) where $u_{03} = 1/u_{00}$, $v_{03} = 1/v_{00}$.

We now focus on examples to provide the key ideas and methodology.

7.3.1 Autonomous qP_{III}

As an example we consider the autonomous difference equation

$$x_{n+1}\, x_{n-1} = \alpha\, \beta\, \frac{(x_n - 1)\,(x_n - \gamma)}{(x_n - \alpha)\,(x_n - \beta)} =: F\big(x_n\big), \tag{7.19}$$

where we assume that 1, α, β and γ are all distinct numbers. If we separate the odd and even iterates

$$x_{2j} = u_j \tag{7.20a}$$
$$x_{2j+1} = v_j \tag{7.20b}$$

and denote $\overline{u} = u_{j+1}$, $\overline{v} = v_{j+1}$, then we can write (7.19) as a first-order system

$$\overline{u}\, u = F(v) \tag{7.21a}$$
$$\overline{v}\, v = F(\overline{u}) \tag{7.21b}$$

which is a mapping in QRT form. As for all QRT systems, this mapping can also be considered as a composition $\iota_2 \circ \iota_1$ of the following two maps:

$$\iota_1 : \begin{pmatrix} u \\ v \end{pmatrix} \mapsto \begin{pmatrix} F(v)/u \\ v \end{pmatrix}, \quad \iota_2 : \begin{pmatrix} u \\ v \end{pmatrix} \mapsto \begin{pmatrix} u \\ F(u)/v \end{pmatrix}. \tag{7.22}$$

This mapping has an invariant K defined by the equation

$$\begin{aligned} P(u,v) := u^2\, v^2 - (\alpha+\beta)\,(u^2\, v + u\, v^2) + \alpha\,\beta\,(u^2+v^2) \\ + 2\,K\,u\,v - \alpha\,\beta\,(1+\gamma)\,(u+v) + \alpha\,\beta\,\gamma = 0 \end{aligned} \tag{7.23}$$

which is a special case of (6.34). Note that this is in the form

$$\mathcal{P}_1(u,v) + 2K\,\mathcal{P}_2(u,v) = 0 \tag{7.24}$$

where $\mathcal{P}_i(u,v)$, $i = 1, 2$ are polynomials in (u, v). Equation (7.24) describes a one-parameter family of curves and is commonly referred to as a *pencil* of curves.[4]

A generic initial value (u, v) defines the invariant K and the corresponding curve in the pencil contains the iterates of the map (7.21) for all n. We can uniquely solve for the next iterate, using the curve except at certain problematic points.

7.3.2 The base points

There are two types of problems. Firstly, note that the value of K becomes undefined where $\mathcal{P}_1(u,v) = 0 = \mathcal{P}_2(u,v)$. All curves in the pencil pass through such points (i.e. the value of K is not specified) and such points are called *base points* of the pencil. Secondly, the next iterate may fail to be defined uniquely at some value of (u, v) where the gradient of $P(u, v)$ vanishes. In this section we will focus on the first problem.

At base points the first equation $\mathcal{P}_2(u,v) = uv = 0$ implies that either u or v is zero. When $v = 0$, the other equation gives $\mathcal{P}_1(u,0) = \alpha\beta(u^2 - (1+\gamma)u + \gamma) = 0$ that is $u = 1$ or γ. On the other hand if $u = 0$ we find $v = 1$ or γ. These give four base points:

$$p_1 = (1,0),\ p_2 = (\gamma,0),\ p_3 = (0,1),\ p_4 = (0,\gamma).$$

[4] Such a pencil of curves exists because in this autonomous case we have an invariant K. In the nonautonomous case we no longer have an invariant; nevertheless the geometric point of view explained in the following subsections still applies.

There are more base points at infinity. In order to see them we rewrite the invariant in terms of the coordinates $(u_{01}, v_{01}) = (1/u, v)$. In this way the line at infinity in the original coordinates ($u = \infty$) becomes the line $u_{01} = 0$. In this case equation (7.23) becomes

$$v_{01}^2 - (\alpha + \beta)\left(v_{01} + u_{01}\, v_{01}^2\right) + \alpha\, \beta\, (1 + u_{01}^2 v_{01}^2)$$
$$+ 2\, K\, u_{01}\, v_{01} - \alpha\, \beta\, (1 + \gamma)\, (u_{01} + u_{01}^2 v_{01}) + \alpha\, \beta\, \gamma u_{01}^2 = 0.$$

At a base point we must have $u_{01} v_{01} = 0$, and if we choose $u_{01} = 0$ we get from the equation $v_{01} = \alpha$ or β and thus we find additional base points

$$p_5 = (0, \alpha), \;\; p_6 = (0, \beta)$$

in the coordinates $(u_{01}, v_{01}) = (1/u, v)$. The choice $v_{01} = 0$ corresponds to $v = 0$, which reproduces p_1, p_2.

To investigate the possibility that v may be infinite, we use new coordinates $(u_{02}, v_{02}) = (u, 1/v)$, and now the line $v = \infty$ becomes the line $v_{01} = 0$. A similar analysis as above now produces new base points

$$p_7 = (\alpha, 0), \;\; p_8 = (\beta, 0)$$

in these new coordinates. The base points are drawn schematically in Figure 7.5.

There are no further base points, because the generic biquadratic polynomial curve is uniquely determined by nine coefficients, one of which can be scaled to 1, and thus there remain only eight essential parameters, determined by eight points through which the curve passes.[5]

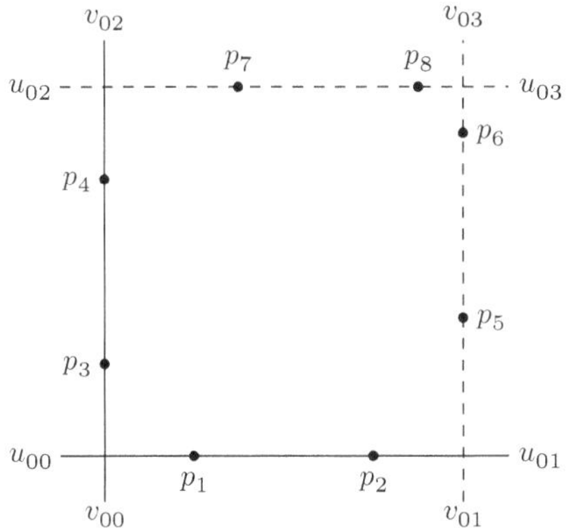

Figure 7.5 The eight base points of the mapping (7.21). We have indicated the lines at infinity by dashes. The coordinates are defined in (7.25).

[5] It turns out that for the discrete Painlevé equations, which do not have such a biquadratic invariant, the number of base points in $\mathbb{P}^1 \times \mathbb{P}^1$ is still eight (Sakai, 2001).

Above, we introduced transformations in cases where either u or v was infinite, but there is the remaining possibility that they are infinite simultaneously, which is captured with the transformation $(u_{03}, v_{03}) = (1/u, 1/v)$. However, this is not needed because no finite base points appear in this coordinate chart.

The eight base points we have obtained correspond precisely to the eight possible ambiguities in the iteration of equation (7.19). From this it follows, that if one does not know the invariant, one can construct the base points by looking at the points where the iteration is ambiguous due to expression such as $0/0$.

In summary, we need to take four coordinate charts:

$$(u_{00}, v_{00}) := (u, v), \tag{7.25a}$$

$$(u_{01}, v_{01}) := (1/u, v), \tag{7.25b}$$

$$(u_{02}, v_{02}) := (u, 1/v), \tag{7.25c}$$

$$(u_{03}, v_{03}) := (1/u, 1/v). \tag{7.25d}$$

These coordinate charts and the corresponding equations are indicated in the schematic diagram of $\mathbb{P}^1 \times \mathbb{P}^1$ in Figure 7.6.

7.3.3 Resolving the singularities

We start with the base point $p_1 = (1, 0)$ and resolve the ambiguity of the map ι_2 (7.22) there. This ambiguity arises because when $u = 1$ we have $F(u) = 0$ and since $v = 0$ the iterate of v becomes $0/0$. We use an invertible rational change of variables that allows us to study it in greater detail. This is the standard technique of resolving a singularity called *blowing up*.

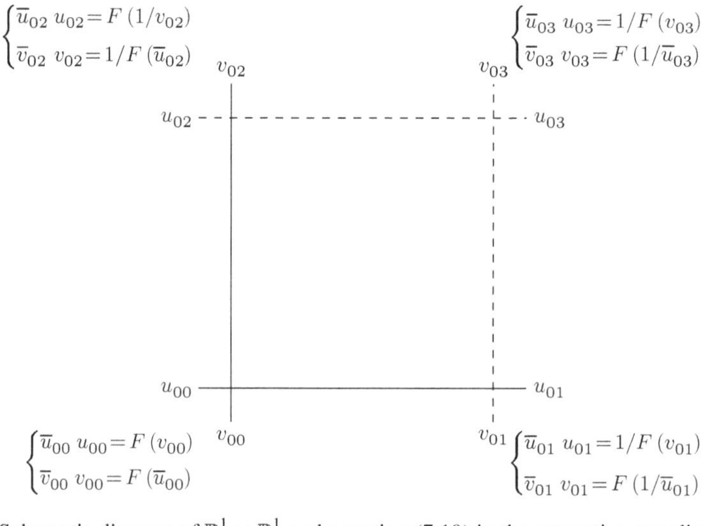

Figure 7.6 Schematic diagram of $\mathbb{P}^1 \times \mathbb{P}^1$ and equation (7.19) in the respective coordinate charts.

Explicitly, we take new coordinates given by

$$u_{11} = \frac{u-1}{v}, \quad v_{11} = v \Rightarrow u = 1 + u_{11}v_{11}, v = v_{11}.$$

(Notice that each value of u_{11} is the slope of the corresponding line $u = u_{11}v + 1$ which is a tangent to a curve of the pencil going through p_1.) In these new coordinates the map ι_2 becomes

$$\begin{cases} \overline{u}_{11}\overline{v}_{11} = u_{11}\, v_{11} \\ \overline{v}_{11} = \alpha\,\beta \dfrac{u_{11}\,(1-\gamma+u_{11}\,v_{11})}{(1-\alpha+u_{11}\,v_{11})(1-\beta+u_{11}\,v_{11})}. \end{cases} \tag{7.26}$$

Notice that a factor of v_{11} has canceled and that there are no longer any ambiguities remaining in this system. However, it is still possible that u_{11} becomes infinite (the line $u = u_{11}v + 1$ becomes vertical). To take care of this case we use the second chart

$$u_{12} = u - 1, \quad v_{12} = \frac{v}{u-1} \Rightarrow u = 1 + u_{12}, v = u_{12}\, v_{12}$$

and then the map becomes

$$\begin{cases} \overline{u}_{12} = u_{12} \\ \overline{u}_{12}\,\overline{v}_{12} = \alpha\,\beta \dfrac{1-\gamma+u_{12}}{(1-\alpha+u_{12})(1-\beta+u_{12})v_{12}}. \end{cases}$$

Here, no cancellation has occurred but nevertheless, no ambiguity of the form $0/0$ arises, because when u_{12} or v_{12} vanishes in the denominator, no terms in the numerator of F vanishes, and vice versa.

In general, if the mapping becomes $0/0$ at say $(u, v) = (\mu, \nu)$, then we blow up this base point by taking two overlapping sets of new coordinates

$$(U_1, V_1) = \left(\frac{u-\mu}{v-\nu}, v-\nu\right),$$
$$(U_2, V_2) = \left(u-\mu, \frac{v-\nu}{u-\mu}\right).$$

We need two coordinate charts, because we may have instances where one set of coordinates becomes infinite and we need to cover this case.

Note that in the original coordinates p_1 and p_2 are singularities of ι_2, while p_3 and p_4 are those of ι_1. Moreover, p_5 and p_6 are singularities of ι_1 in the chart (u_{02}, v_{02}), while p_6 and p_7 are those of ι_2 in the chart (u_{01}, v_{01}).

Similar explicit changes of variables may be carried out for each of the base points and the corresponding map. The results show that no further base points arise and therefore we have resolved the singularities at all of base points.

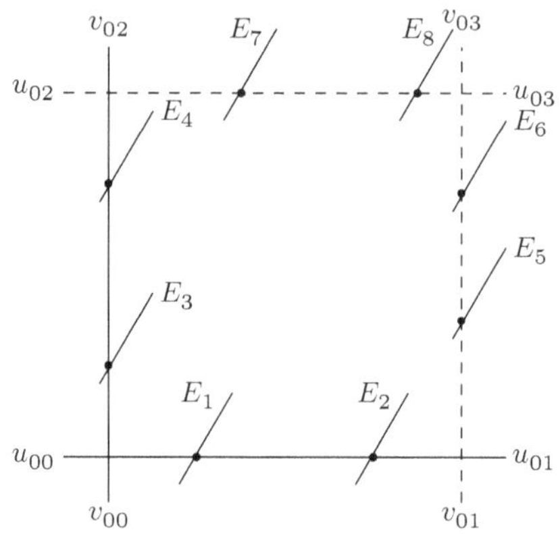

Figure 7.7 The eight exceptional lines that arise after resolving the base points of the mapping (7.21).

Each time we "blow-up"a base point, we replace it with a line, called the "exceptional line", which is parameterized by the slopes of the tangents of curves passing through that point. In the case of the blow-up of p_1 provided above, the point is replaced by the line E_1 which is given by $v_{11} = 0$. We can now find out how this line changes when it is iterated under the mapping ι_2 given by equation (7.26). The equation shows that $v_{11} = 0 \mapsto \overline{v}_{11} \sim \alpha\,\beta\,(1-\gamma)(1-\alpha)u_{11}/(1-\beta)$, for $v_{11} \ll 1$. In other words, the exceptional line E_1 is mapped to an ordinary line coordinatized by u_{11}. Similar results can be found for each of the remaining exceptional lines E_j, replacing each base point p_j when it is blown up.

In summary, after resolution of all singularities, we have a regularized space in which the base points have been replaced by exceptional lines that are indicated schematically in Figure 7.7. In this construction it is important to keep track of the self-intersection numbers of all lines in the space. The coordinate lines H_u, defined to be $\{u = \text{constant}\}$, and H_v, which is $\{v = \text{constant}\}$, are initially ordinary lines with zero self-intersection number. However, each time a point on such a line is resolved or "blown up", its self-intersection number decreases by unity. So at the end of the process, we find four curves with self-intersection number equal to two, namely $L_1 = H_v - E_1 - E_2$, $L_2 = H_u - E_3 - E_4$, $L_3 = H_v - E_7 - E_8$ and $L_4 = H_u - E_5 - E_6$. Notice that L_1 intersects with L_2, which in turn intersects with L_3, which intersects with L_4. If we were to replace each L_k, $1 \le k \le 4$, with a node and draw a line connecting a pair of nodes if the corresponding lines intersect, we would obtain the intersection diagram shown in Figure 7.8. These lines L_k, $1 \le k \le 4$ span the homology class of the now resolved initial-value space. It can be shown that the action of the mapping can be lifted to a linear mapping on the Picard group of this space. For further details, we refer the reader to Duistermaat (2010).

7.3.4 Integrability

This seemingly abstract construction has a crucial and miraculous connection with integrability. For integrable systems, it turns out that, as for QRT maps, all singularities

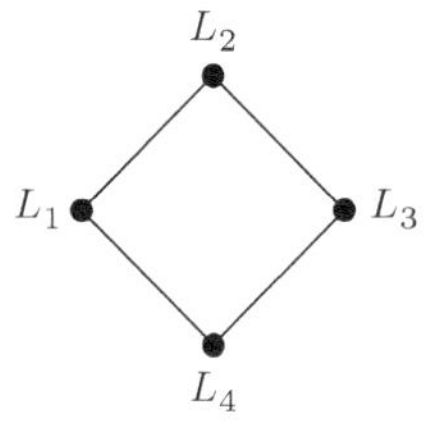

Figure 7.8 Intersection diagram of (7.21).

in the initial value space can be resolved after blowing-up a minimum of eight points in $\mathbb{P}^1 \times \mathbb{P}^1$ or nine points in $\mathbb{P}^2$ (depending on the choice of coordinates). This remarkable observation was first made by Okamoto (1979) for the continuous Painlevé equations. Sakai (2001) used this property as a starting point in $\mathbb{P}^2$ to find and classify all second-order integrable equations that generalize the Painlevé equations.

This connection with algebraic geometry also provides us with another way to understand complexity and algebraic entropy (cf. Section 7.2). As mentioned above, the time evolution of the integrable system can be lifted to a linear action on the Picard group. For mappings, this linear action is expressible in terms of a matrix which gives the action on the union of equivalence classes of the lines H_u, H_v, E_j, $1 \leq j \leq 8$. It turns out that the modulus of the largest eigenvalue of this matrix gives the quantity ε defined in equation (7.17) and consequently the algebraic entropy $E = \log(\varepsilon)$. In the integrable case (with a minimal eight or nine blow-ups, depending on choice of the projective space) the eigenvalues turn out to have modulus one and therefore $\varepsilon = 1$; that is, $E = 0$.

The nonintegrable but singularity confining case (7.9) was shown in Takenawa (2001a,b) to need many more resolutions (in fact, a minimum of 14 blow-ups are needed) and the maximal eigenvalue of the linear action on the Picard group has modulus $\frac{1}{2}(3 + \sqrt{5})$, which was obtained by other means in Section 7.2.2.

In summary, the abstract theory of resolution of singularities (Hartshorne, 1977) encodes and provides many of the important concepts discovered separately through the theory of integrability. The investigation of the initial value space offers the tantalizing prospect that many more discoveries about integrable systems remain to be made through the geometric study of the resulting surface.

7.4 Notes

The singularity confinement test for discrete equations was introduced by Grammaticos, Ramani and Papageorgiou (1991) for both PΔEs as well as OΔEs. As can be seen in Section 7.1.2 it is a rather algorithmic procedure and as such it became popular immediately and has since then been applied on numerous equations as the first indicator of integrability; see e.g. Ramani et al. (1996); Lafortune et al. (2001). The discovery of the Hietarinta–Viallet counterexample (Hietarinta and Viallet, 1998) showed that is necessary but sufficient, but it has nevertheless turned out to be a very efficient method for de-autonomizing integrable maps and has produced many PΔEs that can be considered as discrete versions of Painlevé equations (Grammaticos et al., 2000; Ramani et al., 1991). This approach has been further developed in Ramani et al. (2015); Grammaticos et al. (2015).

The concept of algebraic entropy arose from the study of rational mappings (Arnold, 1990; Veselov, 1992; Falqui and Viallet, 1993; Boukraa et al., 1994; Boukraa and Maillard, 1995) and was specifically defined in Bellon and Viallet (1999). Degree growth, entropy and complexity has been studied further from various directions; see e.g. Abarenkova et al. (1999); d'Auriac et al. (2006); Hasselblatt and Propp (2007). As was shown in Section 7.2, this method is a refinement of singularity confinement; this connection has been further analyzed in Lafortune et al. (2001); Ohta et al. (1999). For lattice equations, algebraic entropy has been applied in Tremblay et al. (2001); Viallet (2006).

Sometimes a map looks nonlinear but can be linearized with a suitable transformation. Algebraic entropy analysis, however, will indicate linearizability by exhibiting at most linear growth in the degree d_n defined in 7.2.2 (Ramani et al., 2000).

In this chapter we saw that the study of growth of solutions played a major role in the identification of integrable systems. In the theory of complex analytic functions the Finnish mathematician Rolf Nevanlinna's (1895–1980) developed a method to characterize the growth of meromorphic functions, which has been used to study the solutions of ODEs since the 1920s; see the monograph by Gromak et al. (2002). Nevanlinna theory was first applied to difference equations by Ablowitz, Halburd and Herbst (2000), who found that for equations of the type $x_{n+1} + x_{n-1} = R(x_n)$ or $x_{n+1}x_{n-1} = R(x_n)$ the rational function R can at most be a ratio of two quadratic polynomials in x_n. The method has been developed further in e.g. Ablowitz et al. (2000); Ramani et al. (2003); Halburd and Korhonen (2006, 2007b,a); Barnett et al. (2007).

The integrability test provided by algebraic entropy analysis and by Nevanlinna theory are computationally or theoretically demanding and therefore easier methods related to them have been developed. One of these is *diophantine integrability* and the connection is provided by the so-called "Vojta's dictionary"; see Halburd (2005). This idea connects the Nevanlinna growth to the "height" of the iterates and since the computations are done for rational numbers this is a computationally effective method.

Another computation simplification is to use finite fields. The dependent variables are assumed to have values in a finite field; that is, all computations are performed modulo a prime number p. In this way the analysis of integrability becomes the study of periodic orbits (Roberts and Vivaldi, 2003). This has been applied to Painlevé equations in Kanki (2013).

The geometric method of *resolution of singularities* discussed in Section 7.3 can be seen as a rigorous description of singularity confinement. This idea was used by Sakai to classify discrete Painlevé equations (Sakai, 2001; Takenawa, 2004); see also Duistermaat (2010); Duistermaat and Joshi (2011); Joshi and Lobb (2016). One particular advantage of this method is that it allows for the rigorous computation of algebraic entropy (Takenawa,

2001a,b). Extension to linearizable mappings works as well (Takenawa et al., 2003). The connection between singularity structure and dynamics is further discussed e.g. in Diller and Favre (2001) and McMullen (2007).

Exercises

7.1 (i) Show that the following nonautonomous generalization of the McMillan map (1.19)

$$x_{n+1} + x_{n-1} = \frac{a_n + b_n\, x_n}{1 - x_n^2}, \tag{7.27}$$

passes the singularity confinement test if a_n, b_n are constants. (Note that there are two values that can generate problems: $x_n = \pm 1$.)

(ii) De-autonomize (7.27) by requiring confinement at the same stage as for the autonomous map.

7.2 Consider the singularity confinement analysis of (7.1) and extend it in the following two ways:

(i) Consider the initial values $x_{-1} = \epsilon$, $x_0 = \epsilon$. Find x_4 and show that the singularity is not confined. However, show also that by iterating backward the above intial values cannot be reached from nonsingular earlier iterates.

(ii) Now assume the initial values $x_{-1} = u$, $x_0 = \epsilon$ and the equation is (7.6). What happens if x_4 is allowed to be singular but x_8 is regular?

7.3 Determine the continuum limit of qP$_{\rm III}$ (7.8). (Hint: take $c = 1/\epsilon + \alpha/2$, $d = -1/\epsilon + \alpha/2$, $a = \beta\epsilon - \gamma\epsilon^2/2$, $b = -\beta\epsilon - \gamma\epsilon^2/2$ and then for the new variable z defined by $\lambda^n = (1+\epsilon)^n =: e^z$ one gets P$_{\rm III}$.)

7.4 Analyze the singularity pattern for the equation (3.90) arising from an initial value $W = \gamma$ at a given vertex of the lattice, and show that singularity confinement takes place. In this case singularities appear for $W = \gamma$ or $W = 1/\gamma$ and the singularity pattern now is in Figure 7.9. The computations go in a similar way, starting from a small deviation from the singularity, i.e. $\gamma_1 = \gamma + \varepsilon$

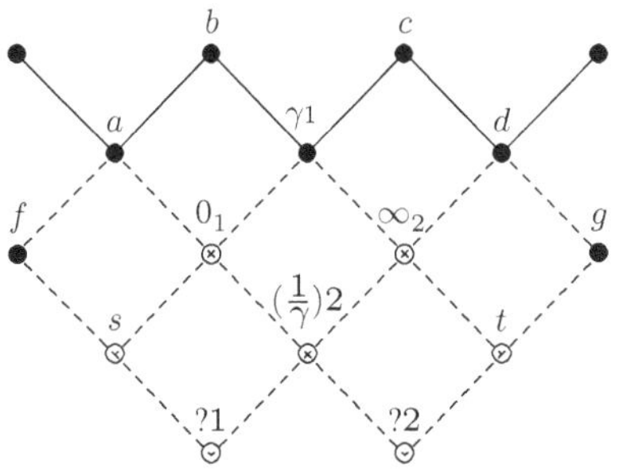

Figure 7.9 Propagation of singularities in the 2D map (2.76). Here n grows in the SE direction and m in the SW direction.

7.5 Consider the q-discrete first Painlevé equation

$$\overline{f} f^2 \underline{f} = t(1-f)$$

where $f = f(t)$, $t = t_0 q^n$ and $\overline{f} = f(qt)$, $\underline{f} = f(t/q)$. Investigate the iterates of the initial conditions $f(t_0) = b$, $f(q\,t_0) = \epsilon$ and show that the resulting singularity is not confined. To overcome this paradox, consider the backward map and show that these initial conditions can only be reached if $b = 1 + \mathcal{O}(\epsilon)$. Restart the investigation of the forward map with $f(t_0) = 1 + \beta\epsilon$, $f(q\,t_0) = \epsilon$ and show that the singularities are now confined.

8

Hirota's bilinear method

In this chapter we will describe Hirota's bilinear approach (Hirota, 2004) for integrable systems. The key observation of Ryogo Hirota (1932–2015) – see Hirota (1971) – was that one can find new dependent variables so that soliton solutions become expressible simply as sums of exponentials. This observation turned out to be universal for integrable PDEs and moreover provided a systematic method for constructing soliton solutions. Hirota's method actually extends to other types of equations, such as partial difference equations and ordinary differential equations.

Figure 8.1 R. Hirota

Hirota's new variables, nowadays called τ-functions, are key ingredients of a deep and extensive theory which was developed in the eighties by the "Kyoto school" of M. Sato and his collaborators, including E. Date, M. Jimbo, M. Kashiwara and T. Miwa (Jimbo and Miwa, 1983; Miwa et al., 2000). In this approach the underlying algebraic structures are given in terms of the representation theory of infinite-dimensional Lie algebras (Kac, 1994). This intriguing connection has been the subject of many studies. Discrete equations are an integral part of this approach (Date et al., 1982a,b, 1983a,b,c) and these led to the discovery of two key three-dimensional equations, Hirota's "Discrete Analogue of the Generalized Toda Equation" (DAGTE) and Miwa's equation, from which many other integrable equations may be derived.

In this chapter we first introduce Hirota's bilinear method in the context of PDEs and then show how it applies to PΔEs.

8.1 Introduction

8.1.1 Sketch of the method in the continuous case

By the end of the 1960s the Korteweg–de Vries equation

$$u_t = u_{xxx} + 6uu_x, \tag{8.1}$$

was already well studied and in particular it was known that its multi-soliton solutions have the form[1]

$$u = 2\partial_x^2 \log \det(\mathbf{I} + \mathbf{M}), \tag{8.2}$$

where the matrix elements of **M** are sums of exponentials of the form e^{ax+bt}. Motivated by this Hirota (1971) proposed a change of the dependent variable[2]

$$u = 2\partial_x^2 \log f, \tag{8.3}$$

because then the new dependent variable should be a finite sum of exponentials.

If the substitution (8.3) is applied to (8.1) and the result is integrated once then one finds the following equation that is quadratic in f and its derivatives

$$f_{xt} f - f_x f_t = f\, f_{xxxx} - 4 f_x f_{xxx} + 3 f_{xx}^2 \,. \tag{8.4}$$

At first glance it seems that we may not have gained anything by writing the equation in this form, but it turns out that the equation can be written in a more transparent form by introducing the **Hirota derivative operator**, defined by

$$D_x^i D_t^j\, f \cdot g \equiv (\partial_x - \partial_{x'})^i\, (\partial_t - \partial_{t'})^j\, f(x,t) g(x',t')\Big|_{x'=x,t'=t} \tag{8.5a}$$

$$\equiv \partial_{x'}^i \partial_{t'}^j\, f(x+x',t+t') g(x-x',t-t')\big|_{x'=0,t'=0} \tag{8.5b}$$

where i, j are nonnegative integers, and where the dot between the functions f and g is used as a separator. These operators are linear in each variable, and are therefore also called **bilinear** operators. Using these we can write (8.4) as

$$\left(D_x D_t - D_x^4\right) f \cdot f = 0. \tag{8.6}$$

Note the sign difference between the Hirota derivative and the usual derivative when applied to products of functions (Leibniz product rule):

$$\partial_x^j\, (fg) \equiv (\partial_x + \partial_{x'})^j\, f(x) g(x')\Big|_{x'=x} ,$$

$$D_x^j\, f \cdot g \equiv (\partial_x - \partial_{x'})^j\, f(x) g(x')\Big|_{x'=x} .$$

Furthermore, Hirota's bilinear operator acts on direct products of spaces of differentiable functions, while the ordinary derivative operator acts on one copy of the space. Thus, for sufficiently differentiable functions we can compose the ordinary derivative, for example

[1] The multi-soliton solutions to the KdV equation were given in the seminal paper by Gardner et al. (1967), based on earlier results by Kay and Moses (1956). For an explicit construction see also Wadati and Toda (1972).

[2] We note that **M** has Cauchy matrix structure, and in the next chapter we will further elaborate on this; in fact, there we use it as a starting point to derive integrable equations. In this chapter this structure is only used to motivate the change of variables in the equation.

$\partial^n(\partial^m f) = \partial^{n+m} f$, while an expression such as $D^n(D^m f \cdot g)$ is not even defined. However, $D^n(D^m f \cdot g) \cdot (D^k f \cdot g)$, for example, does make sense. In particular, from the observation that $D_x f \cdot f = 0$ (which holds due to antisymmetry), one cannot conclude that $D_x^2 f \cdot f$ or $D_x D_t f \cdot f$ vanishes.

The Hirota bilinear derivative obeys several rules which follow directly from the definition. The most important are those of *symmetry*, *linearity* and *bilinearity*; that is, for all differentiable functions f, g, h and constants α, β we have:

$$\begin{aligned} D_x^j f \cdot g &= (-1)^j D_x^j g \cdot f\,, \\ (\alpha D_x^j + \beta D_x^k)\, f \cdot g &= \alpha D_x^j f \cdot g + \beta D_x^k f \cdot g, \\ D_x^j (\alpha f + \beta g) \cdot h &= \alpha D_x^j f \cdot h + \beta D_x^j g \cdot h. \end{aligned}$$

The bilinear derivatives operate on exponentials in a simple manner:

$$P(D_{\vec{x}}) e^{\vec{x}\cdot\vec{p}} \cdot e^{\vec{x}\cdot\vec{q}} = P(\vec{p} - \vec{q}) e^{\vec{x}\cdot(\vec{p}+\vec{q})}, \tag{8.7}$$

where P is any function expandable in a power series of Hirota derivatives. Here we use the arrow to denote multicomponent vectors of arbitrary but fixed dimension, and the dot between vectors denotes inner product. This formula turns out to be quite useful in constructing soliton solutions.

There are also various formulae for expressing combinations of ordinary derivatives in terms of bilinear ones, such as

$$2\partial_x^2 \log f = \frac{D_x^2 f \cdot f}{f^2}.$$

See e.g. the appendix of Hirota and Satsuma (1976).

8.1.2 Discrete Hirota operator

Note first that the usual Taylor series can be (formally) expressed as:

$$f(x+a) = \sum_{j=0}^{\infty} \frac{a^j}{j!} \partial_x^j f(x) = e^{a\partial_x} f(x).$$

Replacing now the derivative ∂_x by the Hirota operator D_x and doing the expansions, one finds

$$e^{aD_x} f(x) \cdot g(x) = e^{a(\partial_x - \partial_{x'})} f(x) g(x')\Big|_{x'=x} = f(x+a)g(x-a). \tag{8.8}$$

Furthermore, as a consequence of the symmetry property of the Hirota operator, we have its discrete analogue:

$$e^{aD_x} f \cdot g = e^{-aD_x} g \cdot f.$$

Thus, if we have a discrete equation in which every product has a sum of shifts adding to the same value (cf. (8.8)), we can write it in Hirota form.

8.1.3 Gauge invariance

The fundamental characterization of the Hirota bilinear form is its invariance under a *linear gauge transformation* $(f, g, \ldots) \to (e^{\vec{x}\cdot\vec{p}} f, e^{\vec{x}\cdot\vec{p}} g, \ldots)$; in other words, for any function P of only Hirota derivatives, we have

$$P(D_{\vec{x}})(e^{\vec{x}\cdot\vec{p}} f) \cdot (e^{\vec{x}\cdot\vec{p}} g) = e^{2\vec{x}\cdot\vec{p}}\, P(D_{\vec{x}}) f \cdot g. \tag{8.9}$$

In fact, the converse statement is also true: any bilinear equation having gauge invariance of the above type can be written in terms of Hirota's bilinear derivatives (Grammaticos et al., 1994). This gauge principle can be used as a guide in generalizing the operator. Note also that gauge factors would drop out when computing u from (8.3).

Therefore we define the discrete Hirota bilinear form by requiring invariance under

$$f_j(n, m) \to f_j'(n, m) = A^n B^m\, f_j(n, m).$$

This implies that an equation is in the **discrete Hirota bilinear form** if it can be written as

$$\sum_j c_j\, f_j(n + \nu_j^+, m + \mu_j^+)\, g_j(n + \nu_j^-, m + \mu_j^-) = 0 \tag{8.10}$$

where the sums of shifts $\nu_j^+ + \nu_j^- = \nu^s$, $\mu_j^+ + \mu_j^- = \mu^s$ do not depend on j.

The above result can be described geometrically as follows. For each j in the sum (8.10) we can draw a line connecting the coordinates of f_j and g_j. For an equation in Hirota form all these lines intersect at one point, given by the average value of the coordinates $(n + \frac{1}{2}\nu^s, m + \frac{1}{2}\mu^s)$; see Figure 8.2. Thus, although lattice potential mKdV (3.11) is quadratic in v, it is not in Hirota bilinear form.

8.2 Soliton solutions

Let us consider a bilinear equation of the general form

$$P(D_{\vec{x}}) f \cdot f = 0 \tag{8.11}$$

where P is some *even* function in D_x, D_y, D_z, ... In the continuous case any even polynomial P is acceptable; in the discrete case P should be an even function built up from exponentials of Hirota derivative. For example, a sum containing terms like $\cosh(D_x)$ or $\sinh(D_x)\sinh(D_t)$ etc. is acceptable.

We are now going to build the soliton solutions in a step-by-step manner without a priori specifying the function P. We note that according to (8.2) the solution $f = 1$ of the bilinear

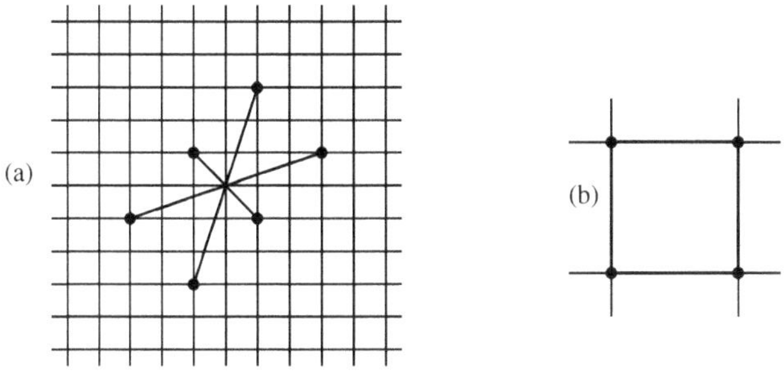

Figure 8.2 A geometric representation of (a) Hirota's bilinear discrete KdV equation (8.22b) and (b) quadratic lattice potential mKdV (3.11). Both equations are quadratic but in (a) the lines connecting the coordinates of the pair of terms in each product intersect at a common point while in (b) they do not.

equation (8.6) corresponds to the trivial solution $u = 0$ of (8.1). Thus we take $f \equiv 1$ as the first step in constructing the soliton solutions. This works, provided that $P(\vec{0}) = 0$.[3]

The next step is to construct the one-soliton solution (1SS)

$$f = 1 + \varepsilon e^{\eta_1}, \tag{8.12}$$

where we have introduced $\eta_i := \vec{x}\cdot\vec{p}_i + \eta_i^0$, for all i, and ε is a parameter. (Here $\vec{x}$ represents a vector of coordinates and $\vec{p}_i$ a vector of parameters.) When this is substituted into (8.11) one finds the constraint $P(\vec{p}_1) = 0$ and below we assume that all parameters $\vec{p}_i$ satisfy the *dispersion relation*

$$P(\vec{p}_i) = 0, \ \forall i. \tag{8.13}$$

Continuing in this manner the ansatz for a two-soliton solution (2SS) is given by

$$f = 1 + \varepsilon e^{\eta_1} + \varepsilon e^{\eta_2} + \varepsilon^2 A_{12} e^{\eta_1+\eta_2}. \tag{8.14}$$

A direct calculation shows that this works for the generic equation (8.11), provided that

$$A_{ij} = -\frac{P(\vec{p}_i - \vec{p}_j)}{P(\vec{p}_i + \vec{p}_j)}, \quad \forall i, j. \tag{8.15}$$

A_{ij} is called the *phase factor*. Thus we have found that the generic equation (8.11) has 2SS, where each $\vec{p}_i$ is restricted by the dispersion relation (8.13), which already appeared in constructing the 1SS, but there are no additional restrictions. The above works equally well for the continuous and the discrete case, the only difference being that in the continuous case the function P is a polynomial; in the discrete case it is a sum of exponentials.

[3] If $P(\vec{0}) \neq 0$ then one obtains periodic solutions expressible in terms of theta-functions (Nakamura, 1979; Hirota and Ito, 1981). For theta-functions, see Appendix B.

In the above computations a crucial role is played by the minus sign in the definition of Hirota's derivative operator, which implies (8.7). In practice this means that the first *and* the last terms in the ε-expansion, such as the 1SS (8.12) and 2SS (8.14), automatically satisfy the equation (since $P(\vec{0}) = 0$).

In order to understand why (8.14) is called a *two*-soliton solution we consider it in two different limits. First suppose η_1 is fixed while $\eta_2 \to -\infty$; then we are left with the 1SS (8.12). On the other hand, when η_2 is fixed while $\eta_1 \to -\infty$ then we are again left with the 1SS but η_1 replaced by η_2 in (8.12). We identify these two limiting cases as individual solitons within the full 2SS (8.14). Next consider the case when η_1 is fixed while the $\eta_2 \to +\infty$. In this limit the leading term is

$$f \sim \varepsilon e^{\eta_2}(1 + \varepsilon A_{12} e^{\eta_1}).$$

However, this is equivalent to the 1SS (8.12) under gauge equivalence and redefining $\eta_1^0 \mapsto \eta_1^0 - \log A_{12}$. The same applies when $\eta_1 \to +\infty$ with η_2 fixed. Therefore we can still see two individual solitons in this limit but each has undergone a phase shift.

The results obtained so far show that any bilinear equation of type (8.11) can have a two-soliton solution. However, we will show that the existence of N-soliton solutions (NSS) for any integer $N \geq 3$ (with generic parameters) is rare and can be used as a definition of integrability.

Assume that the bilinear equation (8.11) has 1SSs of the form

$$f = 1 + \varepsilon e^{\eta_j}, \quad \eta_j = \vec{x} \cdot \vec{p}_j + \eta_j^0,$$

where the parameters $\vec{p}_j$ are only restricted by the dispersion relation $P(\vec{p}_j) = 0$. The equation is said to be **Hirota-integrable** if for all positive integers N it has N-soliton solutions of the form

$$f = 1 + \varepsilon \sum_{j=1}^{N} e^{\eta_j} + (\text{finite number of higher-order terms in } \varepsilon)$$

without any further conditions on the parameters $\vec{p}_j$, beyond the dispersion relation (8.13).

The assumption that there are no further restrictions on the parameters is essential, because any equation may have multi-soliton solutions for some restricted set of parameters. (As an example see Hirota (1973) for a restricted three-soliton solution of the (2+1)-dimensional sine-Gordon equation.)

If we now apply the above principle to the 3SS we start with the ansatz

$$\begin{aligned} f = 1 &+ \varepsilon e^{\eta_1} + \varepsilon e^{\eta_2} + \varepsilon e^{\eta_3} \\ &+ \varepsilon^2 A_{12} e^{\eta_1+\eta_2} + \varepsilon^2 A_{23} e^{\eta_2+\eta_3} + \varepsilon^2 A_{13} e^{\eta_1+\eta_3} \\ &+ \varepsilon^3 A_{12} A_{23} A_{13} e^{\eta_1+\eta_2+\eta_3}. \end{aligned} \tag{8.16}$$

This form is fixed by the requirement that if any one of the solitons in an N-soliton solution goes far away, then the rest should look like an $(N-1)$-soliton solution. As a consequence there is no freedom left in f: the parameters are restricted only by the dispersion relation (8.13) and the phase factors A_{ij} were given already in (8.15). Thus the existence of three-soliton solutions is not automatic, but rather it imposes severe requirements on the polynomial P.

The above definition of integrability can be used as *a method for searching for and classifying new integrable equations*. In practice it means that one has to verify the "three-soliton condition" which follows when the ansatz (8.16) is substituted into (8.11), namely

$$\sum_{\sigma_i=\pm} P(\sigma_1\vec{p}_1+\sigma_2\vec{p}_2+\sigma_3\vec{p}_3)P(\sigma_1\vec{p}_1-\sigma_2\vec{p}_2) \times P(\sigma_2\vec{p}_2-\sigma_3\vec{p}_3)P(\sigma_1\vec{p}_1-\sigma_3\vec{p}_3)=0, \tag{8.17}$$

on the manifold $P(\vec{p}_i) = 0, \forall i$. This has been used to classify continuous equations (Hietarinta, 1987a) and some restricted discrete cases (Hietarinta and Zhang, 2013).[4]

If the equation passes the "three-soliton test" it is conjectured to be integrable by any other test, and the corresponding NSS is given by

$$f=\sum_{\mu_i\in\{0,1\}} \exp\left[\sum_{\substack{i,j=1\\ i<j}}^{N} a_{ij}\,\mu_i\,\mu_j+\sum_{i=1}^{N}\mu_i\,\eta_i\right], \tag{8.18}$$

where $\exp(a_{ij}) = A_{ij}$.

8.3 Hirota's and Miwa's equations

The first work on the discrete bilinear method was by Hirota (1977a). In this section we explain how Hirota approached the question of discretizing soliton equations by using the KdV equation as an example. Hirota's approach led to two fundamental new equations, which are now called Hirota's and Miwa's equations. We will also discuss these equations and some of their properties.

8.3.1 Hirota's method of discretization

In a series of papers Hirota (1977a,b,c) discretized the bilinear forms of KdV, Toda and sG equations, constructed their NSS and also proposed discrete nonlinear forms for these equations. In order to illustrate Hirota's astute observations that led to these results, we

[4] Letting $\vec{p} = (p, q, \omega)$ it is easy to show that the cases $P = p^4 - p\omega$ (KdV) and $P = p^4 - 4p\omega + 3q^2$ (KP) solve (8.17). Note that (8.17) needs to hold only when (8.13) holds. For these examples (which are evolution equations and therefore first-order in time) one can easily do this by solving $P(\vec{p}_i) = 0$ for ω_i. For other equations this step may present difficulties.

reproduce the step-by-step derivation in the KdV case, but use simpler notation in order to make this more accessible. Starting from the famous Toda-lattice equation (Toda, 1967) Hirota and Satsuma (1976) proposed a semi-discrete KdV equation:

$$\frac{d}{dt}\frac{w_n}{1+w_n} = w_{n-2} - w_{n+2}. \tag{8.19}$$

This is equivalent to the modified Volterra equation (5.62) with $a_n \propto 1 + w_{n'}$, together with a scaling of the independent variable. Equation (8.19) can be bilinearized using the substitution

$$w_n = \frac{f_{n-2} f_{n+2}}{f_n^2} - 1. \tag{8.20}$$

Indeed, if one substitutes (8.20) into (8.19) the result can be written as

$$\frac{E_{n+1}}{f_{n+2} f_n} - \frac{E_{n-1}}{f_n f_{n-2}} = 0,$$

with

$$E_n := D_t\, f_{n+1} \cdot f_{n-1} + f_{n+3} f_{n-3} + \sigma f_{n+1} f_{n-1} = 0, \tag{8.21a}$$

where σ is a separation constant. For $\sigma = -1$ this can also be written as

$$[\sinh(D_n)(D_t + 2\sinh(2D_n))] f_n \cdot f_n = 0. \tag{8.21b}$$

Hirota then made the crucial step of proposing the fully discrete *symmetric* form (Hirota, 1977a):

$$[\sinh(D_n + D_m)(2\delta^{-1}\sinh(2D_m) + 2\sinh(2D_n))] f_{n,m} \cdot f_{n,m} = 0, \tag{8.22a}$$

which has the continuum limit (8.21b) if the m-shift corresponds to δ-shift in time. Then writing sinh and cosh in terms of exponentials and using (8.8) we can rewrite (8.22a) as

$$f_{n+3,m+1} f_{n-3,m-1} + \delta^{-1} f_{n+1,m+3} f_{n-1,m-3} - (1+\delta^{-1}) f_{n+1,m-1} f_{n-1,m+1} = 0. \tag{8.22b}$$

Next, in order to get the integrable nonlinear fully discrete form one uses the fully discretized form of (8.20), namely

$$W_{n,m} = \frac{f_{n+2,m} f_{n-2,m}}{f_{n,m+2} f_{n,m-2}} - 1. \tag{8.23}$$

It is then easy to verify that if f satisfies (8.22b) then W satisfies

$$\frac{1}{1 + W_{n,m+2}} - \frac{1}{1 + W_{n,m-2}} = \delta(W_{n+2,m} - W_{n-2,m}). \tag{8.24}$$

This is then Hirota's discretization of the KdV equation; it can be easily transformed into the form given in (1.45).

8.3.2 Hirota's DAGTE

Most of the bilinear equations that Hirota discretized using the above technique were later unified under one culminating equation, which Hirota called "Discrete Analogue of a Generalized Toda Equation" (DAGTE) (Hirota, 1981):

$$[Z_1 \exp(D_1) + Z_2 \exp(D_2) + Z_3 \exp(D_3)]\, f \cdot f = 0, \tag{8.25}$$

which was, in fact, the first fully discrete equation in three dimensions. Here Z_i are arbitrary constants and D_i are Hirota derivatives in arbitrary directions.

This is a remarkable equation in that with suitable choices for the derivatives D_i this equation seems to contain most of the known integrable soliton equations. For example, the bilinear dKdV is obtained with $D_1 = \frac{1}{4}(3\delta D_t + D_n)$, $D_2 = \frac{1}{4}(\delta D_t + 3D_n)$, $D_3 = \frac{1}{4}(\delta D_t - D_n)$, $Z_1 = 1$, $Z_2 = \delta$, $Z_3 = -(1+\delta)$ and Hirota (1981) gave similar choices for many other equations, including KP, mKdV, sG, Benjamin-Ono etc. Equation (8.25) was also shown to satisfy the three-soliton condition and conjectured to have NSS of the type (8.18), futhermore Hirota gave the bilinear BT for (8.25), from which a Lax pair can be derived.

If we take DAGTE in its most general form and assume that D_i are derivatives in different coordinates ($D_1 = \delta D_x$ etc.) and use (8.8) we arrive at the three-dimensional discrete equation

$$\begin{aligned} Z_1 f(x+\delta, y, t) f(x-\delta, y, t) + Z_2 f(x, y+\varepsilon, t) f(x, y-\varepsilon, t) \\ + Z_3 f(x, y, t+\kappa) f(x, y, t-\kappa) = 0\,, \end{aligned}$$

which can also be written as

$$Z_1\, f_{n+1,m,k} f_{n-1,m,k} + Z_2\, f_{n,m+1,k} f_{n,m-1,k} + Z_3\, f_{n,m,k+1} f_{n,m,k-1} = 0. \tag{8.26}$$

By the change of independent variables $n' = \frac{1}{2}(-n+m+k)$, $m' = \frac{1}{2}(n-m+k)$, $k' = \frac{1}{2}(n+m-k)$, this can also be written (after a $\frac{1}{2}$ shift in all coordinates) as

$$\begin{aligned} Z_1\, f_{n'+1,m',k'} f_{n',m'+1,k'+1} + Z_2\, f_{n',m'+1,k'} f_{n'+1,m',k'+1} \\ + Z_3\, f_{n',m',k'+1} f_{n'+1,m'+1,k'} = 0, \end{aligned} \tag{8.27}$$

or

$$Z_1\, \widetilde{f}\,\widehat{\overline{f}} + Z_2\, \widehat{f}\,\widetilde{\overline{f}} + Z_3\, \overline{f}\,\widehat{\widetilde{f}} = 0. \tag{8.28}$$

To compare these two equations it is informative to look at the diagram connecting the lattice points of each term. In the case of (8.26) the diagram takes the form of a

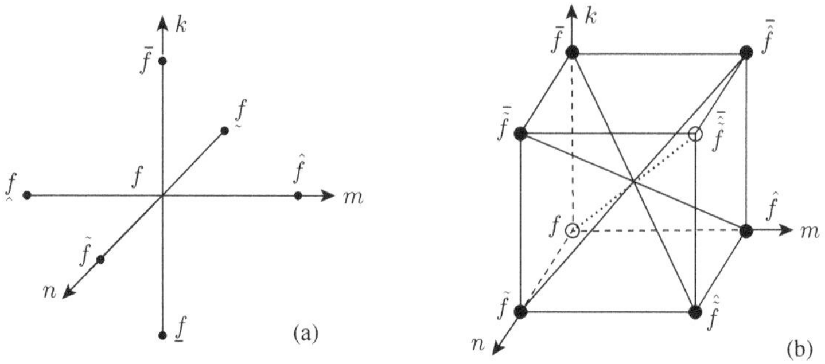

Figure 8.3 Diagrammatic representation of equation (8.26) in (a) and equation (8.27) in (b). The dotted line corresponds to the fourth term that is included in Miwa's equation (8.31).

three-dimensional star, with center at the origin; see Figure 8.3a, while equation (8.28) connects six of the eight corners of the elementary cube as in Figure 8.3b.

Recalling that the construction of multi-soliton solutions in Section 8.2 starts with $f \equiv 1$ which implies $Z_1 + Z_2 + Z_3 = 0$, Miwa proposed in Miwa (1982) the parameterization

$$a(b-c)\,\tilde{f}\,\hat{\bar{f}} + b(c-a)\,\hat{f}\,\tilde{\bar{f}} + c(a-b)\,\bar{f}\,\hat{\tilde{f}} = 0 \tag{8.29}$$

which leads to particularly elegant soliton solutions of the form (8.18) with

$$e^{\eta_i} = \left(\frac{1-aq_i}{1-ap_i}\right)^n \left(\frac{1-bq_i}{1-bp_i}\right)^m \left(\frac{1-cq_i}{1-cp_i}\right)^k, \qquad A_{i,j} = \frac{(p_i-p_j)(q_i-q_j)}{(p_i-q_j)(q_i-p_j)}. \tag{8.30}$$

Equation (8.29) is often referred to in the literature as the *Hirota–Miwa equation*. However Hirota's original form (8.25) (which we will refer to as the *Hirota equation*, or *bilinear lattice KP* equation) is more general as it has three free parameters Z_i. It is still integrable, and in order to construct more general solutions (elliptic or periodic solutions) this freedom in parameters is needed; see e.g. Krichever et al. (1998); Yoo-Kong and Nijhoff (2013). Furthermore, in Ramani et al. (1992) it was shown that (8.25) passes the singularity confinement test for generic Z_i, as expected. We note that there are still other transformed versions of the DAGTE in the literature; see e.g. Zabrodin (1997a).

8.3.3 Miwa's equations

Inspired by Hirota's discretization of the bilinear KP equation, Miwa (1982) proposed a generalization to the BKP case:

$$\begin{aligned}&(a+b)(a+c)(b-c)\,\tilde{\tau}\hat{\bar{\tau}} + (b+c)(b+a)(c-a)\,\hat{\tau}\tilde{\bar{\tau}}\\ &+(c+a)(c+b)(a-b)\,\bar{\tau}\hat{\tilde{\tau}} + (a-b)(b-c)(c-a)\,\tau\hat{\tilde{\bar{\tau}}} = 0.\end{aligned} \tag{8.31}$$

This generalization is natural from the geometric point of view, because it now connects all vertex pairs of the cube displayed in Figure 8.3b. Equation (8.31) is commonly referred to as the *Miwa equation* or also as the bilinear lattice BKP equation (whereas (8.29) is sometimes also referred to as the bilinear lattice AKP equation – this terminology being inspired by connections with the classification of root systems of semi-simple Lie algebras).

Also, this equation has soliton solutions of the form (8.18), but now (Miwa, 1982; Date et al., 1983c)

$$\begin{aligned} e^{\eta_i} &= \left[\frac{(1-ap_i)(1-aq_i)}{(1+ap_i)(1+aq_i)}\right]^n \left[\frac{(1-bp_i)(1-bq_i)}{(1+bp_i)(1+bq_i)}\right]^m \\ &\qquad \times \left[\frac{(1-cp_i)(1-cq_i)}{(1+cp_i)(1+cq_i)}\right]^k, \end{aligned} \tag{8.32a}$$

$$A_{ij} = \frac{(p_i-p_j)(p_i-q_j)(q_i-p_j)(q_i-q_j)}{(p_i+p_j)(p_i+q_j)(q_i+p_j)(q_i+q_j)}. \tag{8.32b}$$

Equation (8.31) is also integrable in the sense that it arises from a Lax pair or actually a Lax triplet (Nimmo and Schief, 1997)

$$\widehat{\widetilde{\psi}} - \psi = \frac{\widetilde{\tau}\widehat{\tau}}{\tau\widehat{\widetilde{\tau}}}(\widehat{\psi} - \widetilde{\psi}), \quad \overline{\widetilde{\psi}} - \psi = \frac{\widetilde{\tau}\overline{\tau}}{\tau\overline{\widetilde{\tau}}}(\overline{\psi} - \widetilde{\psi}), \quad \overline{\widehat{\psi}} - \psi = \frac{\widehat{\tau}\overline{\tau}}{\tau\overline{\widehat{\tau}}}(\overline{\psi} - \widehat{\psi}), \tag{8.33}$$

in the sense that the compatibility conditions

$$(\widehat{\widetilde{\psi}})^{-} = (\overline{\widetilde{\psi}})^{\sim} = (\overline{\widehat{\psi}})^{\wedge}$$

yield

$$\overline{\tau}\widehat{\widetilde{\tau}} + \widetilde{\tau}\overline{\widehat{\tau}} - \widehat{\tau}\overline{\widetilde{\tau}} - \tau\overline{\widehat{\widetilde{\tau}}} = 0. \tag{8.34}$$

This last equation follows by the scaling transformation

$$\tau \mapsto \left(\frac{a+b}{a-b}\right)^{nm} \left(\frac{b+c}{b-c}\right)^{mk} \left(\frac{a+c}{a-c}\right)^{kn} \tau,$$

from (8.31).

8.4 Reductions of the Hirota-Miwa equation

Hirota's DAGTE is fundamental in the sense that many other equations can be obtained from it by reductions. In this section we illustrate this by showing how to obtain the bilinear dKdV (8.22b) and dmKdV (8.51) equations as dimensional reductions. We will also study what happens to the solution (8.30) in these reductions.

8.4.1 Reduction to dKdV

Reduction of equation

In order to get the dKdV equation we impose on (8.29) the diagonal reduction

$$f_{n,m,k+1} = f_{n-1,m,k} \tag{8.35}$$

for all n, m, k. Using this we can eliminate all forward shifts of k and thereafter omit that index. The resulting equation is

$$a(b-c)\, f_{n+1,m} f_{n-1,m+1} + c(a-b)\, f_{n-1,m} f_{n+1,m+1} + b(c-a)\, f_{n,m+1} f_{n,m} = 0, \tag{8.36}$$

which agrees with (8.22b) after a coordinate change $[(\frac{1}{4}(n+m), \frac{1}{4}(n-m+2))$ in (8.22b) corresponds to (n, m) in (8.36)$]$ and defining $\delta = b(c-a)/(a(b-c))$.

Reduction of solution

The reduction of the equations also allows us to construct solutions. But note that because the reduction lowers the dimension of the equation we will automatically get conditions on the parameters of the solution. In this case imposing the reduction (8.35) directly on the NSS (8.18) and (8.30) leads to

$$\begin{aligned} &\left(\frac{1-aq_i}{1-ap_i}\right)^n \left(\frac{1-bq_i}{1-bp_i}\right)^m \left(\frac{1-cq_i}{1-cp_i}\right)^{k+1} \\ &\quad = \left(\frac{1-aq_i}{1-ap_i}\right)^{n-1} \left(\frac{1-bq_i}{1-bp_i}\right)^m \left(\frac{1-cq_i}{1-cp_i}\right)^k, \end{aligned}$$

and therefore

$$\frac{1-cq_i}{1-cp_i} = \frac{1-ap_i}{1-aq_i}.$$

In addition to the trivial solution $p_i = q_i$ this equation has the solution

$$q_i = -p_i + \frac{1}{a} + \frac{1}{c}. \tag{8.37}$$

We will now use this solution in the plane-wave factor and the phase factor (8.30). In order to get a more symmetric form for A_{ij} we change to a new parameter P_i by taking $p_i = P_i + (a+c)/(2ac)$, which yields the phase-factor in the standard form for the KdV hierarchy

$$A_{i,j} = \left(\frac{P_i - P_j}{P_i + P_j}\right)^2. \tag{8.38}$$

Next, in order to symmetrize the plane-wave factor we introduce new parameters α, β by

$$\frac{1}{b} = \frac{1}{a} + \alpha - \beta, \quad \frac{1}{c} = \frac{1}{a} + 2\alpha$$

after which the plane-wave factor gets the simple form

$$e^{\eta_i} = \left(\frac{\alpha - P_i}{\alpha + P_i}\right)^n \left(\frac{\beta - P_i}{\beta + P_i}\right)^m, \tag{8.39}$$

and we can write equation (8.36) as

$$(\alpha + \beta)\, \underset{\sim}{\widehat{f}}\, \widetilde{f} + (\alpha - \beta)\, \underset{\sim}{f}\, \widehat{\widetilde{f}} = 2\alpha\, f\, \widehat{f}. \tag{8.40a}$$

Second reduction for the equation and consistency

A reduction like (8.35) can also be performed in other directions, for example in the m, k direction by using $f_{n,m,k+1} = f_{n,m-1,k}$. It turns out that we can parameterize this case so that we end up with the exactly same phase factor (8.38) and plane-wave factor (8.39), and then we arrive at the equation

$$(\alpha + \beta) \underset{\wedge}{\widetilde{f}} \widehat{f} + (\beta - \alpha) \underset{\wedge}{f} \widehat{\widetilde{f}} = 2\beta\, f\, \widetilde{f}. \tag{8.40b}$$

Therefore we have found, surprisingly, that both of the equations (8.40) have the same multi-soliton solutions. We may then expect that these six-point equations hold simultaneously for the function f, and thus there is an issue of the consistency of these two equations.

In this case the consistency condition is described in Figure 8.4c: we will be using (8.40a) and its hat downshift as well as (8.40b) and its tilde downshift. In this figure the values at black disks are given. The equations then provide unique values at open circles, but the value of $\widehat{\widetilde{f}}$ at the upper-right hand corner can be computed in two ways. In this case the two ways give the same result. In fact, expressing $\widehat{\widetilde{f}}$ in terms of the values of the initial data, the following five-point equation emerges:

$$(\alpha - \beta)^2 \underset{\approx}{f}\, \widehat{\widetilde{f}} - (\alpha + \beta)^2 \underset{\wedge}{\widetilde{f}}\, \underset{\sim}{\widehat{f}} + 4\alpha\beta\, f^2 = 0, \tag{8.41}$$

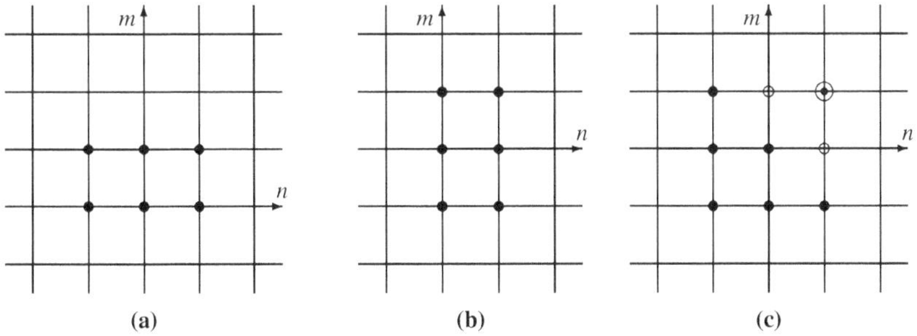

Figure 8.4 Points involved in the equations (8.40a) (a) and (8.40b) (b), and points used in their consistency analysis (c).

which is the famous *discrete-time Toda equation* (Hirota, 1977a) (see Section 3.8.4: equation (3.93) is obtained from this by a rotation).

In this section we focused on a reduction of the form (8.35) but there are other reductions of the type $f(n, m, k+1) = f(n+\nu, m+\mu, k)$ leading to other equations (see e.g. Date et al. (1983a) and Zabrodin (1997a), Section 8).

8.4.2 Reduction to the dmKdV equations

Reduction of equation

Now consider the period-two reduction of (8.29) in the k-variable:

$$f_{n,m,k} = \begin{cases} F_{n,m} & \text{if} \quad k \text{ is even}, \\ G_{n,m} & \text{if} \quad k \text{ is odd}. \end{cases} \tag{8.42}$$

We then get the two equations

$$a(b-c)\,\widetilde{F}\widehat{G} + b(c-a)\,\widehat{F}\widetilde{G} + c(a-b)\,\widehat{\widetilde{F}}\,G = 0, \tag{8.43a}$$
$$a(b-c)\,\widehat{F}\widetilde{G} + b(c-a)\,\widetilde{F}\widehat{G} + c(a-b)\,F\widehat{\widetilde{G}} = 0. \tag{8.43b}$$

We will next show that this equation is the Hirota-bilinear version of the lattice potential mKdV equation. When Hirota nonlinearized the continuous mKdV equation, he used the substitution

$$v := G/F, \tag{8.44}$$

which was motivated by the structure of the soliton solutions. Since we expect the soliton structure to be the same, we let $G = vF$, and find

$$\frac{\widehat{\widetilde{F}}\,F}{\widetilde{F}\widehat{F}}(1-\gamma) = \frac{\widehat{v} - \gamma\widetilde{v}}{v} = \frac{\widetilde{v} - \gamma\widehat{v}}{\widehat{\widetilde{v}}},$$

where we have used $\gamma = (b(a-c))/(a(c-b))$. The second equality is equation (3.11) with $\gamma = p/q$, as expected.

We will see later that there is another natural combination of variables that we can use to nonlinearize equations (8.43), namely

$$W := \frac{\widetilde{F}\widehat{G}}{\widehat{F}\widetilde{G}}. \tag{8.45}$$

Dividing both equations (8.43) by $\widehat{F}\widetilde{G}$ we find that they can be rewritten as

$$W - \gamma = (1-\gamma)\frac{\widehat{\widetilde{F}}\,G}{\widehat{F}\,\widetilde{G}}, \quad 1 - \gamma W = (1-\gamma)\frac{F\,\widehat{\widetilde{G}}}{\widehat{F}\,\widetilde{G}}.$$

We can eliminate F, G between these two equations by taking shifts and ratios and thereby obtain (3.90). The substitutions (8.44) and (8.45) are related by $W = \widetilde{v}/\widehat{v}$ and the relationship between the corresponding nonlinear equations is discussed in Exercise 3.10.

Reduction of solution

The requirement of periodicity (8.42) implies for the soliton solution (8.30) that $((1 - cq_i)/(1 - cp_i))^2 = 1$, i.e.

$$\frac{1 - cq_i}{1 - cp_i} = \pm 1. \tag{8.46}$$

The nontrivial solution comes from the minus sign (otherwise the plane-wave factor is trivially equal to one). Therefore, the even-k case F and the odd-k case G differ by the signs of the plane-wave factor. Writing F as in (8.18) we can write G as

$$G = \sum_{\mu_i \in \{0,1\}} \exp\left[\sum_{\substack{i,j=1 \\ i<j}}^{N} a_{ij}\, \mu_i\, \mu_j + \sum_{i=1}^{N} \mu_i\,(\eta_i + i\pi) \right]. \tag{8.47}$$

The condition (8.46) is solved by

$$q_i = -p_i + \frac{2}{c}, \tag{8.48a}$$

and if in addition we reparameterize as follows:

$$p_i = P_i + \frac{1}{c}, \quad \frac{1}{a} = \frac{1}{c} - \alpha, \quad \frac{1}{b} = \frac{1}{c} - \beta, \tag{8.48b}$$

then the plane-wave factors and the phase factors are respectively as in (8.38) and (8.39) and the equations become

$$\alpha(\widehat{\widetilde{\widetilde{F}}} G - \widehat{F}\widetilde{G}) = \beta(\widehat{\widetilde{\widetilde{F}}} G - \widetilde{F}\widehat{G}), \quad \alpha(\widehat{\widetilde{\widetilde{G}}} F - \widehat{G}\widetilde{F}) = \beta(\widehat{\widetilde{\widetilde{G}}} F - \widetilde{G}\widehat{F}). \tag{8.49a}$$

Other reductions and their consistency

Equation (8.49a) is an equation on the elementary square in the n, m-plane, because we eliminated the k-variable by the reduction. But we could do the period two reduction equally well in the n or in the m directions, resulting in equations defined on the m, k-plane and n, k-plane, respectively. In addition to the above equations we get further equations by cyclic permutation of (a, b, c), (n, m, k) and (α, β, γ):

$$\beta(\bar{\bar{\widehat{F}}} G - \bar{F}\widehat{G}) = \gamma(\bar{\bar{\widehat{F}}} G - \widehat{F}\bar{G}), \quad \beta(\bar{\bar{\widehat{G}}} F - \bar{G}\widehat{F}) = \gamma(\bar{\bar{\widehat{G}}} F - \widehat{G}\bar{F}), \tag{8.49b}$$

$$\gamma(\widetilde{\bar{\bar{F}}} G - \widetilde{F}\bar{G}) = \alpha(\widetilde{\bar{\bar{F}}} G - \bar{F}\widetilde{G}), \quad \gamma(\widetilde{\bar{\bar{G}}} F - \widetilde{G}\bar{F}) = \alpha(\widetilde{\bar{\bar{G}}} F - \bar{G}\widetilde{F}). \tag{8.49c}$$

This is similar to the study of multidimensional consistency in Chapter 3: the above triplet of pairs of equations can be interpreted as equations posed on the bottom, back and left sides of a cube in the n, m, k space. By taking suitable shifts we get equations on the top, front and right sides of the same cube. These equations allow us to compute $\widehat{\widetilde{F}}$, $\widehat{\widetilde{G}}$, $\bar{\widetilde{F}}$, $\widetilde{\bar{G}}$, $\bar{\widehat{F}}$, $\bar{\widehat{G}}$ uniquely, and then $\bar{\widetilde{\widehat{F}}}$, $\bar{\widetilde{\widehat{G}}}$ in three different ways, but they all give the same result:

$$\bar{\widetilde{\widehat{F}}} = \frac{\gamma(\alpha^2-\beta^2)\widehat{F}\widetilde{F}\bar{G} + \alpha(\beta^2-\gamma^2)\bar{F}\widehat{F}\widetilde{G} + \beta(\gamma^2-\alpha^2)\widetilde{F}\bar{F}\widehat{G}}{(\alpha-\beta)(\beta-\gamma)(\gamma-\alpha)FG}, \tag{8.50a}$$

$$\bar{\widetilde{\widehat{G}}} = \frac{\gamma(\alpha^2-\beta^2)\widehat{G}\widetilde{G}\bar{F} + \alpha(\beta^2-\gamma^2)\bar{G}\widehat{G}\widetilde{F} + \beta(\gamma^2-\alpha^2)\widetilde{G}\bar{G}\widehat{F}}{(\alpha-\beta)(\beta-\gamma)(\gamma-\alpha)FG}. \tag{8.50b}$$

Thus we have found a two-component system of equations that (1) is in the Hirota bilinear form, (2) has the CAC property, and 3) does not satisfy the tetrahedron property. In contrast, in the case of one-component Hirota bilinear equations, there is only one quadrilateral case, namely $\widehat{\widetilde{F}}F - \widehat{F}\widetilde{F} = 0$, which is linearizable.

8.5 Bilinearization of a lattice equation

In order to apply the Hirota bilinear method for constructing NSS the original nonlinear equation must first be converted into Hirota bilinear form. The bilinearization process requires some care and insight that may change with the equation because a priori it is not clear how many dependent or even independent variables are needed for this purpose. Hirota (2004) has showed that one dependent variable is enough for bilinearizing the KdV equation, but for the mKdV, sG and NLS equations one needs two real functions (or one complex).[5]

In this section, we give two methods for bilinearization based on singularity confinement and construction of soliton solutions, respectively.

8.5.1 Using singularity confinement

The transformation to Hirota bilinear form can be seen as a regularization of movable singularities in the solutions. As an example (8.3) transforms movable double poles in u to movable simple zeroes in f. We use this crucial idea here to construct Hirota bilinear forms for lattice equations.

The idea of transforming to more regular functions is a familiar technique in mathematics. Recall that the solutions to Painlevé equations are meromorphic functions with movable poles, but Painlevé himself noted (Painlevé, 1902; cf. Hietarinta and Kruskal, 1992) that the solutions of his new equations can be written in terms of entire functions.

For lattice equations, the analysis of singularity confinement, as discussed in Section 7.1.3, provides us a singularity pattern for the original variable, which as we will show, can

[5] After these observations it is perhaps not surprising that the literature contains examples with unacceptable bilinearizations; for example "over-bilinearization", in which a multilinear equation is broken down into too many bilinear equations.

be generated from a simple zero of one or more function(s). The equation in terms of the new dependent variable(s) may not immediately be in Hirota bilinear form but rather be in a multilinear form. Further work is then required to split it into reasonable bilinear form, but this step may be difficult.

First application of this method for discrete Painlevé equations was in Ramani et al. (1995) and for soliton equations in Maruno et al. (1997); Kajiwara et al. (2000).

Bilinearizing discrete KdV equation

As the first example let us recall the singularity analysis of the dKdV equation (7.4) in Section 7.1.3, where we found the singularity pattern $w_{n,m}=0$, $w_{n+1,m}=\infty$, $w_{n,m+1}=\infty$, $w_{n+1,m+1}=0$. All these values of w can be generated from a single zero of a new function τ at the point n,m. The zeros of w can be obtained from the product $\tau_{n,m}\tau_{n-1,m-1}$. Moreover the infinite values of w could arise from $1/(\tau_{n-1,m}\tau_{n,m-1})$. Putting these together we propose the transformation

$$w_{n,m}=\frac{\tau_{n,m}\tau_{n-1,m-1}}{\tau_{n-1,m}\tau_{n,m-1}}.$$

When this is substituted into (7.4) we get a trilinear equation

$$\begin{aligned}\tau_{n-1,m-1}\tau_{n+1,m}\tau_{n,m+1}-\tau_{n-1,m}\tau_{n+1,m+1}\tau_{n,m-1}&=\\ \kappa(\tau_{n-1,m+1}\tau_{n+1,m}\tau_{n,m-1}-\tau_{n-1,m}\tau_{n+1,m-1}\tau_{n,m+1})&.\end{aligned}$$

Now we face the usual problem in the bilinearization process: how can we split a trilinear equation into bilinear ones? Dividing the equation with $\tau_{n-1,m}\tau_{n+1,m}\tau_{n,m}$ we can group the terms as

$$\frac{\tau_{n-1,m-1}\,\tau_{n,m+1}-\kappa\tau_{n-1,m+1}\,\tau_{n,m-1}}{\tau_{n-1,m}\,\tau_{n,m}}=\frac{\tau_{n,m-1}\,\tau_{n+1,m+1}-\kappa\tau_{n,m+1}\,\tau_{n+1,m-1}}{\tau_{n,m}\,\tau_{n+1,m}}.$$

Clearly the right-hand side is an $n\to n+1$ shift of the left-hand side, so we can separate the equations as

$$\tau_{n-1,m-1}\,\tau_{n,m+1}-\kappa\tau_{n-1,m+1}\,\tau_{n,m-1}=\lambda\tau_{n-1,m}\,\tau_{n,m}$$

where the separation "constant" λ could in principle depend on m but we take $\lambda=1-\kappa$ because we want the equation to have soliton solutions that start as $\tau_{nm}=1+\dots$.

Bilinearizing discrete mKdV equation

As the second example consider the dmKdV equation (3.90)

$$\frac{\widehat{\widetilde{W}}}{W}=\frac{(\gamma\widehat{W}-1)}{(\gamma-\widehat{W})}\frac{(\gamma-\widetilde{W})}{(\gamma\widetilde{W}-1)}. \tag{8.51}$$

It was shown in Exercise 7.4 that this equation has two kinds of singularity patterns:

$$W_{n,m} = \gamma,\ W_{n,m+1} = 0,\ W_{n+1,m} = \infty,\ W_{n+1,m+1} = 1/\gamma, \tag{8.52a}$$

$$W_{n,m} = 1/\gamma,\ W_{n,m+1} = \infty,\ W_{n+1,m} = 0,\ W_{n+1,m+1} = \gamma. \tag{8.52b}$$

Since there are two patterns we need two dependent functions (Maruno et al., 1997), so that each pattern is generated by the zero of its own τ-function. Let the pattern (8.52a) be given by zeroes of $F_{n,m}$ and the pattern (8.52b) by zeroes of $G_{n,m}$. The value $W = \gamma$ is generated for both patterns by

$$W_{n,m} = \gamma + \alpha' \frac{F_{n,m} G_{n-1,m-1}}{F_{n-1,m} G_{n,m-1}}, \tag{8.53a}$$

and similarly the value $W = 1/\gamma$ by

$$W_{n,m} = \frac{1}{\gamma} + \beta' \frac{F_{n-1,m-1} G_{n,m}}{F_{n-1,m} G_{n,m-1}}\ . \tag{8.53b}$$

On the other hand the zeroes of W suggest

$$W_{n,m} = \frac{F_{n,m-1} G_{n-1,m}}{F_{n-1,m} G_{n,m-1}}. \tag{8.53c}$$

Note that the denominators of all three expression (8.53) generate the infinities of W at $(n+1, m)$ and $(n, m+1)$.

Since all the above representations (8.53) of W in terms of F, G must be equal we get two equations for two functions, namely

$$\gamma F_{n-1,m} G_{n,m-1} + \alpha' F_{n,m} G_{n-1,m-1} - F_{n,m-1} G_{n-1,m} = 0, \tag{8.54a}$$

$$F_{n-1,m} G_{n,m-1} + \gamma \beta' F_{n-1,m-1} G_{n,m} - \gamma F_{n,m-1} G_{n-1,m} = 0. \tag{8.54b}$$

These equations yield (8.49a) after up-shifting them in n and m and taking $\gamma = \alpha/\beta$, $\alpha' = 1 - \alpha/\beta$, $\beta' = 1 - \beta/\alpha$.

8.5.2 Using the 1SS

In this section we show how to use the soliton solution to find a transformation to τ-functions. We will construct a one-soliton solution and use it to suggest a transformation.

As an example let us consider the H1 equation of the ABS list, namely

$$(u - \widehat{\widetilde{u}})(\widetilde{u} - \widehat{u}) = p - q. \tag{8.55}$$

Recall that in the continuous case the trivial solution $u = 0$ corresponded to the solution $f = 1$ for the τ-function. However, for this equation $u = 0$ is not a solution, and to circumvent this problem we start by constructing an explicit solution around which to expand the variable u. For (8.55) an explicit solution is given by

$$u^{(0)}_{n,m} = an + bm + \gamma, \tag{8.56}$$

where we have reparameterized $p = c^2 - a^2$, $q = c^2 - b^2$. Transforming variables from u to v by

$$u_{n,m} = an + bm + \gamma + v_{n,m}, \tag{8.57}$$

and then $v = 0$ is a solution. However, the equation for v is still not bilinear. In Section 3.3.2 we derived the 1SS (3.33). Motivated by its form we attempt a bilinearizing transformation

$$u_{n,m} = an + bm + \gamma - \frac{g_{n,m}}{f_{n,m}}. \tag{8.58}$$

One can show (Hietarinta and Zhang, 2009) that for this u the following identity holds

$$(u - \widehat{\widetilde{u}})(\widetilde{u} - \widehat{u}) - p + q = -\big[\mathfrak{H}_1 + (a-b) f \widehat{\widetilde{f}}\big]\big[\mathfrak{H}_2 + (a+b)\widehat{f}\widetilde{f}\big]/(f\widehat{f}\widetilde{f}\widehat{\widetilde{f}}) + (a^2 - b^2),$$

where

$$\mathfrak{H}_1 \equiv \widehat{g}\widetilde{f} - \widetilde{g}\widehat{f} + (a-b)(\widehat{f}\widetilde{f} - f\widehat{\widetilde{f}}), \tag{8.59a}$$
$$\mathfrak{H}_2 \equiv g\widehat{\widetilde{f}} - \widehat{\widetilde{g}}f + (a+b)(f\widehat{\widetilde{f}} - \widehat{f}\widetilde{f}). \tag{8.59b}$$

Therefore it follows that if $\mathfrak{H}_1 = 0$ and $\mathfrak{H}_2 = 0$ then equation (8.55) is satisfied. These equations then give the bilinear forms of (8.55).

The method presented here works also for the other equations in the ABS list, but the solutions for higher members are progressively more complicated (Nijhoff et al., 2009; Hietarinta and Zhang, 2009).

8.6 Solutions in matrix form

In this section we assume that a bilinear equation is given and we describe a method of deriving its solutions in terms of determinants of certain types of matrices. The reason why this works is that bilinear equations are related to determinantal identities. There are various determinants that have been used for this purpose, including Wronskians, Casoratians, Grammians and Pfaffians. In the next chapter we will study the Cauchy matrix approach in detail. In this section we will first briefly discuss the Wronskian and then focus on its discrete analogue, the Casoratian.

8.6.1 Wronskian solution for the continuous KP equation

As a prelude to the discrete case, we will consider the Wronskian form of the NSS for the continuous potential KP equation (3.101) following Freeman and Nimmo (1983). The Hirota bilinear form for this equation is

$$(D_x^4 - 4D_xD_t + 3D_y^2)F \cdot F = 0, \tag{8.60}$$

and it is easy to verify that it has the 1SS

$$F = 1 + \exp[(p-q)x + (p^2-q^2)y + (p^3-q^3)t + \eta_0].$$

First of all note that, due to gauge invariance (see Section 8.1.3), we can write the 1SS as

$$\Psi_i(x,y,t) := c_i e^{q_i x + q_i^2 y + q_i^3 t} + d_i e^{p_i x + p_i^2 y + p_i^3 t}. \tag{8.61}$$

We will see these Ψ_i, $i = 1, \ldots, N$ will be the building blocks for constructing the NSS. The Ψ_i must be different so we require that $(p_i, q_i) \neq (p_j, q_j)$ for $i \neq j$. In this construction we do not need the explicit form (8.61) but only the properties

$$\partial_y \Psi_i(x,y,t) = \partial_x^2 \Psi_i(x,y,t), \quad \partial_t \Psi_i(x,y,t) = \partial_x^3 \Psi_i(x,y,t). \tag{8.62}$$

We construct the NSS by starting with the column vector

$$\Psi = (\Psi_1(x,y,t), \ldots, \Psi_N(x,y,t))^T \tag{8.63}$$

and the Wronskian is then the determinant

$$F = |\Psi, \partial_x \Psi, \ldots, \partial_x^{N-1}\Psi|. \tag{8.64}$$

The bilinear equation (8.60) expands to give

$$F_{xxxx}F - 4F_{xxx}F_x + 3F_{xx}^2 - 4F_{xt}F + 4F_xF_t + 3F_{yy}F - 3F_y^2 = 0. \tag{8.65}$$

Therefore we need to compute several partial derivatives of the Wronskian F. In order to simplify the notation we introduce the following short-hand notation to keep track of the number of derivatives

$$|\partial_x^{k_0}\Psi, \partial_x^{k_1}\Psi, \ldots, \partial_x^{k_{N-1}}\Psi| \equiv |k_0, k_1, \ldots, k_{N-1}|.$$

Furthermore, if there is an unbroken sequence of derivatives from 0 to K we condense the notation as follows

$$|0, 1, \ldots, K-1, K, L \ldots| \equiv |\widehat{K}, L, \ldots|,$$

thus for example (8.64) can be written as $F = |\widehat{N-1}|$.

Since derivation is distributive, its action on a determinant is given by the sum of the derivative operating on each column separately. Since the determinant vanishes when two columns are identical, we see that in $\partial_x F$ the derivative produces a nonzero result only when acting on the last column. Thus we find

$$\partial_x F = |\widehat{N-2}, N|.$$

Similarly using (8.62) to express the y- and t-derivatives in terms of x-derivatives one finds

$$\begin{aligned}\partial_y F &= |\widehat{N-2}, N+1| + |\widehat{N-3}, N, N-1|,\\ \partial_t F &= |\widehat{N-2}, N+2| + |\widehat{N-3}, N+1, N-1| + |\widehat{N-4}, N-2, N-1, N|,\end{aligned}$$

and analogous formulas for higher derivatives. After computing all the necessary derivatives (for details, see Freeman and Nimmo, 1983) and substituting into (8.65) it becomes

$$\begin{aligned}&|\widehat{N-3}, N-2, N-1||\widehat{N-3}, N, N+1|\\ &-|\widehat{N-3}, N-2, N||\widehat{N-3}, N-1, N+1|\\ &+|\widehat{N-3}, N-2, N+1||\widehat{N-3}, N-1, N| = 0.\end{aligned} \tag{8.66}$$

This result can be proved using Laplace expansion in the form (D.6). For if we choose $\mathbf{M}$ to contain the columns $0, \dots, N-3$ and $\mathbf{a}, \mathbf{b}, \mathbf{c}, \mathbf{d}$ to be columns $N-2, N-1, N, N+1$, respectively, then after some column permutations the identity (D.6) becomes (8.66). This shows that F defined by (8.64) with (8.62) solves the bilinear KP equation (8.60).

8.6.2 Wronskian solution for the continuous KdV equation

We can apply the same technique to construct the NSS for the KdV equation except that we have to eliminate the y-dependence in the solution F by the "2-reduction" (Hirota, 1986) by which KdV is obtained from KP; that is, $p^2 - q^2 = 0$, $p \neq q$. This means that the basic matrix element is

$$\Psi_i(x,t) := c_i e^{-p_i x - p_i^3 t} + d_i e^{p_i x + p_i^3 t}. \tag{8.67}$$

However, the absence of y derivatives requires a modification of the procedure; thus, instead of the ∂_y-equation of (8.62) we will use

$$\partial_x^2 \Psi_i(x,t) = p_i^2 \Psi_i(x,t)$$

in the derivation. In order to use this property we need the identity (D.8). Using this and other identities in Appendix D one can show that the Wronskian (8.64) solves the KdV equation.

8.6.3 Casoratian NSS for the Hirota–Miwa equation

In this section we consider the discrete case (Ohta et al., 1993). The Wronskian is replaced with the Casoratian, where columns differ by shifts instead of derivatives; in this case the shift is indicated by the last index s of the column vector $\Psi(n, m, k; s)$

$$\tau(n, m, k) := |\Psi(n, m, k; 0), \Psi(n, m, k; 1), \dots, \Psi(n, m, k; N-1)|.$$

Recall the plane-wave factor (8.30), from which the 1SS is constructed by $F = 1 + e^{\eta_i}$. We choose as components ψ_i of Ψ

$$\begin{aligned}\psi_i(n, m, k; s) = {} & \sigma_i p_i^s (1 - ap_i)^{-n}(1 - bp_i)^{-m}(1 - cp_i)^{-k} \\ & + \rho_i q_i^s (1 - aq_i)^{-n}(1 - bq_i)^{-m}(1 - cq_i)^{-k},\end{aligned} \tag{8.68}$$

which are gauge equivalent (recall Section 8.1.3) to F. Note the negative exponents, the extra factors p_i^s, q_i^s which are needed to define the shifts, and σ_i, ρ_i which take the role of η_i^0.

We will now derive the determinantal identity which allows us to solve the Hirota–Miwa equation (8.29). For this purpose, we need PΔEs that follow from (8.68)

$$\Psi(n+1, m, k; s) - a\Psi(n+1, m, k; s+1) = \Psi(n, m, k; s), \tag{8.69a}$$

$$\Psi(n, m+1, k; s) - b\Psi(n, m+1, k; s+1) = \Psi(n, m, k; s), \tag{8.69b}$$

$$\Psi(n, m, k+1; s) - c\Psi(n, m, k+1; s+1) = \Psi(n, m, k; s). \tag{8.69c}$$

Note that PDEs of the continuous case (8.62) have now been replaced by PΔEs.

One difference with respect to the continuous case is that, while the derivative operated on the columns one-by-one, the shift operates on all columns at once. Using (8.69a) (suppressing for the moment the unchanged m, k indices) we have

$$\begin{aligned}\tau(n+1) &= |\Psi(n+1; 0), \Psi(n+1; 1), \Psi(n+1; 2), \dots| \\ &= |\Psi(n; 0) + a\Psi(n+1; 1), \Psi(n+1; 1), \Psi(n+1; 2), \dots| \\ &= |\Psi(n; 0), \Psi(n+1; 1), \Psi(n+1; 2), \dots|\end{aligned}$$

where in the last step we have subtracted $a\times$(second column) from the first. Clearly this can be propagated up to the last column, and thus we get

$$\tau(n+1, m, k) = |\widehat{N-2}, \Psi(n+1, m, k; N-1)|, \tag{8.70}$$

where the hat-notation is here used to represents a sequence of columns $\Psi(n, m, k; s)$ with shifts in s from 0 to $N-2$. By using (8.69a) once more we can lower the s-index of the last column and obtain

$$a\tau(n+1, m, k) = |\widehat{N-2}, \Psi(n+1, m, k; N-2)|. \tag{8.71}$$

Starting with (8.69b, 8.69c) we can derive formulas for the m, k shifts, similar to (8.71).

We also need formulas for double shifts. Starting from (8.70) and using the same method for the m-shift we get first

$$\tau(n+1, m+1, k) = |\widehat{N-3}, \Psi(n, m+1, k; N-2), \Psi(n+1, m+1, k; N-1)|.$$

Next using (8.69a, 8.69b) we can also derive

$$(a-b)\Psi(n+1, m+1, k; s+1) = \Psi(n+1, m, k; s) - \Psi(n, m+1, k; s), \tag{8.72}$$

and thus we get finally

$$\begin{aligned}(a-b)&\tau(n+1, m+1, k)\\ &= |\widehat{N-3}, \Psi(n, m+1, k; N-2), \Psi(n+1, m, k; N-2)|.\end{aligned} \tag{8.73}$$

With the above shift formulas we can use the Laplace expansion (D.6) with $\mathbf{M} = |\widehat{N-3}|$, $\mathbf{a} = \Psi(n, m, k; N-2)$, $\mathbf{b} = \Psi(n+1, m, k; N-2)$, $\mathbf{c} = \Psi(n, m+1, k; N-2)$, $\mathbf{d} = \Psi(n, m, k+1; N-2)$ and it yields (8.29) with $f = \tau$.

8.6.4 Casoratian NSS for the discrete KdV equation

In Section 8.4.1 we showed that the bilinear discrete KdV equation (8.40a) can be obtained from the Hirota–Miwa equation (8.29) by the reduction (8.35). We will now show that this also works for the Casoratian form of the τ-function (Ohta et al., 1993).

When we apply the parameter condition (8.37) to the matrix entry (8.68) we find that the result, which we call ψ', satisfies the periodicity property

$$\psi_i'(n+1, m, k+1; s) = \frac{1}{(1-ap_i)(1-cp_i)}\,\psi_i'(n, m, k; s),$$

and applying this to the Casoratian constructed the usual way we get

$$\tau'(n+1, m, k+1) = \frac{1}{\prod_{i=1}^{N}(1-ap_i)(1-cp_i)}\,\tau'(n, m, k).$$

Since the equation is in Hirota bilinear form the overall factor cancels out and we find that τ' solves the bilinear discrete KdV equation (8.40a).

8.7 Notes

In hindsight we can find instances of Hirota's bilinear forms throughout the mathematical literature going back to the nineteenth century, for example by Frobenius (relations between Hankel determinants), Plücker (Plücker relations for the coordinates of Grassmannians), Baker (in the context of Abelian functions), Painlevé (bilinear forms of Painlevé equations), and in general in the theory of determinant, Pfaffians etc. (see also Appendix D). Hirota brought this method to the forefront and started using it as a solution method for integrable equations. A testimony of the success of this method is that it allowed Hirota to be a pioneer in the world of DIS through a sequence of papers (Hirota, 1977a,b,c).

Although Hirota initially developed and used the bilinear method to construct and prove multi-soliton solutions, other important properties of integrable systems were soon formulated in that language. For example, Bäcklund transformations were given in bilinear form in Hirota (1974).

Underlying Hirota's bilinear approach is the elegant theory of Kac-Moody Lie algebras. For representation of these one needs vertex operators, which are in fact infinitesimal BT (Date et al., 1981). Furthermore it was found by M. Sato and Y. Sato that the space of τ-functions (as Hirota's function was called) is the infinite-dimensional Grassmannian. For an introduction to Sato theory see e.g. Ohta et al. (1988); Willox and Satsuma (2004); Miwa et al. (2000). From the point of view of representation theory and vertex operators the subject has been discussed in Dickey (2003).

Most of these early developments were for PDEs (see e.g. Jimbo and Miwa (1983) and Miwa et al. (2000)) but a method of discretization at the vertex operator level was then provided by Miwa's transformation (see equation (2.3) in Miwa (1982)). Many discrete equations were then provided by Date, Jimbo and Miwa in a series of papers (Date et al., 1982a,b, 1983a,b,c). This led to growing interest in the bilinear approach to discrete equations and, for example, the Hirota–Miwa equation, i.e. discrete KP equation, has thereafter been studied from many different points of view; see e.g. Ohta et al. (1993); Zabrodin (1997a); Willox et al. (1997); Adler and van Moerbeke (1999); Schief (2003); Doliwa (2010).

In Atkinson et al. (2007); Nijhoff et al. (2009); Hietarinta and Zhang (2009) several equations in the ABS list have been bilinearized and their solutions constructed. Non-autonomous bilinear equations have been discussed in Kajiwara and Ohta (2008); Shi et al. (2014). The bilinear method has also been generalized to the non-abelian direction and the non-abelian (i.e. matrix) form of the Hirota–Miwa equation was presented in Nimmo (2006). The solutions to this equation are given in terms of quasideterminants; see e.g. Gilson et al. (2007). A classification of discrete Hirota-type equations in three-dimension is given in Ferapontov et al. (2014).

Exercises

8.1 Show that

$$\partial_x^4 \log(f) = [f^2 (D_x^4 f \cdot f) - 3 (D_x^2 f \cdot f)^2]/(2f^4).$$

8.2 Verify by explicit calculation that (8.14) solves (8.11) under the assumption of the dispersion relation (8.13) and phase factor (8.15).

8.3 Simplify the phase factor (8.15) in the KdV case, where the function P is given by $P(X, T) = XT - X^4$. (Hint: use also the dispersion relation.)

8.4 The bilinear forms for the continuous sine-Gordon and mKdV equations are of the type

$$\begin{cases} A(D_{\vec{x}})\, g \cdot f = 0, \\ B(D_{\vec{x}})(f \cdot f - g \cdot g) = 0. \end{cases} \tag{8.74}$$

where for sine-Gordon A and B are even functions of Hirota derivatives and for mKdV A is odd and B is even (Hietarinta, 1987b,c).

Show that these equations have, generically, 1SS of the type

$$f = 1,\ g = e^{\eta_i}, \tag{8.75}$$

and 2SS

$$f = 1 + K_{12} e^{\eta_1+\eta_2},\ g = e^{\eta_1} + e^{\eta_2}, \tag{8.76a}$$

where the parameters satisfy the dispersion relations $A(\vec{p}_i) = 0$, and phase factor is given by

$$K_{ij} = \frac{B(\vec{p}_i - \vec{p}_j)}{B(\vec{p}_i + \vec{p}_j)}. \tag{8.76b}$$

8.5 The bilinear form of the continuous sine-Gordon equation is

$$\begin{cases} (D_x D_t - 1)\, g \cdot f = 0, \\ D_x D_t\, (f \cdot f - g \cdot g) = 0. \end{cases} \tag{8.77}$$

Hirota (1977c) proposed to discretize the continuous bilinear form (8.77) by

$$\begin{cases} \sinh(\delta D_x) \sinh(\delta D_t)\, g \cdot f - \delta^2 \cosh(\delta D_x) \cosh(\delta D_t)\, g \cdot f = 0, \\ \sinh(\delta D_x) \sinh(\delta D_t)\, (f \cdot f - g \cdot g) = 0. \end{cases} \tag{8.78a}$$

(a) Using the conversion to discrete variables by $g(x, t) = g(x_0{+}n\delta, t_0{+}m\delta) = g_{n,m}$ show that (8.78a) can also be written as

$$\begin{cases} (\delta^2 - 1)(g_{n+1,m+1} f_{n-1,m-1} + g_{n-1,m-1} f_{n+1,m+1}) \\ \qquad + (\delta^2 + 1)(g_{n+1,m-1} f_{n-1,m+1} + g_{n-1,m+1} f_{n+1,m-1}) = 0, \\ f_{n+1,m+1} f_{n-1,m-1} - f_{n-1,m+1} f_{n+1,m-1} \\ \qquad - g_{n+1,m+1} g_{n-1,m-1} + g_{n-1,m+1} g_{n+1,m-1} = 0. \end{cases} \tag{8.78b}$$

(b) Up-shift both equations in n and m, and use $f_{n+2,m} = \widetilde{f}$, $f_{n,m+2} = \widehat{f}$ etc. to write (8.78b) as a coupled system on the quadrilateral:

$$\begin{cases} (\delta^2 - 1)(\widehat{\widetilde{g}} f + g \widehat{\widetilde{f}}) + (\delta^2 + 1)(\widetilde{g} \widehat{f} + \widehat{g} \widetilde{f}) = 0, \\ \widehat{\widetilde{f}} f - \widetilde{f} \widehat{f} - \widehat{\widetilde{g}} g + \widetilde{g} \widehat{g} = 0. \end{cases} \tag{8.78c}$$

(c) Use the transformation

$$f = e^{\rho/4}\cos(\phi/4), \qquad g = e^{\rho/4}\sin(\phi/4) \tag{8.79}$$

on (8.78c) and, by eliminating ρ, derive

$$\sin\{\tfrac{1}{4}[\widehat{\widetilde{\phi}} + \phi - \widetilde{\phi} - \widehat{\phi}]\} = \delta^2 \sin\{\tfrac{1}{4}[\widehat{\widetilde{\phi}} + \phi + \widetilde{\phi} + \widehat{\phi}]. \tag{8.80}$$

Compare this to (2.93). What is the transformation that connects these two equations?

8.6 Verify the derivation of (8.24), i.e.

(1) Show that substituting (8.20) into (8.19) yields an equation that is a combination of two copies of (8.21a) (for any σ), shifted by $n \to n \pm \frac{1}{4}$.

(2) Show that when (8.22a) is written out it gives (8.22b).

(3) Show that substituting (8.23) into (8.24) yields an equation that is a combination of two copies of (8.22b), shifted by $(n, m) \to (n \pm \frac{1}{4}, m \to \mp\frac{1}{4})$.

(4) Do the transformation that takes (8.24) into the form (3.87).

8.7 Consider a Hirota's DAGTE equation (8.29) of the following form:

$$a_{ij} f_k f_{ij} + a_{jk} f_i f_{jk} + a_{ki} f_j f_{ik} = 0 \tag{8.81}$$

where f denotes the τ-function and the suffices denote shifts in corresponding directions in a multidimensional lattice, i.e. $f_i = T_i f$, etc. Here a_{ij} are constant coefficients *antisymmetric* in the indices: $a_{ij} = -a_{ji}$.

(a) Derive from (8.81) the four-dimensional equation

$$a_{jk}a_{jl}a_{kl} f_i f_{jkl} - a_{ik}a_{il}a_{kl} f_j f_{ikl} + a_{ij}a_{il}a_{jl} f_k f_{ijl} - a_{ij}a_{ik}a_{jk} f_l f_{ijk} = 0, \tag{8.82}$$

by performing the following sequence of intermediate steps. First, by shifting Equation (8.81) in the fourth direction indicated by suffix l to obtain

$$a_{ij} f_{kl} f_{ijl} + a_{jk} f_{il} f_{jkl} + a_{ki} f_{jl} f_{ikl} = 0, \tag{8.83}$$

and by similarly considering Equation (8.81) in the ijl directions shifted in the direction k, namely

$$a_{ij} f_{kl} f_{ijk} + a_{jl} f_{ik} f_{jkl} + a_{li} f_{jk} f_{ilk} = 0, \tag{8.84}$$

derive (8.82) by subtracting $a_{lj}a_{ki}^{-1} f_k \times$(8.83) and $a_{kj}a_{li}^{-1} f_l \times$(8.84) and using (8.81) in the ikl and jkl directions in order to eliminate f_{il} and f_{jl}.

(b) Imposing now the condition that solution f is a constant in the single fixed direction indicated by l_0, i.e., $f_{l_0} = T_{l_0} f = f$ and using that direction for l in (8.82), show that it reduces to the following equation

$$b_{jk} f_i f_{jk} - b_{ik} f_j f_{ik} + b_{ij} f_k f_{ij} + c_{ijk} f\, f_{ijk} = 0, \tag{8.85}$$

with $b_{ij} = a_{ij}a_{il_0}a_{jl_0}$, $c_{ijk} = a_{ij}a_{jk}a_{ki}$. Note that this equation resembles Miwa's equation (8.31) but has different parameters.

(c) By imposing the fixed point condition $T_{l_0}f = f$ directly on (8.81) in directions with labels i, j and replacing k by l_0, deduce the following two-dimensional PΔE

$$a_{ij}ff_{ij} = (a_{il_0} - a_{jl_0})f_i f_j.$$

Show that this equation can be linearized and in three dimensions leads to the covariant solution

$$f \equiv f(n_i, n_j, n_k) = \tau(n_i, n_j, n_k) \times \left(\frac{a_{il_0} - a_{jl_0}}{a_{ij}}\right)^{n_i n_j} \left(\frac{a_{il_0} - a_{kl_0}}{a_{ik}}\right)^{n_i n_k} \left(\frac{a_{jl_0} - a_{kl_0}}{a_{jk}}\right)^{n_j n_k}, \tag{8.86}$$

of (8.85), if $\tau(n_i, n_j, n_k) = \tau_0$ (a constant).

(d) Now consider (8.86) as a *nonlinear* gauge transformation (cf. Section 8.1.3) from f to τ and apply it to (8.85). Show that one then obtains Miwa's equation (8.31) if one chooses

$$a_{il_0} = b + c, \quad a_{jl_0} = a + c, \quad a_{kl_0} = a + b.$$

9

Multi-soliton solutions and the Cauchy matrix scheme

There are deep mathematical structures underlying integrable systems. In this chapter we explore one of these, namely the Cauchy matrix structure. By a Cauchy matrix we mean a matrix $\boldsymbol{A}$ with entries

$$A_{i,j} = \frac{1}{k_i - l_j}, \tag{9.1}$$

where the indices i, j run over some index sets and where k_i, l_j are chosen so that the matrix is non-singular. In the case of square matrices, the determinant $|\boldsymbol{A}|$ has the factorizing property

$$\det(\boldsymbol{A}) = \frac{\prod_{i<j}(k_i - k_j)(l_j - l_i)}{\prod_{i,j}(k_i - l_j)} \tag{9.2}$$

which is essential for the structure we will develop in this chapter.

The Cauchy matrix structure can be used to not only derive soliton solutions but also new equations and their properties. Following this idea we will develop a recursive structure which will provide us with an infinite class of solutions of the lattice equations, and also various interrelations between different but similar difference equations. This approach unifies the solutions of different equations and puts them in coherent framework. In the structure that emerges, there are several functions that obey different equations, and amongst these are the τ-functions mentioned in the previous chapter.

In this chapter we will consider two-dimensional as well as higher-dimensional equations. The recursive structure transcends the strict setting of soliton solutions and can be used to construct other types of solutions that extend the soliton initial value problem.

9.1 Cauchy matrix structure for KdV-type equations

In this section we will construct the framework built on symmetric Cauchy matrices following Nijhoff et al. (2009). As we shall see this will lead to solutions of KdV-type lattice equations.

9.1.1 Basic ingredients

Before considering the general case we consider symmetric $N \times N$ Cauchy matrices $\mathcal{A}$ with $l_j = -k_j$ in (9.1), i.e. with entries $\mathcal{A}_{i,j} = 1/(k_i + k_j)$, where k_i are arbitrary except that $k_i + k_j \neq 0, \forall i, j = 1, \dots, N$. Using this we construct the main object of the present theory, namely the matrix $\boldsymbol{M}$, which we also call a Cauchy matrix, defined by

$$\boldsymbol{M} = \boldsymbol{R}\mathcal{A}\boldsymbol{C},$$

where $\boldsymbol{R} = \operatorname{diag}(\rho(k_1), \dots, \rho(k_N))$ and $\boldsymbol{C} = \operatorname{diag}(c_1, \dots, c_N)$ are diagonal matrices. Therefore the entries of $\boldsymbol{M}$ are given by

$$M_{i,j} = \frac{\rho(k_i)c_j}{k_i + k_j}. \tag{9.3}$$

Here c_i are constants but $\rho(k_i)$ will be functions that give the dynamics. For most of this chapter we use the notation:

$$\rho(x) \equiv \rho_{n,m}(p,q;x) := \left(\frac{p+x}{p-x}\right)^n \left(\frac{q+x}{q-x}\right)^m \rho^0(x), \tag{9.4}$$

where n, m are the lattice variables, x is a generic variable and ρ^0 is a constant in n, m but may depend on other parameters. This definition of ρ dependence implies the following one step evolutions in n, m

$$\widetilde{\rho}(k_i) = \frac{p+k_i}{p-k_i}\,\rho(k_i), \quad \widehat{\rho}(k_i) = \frac{q+k_i}{q-k_i}\,\rho(k_i), \tag{9.5}$$

which will be used often in the following.

The $\rho_{n,m}(p,q;k_i)$ are the discrete plane-wave factors, which we already met in Chapters 5 and 8 where we analyzed equations in the KdV class. So it is natural that the equations we will derive here turn out to be connected to the KdV hierarchy.

For convenience, let

$$\boldsymbol{K} := \begin{pmatrix} k_1 & & & \\ & k_2 & & \\ & & \ddots & \\ & & & k_N \end{pmatrix}, \quad \boldsymbol{r} := \begin{pmatrix} \rho(k_1) \\ \rho(k_2) \\ \vdots \\ \rho(k_N) \end{pmatrix}, \quad \boldsymbol{c} := \begin{pmatrix} c_1 \\ c_2 \\ \vdots \\ c_N \end{pmatrix}. \tag{9.6}$$

From the definition (9.3) we immediately have the relation:

$$\boldsymbol{MK} + \boldsymbol{KM} = \boldsymbol{r}\boldsymbol{c}^T, \tag{9.7}$$

where $\boldsymbol{c}^T$ stands for the transpose of $\boldsymbol{c}$. This equation is known as the Sylvester equation. Note that the *dyadic* on the right-hand side is a matrix of rank 1.

9.1.2 Linear dynamics

We are now in a position to study the dynamics of the matrix $\boldsymbol{M}$, i.e. the behavior of $\boldsymbol{M}$ as a function of the lattice variables n and m. Note that all n, m dependence comes from $\rho(k_i)$ which appears in $\boldsymbol{M}$ and in $\boldsymbol{r}$. As a consequence of (9.5) we have

$$
\begin{aligned}
\widetilde{M}_{i,j} &= \frac{\widetilde{\rho}(k_i)c_j}{k_i+k_j} = \frac{p+k_i}{p-k_i}\,\frac{\rho(k_i)c_j}{k_i+k_j} \\
&= \left(\frac{1}{p-k_i}+\frac{1}{k_i+k_j}\right)\frac{p+k_i}{p+k_j}\,\rho(k_i)c_j \\
&= \frac{1}{p+k_j}\widetilde{\rho}(k_i)c_j + \frac{p+k_i}{p+k_j}M_{i,j} \\
\Rightarrow \qquad \widetilde{M}_{i,j}(p+k_j) &= \widetilde{\rho}(k_i)c_j + (p+k_i)M_{i,j},
\end{aligned}
$$

which leads to the matrix relation:

$$
\widetilde{\boldsymbol{M}}\,(p\boldsymbol{I}+\boldsymbol{K}) - (p\boldsymbol{I}+\boldsymbol{K})\,\boldsymbol{M} = \widetilde{\boldsymbol{r}}\,\boldsymbol{c}^T, \tag{9.8a}
$$

where $\boldsymbol{I}$ is the $N \times N$ unit matrix. (Subsequently we will adopt the shorthand notation where the unit matrix symbol will be omitted when it multiplies a scalar.)

Conversely, we can do the calculation the other way around, namely by writing:

$$
\begin{aligned}
M_{i,j} &= \frac{p-k_i}{p+k_i}\,\frac{\widetilde{\rho}(k_i)c_j}{k_i+k_j} \\
&= \left(-\frac{1}{p+k_i}+\frac{1}{k_i+k_j}\right)\frac{p-k_i}{p-k_j}\,\widetilde{\rho}(k_i)c_j \\
&= -\frac{1}{p-k_j}\rho(k_i)c_j + \frac{p-k_i}{p-k_j}\widetilde{M}_{i,j} \\
\Rightarrow \qquad M_{i,j}(p-k_j) &= -\rho(k_i)c_j + (p-k_i)\widetilde{M}_{i,j},
\end{aligned}
$$

leading to the matrix equation:

$$
(p-\boldsymbol{K})\,\widetilde{\boldsymbol{M}} - \boldsymbol{M}\,(p-\boldsymbol{K}) = \boldsymbol{r}\,\boldsymbol{c}^T. \tag{9.8b}
$$

Similar relations can be derived for the dynamics in the other variable m, by interchanging the roles of the parameters p and q and of the variables n and m. This yields

$$
\widehat{\boldsymbol{M}}\,(q+\boldsymbol{K}) - (q+\boldsymbol{K})\,\boldsymbol{M} = \widehat{\boldsymbol{r}}\,\boldsymbol{c}^T, \tag{9.8c}
$$

$$
(q-\boldsymbol{K})\,\widehat{\boldsymbol{M}} - \boldsymbol{M}\,(q-\boldsymbol{K}) = \boldsymbol{r}\,\boldsymbol{c}^T. \tag{9.8d}
$$

Equations (9.8a)–(9.8d) encode all the information on the dynamics of $\boldsymbol{M}$.

9.1.3 Nonlinear dynamics

Let us now introduce the following quantities:

$$\boldsymbol{u}_\alpha = (\boldsymbol{I}+\boldsymbol{M})^{-1}\,\boldsymbol{K}^\alpha\,\boldsymbol{r} \tag{9.9a}$$

$$^t\boldsymbol{u}_\beta = \boldsymbol{c}^T\,\boldsymbol{K}^\beta\,(\boldsymbol{I}+\boldsymbol{M})^{-1} \tag{9.9b}$$

$$U_{\alpha,\beta} = \boldsymbol{c}^T\,\boldsymbol{K}^\beta\,(\boldsymbol{I}+\boldsymbol{M})^{-1}\,\boldsymbol{K}^\alpha\,\boldsymbol{r} \tag{9.9c}$$

for integers α, β. (Note that here α in the subscript is an index while as a superscript it denotes a power, which, if none of the parameters k_i is zero, can be negative.) Thus, we obtain an infinite sequence of column vectors $\boldsymbol{u}_\alpha$, row vectors $^t\boldsymbol{u}_\beta$, and scalar quantities $U_{\alpha,\beta}$. Note that we can rewrite definition (9.9c) using (9.9a) and (9.9b) in two alternative ways

$$U_{\alpha,\beta} = {}^t\boldsymbol{u}_\beta\,\boldsymbol{K}^\alpha\,\boldsymbol{r} = \boldsymbol{c}^T\,\boldsymbol{K}^\beta\,\boldsymbol{u}_\alpha. \tag{9.10}$$

We note that U as defined above is symmetric

$$U_{\alpha,\beta} = U_{\beta,\alpha}. \tag{9.11}$$

This symmetry property is shown in Exercise 9.1. It is characteristic of the KdV class of lattice equations and will be used in the following.

Furthermore, it is easy to show that generically the matrix $\boldsymbol{I}+\boldsymbol{M}$ is invertible, and hence all quantities (9.9) are well defined. These are the quantities that play the main role in the construction that follows.

We shall now derive, starting from the relations (9.8), a system of recurrence relations which describe the dynamics for the quantities (9.9). Once this recursive structure is in place, we will then single out specific choices among the latter for which we can derive closed-form discrete equations.

First, let us do some simple steps. From the definition (9.9a) we have

$$\boldsymbol{K}^\alpha\,\boldsymbol{r} = (\boldsymbol{I}+\boldsymbol{M})\,\boldsymbol{u}_\alpha\,,$$

and then using the fact that $\boldsymbol{K}$ is a constant matrix, we obtain

$$\boldsymbol{K}^\alpha\,\widetilde{\boldsymbol{r}} = (\boldsymbol{I}+\widetilde{\boldsymbol{M}})\,\widetilde{\boldsymbol{u}}_\alpha\,.$$

Next from (9.5) and (9.6) we get

$$(p-\boldsymbol{K})\widetilde{\boldsymbol{r}} = (p+\boldsymbol{K})\boldsymbol{r}, \tag{9.12}$$

therefore

$$\begin{aligned}
\boldsymbol{K}^\alpha\,(p+\boldsymbol{K})\,\boldsymbol{r} &= (p-\boldsymbol{K})\,(\boldsymbol{I}+\widetilde{\boldsymbol{M}})\,\widetilde{\boldsymbol{u}}_\alpha \\
&= (p-\boldsymbol{K})\,\widetilde{\boldsymbol{u}}_\alpha + (p-\boldsymbol{K})\,\widetilde{\boldsymbol{M}}\,\widetilde{\boldsymbol{u}}_\alpha \\
&= (p-\boldsymbol{K})\,\widetilde{\boldsymbol{u}}_\alpha + \boldsymbol{M}\,(p-\boldsymbol{K})\,\widetilde{\boldsymbol{u}}_\alpha + \boldsymbol{r}\,\boldsymbol{c}^T\widetilde{\boldsymbol{u}}_\alpha,
\end{aligned}$$

where we have used (9.8b) in the last step.

Recalling (9.10) we conclude that

$$p\boldsymbol{K}^{\alpha}\,\boldsymbol{r}+\boldsymbol{K}^{\alpha+1}\,\boldsymbol{r}=(\boldsymbol{I}+\boldsymbol{M})\,(p-\boldsymbol{K})\,\widetilde{\boldsymbol{u}}_{\alpha}+\widetilde{U}_{\alpha,0}\boldsymbol{r}\,.$$

Multiplying both sides by the inverse matrix $(\boldsymbol{I}+\boldsymbol{M})^{-1}$ we get

$$\begin{aligned}(p-\boldsymbol{K})\,\widetilde{\boldsymbol{u}}_{\alpha}&=(\boldsymbol{I}+\boldsymbol{M})^{-1}\left[p\boldsymbol{K}^{\alpha}\,\boldsymbol{r}+\boldsymbol{K}^{\alpha+1}\,\boldsymbol{r}-\widetilde{U}_{\alpha,0}\boldsymbol{r}\right]\\&=p\boldsymbol{u}_{\alpha}+\boldsymbol{u}_{\alpha+1}-\widetilde{U}_{\alpha,0}\boldsymbol{u}_{0},\end{aligned}\tag{9.13}$$

where in the last step we have used (9.9a).

Thus, we have obtained a linear recursion relation between the objects $\boldsymbol{u}_{\alpha}$ with the objects $U_{\alpha,\beta}$ acting as coefficients. In similar fashion we can derive, starting with (9.8a), the relation

$$(p+\boldsymbol{K})\,\boldsymbol{u}_{\alpha}=p\widetilde{\boldsymbol{u}}_{\alpha}-\widetilde{\boldsymbol{u}}_{\alpha+1}+U_{\alpha,0}\widetilde{\boldsymbol{u}}_{0}.\tag{9.14}$$

Equation (9.14) can be thought of as an inverse relation to (9.13), noting that the tilde-shifted objects are now at the right-hand side of the equation. Multiplying both sides of (9.13) from the left by the row vector $\boldsymbol{c}^{T}\,\boldsymbol{K}^{\beta}$ we get

$$\boldsymbol{c}^{T}\,\boldsymbol{K}^{\beta}\,(p-\boldsymbol{K})\,\widetilde{\boldsymbol{u}}_{\alpha}=\boldsymbol{c}^{T}\,\boldsymbol{K}^{\beta}\,\left[p\boldsymbol{u}_{\alpha}+\boldsymbol{u}_{\alpha+1}-\widetilde{U}_{\alpha,0}\boldsymbol{u}_{0}\right].$$

Then using (9.10) we obtain a relation purely in terms of the objects $U_{\alpha,\beta}$, namely:

$$p\widetilde{U}_{\alpha,\beta}-\widetilde{U}_{\alpha,\beta+1}=pU_{\alpha,\beta}+U_{\alpha+1,\beta}-\widetilde{U}_{\alpha,0}U_{0,\beta}.\tag{9.15}$$

Thus, we have obtained a *nonlinear* recursion relation between the $U_{\alpha,\beta}$ and its tilde-shifted counterparts.

In a similar way, multiplying equation (9.14) by the row vector $\boldsymbol{c}^{T}\,\boldsymbol{K}^{\beta}$ we obtain the complementary relation:

$$pU_{\alpha,\beta}+U_{\alpha,\beta+1}=p\widetilde{U}_{\alpha,\beta}-\widetilde{U}_{\alpha+1,\beta}+U_{\alpha,0}\widetilde{U}_{0,\beta}.\tag{9.16}$$

This relation can also be obtained from (9.15) by interchanging α, β and using symmetry (9.11).

We can also derive the following algebraic relations

$$\boldsymbol{K}^{2}\boldsymbol{u}_{\alpha}=\boldsymbol{u}_{\alpha+2}+U_{\alpha,1}\,\boldsymbol{u}_{0}-U_{\alpha,0}\,\boldsymbol{u}_{1},\tag{9.17}$$

which we leave as an exercise. By multiplying (9.17) by $\boldsymbol{c}^{T}\,\boldsymbol{K}^{\beta}$ from the left we also obtain

$$U_{\alpha,\beta+2}=U_{\alpha+2,\beta}+U_{\alpha,1}\,U_{0,\beta}-U_{\alpha,0}\,U_{1,\beta}.\tag{9.18}$$

Obviously, all relations that we have derived for the tilde-shifts (involving lattice parameter p and lattice variable n) hold also for the hat-shifts, simply by replacing p by q and interchanging the roles of n and m.

To summarize, starting from the matrix $\boldsymbol{M}$ given in (9.3) depending dynamically on the lattice variables through the plane-wave factors $\rho(k_i)$ given in (9.4), we have defined a collection of objects, namely column and row vectors $\boldsymbol{u}_\alpha$ and ${}^t\boldsymbol{u}_\alpha$, and scalar functions $U_{\alpha,\beta}$, all related through a system of recurrence relations involving dynamical shifts in the lattice variables. In particular, for the scalar objects $U_{\alpha,\beta}$ we have the following set of coupled recurrence relations:

$$p\widetilde{U}_{\alpha,\beta} - \widetilde{U}_{\alpha,\beta+1} = pU_{\alpha,\beta} + U_{\alpha+1,\beta} - \widetilde{U}_{\alpha,0}U_{0,\beta}, \tag{9.19a}$$
$$pU_{\alpha,\beta} + U_{\alpha,\beta+1} = p\widetilde{U}_{\alpha,\beta} - \widetilde{U}_{\alpha+1,\beta} + U_{\alpha,0}\widetilde{U}_{0,\beta}, \tag{9.19b}$$
$$q\widehat{U}_{\alpha,\beta} - \widehat{U}_{\alpha,\beta+1} = qU_{\alpha,\beta} + U_{\alpha+1,\beta} - \widehat{U}_{\alpha,0}U_{0,\beta}, \tag{9.19c}$$
$$qU_{\alpha,\beta} + U_{\alpha,\beta+1} = q\widehat{U}_{\alpha,\beta} - \widehat{U}_{\alpha+1,\beta} + U_{\alpha,0}\widehat{U}_{0,\beta}. \tag{9.19d}$$

It is not yet apparent that this system yields closed-form lattice equations, because each iteration in α, β links to a new object. We show in the next section how to achieve this.

9.2 Closed-form lattice equations

Now that we have the full set of recurrence relations (9.19), the next step is to derive closed-form equations from this system for individual elements. Let us first, as a warming-up exercise, derive an equation for the variable $U_{0,0}$. In fact, subtracting (9.19c) from (9.19a) we obtain

$$p\widetilde{U}_{\alpha,\beta} - q\widehat{U}_{\alpha,\beta} - \widetilde{U}_{\alpha,\beta+1} + \widehat{U}_{\alpha,\beta+1} = (p-q)U_{\alpha,\beta} - (\widetilde{U}_{\alpha,0} - \widehat{U}_{\alpha,0})U_{0,\beta}. \tag{9.20}$$

On the other hand, taking the hat-shift of (9.19b), and subtracting from it the tilde-shift of (9.19d), we obtain

$$p\widehat{U}_{\alpha,\beta} - q\widetilde{U}_{\alpha,\beta} + \widehat{U}_{\alpha,\beta+1} - \widetilde{U}_{\alpha,\beta+1} = (p-q)\widehat{\widetilde{U}}_{\alpha,\beta} + (\widehat{U}_{\alpha,0} - \widetilde{U}_{\alpha,0})\widehat{\widetilde{U}}_{0,\beta}. \tag{9.21}$$

Subtracting (9.21) from (9.20), the terms which have a shift in their second index drop out and we obtain the equation:

$$(p+q)(\widetilde{U}_{\alpha,\beta} - \widehat{U}_{\alpha,\beta}) = (p-q)(U_{\alpha,\beta} - \widehat{\widetilde{U}}_{\alpha,\beta}) + (\widehat{U}_{\alpha,0} - \widetilde{U}_{\alpha,0})(U_{0,\beta} - \widehat{\widetilde{U}}_{0,\beta}). \tag{9.22}$$

More generally, we can choose certain elements among the $U_{\alpha,\beta}$, or (linear) combinations of them, and then systematically investigate what equations these choices satisfy by exploring the system of recurrence relations (9.19).

9.2.1 Lattice potential KdV equation

Consider the special case $\alpha = \beta = 0$ in (9.22). Then we get a closed-form equation in terms of u defined by

$$u := U_{0,0}, \tag{9.23}$$

that is

$$(p + q + u - \widehat{\widetilde{u}})(p - q + \widehat{u} - \widetilde{u}) = p^2 - q^2, \tag{9.24}$$

which is the lattice potential KdV equation, introduced earlier (1.46).

By construction from (9.9c) we find an infinite family of *exact* solutions of the nonlinear PΔE (9.24), of the form $u = \boldsymbol{c}^T\ (\boldsymbol{I} + \boldsymbol{M})^{-1}\boldsymbol{r}$. Note that for each N, this provides us with a solution where the parameters k_i, c_i, ρ_i^0, $i = 1, \ldots, N$ are all free. The corresponding solutions are N-soliton solutions of the lattice potential KdV equation.

9.2.2 Lattice potential mKdV equation

Now considering $\alpha = 0$, $\beta = -1$ in (9.20) and introducing the variable v by

$$v := 1 - U_{0,-1}, \tag{9.25}$$

we obtain the following relation (with $u = U_{0,0}$):

$$p - q + \widehat{u} - \widetilde{u} = \frac{p\widetilde{v} - q\widehat{v}}{v}. \tag{9.26}$$

In addition we need another equation to close the system. We obtain this by adding the hat-shift of (9.19a) to (9.19d)

$$p\widehat{\widetilde{U}}_{\alpha,\beta} + qU_{\alpha,\beta} - \widehat{\widetilde{U}}_{\alpha,\beta+1} + U_{\alpha,\beta+1} = (p+q)\widehat{U}_{\alpha,\beta} + (U_{\alpha,0} - \widehat{\widetilde{U}}_{\alpha,0})\widehat{U}_{0,\beta}, \tag{9.27}$$

and then taking $\alpha = 0$, $\beta = -1$, which yields:

$$p + q + u - \widehat{\widetilde{u}} = \frac{p\widehat{\widetilde{v}} + qv}{\widehat{v}}. \tag{9.28}$$

Clearly, in (9.28), interchanging p and q and the tilde-shift and the hat-shift should not make a difference, since the left-hand side is invariant under this change. Thus, the right-hand side must be invariant as well, leading to the relation:

$$p\left(v\widehat{v} - \widetilde{v}\widehat{\widetilde{v}}\right) = q\left(v\widetilde{v} - \widehat{v}\widehat{\widetilde{v}}\right). \tag{9.29}$$

Alternatively, the left-hand sides of (9.26) and (9.28) appear as factors in (9.24), so after substitution we get (9.29).

This PΔE for the variable v is the lattice potential mKdV equation (3.11). Once again, by construction this equation has an infinite family of solutions given by

$$v = 1 - U_{0,-1} = 1 - \boldsymbol{c}^T \boldsymbol{K}^{-1} (\boldsymbol{I} + \boldsymbol{M})^{-1} \boldsymbol{r} .$$

Note also that in this way the Cauchy matrix approach has led to the Miura transform (3.12).

9.2.3 Lattice SKdV equation

Another choice we can consider is the variable $U_{-1,-1}$; that is, we can consider (9.19a) for $\alpha = \beta = -1$, that is

$$p\left(\widetilde{U}_{-1,-1} - U_{-1,-1}\right) = 1 - \left(1 - \widetilde{U}_{-1,0}\right)\left(1 - U_{0,-1}\right),$$

and a similar relation for p replaced by q and the tilde-shift replaced by the hat-shift obtained from (9.19c). Using the fact that $U_{-1,0} = U_{0,-1} = 1 - v$ and introducing the new variable $z = U_{-1,-1} - \frac{n}{p} - \frac{m}{q}$, the above two relations reduce to:

$$p(z - \widetilde{z}) = \widetilde{v}v, \quad q(z - \widehat{z}) = \widehat{v}v. \tag{9.30}$$

This is a Bäcklund transformation, because on the one hand, these two equations lead back to the equation (9.29) by eliminating the variable z (considering in addition the tilde- and hat-shifts of the two relations), while on the other hand, by eliminating the variable v we obtain a PΔE for z which reads:

$$\frac{(z - \widetilde{z})(\widehat{z} - \widehat{\widetilde{z}})}{(z - \widehat{z})(\widetilde{z} - \widehat{\widetilde{z}})} = \frac{q^2}{p^2}, \tag{9.31}$$

which is the lattice SKdV equation (3.13). Thus, an infinite family of solutions of equation (9.31) is given by the formula:

$$z = z_0 - \frac{n}{p} - \frac{m}{q} + \boldsymbol{c}^T \boldsymbol{K}^{-1} (\boldsymbol{I} + \boldsymbol{M})^{-1} \boldsymbol{K}^{-1} \boldsymbol{r},$$

in which z_0 is an arbitrary constant.

In general, we have shown that we can obtain many closed sets of equations together with their N-soliton solutions from the recurrence relations (9.19), following from the Cauchy matrix (9.3).

9.3 Derivation of Lax pairs

We will now show how the infinite recurrence structure we developed in Section 9.2 can also be used to systematically derive Lax pairs for the lattice equations that emerge from the system. Our starting point is the pair of linear recurrence relations (9.13), (9.14) for the vectors $\boldsymbol{u}_i$, as well their counterparts involving the hat-shift (with lattice parameter q).

9.3.1 Lax pair for the lattice potential KdV equation

Recalling the set of linear equations (9.13, 9.14) for $\alpha = 0$, we immediately get from (9.13)

$$(p - \boldsymbol{K})\widetilde{\boldsymbol{u}}_0 = p\boldsymbol{u}_0 + \boldsymbol{u}_1 - \widetilde{U}_{0,0}\,\boldsymbol{u}_0 = (p - \widetilde{u})\boldsymbol{u}_0 + \boldsymbol{u}_1.$$

(Note that $U_{0,0} = u$.) On the other hand, from (9.14) we obtain:

$$\begin{aligned} &(p + \boldsymbol{K})\boldsymbol{u}_0 = p\widetilde{\boldsymbol{u}}_0 - \widetilde{\boldsymbol{u}}_1 + U_{0,0}\,\widetilde{\boldsymbol{u}}_0 = (p + u)\widetilde{\boldsymbol{u}}_0 - \widetilde{\boldsymbol{u}}_1 \\ \Rightarrow\quad &(p^2 - \boldsymbol{K}^2)\boldsymbol{u}_0 = (p + u)\left[(p - \widetilde{u})\boldsymbol{u}_0 + \boldsymbol{u}_1\right] - (p - \boldsymbol{K})\widetilde{\boldsymbol{u}}_1. \end{aligned}$$

Solving for $(p - \boldsymbol{K})\widetilde{\boldsymbol{u}}_1$ from the last relation, we obtain a system of two coupled linear equations giving the discrete evolution of the vectors $\boldsymbol{u}_0$ and $\boldsymbol{u}_1$. To express this succinctly, we introduce the vector

$$\boldsymbol{\phi} = \begin{pmatrix} \boldsymbol{u}_0 \\ \boldsymbol{u}_1 \end{pmatrix}$$

which is a $2N$-component vector, consisting of two N-component parts. In terms of $\boldsymbol{\phi}$, we can rewrite the system in the following 2×2 block matrix form:

$$(p - \boldsymbol{K})\widetilde{\boldsymbol{\phi}} = \begin{pmatrix} (p - \widetilde{u})\boldsymbol{I} & \boldsymbol{I} \\ \boldsymbol{K}^2 + [-p^2 + (p - \widetilde{u})(p + u)]\boldsymbol{I} & (p + u)\boldsymbol{I} \end{pmatrix} \boldsymbol{\phi}. \tag{9.32a}$$

By a similar derivation, we obtain the following matrix equation involving the hat-shift:

$$(q - \boldsymbol{K})\widehat{\boldsymbol{\phi}} = \begin{pmatrix} (q - \widehat{u})\boldsymbol{I} & \boldsymbol{I} \\ \boldsymbol{K}^2 + [-q^2 + (q - \widehat{u})(q + u)]\boldsymbol{I} & (q + u)\boldsymbol{I} \end{pmatrix} \boldsymbol{\phi}, \tag{9.32b}$$

which must hold simultaneously.

Notice that although (9.32) involve $2N \times 2N$ matrices, they are actually 2×2 block matrices where each block only involves a scalar quantity multiplying an $N \times N$ unit matrix or the diagonal matrix $\boldsymbol{K}$. Thus, this $2N \times 2N$ matrix system can be interpreted as a family of N decoupled 2×2 matrix systems, each acting on a two-component vector, which can be written as

$$\boldsymbol{\phi}_i = \begin{pmatrix} (\boldsymbol{u}_0)_i \\ (\boldsymbol{u}_1)_i \end{pmatrix}, \quad i = 1, \dots, N,$$

where $(\boldsymbol{u}_j)_i$ is the ith component of the N-component vector $\boldsymbol{u}_j$. In each of these equations, the diagonal matrix $\boldsymbol{K}$ is then simply replaced by its ith entry, k_i, and this parameter plays the role of the spectral parameter in the Lax pair. Thus, these N matrix systems take the form:

$$(p-k_i)\widetilde{\boldsymbol{\phi}}_i = L(k_i)\boldsymbol{\phi}_i, \quad (q-k_i)\widehat{\boldsymbol{\phi}}_i = M(k_i)\boldsymbol{\phi}_i, \quad i=1,\ldots,N \tag{9.33}$$

with the Lax matrices:

$$L(k) = \begin{pmatrix} p-\widetilde{u} & 1 \\ k^2-p^2+(p-\widetilde{u})(p+u) & p+u \end{pmatrix},$$

$$M(k) = \begin{pmatrix} q-\widehat{u} & 1 \\ k^2-q^2+(q-\widehat{u})(q+u) & q+u \end{pmatrix}.$$

These are related by a simple transformation to (3.17b).

9.3.2 Lax pair for the lattice potential mKdV equation

A second example is the derivation of the Lax pair for the lattice potential mKdV equation (9.29), which proceeds along similar lines, but with a slightly different choice of vectors from among the $\boldsymbol{u}_i$. In this case we select the vectors $\boldsymbol{u}_0$ (again) and $\boldsymbol{u}_{-1}$ instead of $\boldsymbol{u}_1$. Setting $\alpha=-1$ in (9.13) we now obtain

$$(p-\boldsymbol{K})\widetilde{\boldsymbol{u}}_{-1} = p\boldsymbol{u}_{-1} + \boldsymbol{u}_0 - \widetilde{U}_{-1,0}\,\boldsymbol{u}_0 = p\boldsymbol{u}_{-1} + \widetilde{v}\boldsymbol{u}_0,$$

recalling that $1-U_{-1,0} = 1-U_{0,-1} = v$. On the other hand, from (9.14) we get

$$\begin{aligned} &(p+\boldsymbol{K})\boldsymbol{u}_{-1} = p\widetilde{\boldsymbol{u}}_{-1} - \widetilde{\boldsymbol{u}}_0 + U_{-1,0}\widetilde{\boldsymbol{u}}_0 = p\widetilde{\boldsymbol{u}}_{-1} - v\widetilde{\boldsymbol{u}}_0 \\ \Rightarrow\quad &(p^2-\boldsymbol{K}^2)\boldsymbol{u}_{-1} = p\left[p\boldsymbol{u}_{-1} + \widetilde{v}\boldsymbol{u}_0\right] - (p-\boldsymbol{K})v\widetilde{\boldsymbol{u}}_0. \end{aligned}$$

Solving the last relation for $(p-\boldsymbol{K})\widetilde{\boldsymbol{u}}_0$ we get a closed system of linear equations describing the dynamics of $\boldsymbol{u}_0$ and $\boldsymbol{u}_{-1}$ in the tilde-direction.

Introducing now the $2N$-component vector

$$\boldsymbol{\psi} = \begin{pmatrix} \boldsymbol{u}_{-1} \\ \boldsymbol{u}_0 \end{pmatrix},$$

we can write this system in a $2N \times 2N$ matrix form, namely as follows:

$$(p-\boldsymbol{K})\widetilde{\boldsymbol{\psi}} = \begin{pmatrix} p\boldsymbol{I} & \widetilde{v}\boldsymbol{I} \\ \frac{\boldsymbol{K}^2}{v} & p\frac{\widetilde{v}}{v}\boldsymbol{I} \end{pmatrix}\boldsymbol{\psi}, \tag{9.34a}$$

accompanied by the analogous equation for the hat-shift:

$$(q-\boldsymbol{K})\widehat{\boldsymbol{\psi}} = \begin{pmatrix} q\boldsymbol{I} & \widehat{v}\boldsymbol{I} \\ \frac{\boldsymbol{K}^2}{v} & q\frac{\widehat{v}}{v}\boldsymbol{I} \end{pmatrix}\boldsymbol{\psi}. \tag{9.34b}$$

As in the previous case, this $2N \times 2N$ matrix system decouples into N similarly looking 2×2 matrix systems for each of the components of the vectors $\boldsymbol{u}_0$ and $\boldsymbol{u}_{-1}$, replacing $\boldsymbol{K}$ by k_i and $\boldsymbol{\psi}$ by $\psi_i = ((\boldsymbol{u}_{-1})_i\,,\,(\boldsymbol{u}_0)_i)^T$, each of which will take the form:

$$(p-k_i)\widetilde{\boldsymbol{\psi}} = \mathcal{L}(k_i)\boldsymbol{\psi}\,, \quad (q-k_i)\widehat{\boldsymbol{\psi}} = \mathcal{M}(k_i)\boldsymbol{\psi}, \tag{9.35}$$

with Lax matrices:

$$\mathcal{L}(k) = \begin{pmatrix} p & \widetilde{v} \\ \frac{k^2}{v} & p\frac{\widetilde{v}}{v} \end{pmatrix}, \quad \mathcal{M}(k) = \begin{pmatrix} q & \widehat{v} \\ \frac{k^2}{v} & q\frac{\widehat{v}}{v} \end{pmatrix},$$

which is related by a simple transformation to (3.21).

As a byproduct, we also obtain the connection between the two Lax pairs (9.33) and (9.35), namely by exploiting the algebraic recurrence (9.17) (which does not involve the discrete dynamics). In fact, setting $\alpha = -1$ in the latter relation we obtain:

$$\boldsymbol{K}^2\boldsymbol{u}_{-1} = \boldsymbol{u}_1 + U_{-1,1}\,\boldsymbol{u}_0 - U_{-1,0}\,\boldsymbol{u}_1 = v\boldsymbol{u}_1 + U_{-1,1}\boldsymbol{u}_0\,,$$

where the new object $U_{-1,1}$ obeys dynamical relations that can be obtained by setting $\alpha = -1$, $\beta = 0$ directly in equations (9.19). Thus, we can relate the two Lax pair through the $2N \times 2N$ matrix relation

$$\boldsymbol{K}^2\boldsymbol{\psi} = \begin{pmatrix} U_{-1,1}\boldsymbol{I} & v\boldsymbol{I} \\ \boldsymbol{K}^2 & \boldsymbol{0} \end{pmatrix}\boldsymbol{\phi}. \tag{9.36}$$

Equation (9.36) decouples into a set of N 2×2 matrix relations which give a gauge transformation between $\boldsymbol{\phi}$ and $\boldsymbol{\psi}$

$$k^2\boldsymbol{\psi} = G\boldsymbol{\phi}, \quad G(k) \equiv \begin{pmatrix} U_{-1,1} & v \\ k^2 & 0 \end{pmatrix}, \tag{9.37}$$

implying that the Lax matrices for the two Lax pairs are related through the formulae:

$$\widetilde{G}(k)\,L(k) = \mathcal{L}(k)\,G(k), \quad \widehat{G}(k)\,M(k) = \mathcal{M}(k)\,G(k). \tag{9.38}$$

To verify these equations it is necessary to use all equations of (9.19) for $\alpha = -1$, $\beta = 0$.

9.3.3 Lax pair for the lattice SKdV equation

Finally, let us observe that from the Lax pair (9.34) one can easily obtain a Lax pair for the lattice SKdV equation, namely by exploiting the relations (9.30). In fact, multiplying the tilde equation of (9.35) from the left by the diagonal matrix $\mathrm{diag}(1, 1/\widetilde{v})$, we obtain

$$(p-k)\begin{pmatrix} 1 & 0 \\ 0 & 1/\widetilde{v} \end{pmatrix}\widetilde{\boldsymbol{\psi}} = \begin{pmatrix} p & \widetilde{v}v \\ \frac{k^2}{\widetilde{v}v} & p \end{pmatrix}\begin{pmatrix} 1 & 0 \\ 0 & 1/v \end{pmatrix}\boldsymbol{\psi}.$$

This suggests a gauge transformation to a new vector $\boldsymbol{\chi} := \mathrm{diag}(1, 1/v)\,\boldsymbol{\psi}$ and then using (9.30) we obtain an equation expressed entirely in terms of z and $\widetilde{z}$, namely

$$(1-k/p)\widetilde{\boldsymbol{\chi}} = \mathfrak{L}\boldsymbol{\chi}, \quad \mathfrak{L} := \begin{pmatrix} 1 & z-\widetilde{z} \\ \frac{k^2/p^2}{z-\widetilde{z}} & 1 \end{pmatrix}. \tag{9.39}$$

It is an easy exercise to show that the consistency relation between (9.39) and its counterpart for the hat-shift (making the usual replacements) gives rise the lattice equation (9.31) for the variable z. We also note that the previously obtained Lax matrix $\boldsymbol{L}$ of (3.23) is related to $\mathfrak{L}$ in (9.39) by the gauge transformation

$$\boldsymbol{L} = \mathfrak{G}^{-1}(\widetilde{z})\,\mathfrak{L}\,\mathfrak{G}(z), \quad \mathfrak{G}(z) = \begin{pmatrix} 1 & -z \\ 0 & 1 \end{pmatrix},$$

and correspondingly $\boldsymbol{M} = \mathfrak{G}^{-1}(\widehat{z})\,\mathfrak{M}\,\mathfrak{G}(z)$, where $\mathfrak{M}$ is the same as $\mathfrak{L}$ in (9.39) with p replaced by q and tilde-shifts by hat-shifts.

9.4 Bilinear form from soliton solutions

Our starting point in this chapter was the Cauchy matrix (9.3), but the same matrix also motivated the definition of the τ-function in Chapter 8. This connection is further elaborated in this section.

9.4.1 Dynamics of the τ-function

Motivated by (8.2) we define

$$\tau_{n,m} = \det(\boldsymbol{I}+\boldsymbol{M}), \tag{9.40}$$

where the matrix $\boldsymbol{M}$ is given by (9.3) and the n, m dependence enters through $\rho_{n,m}(k_i)$ as defined in (9.4). We will show that this τ-function satisfies the bilinear equations (8.40), by using a variation of the technique from Section 9.1.

From this definition, using the relation (9.8a), we can perform the following straightforward calculation:

$$\begin{aligned}
\widetilde{\tau} = \det(\boldsymbol{I}+\widetilde{\boldsymbol{M}}) &= \det\left\{\boldsymbol{I} + \left[(p+\boldsymbol{K})\boldsymbol{M} + \widetilde{\boldsymbol{r}}\boldsymbol{c}^T\right](p+\boldsymbol{K})^{-1}\right\} \\
&= \det\left\{(p+\boldsymbol{K})\left[\boldsymbol{I}+\boldsymbol{M}+(p+\boldsymbol{K})^{-1}\widetilde{\boldsymbol{r}}\boldsymbol{c}^T\right](p+\boldsymbol{K})^{-1}\right\} \\
&= \det\left\{(\boldsymbol{I}+\boldsymbol{M})\left[\boldsymbol{I}+(\boldsymbol{I}+\boldsymbol{M})^{-1}(p+\boldsymbol{K})^{-1}\widetilde{\boldsymbol{r}}\boldsymbol{c}^T\right]\right\} \\
&= \tau\,\det\left\{\boldsymbol{I}+(\boldsymbol{I}+\boldsymbol{M})^{-1}(p+\boldsymbol{K})^{-1}\widetilde{\boldsymbol{r}}\boldsymbol{c}^T\right\},
\end{aligned}$$

from which, using also (9.12) we have

$$\frac{\widetilde{\tau}}{\tau} = 1 + \boldsymbol{c}^T \, (\boldsymbol{I} + \boldsymbol{M})^{-1} \, (p - \boldsymbol{K})^{-1} \, \boldsymbol{r}, \tag{9.41}$$

where in the last step we have used the Weinstein–Aronszajn formula (D.3) with $\boldsymbol{b} = (\boldsymbol{I} + \boldsymbol{M})^{-1} \, (p - \boldsymbol{K})^{-1} \, \boldsymbol{r}$.

9.4.2 Derivation of the NQC equation

The object created on the right-hand side of equation (9.41) is new and we need to investigate what relations it satisfies.

Introduce the function

$$V(a) := 1 - \boldsymbol{c}^T \, (\boldsymbol{I} + \boldsymbol{M})^{-1} \, (a + \boldsymbol{K})^{-1} \, \boldsymbol{r} \tag{9.42a}$$
$$= 1 - {}^t\boldsymbol{u}_0 \, (a + \boldsymbol{K})^{-1} \, \boldsymbol{r} \tag{9.42b}$$
$$= 1 - \boldsymbol{c}^T \, (a + \boldsymbol{K})^{-1} \boldsymbol{u}_0. \tag{9.42c}$$

From the calculation above, we have immediately:

$$\frac{\widetilde{\tau}}{\tau} = V(-p). \tag{9.43}$$

In order to derive equations for this function $V(a)$ we need to introduce some further objects, namely:

$$\boldsymbol{u}(a) = (\boldsymbol{I} + \boldsymbol{M})^{-1} (a + \boldsymbol{K})^{-1} \boldsymbol{r}, \tag{9.44a}$$
$${}^t\boldsymbol{u}(b) = \boldsymbol{c}^T \, (b + \boldsymbol{K})^{-1} (\boldsymbol{I} + \boldsymbol{M})^{-1}, \tag{9.44b}$$
$$S(a, b) = \boldsymbol{c}^T \, (b + \boldsymbol{K})^{-1} (\boldsymbol{I} + \boldsymbol{M})^{-1} (a + \boldsymbol{K})^{-1} \boldsymbol{r}, \tag{9.44c}$$

in which a, b are arbitrary (real or complex valued) parameters. All of these new functions depend on n, m through ρ_{nm} appearing in $\boldsymbol{r}$ and $\boldsymbol{M}$. Comparing this with quantities defined in (9.9), notice that instead of powers of $\boldsymbol{K}$ we now have additive shifts in $\boldsymbol{K}$. Note that $S(a, b)$ can also be expressed in two further ways:

$$S(a, b) = \boldsymbol{c}^T \, (b + \boldsymbol{K})^{-1} \boldsymbol{u}(a) = {}^t\boldsymbol{u}(b) \, (a + \boldsymbol{K})^{-1} \boldsymbol{r}. \tag{9.45}$$

Following a similar derivation as the one leading to (9.13) and (9.14), one can derive the following relations:

$$(p - \boldsymbol{K}) \widetilde{\boldsymbol{u}}(a) = \widetilde{V}(a) \boldsymbol{u}_0 + (p - a) \boldsymbol{u}(a), \tag{9.46a}$$
$$(p + \boldsymbol{K}) \boldsymbol{u}(a) = -V(a) \widetilde{\boldsymbol{u}}_0 + (p + a) \widetilde{\boldsymbol{u}}(a). \tag{9.46b}$$

By multiplying (9.46) from the left by the row vector $\boldsymbol{c}^T\,(b+\boldsymbol{K})^{-1}$, one also finds that (9.46) leads to

$$1-(p+b)\,\widetilde{S}(a,b)+(p-a)\,S(a,b)=\widetilde{V}(a)\,V(b). \tag{9.47}$$

It can be shown that as a consequence of (9.11) we have that $S(a,b)$ is symmetric under the exchange of the parameters a and b:

$$S(a,b)=S(b,a), \tag{9.48}$$

which is easily seen from the explicit expression (9.44c) by expanding the diagonal matrices $(a+\boldsymbol{K})^{-1}$ and $(b+\boldsymbol{K})^{-1}$ in powers of a and b. Using this property and the identity

$$\frac{\left[\widetilde{V}(a)\,V(b)\right]^{\widehat{}}}{\left[\widehat{V}(a)\,V(b)\right]^{\widetilde{}}}=\frac{\widehat{V}(b)\,V(a)}{\widetilde{V}(b)\,V(a)}$$

by inserting (9.47) and its counterpart, with p replaced by q and the tilde-shift replaced by the hat-shift, as well as the relations with a and b interchanged, we obtain the following closed-form equation for $S(a,b)$:

$$\frac{1-(p+b)\,\widehat{\widetilde{S}}(a,b)+(p-a)\,\widehat{S}(a,b)}{1-(q+b)\,\widehat{\widetilde{S}}(a,b)+(q-a)\,\widetilde{S}(a,b)}=\frac{1-(q+a)\,\widehat{S}(a,b)+(q-b)\,S(a,b)}{1-(p+a)\,\widetilde{S}(a,b)+(p-b)\,S(a,b)}, \tag{9.49}$$

which is the NQC-equation (Nijhoff et al., 1983a). It is a quite general PΔE which contains many special subcases for different choices of the parameters a and b, including lattice potential KdV, mKdV and SKdV equations. We will return to it in Section 9.5. The conservation laws of (9.49) have been discussed in Rasin and Hydon (2006).

9.4.3 Relation for the τ-functions

Making the special choice $a=p$, $b=-p$ in (9.47), we have the relation:

$$\widetilde{V}(p)V(-p)=1 \quad \text{and similarly,} \quad \widehat{V}(q)V(-q)=1. \tag{9.50}$$

Furthermore, introducing the variable

$$W(a)\equiv a-\boldsymbol{c}^T\,\boldsymbol{K}\,\boldsymbol{u}(a)=a-\boldsymbol{c}^T\,(a+\boldsymbol{K})^{-1}\boldsymbol{u}_1, \tag{9.51}$$

and multiplying equations (9.46) from the left by $\boldsymbol{c}^T$ we find:

$$W(a)=(p+a)\widetilde{V}(a)-(p-\widetilde{u})V(a)\,, \tag{9.52a}$$
$$\widetilde{W}(a)=-(p-a)V(a)+(p+u)\widetilde{V}(a)\,. \tag{9.52b}$$

(Recalling the definition of u in (9.10).)

A similar set of equations can be obtained for iterates in the m-direction (with hat-shifts and lattice parameter q):

$$W(a) = (q+a)\widehat{V}(a) - (q-\widehat{u})V(a), \tag{9.52c}$$

$$\widehat{W}(a) = -(q-a)V(a) + (q+u)\widehat{V}(a). \tag{9.52d}$$

Eliminating $W(a)$ leads to several relations, for instance by subtracting (9.52c) from (9.52a), or the tilde-shift of (9.52d) from the hat-shift of (9.52b), we obtain the set of relations:

$$p - q + \widehat{u} - \widetilde{u} = (p+a)\frac{\widetilde{V}(a)}{V(a)} - (q+a)\frac{\widehat{V}(a)}{V(a)} \tag{9.53a}$$

$$= (p-a)\frac{\widehat{V}(a)}{\widehat{\widetilde{V}}(a)} - (q-a)\frac{\widetilde{V}(a)}{\widehat{\widetilde{V}}(a)}. \tag{9.53b}$$

Alternatively, by combining (9.52b) with the tilde-shift of (9.52c), or (9.52d) with the hat-shift of (9.52a), one obtains:

$$p + q + u - \widehat{\widetilde{u}} = (p+a)\frac{\widehat{\widetilde{V}}(a)}{\widehat{V}(a)} + (q-a)\frac{V(a)}{\widehat{V}(a)} \tag{9.54a}$$

$$= (p-a)\frac{V(a)}{\widetilde{V}(a)} + (q+a)\frac{\widehat{\widetilde{V}}(a)}{\widetilde{V}(a)}. \tag{9.54b}$$

Notice that the right-hand-sides of each of (9.53) and (9.54) being equal give us a PΔE for $V(a)$. In particular, choosing $a = p$ in (9.53) we get the PΔE for $V(p)$ which takes the form

$$2p\frac{\widetilde{V}(p)}{V(p)} = (p+q)\frac{\widehat{V}(p)}{V(p)} + (p-q)\frac{\widetilde{V}(p)}{\widehat{\widetilde{V}}(p)}. \tag{9.55}$$

Now we are in the position to complete our aim of deducing bilinear equations for τ. For this we need the definition (9.43) and the relation (9.50), which yield $V(p) = \underset{\sim}{\tau}/\tau$ and similarly $V(q) = \underset{\wedge}{\tau}/\tau$. Equation (9.55) and the analogous equation for $V(q)$ lead to the bilinear equations

$$(p+q)\,\widehat{\underset{\sim}{\tau}}\,\widetilde{\tau} + (p-q)\,\underset{\sim}{\tau}\,\widehat{\widetilde{\tau}} = 2p\tau\widehat{\tau}, \tag{9.56a}$$

$$(p+q)\widetilde{\underset{\wedge}{\tau}}\widehat{\tau} + (q-p)\underset{\wedge}{\tau}\widehat{\widetilde{\tau}} = 2q\tau\widetilde{\tau}, \tag{9.56b}$$

which were discussed in Section 8.4.1.

9.4.4 Explicit form of the N-soliton solutions

We will now further elaborate the connection of the τ-function defined by equation (9.40) and in Chapter 8. For this purpose we need two explicit formulae for determinants.

The first formula is the expansion formula for the determinants of matrices of the form $\lambda \boldsymbol{I} + \boldsymbol{M}$, where λ is a scalar:

$$\det(\lambda \boldsymbol{I} + \boldsymbol{M}) = \lambda^N + \lambda^{N-1} \sum_{i=1}^{N} |M_{i,i}| + \lambda^{N-2} \sum_{i<j} \begin{vmatrix} M_{i,i} & M_{i,j} \\ M_{j,i} & M_{j,j} \end{vmatrix}$$
$$+ \lambda^{N-3} \sum_{i<j<k} \begin{vmatrix} M_{i,i} & M_{i,j} & M_{i,k} \\ M_{j,i} & M_{j,j} & M_{j,k} \\ M_{k,i} & M_{k,j} & M_{k,k} \end{vmatrix} + \cdots + \det(\boldsymbol{M}), \tag{9.57}$$

where $M_{i,j}$ are entries of the matrix $\boldsymbol{M}$. The second formula is for determinants of Cauchy matrices $\boldsymbol{A}$ as defined in (9.1), namely the famous formula (9.2).

Noting that the matrix $\boldsymbol{M}$ defined in (9.3) can be written using a Cauchy matrix of the form (9.1) with $l_j = -k_j$ multiplied from the left and right by diagonal matrices we have:

$$\det\left(\frac{\rho_i c_j}{k_i + k_j}\right) = \left(\prod_i \frac{\rho_i c_i}{2k_i}\right) \prod_{i<j} \left(\frac{k_i - k_j}{k_i + k_j}\right)^2 . \tag{9.58}$$

Since all terms in the expansion (9.57) are given by determinants with a selection of entries from $\boldsymbol{M}$, each term is expressible in a form similar to (9.58). Introducing for convenience the notations

$$e^{a_{ij}} := \left(\frac{k_i - k_j}{k_i + k_j}\right)^2, \quad e^{\eta_i} := \frac{\rho_i c_i}{2k_i}, \tag{9.59}$$

the formula for the τ-function thus takes the form:

$$\tau = 1 + \sum_{i=1}^{N} e^{\eta_j} + \sum_{1\le i<j\le N} e^{\eta_i + \eta_j + a_{ij}} +$$
$$+ \sum_{1\le i<j<k\le N} e^{\eta_i+\eta_j+\eta_k+a_{ij}+a_{ik}+a_{jk}} + \cdots . \tag{9.60}$$

This is in fact the same expression as was given in (8.18). Using the definitions (9.59) we obtain N-soliton solutions to the PΔEs discussed so far in this chapter. Note that the phase factor $e^{a_{jk}}$ above is special case of the one given in (8.15), so for other classes of equations some modifications of the present approach are necessary.

9.5 The NQC and Q3 equations

In this section we show that the results obtained in the previous section enable us to solve the Q3 equation (3.42c) introduced in Section 3.5. We consider the two cases $(\mathrm{Q3})_{\delta=0}$ and $(\mathrm{Q3})_{\delta}$ separately. We will see that the connections between (9.49) and (3.42c) provide multi-soliton solutions and a bilinearization of Q3.

9.5.1 Connection between the NQC and the $(Q3)_{\delta=0}$ equations

We will first establish the connection between $(\mathrm{Q3})_{\delta=0}$, i.e. the equation (3.42c) in which we set the parameter $\delta = 0$ and the NQC equation (9.49). We introduce the variable

$$\mathrm{u}^0_{n,m} := [\rho_{n,m}(a)\rho_{n,m}(b)]^{1/2}\left[1 - (a+b)S_{n,m}(a,b)\right], \tag{9.61}$$

where the ρ-functions were defined in (9.4). Now solving for $S(a,b)$ from this and substituting into (9.49) we obtain for $\mathrm{u}^0 = \mathrm{u}^0_{n,m}$ the following equation

$$P(\mathrm{u}^0\widehat{\mathrm{u}}^0 + \widetilde{\mathrm{u}}^0\widehat{\widetilde{\mathrm{u}}}^0) - Q(\mathrm{u}^0\widetilde{\mathrm{u}}^0 + \widehat{\mathrm{u}}^0\widehat{\widetilde{\mathrm{u}}}^0) = (p^2 - q^2)(\widehat{\mathrm{u}}^0\widetilde{\mathrm{u}}^0 + \mathrm{u}^0\widehat{\widetilde{\mathrm{u}}}^0), \tag{9.62}$$

which is equivalent to (3.42c) for $\delta = 0$. In (9.62) we have new parameters P and Q, associated with the lattice parameters p and q respectively, given by the formulae:[1]

$$P^2 = (p^2 - a^2)(p^2 - b^2), \quad Q^2 = (q^2 - a^2)(q^2 - b^2). \tag{9.63}$$

9.5.2 Connection between the NQC and the general Q3 equations

Now consider the case when $\delta \neq 0$ in Q3 equation. Our starting point is the observation that we get a solution to (9.62) for all choices of sign of a, b. Thus if we define

$$\mathrm{u}^{\sigma_1\sigma_2}_{n,m} := [\rho_{n,m}(\sigma_1 a)\rho_{n,m}(\sigma_2 b)]^{1/2}\left[1 - (\sigma_1 a + \sigma_2 b)S_{n,m}(\sigma_1 a, \sigma_2 b)\right], \tag{9.64}$$

where $\sigma_1, \sigma_2 \in \{+,-\}$ then each $\mathrm{u}^{\pm\pm}$ solves (9.62). But the remarkable fact is (Atkinson et al., 2008) that a linear combination of these functions

$$\mathrm{u} := A\mathrm{u}^{++} + B\mathrm{u}^{+-} + C\mathrm{u}^{-+} + D\mathrm{u}^{--}, \tag{9.65}$$

solves the Q3 equation in the form

$$P(\mathrm{u}\widehat{\mathrm{u}} + \widetilde{\mathrm{u}}\widehat{\widetilde{\mathrm{u}}}) - Q(\mathrm{u}\widetilde{\mathrm{u}} + \widehat{\mathrm{u}}\widehat{\widetilde{\mathrm{u}}}) = (p^2 - q^2)\left((\widehat{\mathrm{u}}\widetilde{\mathrm{u}} + \mathrm{u}\widehat{\widetilde{\mathrm{u}}}) + \frac{\delta^2}{4PQ}\right), \tag{9.66}$$

[1] In fact, equations (9.63) imply that (p, P), (q, Q) are points on a (Jacobi) elliptic curve with branch points $\{(\pm a, 0), (\pm b, 0)\}$.

provided that the coefficients A, B, C, D in (9.65) are related by

$$AD(a+b)^2 - BC(a-b)^2 = -\frac{\delta^2}{16ab}\,. \tag{9.67}$$

What is quite remarkable about the formula (9.65) is that even though we are dealing with a highly *nonlinear* equation (9.66), nevertheless it seems a kind of linear superposition formula holds. For an outline of the proof of this result see Section 9.6.

9.5.3 Bilinearization

In this section we show how to write the solution (9.65) in terms of the τ-functions. For this purpose we have to extend the lattice from two dimensions to four dimensions. The reason is that, similar to p and q, we can reinterpret the parameters a and b as associated with new lattice directions coordinatized by variables α, β.

Since there are several parameters playing different roles, let us write the dependencies explicitly and extend the definition of the $\rho(x)$ previously given in (9.4) as follows:

$$\rho_{n,m,\alpha,\beta}(p,q,a,b,k_i) := \left(\frac{p+k_i}{p-k_i}\right)^n \left(\frac{q+k_i}{q-k_i}\right)^m \left(\frac{a+k_i}{a-k_i}\right)^\alpha \left(\frac{b+k_i}{b-k_i}\right)^\beta \rho_i^0. \tag{9.68}$$

Then in addition to (9.41, 9.43) we would get, for example

$$\frac{\tau_{n,m,\alpha+1,\beta}}{\tau_{n,m,\alpha,\beta}} = 1 + \boldsymbol{c}^T\,(\boldsymbol{I}+\boldsymbol{M})^{-1}\,(a-\boldsymbol{K})^{-1}\boldsymbol{r} = V(-a), \tag{9.69}$$

and following the derivation of (9.47), also

$$1-(a+b')\,S_{n,m,\alpha+1,\beta}(a',b') + (a-a')\,S_{n,m,\alpha,\beta}(a',b') = V_{n,m,\alpha+1,\beta}(a')\,V_{n,m,\alpha,\beta}(b'), \tag{9.70}$$

where a', b' now play the same role as a, b when introduced in (9.44c), while a is associated with the lattice shift. The n, m, α, β dependence in S comes only through ρ.

If we now take $a' = a$, $b' = b$ and α-downshift the equation we get

$$1-(a+b)\,S_{n,m,\alpha,\beta}(a,b) = V_{n,m,\alpha,\beta}(a)\,V_{n,m,\alpha-1,\beta}(b)$$

while choosing $b' = -a$, $a' = b$ and using symmetry of S yields

$$1-(-a+b)\,S_{n,m,\alpha,\beta}(-a,b) = V_{n,m,\alpha+1,\beta}(b)\,V_{n,m,\alpha,\beta}(-a).$$

Similarly to (9.43) we now have

$$V_{n,m,\alpha,\beta}(a) = \frac{\tau_{n,m,\alpha-1,\beta}}{\tau_{n,m,\alpha,\beta}}, \qquad V_{n,m,\alpha,\beta}(-a) = \frac{\tau_{n,m,\alpha+1,\beta}}{\tau_{n,m,\alpha,\beta}}$$

and analogous equations for β-shifts. Using these and recalling the connection to τ-function (9.69) as well as (9.64) and the relation from V to τ in (9.43), which extends to shifts in α, β, with which we arrive at

$$\begin{aligned} \mathfrak{u}_{n,m} = & \, A\,[\rho(a)\rho(b)]^{1/2}\,\frac{\tau_{n,m,\alpha-1,\beta-1}}{\tau_{n,m,\alpha,\beta}} + B\,[\rho(a)\rho(-b)]^{1/2}\,\frac{\tau_{n,m,\alpha-1,\beta+1}}{\tau_{n,m,\alpha,\beta}} \\ & + C\,[\rho(-a)\rho(b)]^{1/2}\,\frac{\tau_{n,m,\alpha+1,\beta-1}}{\tau_{n,m,\alpha,\beta}} + D\,[\rho(-a)\rho(-b)]^{1/2}\,\frac{\tau_{n,m,\alpha+1,\beta+1}}{\tau_{n,m,\alpha,\beta}}. \end{aligned} \tag{9.71}$$

(Note that the explicit ρ appearing above is still defined by (9.4).) This gives a bilinearization of the Q3 equation, but to prove it one needs several bilinear equations relating various shifts of τ; for details, see Atkinson et al. (2008).

In addition to the solution $\mathfrak{u}$ it turns out that we will also have to make use of an auxiliary quantity $\mathfrak{U}$

$$\begin{aligned} \mathfrak{U}_{n,m} = & \, A(a+b)\,[\rho(a)\rho(b)]^{1/2}\,\frac{\tau_{\alpha-1,\beta}\tau_{\alpha,\beta-1}}{\tau_{\alpha,\beta}^2} + B(a-b)\,[\rho(a)\rho(-b)]^{1/2}\,\frac{\tau_{\alpha-1,\beta}\tau_{\alpha,\beta+1}}{\tau_{\alpha,\beta}^2} \\ & - C(a-b)\,[\rho(-a)\rho(b)]^{1/2}\,\frac{\tau_{\alpha+1,\beta}\tau_{\alpha,\beta-1}}{\tau_{\alpha,\beta}^2} - D(a+b)\,[\rho(-a)\rho(-b)]^{1/2}\,\frac{\tau_{\alpha+1,\beta}\tau_{\alpha,\beta+1}}{\tau_{\alpha,\beta}^2} \end{aligned} \tag{9.72}$$

(where we have suppressed the common dependence of τ on n, m), which solves the linearized version of the equation which is given by

$$\begin{aligned} & P(\mathfrak{u}\widehat{\mathfrak{U}} + \mathfrak{U}\widehat{\mathfrak{u}} + \widetilde{\mathfrak{u}}\widehat{\widetilde{\mathfrak{U}}} + \widetilde{\mathfrak{U}}\widehat{\widetilde{\mathfrak{u}}}) - Q(\mathfrak{u}\widetilde{\mathfrak{U}} + \mathfrak{U}\widetilde{\mathfrak{u}} + \widehat{\mathfrak{u}}\widehat{\widetilde{\mathfrak{U}}} + \widehat{\mathfrak{U}}\widehat{\widetilde{\mathfrak{u}}}) \\ & = (p^2 - q^2)(\widehat{\mathfrak{u}}\widetilde{\mathfrak{U}} + \widehat{\mathfrak{U}}\widetilde{\mathfrak{u}} + \mathfrak{u}\widehat{\widetilde{\mathfrak{U}}} + \mathfrak{U}\widehat{\widetilde{\mathfrak{u}}}). \end{aligned}$$

9.6 Proof of the Q3 N-soliton solution

The proof of the soliton solution (Nijhoff et al., 2009; Nijhoff and Atkinson, 2010), given by (9.71) for the Q3 equation in the form (9.66), relies on a number of relations that we have established before as well as on some determinantal identities. In fact, it exhibits the interplay of many of the quantities that have arrived on the scene so far: the lattice potential KdV variable u, the variables $V(a)$ given in (9.42a) the quantity $S(a,b)$ defined in (9.44c) and also the auxiliary quantity $W(a)$ given in (9.51). In fact, we need the relations (9.47), and both relations (9.52), and their counterparts involving shifts in the other lattice direction. Remarkably, in the proof the lattice potential KdV equation (9.24) makes its appearance as well.

Using the plane-wave factors $\rho(a)$ and $\rho(b)$ as given in (9.4) we can now introduce the two-component vectors:

$$\boldsymbol{v}_a := \begin{pmatrix} \rho(a)^{1/2}V(a) \\ \rho(-a)^{1/2}V(-a) \end{pmatrix}, \quad \boldsymbol{w}_a := \begin{pmatrix} \rho(a)^{1/2}W(a) \\ \rho(-a)^{1/2}W(-a) \end{pmatrix} \tag{9.73}$$

and the quantities:

$$\mathfrak{u} := \left(\rho(a)^{1/2}, \rho(-a)^{1/2}\right)\begin{pmatrix} AS_{a,b} & BS_{a,-b} \\ CS_{-a,b} & DS_{-a,-b} \end{pmatrix}\begin{pmatrix} \rho(b)^{1/2} \\ \rho(-b)^{1/2} \end{pmatrix}, \tag{9.74a}$$

$$\mathfrak{U} := \boldsymbol{v}_a^T \boldsymbol{A}\, \boldsymbol{v}_b, \tag{9.74b}$$

where $S_{a,b} := (a+b)S(a,b) - 1$, and where $\boldsymbol{A}$ is the matrix:

$$\boldsymbol{A} = \begin{pmatrix} (a+b)A & (a-b)B \\ (-a-b)C & (-a+b)D \end{pmatrix}.$$

In fact, the $\mathfrak{u}$ given in (9.74) is the Q3 solution (9.71), while $\mathfrak{U}$ is an auxiliary object, which obeys a linearized version of the Q3 equation. The relations (9.52) lead to the vector relations

$$\widetilde{\boldsymbol{w}}_a = (p+u)\widetilde{\boldsymbol{v}}_a - P_a\boldsymbol{v}_a, \tag{9.75a}$$

$$\boldsymbol{w}_a = P_a\widetilde{\boldsymbol{v}}_a - (p-\widetilde{u})\boldsymbol{v}_a \tag{9.75b}$$

and similar relations for the dynamics in the variable m, with the usual replacements and changing P_a into Q_a, where

$$P_a = \sqrt{p^2 - a^2}, \quad Q_a = \sqrt{q^2 - a^2}.$$

From (9.75) by elimination $\boldsymbol{w}_a$ we obtain

$$P_a\widetilde{\boldsymbol{v}}_a - Q_a\widehat{\boldsymbol{v}}_a = (p - q + \widehat{u} - \widetilde{u})\boldsymbol{v}_a, \tag{9.76a}$$

$$P_a\boldsymbol{v}_a + Q_a\widehat{\widetilde{\boldsymbol{v}}}_a = (p + q + u - \widehat{\widetilde{u}})\widetilde{\boldsymbol{v}}_a, \tag{9.76b}$$

and we will also make use of the quantities $P = P_aP_b$, $Q = Q_aQ_b$ of (9.63).

The relation (9.47) and its counterpart interchanging a and b, now become

$$P_a\mathfrak{u} - P_b\widetilde{\mathfrak{u}} = \widetilde{\boldsymbol{v}}_a^T\boldsymbol{A}\boldsymbol{v}_b, \quad P_b\mathfrak{u} - P_a\widetilde{\mathfrak{u}} = \boldsymbol{v}_a^T\boldsymbol{A}\widetilde{\boldsymbol{v}}_b, \tag{9.77}$$

where it should be noted that the matrix $\boldsymbol{A}$ need *not* be symmetric.

The construction of the Q3 equation from these data now follows the following steps:
(i) Biquadratic identity: Consider the following 2×2 matrix

$$E_{a,b} = \begin{pmatrix} \boldsymbol{v}_a^T \boldsymbol{A} \boldsymbol{v}_b, & \boldsymbol{v}_a^T \boldsymbol{A} \boldsymbol{w}_b \\ \boldsymbol{w}_a^T \boldsymbol{A} \boldsymbol{v}_b, & \boldsymbol{w}_a^T \boldsymbol{A} \boldsymbol{w}_b \end{pmatrix}. \tag{9.78}$$

The determinant of this matrix can be computed in two different ways; on the one hand, we have, using (9.75)

$$\begin{aligned} \det(E_{a,b}) &= P_b \begin{vmatrix} \boldsymbol{v}_a^T \boldsymbol{A} \boldsymbol{v}_b & \boldsymbol{v}_a^T \boldsymbol{A} \widetilde{\boldsymbol{v}}_b \\ \boldsymbol{w}_a^T \boldsymbol{A} \boldsymbol{v}_b & \boldsymbol{w}_a^T \boldsymbol{A} \widetilde{\boldsymbol{v}}_b \end{vmatrix} = P_a P_b \begin{vmatrix} \boldsymbol{v}_a^T \boldsymbol{A} \boldsymbol{v}_b & \boldsymbol{v}_a^T \boldsymbol{A} \widetilde{\boldsymbol{v}}_b \\ \widetilde{\boldsymbol{v}}_a^T \boldsymbol{A} \boldsymbol{v}_b & \widetilde{\boldsymbol{v}}_a^T \boldsymbol{A} \widetilde{\boldsymbol{v}}_b \end{vmatrix} \\ &= P \begin{vmatrix} \mathfrak{U} & P_b \mathfrak{u} - P_a \widetilde{\mathfrak{u}} \\ P_a \mathfrak{u} - P_b \widetilde{\mathfrak{u}} & \widetilde{\mathfrak{U}} \end{vmatrix} \\ &= P \left[\mathfrak{U}\widetilde{\mathfrak{U}} - P(\mathfrak{u}^2 + \widetilde{\mathfrak{u}}^2) + (2p^2 - a^2 - b^2)\mathfrak{u}\widetilde{\mathfrak{u}} \right]. \end{aligned}$$

On the other hand, computing determinant of E directly as a product of three different 2×2 matrices we have

$$E_{a,b} = \begin{pmatrix} v_{a,1} & v_{a,2} \\ w_{a,1} & w_{a,2} \end{pmatrix} \boldsymbol{A} \begin{pmatrix} v_{b,1} & w_{b,1} \\ v_{b,2} & w_{b,2} \end{pmatrix} \equiv (\boldsymbol{v}_a, \boldsymbol{w}_a)^T \boldsymbol{A} \, (\boldsymbol{v}_b, \boldsymbol{w}_b)$$

and therefore

$$\det(E_{a,b}) = \det(\boldsymbol{A}) \det(\boldsymbol{v}_a \, , \, \boldsymbol{w}_a) \det(\boldsymbol{v}_b \, , \, \boldsymbol{w}_b) = \mathfrak{a}\mathfrak{b} \det(\boldsymbol{A}) \, ,$$

where $\mathfrak{a} := \det(\boldsymbol{v}_a \, , \, \boldsymbol{w}_a)$ and $\mathfrak{b} := \det(\boldsymbol{v}_b \, , \, \boldsymbol{w}_b)$ are constants, as can be seen from the chain:

$$\widetilde{\mathfrak{a}} = \det(\widetilde{\boldsymbol{v}}_a \, , \, \widetilde{\boldsymbol{w}}_a) = -P_a \det(\widetilde{\boldsymbol{v}}_a \, , \, \boldsymbol{v}_a) = \det(\boldsymbol{v}_a \, , \, P_a \widetilde{\boldsymbol{v}}_a) = \det(\boldsymbol{v}_a \, , \, \boldsymbol{w}_a) = \mathfrak{a} \, ,$$

using once again the vector relations (9.75), and a similarly for $\mathfrak{b}$ and the same argument holds for the shift in the variable m.

Thus, equating both sides of the evaluation of $\det(E_{a,b})$ we obtain the factorization of the biquadratic

$$\mathfrak{U}\widetilde{\mathfrak{U}} = P(\mathfrak{u}^2 + \widetilde{\mathfrak{u}}^2) - (2p^2 - a^2 - b^2)\mathfrak{u}\widetilde{\mathfrak{u}} + \frac{\mathfrak{a}\mathfrak{b}}{P} \det(\boldsymbol{A}). \tag{9.79}$$

(ii) Q3 equation: To obtain the main equation of interest, Q3, for the variable u we proceed as follows. Consider the relations (9.77) and insert them into the following determinant

$$\begin{aligned} &\begin{vmatrix} \boldsymbol{v}_a^T \boldsymbol{A} \widetilde{\boldsymbol{v}}_b & \boldsymbol{v}_a^T \boldsymbol{A} \widehat{\boldsymbol{v}}_b \\ \widehat{\widetilde{\boldsymbol{v}}}_a^T \boldsymbol{A} \widetilde{\boldsymbol{v}}_b & \widehat{\widetilde{\boldsymbol{v}}}_a^T \boldsymbol{A} \widehat{\boldsymbol{v}}_b \end{vmatrix} = \begin{vmatrix} P_b \mathfrak{u} - P_a \widetilde{\mathfrak{u}} & Q_b \mathfrak{u} - Q_a \widehat{\mathfrak{u}} \\ Q_a \widetilde{\mathfrak{u}} - Q_b \widehat{\widetilde{\mathfrak{u}}} & P_a \widehat{\mathfrak{u}} - P_b \widehat{\widetilde{\mathfrak{u}}} \end{vmatrix} \\ &= P(\mathfrak{u}\widehat{\mathfrak{u}} + \widetilde{\mathfrak{u}}\widehat{\widetilde{\mathfrak{u}}}) - Q(\mathfrak{u}\widetilde{\mathfrak{u}} + \widehat{\mathfrak{u}}\widehat{\widetilde{\mathfrak{u}}}) - (p^2 - q^2)(\widetilde{\mathfrak{u}}\widehat{\mathfrak{u}} + \mathfrak{u}\widehat{\widetilde{\mathfrak{u}}}). \end{aligned} \tag{9.80}$$

On the other hand we have that the starting determinant equals:

$$
\begin{aligned}
&\frac{1}{Q_b}\begin{vmatrix} \boldsymbol{v}_a^T\boldsymbol{A}\widetilde{\boldsymbol{v}}_b & \boldsymbol{v}_a^T\boldsymbol{A}\left[P_b\widetilde{\boldsymbol{v}}_b-(p-q+\widehat{u}-\widetilde{u})\boldsymbol{v}_b\right] \\ \widehat{\widetilde{\boldsymbol{v}}}_a^T\boldsymbol{A}\widetilde{\boldsymbol{v}}_b & \widehat{\widetilde{\boldsymbol{v}}}_a^T\boldsymbol{A}\left[P_b\widetilde{\boldsymbol{v}}_b-(p-q+\widehat{u}-\widetilde{u})\boldsymbol{v}_b\right] \end{vmatrix} \\
&=-\frac{p-q+\widehat{u}-\widetilde{u}}{P_bQ_b}\begin{vmatrix} \boldsymbol{v}_a^T\boldsymbol{A}\boldsymbol{w}_a & \boldsymbol{v}_a^T\boldsymbol{A}\boldsymbol{v}_b \\ \widehat{\widetilde{\boldsymbol{v}}}_a^T\boldsymbol{A}\boldsymbol{w}_b & \widehat{\widetilde{\boldsymbol{v}}}_a^T\boldsymbol{A}\boldsymbol{v}_b \end{vmatrix} \\
&=-\frac{(p-q+\widehat{u}-\widetilde{u})(p+q+u-\widehat{\widetilde{u}})}{P_aP_bQ_aQ_b}\begin{vmatrix} \boldsymbol{v}_a^T\boldsymbol{A}\boldsymbol{w}_b & \boldsymbol{v}_a^T\boldsymbol{A}\boldsymbol{v}_b \\ \boldsymbol{w}_a^T\boldsymbol{A}\boldsymbol{w}_b & \boldsymbol{w}_a\boldsymbol{A}\boldsymbol{v}_b \end{vmatrix} \\
&=\frac{p^2-q^2}{PQ}\mathfrak{a}\mathfrak{b}\det(\boldsymbol{A}).
\end{aligned} \tag{9.81}
$$

In the derivation we successively used (9.76a) with a replaced by b, then (9.76b), subsequently (9.75b), as well as the lattice potential KdV equation for u

$$
(p-q+\widehat{u}-\widetilde{u})(p+q+u-\widehat{\widetilde{u}})=p^2-q^2. \tag{9.82}
$$

Thus by comparing (9.80) and (9.81), we obtain the Q3 equation (9.66) with the identification $\delta^2/4=\mathfrak{a}\mathfrak{b}\det(\boldsymbol{A})$.

Remark: We note that with the help of the auxiliary quantity $\mathfrak{U}$ we have the following system of equations:

$$
\begin{aligned}
p-q+\widehat{u}-\widetilde{u}&=\frac{1}{\mathfrak{U}}\left[P\widetilde{\mathfrak{u}}-Q\widehat{\mathfrak{u}}-(p^2-q^2)\mathfrak{u}\right] &\text{(9.83a)}\\
&=-\frac{1}{\widehat{\widetilde{\mathfrak{U}}}}\left[P\widehat{\mathfrak{u}}-Q\widetilde{\mathfrak{u}}-(p^2-q^2)\widehat{\widetilde{\mathfrak{u}}}\right], &\text{(9.83b)}\\
p+q+u-\widehat{\widetilde{u}}&=\frac{1}{\widehat{\mathfrak{U}}}\left[P\widehat{\widetilde{\mathfrak{u}}}-Q\mathfrak{u}-(p^2-q^2)\widehat{\mathfrak{u}}\right] &\text{(9.83c)}\\
&=-\frac{1}{\widetilde{\mathfrak{U}}}\left[P\mathfrak{u}-Q\widehat{\widetilde{\mathfrak{u}}}-(p^2-q^2)\widetilde{\mathfrak{u}}\right], &\text{(9.83d)}
\end{aligned}
$$

which effectively establishes a Miura relation between Q3 and the lattice potential KdV equation for u.

Remark: We note that establishing the solution for Q3 is important, because it implies that we have effectively the N-soliton solutions for the entire class of ABS equations from Q3 downward. In fact, all the equations: Q1, Q2, H1, H2 and H3 of the ABS list can be obtained by degenerations, i.e. special limits on the parameters a and b, which also can be implemented on the solutions themselves, cf. Nijhoff et al. (2009); Nijhoff and Atkinson (2010), according to the diagram of Figure 3.7.

9.7 Higher-dimensional soliton systems: the KP class

In two dimensions, each soliton solution needs only one parameter k_i to describe its height and speed. In order to describe solitons in three dimensions we will need one more parameter to describe the angle between different planar solitons. For this purpose we will generalize the Cauchy matrix method in this section.

So far all considerations of the soliton structure were based on the form of the underlying Cauchy matrix (9.3). However, that matrix has a special form, while the determinantal identity (9.2) suggests that we could explore a more general form of the matrix with two types of parameters: k_i and l_j, where $i = 1, \dots, N$, $j = 1, \dots, N'$ and N is not necessarily equal to N'.

9.7.1 Generalized Cauchy matrices and higher dimensional solitons

The starting point of our generalization is the $N \times N'$ matrix $\boldsymbol{M}$ given by

$$\boldsymbol{M} = (M_{i,j})_{i=1,\dots,N;j=1,\dots,N'} \quad \text{with} \quad M_{i,j} = \frac{\rho(k_i)\sigma(l_j)}{k_i + l_j}. \tag{9.84}$$

We will see that we can still implement the approach developed in Sections 9.1 and 9.2, except that half of the relations will be lost because we no longer have symmetry between the variables k and l.

Since we are generalizing from two to three dimensions we have to assume that there are three discrete independent variables n, m, h (each associated with its respective lattice parameter, p, q and r, respectively), that is

$$\rho_i = \rho_{n,m,h}(k_i), \quad \sigma_j = \sigma_{n,m,h}(l_j), \tag{9.85}$$

where

$$\rho_{n,m,h}(k) = (p+k)^n \, (q+k)^m \, (r+k)^h \, \rho_{0,0,0}(k) \,, \tag{9.86a}$$

$$\sigma_{n,m,h}(l) = (p-l)^{-n} \, (q-l)^{-m} \, (r-l)^{-h} \, \sigma_{0,0,0}(l) \,, \tag{9.86b}$$

where $\rho_{0,0,0}$, $\sigma_{0,0,0}$ indicate some initial values independent of the discrete variables n, m, h. In what follows, shifts in n, m, and h by one step will be represented by tilde- , hat- and bar-symbols, respectively.

Let $\boldsymbol{K} = \operatorname{diag}(k_1, \dots, k_N)$, $\boldsymbol{L} = \operatorname{diag}(l_1, \dots, l_{N'})$, $\boldsymbol{r}, \boldsymbol{s}$ be column vectors $\boldsymbol{r} = (\rho_1, \dots, \rho_N)^T$, $\boldsymbol{s} = (\sigma_1, \dots, \sigma_{N'})^T$, and $\boldsymbol{I}_M$ be an $M \times M$ unit matrix. Equations (9.86) now give the tilde-shifts as

$$\widetilde{\boldsymbol{r}} = (p\boldsymbol{I}_N + \boldsymbol{K})\boldsymbol{r}, \quad \widetilde{\boldsymbol{s}} = (p\boldsymbol{I}_{N'} - \boldsymbol{L})^{-1}\boldsymbol{s}, \tag{9.87}$$

and similarly for hat- and bar-shifts.

From the definition (9.84) we immediately have the following relation

$$\boldsymbol{K}\boldsymbol{M} + \boldsymbol{M}\boldsymbol{L} = \boldsymbol{r}\,\boldsymbol{s}^T\,. \tag{9.88}$$

For the tilde-shift on the matrix $\boldsymbol{M}$, we then have

$$\widetilde{\boldsymbol{M}}\,(p\boldsymbol{I}_{N'} - \boldsymbol{L}) = (p\boldsymbol{I}_N + \boldsymbol{K})\boldsymbol{M} \quad \Rightarrow \quad \widetilde{\boldsymbol{M}} = \boldsymbol{M} + \boldsymbol{r}\widetilde{\boldsymbol{s}}^T\,, \tag{9.89}$$

and similar formulae for the other shifts.

We want to study similar objects as before, namely row and column vectors and scalars as in (9.9). However, since $\boldsymbol{M}$ in this case is not necessarily a square matrix, we need to introduce an $N' \times N$ constant matrix $\boldsymbol{J}$ to compensate for the possible deviation, so that the products $\boldsymbol{J}\boldsymbol{M}$ and $\boldsymbol{M}\boldsymbol{J}$ are now square $N' \times N'$, $N \times N$ matrices, respectively.

The objects we want to study are the following sets of column- and row-vectors, respectively

$$\boldsymbol{u}_\alpha := (\boldsymbol{I}_N + \boldsymbol{M}\boldsymbol{J})^{-1}\boldsymbol{K}^\alpha\boldsymbol{r}\,, \quad {}^t\boldsymbol{u}_\beta := \boldsymbol{s}^T\,\boldsymbol{L}^\beta\,(\boldsymbol{I}_{N'} + \boldsymbol{J}\boldsymbol{M})^{-1}\,, \quad \alpha, \beta \in \mathbb{Z}, \tag{9.90}$$

where each $\boldsymbol{u}_\alpha$ is an N-component column vector, and each ${}^t\boldsymbol{u}_\beta$ an N'-component row vector. Furthermore, as before, we introduce the bi-infinite set of scalar quantities

$$U_{\alpha,\beta} = \boldsymbol{s}^T\,\boldsymbol{L}^\beta\,(\boldsymbol{I}_{N'} + \boldsymbol{J}\boldsymbol{M})^{-1}\boldsymbol{J}\boldsymbol{K}^\alpha\boldsymbol{r} = {}^t\boldsymbol{u}_\beta\,\boldsymbol{J}\,\boldsymbol{K}^\alpha\boldsymbol{r}\,, \tag{9.91a}$$

$$= \boldsymbol{s}^T\boldsymbol{L}^\beta\boldsymbol{J}(\boldsymbol{I}_N + \boldsymbol{M}\boldsymbol{J})^{-1}\boldsymbol{K}^\alpha\boldsymbol{r} = \boldsymbol{s}^T\boldsymbol{L}^\beta\boldsymbol{J}\,\boldsymbol{u}_\alpha \tag{9.91b}$$

for all α, β integers. Note that U is no longer symmetric in its indices.

We now use the same approach as in Section 9.1.3 to derive the basic recurrence relations for the vectors $\boldsymbol{u}_\alpha$ and ${}^t\boldsymbol{u}_\beta$, this yields

$$\widetilde{\boldsymbol{u}}_\alpha = p\boldsymbol{u}_\alpha + \boldsymbol{u}_{\alpha+1} - \widetilde{U}_{\alpha,0}\,\boldsymbol{u}_0, \tag{9.92a}$$

$${}^t\boldsymbol{u}_\beta = p\,{}^t\widetilde{\boldsymbol{u}}_\beta - {}^t\widetilde{\boldsymbol{u}}_{\beta+1} + {}^t\widetilde{\boldsymbol{u}}_0\,U_{0,\beta}, \tag{9.92b}$$

$$\widehat{\boldsymbol{u}}_\alpha = q\boldsymbol{u}_\alpha + \boldsymbol{u}_{\alpha+1} - \widehat{U}_{\alpha,0}\boldsymbol{u}_0, \tag{9.92c}$$

$$q\,{}^t\boldsymbol{u}_\beta = {}^t\widehat{\boldsymbol{u}}_\beta - {}^t\widehat{\boldsymbol{u}}_{\beta+1} + {}^t\widehat{\boldsymbol{u}}_0\,U_{0,\beta}, \tag{9.92d}$$

$$\overline{\boldsymbol{u}}_\alpha = r\boldsymbol{u}_\alpha + \boldsymbol{u}_{\alpha+1} - \overline{U}_{\alpha,0}\boldsymbol{u}_0, \tag{9.92e}$$

$$r\,{}^t\boldsymbol{u}_\beta = {}^t\overline{\boldsymbol{u}}_\beta - {}^t\overline{\boldsymbol{u}}_{\beta+1} + {}^t\overline{\boldsymbol{u}}_0\,U_{0,\beta}. \tag{9.92f}$$

And from these we can derive, as before, the double-sided recurrence relation for the $U_{\alpha,\beta}$, namely:

$$p\widetilde{U}_{\alpha,\beta} - \widetilde{U}_{\alpha,\beta+1} = pU_{\alpha,\beta} + U_{\alpha+1,\beta} - \widetilde{U}_{\alpha,0}U_{0,\beta}, \tag{9.93a}$$

$$q\widehat{U}_{\alpha,\beta} - \widehat{U}_{\alpha,\beta+1} = qU_{\alpha,\beta} + U_{\alpha+1,\beta} - \widehat{U}_{\alpha,0}U_{0,\beta}, \tag{9.93b}$$

$$r\overline{U}_{\alpha,\beta} - \overline{U}_{\alpha,\beta+1} = rU_{\alpha,\beta} + U_{\alpha+1,\beta} - \overline{U}_{\alpha,0}U_{0,\beta}. \tag{9.93c}$$

We observe that relation (9.93a) coincides with relation (9.15), and relation (9.92a), up to a factor, with (9.13). Since we no longer have any connection between parameters k_i and l_j we do not have the symmetry (9.11). However, to compensate for this deficiency, we have the third variable h and the corresponding relations governing shifts in that variable.

In order to obtain closed-form PΔEs, we now combine these relations by eliminating the shifts in the indices α and β. There are three cases which run parallel to the cases obtained in the two-dimensional case of Section 9.2.

Lattice potential KP equation

As a first example, taking $\alpha = \beta = 0$ in (9.93) and eliminating the variables $U_{1,0}$ and $U_{0,1}$ by combining the relations obtained for different lattice directions, it is straightforward to derive the following three-dimensional lattice equation for the variable $u = U_{0,0}$

$$(p-\widetilde{u})(q-r+\widetilde{\overline{u}}-\widehat{\widetilde{u}})+(q-\widehat{u})(r-p+\widehat{\widetilde{u}}-\widehat{\overline{u}})+(r-\overline{u})(p-q+\widehat{\overline{u}}-\widetilde{\overline{u}})=0, \quad (9.94)$$

which we shall refer to as the *lattice potential Kadomtsev–Petviashvili (lpKP) equation.* Using $\mathfrak{u} := u - pn - qm - rh$ we can write this equation also as

$$\widehat{\widetilde{\mathfrak{u}}}(\widetilde{\mathfrak{u}}-\widehat{\mathfrak{u}})+\widehat{\overline{\mathfrak{u}}}(\widehat{\mathfrak{u}}-\overline{\mathfrak{u}})+\widetilde{\overline{\mathfrak{u}}}(\overline{\mathfrak{u}}-\widetilde{\mathfrak{u}})=0, \quad (9.95)$$

which is equation (3.107c).

Lattice potential MKP equation

A second example is obtained by setting $\alpha = -1, \beta = 0$ or $\alpha = 0, \beta = -1$ in the set of equations (9.93). By combining any two of these equations and eliminating the variables $U_{1,-1}$ and $U_{-1,1}$, we can derive the relations

$$p-q+\widehat{u}-\widetilde{u}=\frac{p\widehat{v}-q\widetilde{v}}{\widehat{\widetilde{v}}}=\frac{p\widetilde{w}-q\widehat{w}}{w}, \quad (9.96)$$

(and similar ones involving a combination of any two of the other lattice shifts), where we have introduced the variables:

$$v = 1 - U_{-1,0}, \qquad w = 1 - U_{0,-1}. \quad (9.97)$$

By combining (9.96) with the other relations obtained by cyclic permutation of the lattice variables and associated parameters, it is straightforward to eliminate the variable u and derive the three-dimensional lattice potential modified KP equation (3.110) for v or (3.111) for w.

Lattice SKP equation

A third example is obtained by considering the choice $\alpha = \beta = -1$ in equations (9.93) and derive the relation:

$$p(z - \widetilde{z}) = \widetilde{v} w , \tag{9.98}$$

where $z = U_{-1,-1} - \frac{n}{p} - \frac{m}{q} - \frac{h}{r}$. By combining (9.98) with two similar equations for the other lattice directions, and substituting the left-hand sides into the identity:

$$\frac{(\widetilde{v}w)\widehat{\ }}{(\widetilde{v}w)^{-}} = \frac{(\widehat{v}w)\widetilde{\ }}{(\overline{v}w)\widetilde{\ }} \, \frac{(\overline{v}w)\widehat{\ }}{(\widehat{v}w)^{-}} ,$$

we obtain the following three-dimensional lattice equation for z:

$$\frac{(\widehat{\widetilde{z}} - \widehat{z})(\widetilde{\overline{z}} - \widetilde{z})(\widehat{\overline{z}} - \overline{z})}{(\widehat{\widetilde{z}} - \widetilde{z})(\widetilde{\overline{z}} - \overline{z})(\widehat{\overline{z}} - \widehat{z})} = 1 . \tag{9.99}$$

We refer to this equation as the *lattice Schwarzian KP (lSKP) equation* (Dorfman and Nijhoff, 1991) (cf. also Nijhoff et al., 1984). This equation was also derived in connection with singularity manifolds (Bogdanov and Konopelchenko, 1998a) and has a connection with *Melenaus' theorem* of classical geometry (Konopelchenko and Schief, 2002a).

9.7.2 Bilinear structure for discrete KP-type equations

The τ-function for the lattice KP soliton family can be defined as:

$$\tau = \tau_{n,m,h} = \det(\boldsymbol{I}_N + \boldsymbol{M}\boldsymbol{J}) = \det(\boldsymbol{I}_{N'} + \boldsymbol{J}\boldsymbol{M}) . \tag{9.100}$$

The latter identity is a consequence of the Sylvester identity (see Appendix D).

Following the procedure described in Section 9.5.3 (but with (9.87, 9.89, 9.90)) we can derive

$$\frac{\widetilde{\tau}}{\tau} = 1 + \boldsymbol{s}^T (p - \boldsymbol{L})^{-1} \boldsymbol{J} \boldsymbol{u}_0 , \quad \frac{\tau}{\underset{\sim}{\tau}} = 1 - {}^t\boldsymbol{u}_0 \boldsymbol{J} (p + \boldsymbol{K})^{-1} \boldsymbol{r} . \tag{9.101}$$

The relations (9.101) suggest the introduction of the quantities, for arbitrary parameter a:

$$V(a) = 1 - {}^t\boldsymbol{u}_0 \boldsymbol{J} (a + \boldsymbol{K})^{-1} \boldsymbol{r} , \quad Y(a) = 1 - \boldsymbol{s}^T (a + \boldsymbol{L})^{-1} \boldsymbol{J} \boldsymbol{u}_0 , \tag{9.102}$$

and, hence

$$\frac{\widetilde{\tau}}{\tau} = Y(-p) = 1/\widetilde{V}(p) . \tag{9.103}$$

Multiplying the difference of (9.92a) and (9.92c) by $\boldsymbol{s}^T (a + \boldsymbol{L})^{-1} \boldsymbol{J}$, we obtain

$$p - q + \widehat{u} - \widetilde{u} = (p - a) \frac{\widehat{V}(a)}{\widehat{\widetilde{V}}(a)} - (q - a) \frac{\widetilde{V}(a)}{\widehat{\widetilde{V}}(a)} \tag{9.104a}$$

$$= (p + a) \frac{\widetilde{Y}(a)}{Y(a)} - (q + a) \frac{\widehat{Y}(a)}{Y(a)} . \tag{9.104b}$$

Setting $a=-p$ in (9.104b) or $a=p$ in (9.104a) and using the relation (9.50), we get the following expression

$$p-q+\widehat{u}-\widetilde{u}=(p-q)\frac{\widehat{\widetilde{\tau}}\tau}{\widehat{\tau}\widetilde{\tau}}, \tag{9.105}$$

which is the same relation as we got in the two-dimensional case. However, in the three-dimensional case the companion relation is absent, and hence we need the third lattice direction to eliminate the terms involving u from the analogous relations in the other directions. This leads to the bilinear relation:

$$(p-q)\widehat{\widetilde{\tau}}\,\overline{\tau}+(q-r)\widehat{\overline{\tau}}\,\widetilde{\tau}+(r-p)\widetilde{\overline{\tau}}\,\widehat{\tau}=0, \tag{9.106}$$

which is another parameterization of (8.29).

Finally, we also derive a higher-dimensional version of (9.49). For that purpose we introduce

$$S(a,b)=\boldsymbol{s}^T\,(b+\boldsymbol{L})^{-1}\boldsymbol{J}\,(\boldsymbol{I}_N+\boldsymbol{M}\boldsymbol{J})^{-1}\,(a+\boldsymbol{K})^{-1}\,\boldsymbol{r}, \tag{9.107}$$

where now in general $S(a,b)\neq S(b,a)$. Similar to the derivation of (9.47) we can find the relation

$$1+(p-a)S(a,b)-(p+b)\widetilde{S}(a,b)=\widetilde{V}(a)Y(b), \tag{9.108}$$

and analogous relations for the other lattice directions. Each of these three equations can be shifted in the two other directions. Taking appropriate ratios of these six equations we find the following three-dimensional equation:

$$\begin{aligned}&\frac{(1+(p-a)\widehat{S}-(p+b)\widehat{\widetilde{S}})}{(1+(q-a)\widetilde{S}-(q+b)\widehat{\widetilde{S}})}\,\frac{(1+(q-a)\overline{S}-(q+b)\widehat{\overline{S}})}{(1+(r-a)\widehat{S}-(r+b)\widehat{\overline{S}})}\\&\quad\times\frac{(1+(r-a)\widetilde{S}-(r+b)\widetilde{\overline{S}})}{(1+(p-a)\overline{S}-(p+b)\widetilde{\overline{S}})}=1.\end{aligned} \tag{9.109}$$

We can use once again the expansion formula for the τ-function:

$$\tau=\det\left(\boldsymbol{1}+\boldsymbol{M}\boldsymbol{J}\right)=1+\sum_{i=1}^{N}\left|N_{i,i}\right|+\sum_{i<j}\begin{vmatrix}N_{i,i}&N_{i,j}\\N_{j,i}&N_{j,j}\end{vmatrix}+\cdots+\det(\boldsymbol{N}), \tag{9.110}$$

where $\boldsymbol{N}=\boldsymbol{M}\boldsymbol{J}$. We can then exploit the Cauchy matrix structure of $\boldsymbol{M}$, using the determinantal property of a general $N\times N$ Cauchy matrix $\boldsymbol{A}$ as in (9.2). However, since the matrix $\boldsymbol{N}$ is a product of two non-square matrices, the determinants in expansion (9.110) can no longer be factorized in the usual fashion. In order to proceed we need the Cauchy–Binet formula (D.4) and, using subsequently the Cauchy identity, we can express the τ-function in explicit form.

9.8 Notes

The treatment of multi-soliton solutions in this chapter using Cauchy matrices has its origins in the direct linearization (DL) method (Nijhoff et al., 1983a; Quispel et al., 1984). The latter approach is a formal version of the inverse scattering method, which is based on singular linear integral equations with arbitrary measures and contours containing a Cauchy-type kernel. Such an approach using singular integrals also appeared in the context of the inverse scattering method, for example, in Fokas and Ablowitz (1981). The treatment of this chapter is a specific case of DL focusing on soliton solutions. In the case of the KdV equation, Cauchy matrices appeared, from the very beginning of soliton theory, as a way to express multi-soliton solutions (Kay and Moses, 1956; Gardner et al., 1967; Hirota, 1971; Wadati and Toda, 1972). The advantage of the DL approach is that it provides a unifying framework for an entire class of equations, e.g. the KdV class, characterized by the same Cauchy kernel. The connection of DL to Sato theory has been discussed in Willox (2000).

In this chapter we have provided N-soliton solutions of Q3, which can be called *rational soliton solutions*, and by degeneration these include solutions of all equations in the ABS list up to Q3. Also other types of solutions exist for these equations, such as elliptic N-soliton solutions and singular boundary solutions (Nijhoff and Atkinson, 2010; Atkinson and Joshi, 2013). For Q4, the situation is different, and in this case the analogue of the N-soliton solutions were constructed in Atkinson and Nijhoff (2010) by a different method, which also provides the singular boundary solution for that case, cf. Atkinson and Joshi (2013). A different, more recent development was the extension to matrix Cauchy kernels and use of Sylvester's theorem to find new classes of solutions, cf. Zhang and Zhao (2013); Xu et al. (2014).

The Cauchy approach was also applied to many other classes of integrable systems, such as the BSQ class (Nijhoff et al., 1984), the NLS class (Quispel et al., 1984; Walker, 2001) and the KP class and its reductions (Nijhoff et al., 1984).

There are still other kinds of solutions, which require different approaches (which are not treated here): periodic solutions of integrable lattice equations, using the techniques of finite-gap integration, have so far been investigated only in a few cases; results were obtained for H1 (Cao and Xu, 2012), pmKdV, (Bobenko and Pinkall, 1996b; Cao and Zhang, 2012), lattice KdV (Nijhoff and Enolskii, 1999) and the lattice SKdV (Hertrich-Jeromin et al., 2001). Furthermore, the inverse scattering formalism (see Section 3.8.1) was extended to lattice equations in Boiti et al. (2001); Butler and Joshi (2010); Butler (2012a,b). Scaling invariant solutions will be the subject of Chapter 10, while their reductions to OΔEs, discrete Painlevé equations, will be treated in their own right in Chapter 11.

Exercises

9.1 Assuming that all $\rho^0(k_\alpha)$ and c_β are nonzero, show that $U_{\alpha,\beta} = U_{\beta,\alpha}$, i.e. (9.11), from the definition (9.9c).

Hint: introduce the matrices $\boldsymbol{C} = \mathrm{diag}(c_1, \ldots, c_N)$ and $\boldsymbol{R} = \mathrm{diag}(\rho(k_1), \ldots, \rho(k_N))$, and rewrite $\boldsymbol{r} = \boldsymbol{R}\,\boldsymbol{e}$, $\boldsymbol{c}^T = \boldsymbol{e}^t\boldsymbol{C}$, where $\boldsymbol{e}^t = (1, 1, \ldots, 1)$, and express the matrix $\boldsymbol{M}$ in terms of the "bare" Cauchy matrix $\boldsymbol{M}^0 = (M^0_{\alpha,\beta}) = \big(1/(k_\alpha + k_\beta)\big)$.

9.2 (a) Combine both equations (9.13) and (9.14), using also (9.16), to derive the algebraic recurrence relation (9.17).
(b) Show that (9.18) can be derived from (9.17) by multiplying from the left by an appropriate vector.

9.3 Show that the PΔE obtained from the equality on the right-hand side of (9.53) is the same as the one obtained from (9.54).

9.4 (a) Show that the original parameters in (3.42c), let us call them p_1, q_1 to avoid confusion, are related to the parameters (p, P) and (q, Q), in (9.62) and (9.63) as follows:

$$p_1^2 = \frac{p^2 - b^2}{p^2 - a^2}, \qquad P = \frac{(b^2 - a^2)p_1}{1 - p_1^2}, \qquad q_1^2 = \frac{q^2 - b^2}{q^2 - a^2}, \qquad Q = \frac{(b^2 - a^2)q_1}{1 - q_1^2}. \tag{9.111}$$

(b) Show that equation (9.66) is equivalent to (3.42c) by using the connection between the parameters as given in (9.111) where, as before, p_1, q_1 are the parameters as in (3.42c), and where the dependent variable u_1 is related to the dependent variable of (9.66) via the scaling $u = (b^2 - a^2)u_1$.

9.5 Derive equation (9.94) from the system (9.93) and show that it can also be written in the equivalent forms:

$$(p+\widehat{\overline{u}})(q-r+\overline{u}-\widehat{u})+(q+\widetilde{\overline{u}})(r-p+\widetilde{u}-\overline{u})+(r+\widehat{\widetilde{u}})(p-q+\widehat{u}-\widetilde{u}) = 0. \tag{9.112}$$

Give also its linearized version.

9.6 Derive from the first equality of (9.104) and its counterparts in two other pairs of lattice directions, a 3D lattice equation for the quantity $V(a)$, for a a fixed parameter, and similarly from the second equality (and its counterparts) a lattice equation for $Y(a)$ with a fixed. Show that these generalize the lattice potential MKP equation (3.110) and (3.111). By choosing $a = p$ show that these lattice equations reduce to a four-term 3D lattice equation (which can be shown to be related to equation (3.107e)).

9.7 Derive the relation (9.108) from the soliton structure, by introducing the parameter-dependent vectors

$$\boldsymbol{I}(a) \equiv (\boldsymbol{u} + \boldsymbol{M}\boldsymbol{J})^{-1}\,(a + \boldsymbol{K})^{-1}\,\boldsymbol{r}, \quad {}^t\boldsymbol{I}(b) \equiv \boldsymbol{s}^T\,(b + \boldsymbol{L})^{-1}\,(\boldsymbol{u} + \boldsymbol{J}\boldsymbol{M})^{-1}, \tag{9.113}$$

and deriving the following relations:

$$\widetilde{\boldsymbol{u}}(a) = (p - a)\boldsymbol{u}(a) + \widetilde{V}(a)\boldsymbol{u}_0\,,$$
$${}^t\boldsymbol{u}(b) = (p + b)\,{}^t\widetilde{\boldsymbol{u}}(b) - Y(b)\,{}^t\widetilde{\boldsymbol{u}}_0,$$

and by multiplying the first from the left by $s^T(b+L)^{-1}J$, or the second from the right by $J(a+K)^{-1}\widetilde{r}$.

Hint: Use can be made of the relations

$$s^T J u(a) = 1 - V(a), \quad {}^t u(b) J r = 1 - Y(b) .$$

9.8 Consider the discrete KP soliton solutions given by (9.91). Set

$$r = R e, \quad s^T = e^t S \quad \text{where} \quad e^t = (1, 1, \ldots, 1), \tag{9.114}$$

with $R = \mathrm{diag}(\rho_1, \ldots, \rho_N)$ and $S = \mathrm{diag}(\sigma_1, \ldots, \sigma_{N'})$ being diagonal matrices containing the plane-wave factors.

(a) Use the Leibniz rule and the fact that

$$\partial M = (\partial R) R^{-1} M + M S^{-1} (\partial S)$$

to derive the formula

$$\partial U_{\alpha,\beta} = {}^t u_\beta \left[S^{-1} (\partial S) J + J (\partial R) R^{-1} \right] u_\alpha , \tag{9.115}$$

where ∂ denotes any differential operator (with regard to a yet unspecified variable) acting only on the plane-wave factors ρ and σ.

(b) Let T denote any shift operator (with regard to a yet unspecified discrete variable) acting only on the plane-wave factors. Derive the following formula

$$T U_{\alpha,\beta} - U_{\alpha,\beta} = (T\, {}^t u_\beta) \left(J (TR) R^{-1} - (TS)^{-1} S J \right) u_\alpha . \tag{9.116}$$

Show that a continuum limit reduces (9.116) to (9.115).

(c) Apply the formula in part (b) to the case where T denotes the shift in the variable n using the dependence given in (9.86).

10

Similarity reductions of integrable PΔEs

Symmetries of equations give us an immense amount of information about their solutions. By *symmetries* we mean transformations of the independent and dependent variables that preserve the form of the equations. When we study symmetries, it should be noted that just because an equation possesses a symmetry, it does not necessarily imply that all solutions possess this symmetry. However, it is natural to search for those special solutions of the equation which do possess this symmetry, and if we find any such solutions we can characterize them by the symmetry of the equation.

In this chapter we will consider the reduction of integrable PDEs and their lattice analogues subject to a special kind of symmetry called *scaling symmetry*. The corresponding symmetry reductions, which we will refer to as *similarity reductions*, lead to lower-dimensional equations. These equations are usually nonautonomous and in many cases they will turn out to be Painlevé equations (see Chapter 11) or their discrete analogues.

We first show how similarity reductions are implemented for the continuous case, and then we discuss how the procedure can be formulated for lattice equations, and how to obtain the reduced equations from this approach in explicit form. As will become clear later, it turns out that some aspects are quite different in the discrete case and require a reformulation, which offers a new point of view even for the continuous case.

10.1 Introduction to dimensional reductions

In this section we provide several introductory examples of how symmetries can be used to obtain dimensional reductions of PDEs.

10.1.1 The heat equation

As the first example, consider the heat equation

$$u_t + u_{xx} = 0. \tag{10.1}$$

This equation has both translational and scaling symmetries. (A detailed discussion of its symmetries can be found in Olver (2000); see Exercise 3.3 on page 190.)

By a *translational* symmetry, we mean that if $u(x,t)$ is a solution of (10.1) then so is $u(x+\alpha, t+\beta)$, for arbitrary constants α, β. Not all solutions are invariant under this symmetry, even though the equation itself is. If we now search for solutions that are invariant, then we are led to consider solutions of the form $u(x,t) = U(x-ct)$, where in terms of the symmetry, we have $c = \alpha/\beta$. Letting $z = x - ct$, we obtain an ODE for $U(z)$ which is of the form

$$-c\,U' + U'' = 0.$$

Notice that we have reduced the dimension (number of independent variables) of the problem from two (in equation (10.1)) to one in the ODE.

Equation (10.1) also has a *scaling* symmetry; that is, if we scale the variables by $x \mapsto x' = \alpha x$ and $t \mapsto t' = \beta t$ and $u(x,t) \mapsto u'(x',t')$, and choose $\beta = \alpha^2$, then the equation becomes

$$u'_{t'} + u'_{x'x'} = 0.$$

So the equation is invariant under this transformation. Searching again for solutions that remain invariant under this change, we find solutions of the form $u(x,t) = U(z)$, where $z := x/\sqrt{t}$, which satisfies the ODE

$$-\tfrac{1}{2}zU' + U'' = 0.$$

Again, this is a dimensional reduction. Note that $u(x,t) = U(x/\sqrt{t})$ satisfies the constraint

$$2t\,u_t + x\,u_x = 0.$$

10.1.2 The KdV equation

Let us next consider the similarity reduction of a nonlinear PDE, namely of the KdV equation (see Olver, 2000, p. 193)

$$u_t + u_{xxx} + 6uu_x = 0.$$

This also has translational symmetry and the traveling wave reduction $u(x,t) = U(x-ct)$ leads to

$$-c\,U' + U''' + 6UU' = 0.$$

The linear part of the KdV equation suggests that we may also try scaling, in this case with $x \mapsto x' = \alpha x$ and $t \mapsto t' = \beta t$ with $\beta = \alpha^3$. However, this is not enough, because we also have to scale $u \mapsto u' = \alpha u$ in order for the equation to remain invariant. Under this symmetry, solutions of the form

$$u(x,t) = t^{-2/3}\,U(z), \qquad \text{where } z := x/t^{1/3}, \tag{10.2}$$

remain invariant. Substituting these into the KdV equation, we obtain the ODE

$$U - \tfrac{1}{3} z U' + U''' + 6UU' = 0. \tag{10.3}$$

This third-order equation is in fact the second Painlevé equation (2.54) in disguise. The connection between soliton equations and Painlevé equations by similarity reductions was discovered in Ablowitz and Segur (1977). This led to the Ablowitz–Ramani–Segur (ARS) conjecture, stating that nonlinear evolution equations which are solvable by the inverse scattering transform, reduce (by similarity reduction) to ODEs of Painlevé type (Ablowitz et al., 1978, 1980a,b).

We note that if u has the form (10.2) then it also satisfies the constraint equation

$$3t u_t + x u_x + 2u = 0,$$

which can also be written in terms of a vector field X as

$$\boldsymbol{X} u = -2u, \quad \text{where} \quad \boldsymbol{X} := 3t\partial_t + x\partial_x,$$

which is the generator of the scaling symmetry.

10.1.3 Similarity reduction of the KdV class of equations

In this section we consider similarity reductions of PDEs, specifically of the following equations belonging to KdV family, namely

$$w_t = w_{xxx} + 3w_x^2, \tag{10.4a}$$

$$v_t = v_{xxx} - 3\frac{v_x v_{xx}}{v}, \tag{10.4b}$$

$$\frac{z_t}{z_x} = \{z, x\} := \frac{z_{xxx}}{z_x} - \frac{3}{2}\frac{z_{xx}^2}{z_x^2}, \tag{10.4c}$$

which are the potential KdV equation, the potential mKdV equation (note that the variable $\bar{v} = (\log v)_x$ obeys the mKdV equation (1.34)), and the SKdV equation, respectively. (Recall that the third-order operator $\{z, x\}$ is called the Schwarzian derivative.) We will consider similarity reductions of these equations from a perspective that will be useful for the lattice case.

In order to carry out a similarity reduction we first need to find a similarity variable ξ, which is invariant under scaling symmetry. Then we impose the condition that the solution of the PDE only depends on this similarity variable.

In the case of (10.4c) the similarity variable is $\xi = xt^{-1/3}$, because every t-derivative is balanced by three x-derivatives in the equation. To find the similarity reduction we therefore assume

$$z(x, t) = t^{\bar{\mu}} Z(\xi), \tag{10.5}$$

where $\bar{\mu}$ is an arbitrary constant. When this is substituted into (10.4c) and everything written in terms of ξ we get

$$\bar{\mu} Z - \tfrac{1}{3}\xi Z' = Z''' - \tfrac{3}{2}\frac{Z''^2}{Z'} \tag{10.6}$$

where the prime denotes derivative with respect to ξ.

The above approach is difficult to generalize to the discrete case because changes of independent discrete variables are difficult to define on the lattice while keeping the lattice structure. We therefore take a different approach, which is based on the observation that (10.5) solves the linear nonautonomous PDE

$$3\bar{\mu} z = x z_x + 3t z_t, \tag{10.7}$$

which we call a **similarity constraint**.

Imposing (10.7) on the solutions of (10.4c) and eliminating the t-derivatives, we obtain the following third-order ODE in x

$$\{z, x\} = \frac{\bar{\mu}}{t}\frac{z}{z_x} - \frac{x}{3t}, \tag{10.8}$$

where t now plays the role of a parameter of the equation. In order to see its relationship to the P_{II} equation, differentiate equation (10.8) and use the Cole–Hopf transformation $\bar{v} = \frac{1}{2} z_{xx}/z_x$. The result is

$$\bar{v}_{xx} = 2\bar{v}^3 - \frac{x}{3t}\bar{v} + \frac{\mu}{3t}, \tag{10.9}$$

where $\mu = (3\bar{\mu} - 1)/2$.

We can apply the similarity constraint approach also directly to (10.1b) but then we must use the constraint equation[1]

$$\mu v = x v_x + 3t v_t. \tag{10.10}$$

Imposing this constraint raises the nontrivial issue of compatibility between the two PDEs for one dependent variable, namely constraint and the equation (see Exercise 10.1).

The two types of symmetries translation and scaling discussed earlier also occur in the discrete setting. However, where the traveling wave reduction can be reinterpreted explicitly as a staircase reduction, there is no direct analogue of the scaling reduction, because ratios of powers of n and m no longer provide integer shifts on a lattice. For example, the translational symmetry $u_{n+2,m} = u_{n,m+1}$ provides a staircase reduction $u_{n,m} = U_{n-2m}$. However, an attempt to introduce scaled independent variables, such as $(n/\sqrt{m})$ would fail, because such variables would not make sense on an integer grid.

[1] The usual method, which starts from mKdV rather than potential mKdV, could be formulated in terms of the constraint $\bar{v} + x\bar{v}_x + 3t\bar{v}_t = 0$, which contains no parameters. This leads first to a third-order ODE in x which then has to be integrated once to yields the P_{II} equation, in which the parameter appears as an integration constant.

Because of this difficulty, we do the next best thing and focus on *similarity constraints* on a lattice, as well as enlarge the search for symmetries to include derivatives with respect to the *lattice parameters*.

10.2 Compatibility of lattice constraint with quad equations

The key idea in finding reductions of lattice equations is to use similarity constraints that are compatible with the given lattice equation (Nijhoff and Papageorgiou, 1991). This is motivated by the continuous examples, but it turns out to be more useful to use constraints themselves rather than seeking explicit similarity variables.

We will first concentrate on quad equations of the form

$$\mathcal{F}(u, \widetilde{u}, \widehat{u}, \widehat{\widetilde{u}}) = 0. \tag{10.11}$$

In this case a possible form for the constraints is

$$\mathcal{G}(u, \widetilde{u}, \underset{\sim}{u}, \widehat{u}, \underset{\frown}{u}) = 0, \tag{10.12}$$

i.e. an equation corresponding to a configuration of vertices forming a cross. Thus we are led to a coupled system which can be symbolically represented as in Figure 10.1.

By posing the two equations (10.11) and (10.12) on the variable u, we effectively reduce the lattice equation to a finite-dimensional system.[2] This can be seen by the fact that all points of the lattice can be computed from a finite set of lattice points, as for example from the one illustrated in Figure 10.2.

In general the iteration using these two equations may lead to multivalued determinations: after some initial steps the vertices can be calculated via different routes, leading to potential inconsistencies. However, this is not acceptable since we are looking for single valued solutions of the lattice equation.

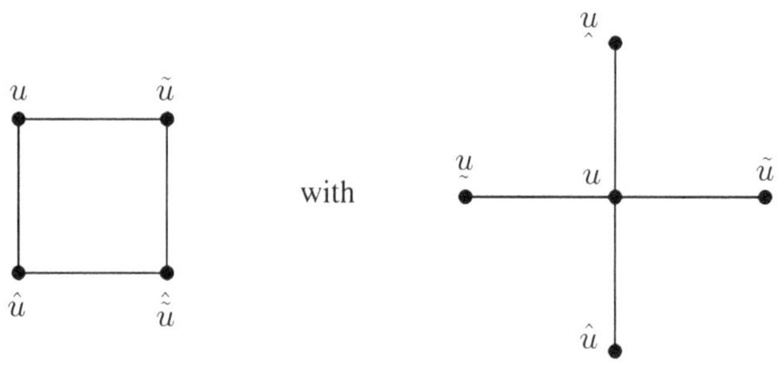

Figure 10.1 The diagram for the equations and its similarity constraint.

[2] This is the lattice analogue of *characteristics* in PDE theory.

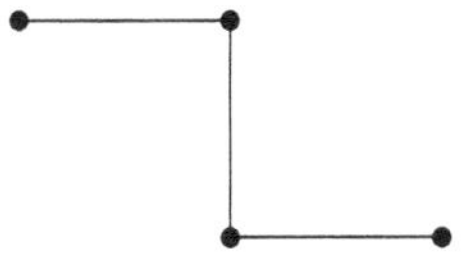

Figure 10.2 The local initial values.

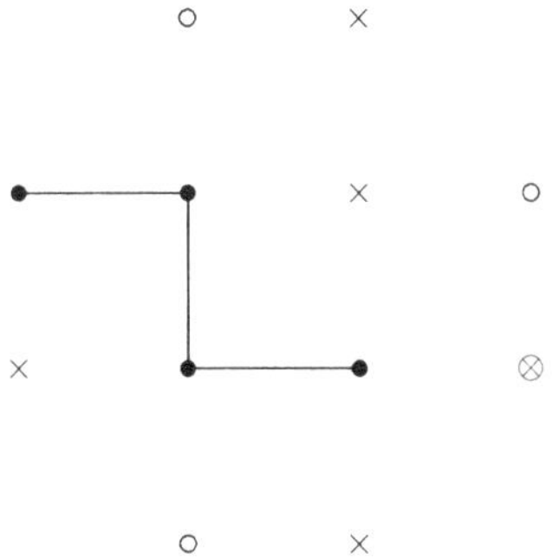

Figure 10.3 Compatibility diagram. $\times$: points that are calculated with the lattice equation, $\circ$: points that are calculated with the similarity constraint and $\otimes$: points at which the evaluation can be multi-valued.

The diagram in Figure 10.3 shows how to check the compatibility. Starting from the initial values at black dots one computes the other values using (10.11) and (10.12) in tandem. At the eighth step, one reaches a point of possible conflict indicated in the diagram by $\otimes$. The equations given by $\mathcal{F}$ and $\mathcal{G}$ must be such that there is no conflict, i.e. both ways of computing the value at this point must coincide.

In the following we show that, starting from integrable quad equations given by $\mathcal{F}$, it is possible to find a function $\mathcal{G}$ such that consistency of the two equations is verified.

10.3 The linear case

In this section we will exhibit the main features of the similarity reduction procedure on a linear multidimensionally consistent system, namely the linear PΔE

$$(p+q)(\widehat{f}-\widetilde{f}) = (p-q)(\widehat{\widetilde{f}}-f). \tag{10.13}$$

A solution of this equation is given by the (discrete Fourier-type) integral representation:

$$f(n,m) = \int_\Gamma \left(\frac{p+k}{p-k}\right)^n \left(\frac{q+k}{q-k}\right)^m d\mu(k), \tag{10.14}$$

where the integration involves an arbitrary measure $d\mu(k)$ (as long as the measure does not depend on the discrete variables n, m or the continuous variables p, q) and the integral is taken over an arbitrary contour in the complex k-plane.[3] (The plane-wave factors considered as discrete exponentials form a basis for discrete holomorphic functions (Mercat, 2004).)

We note that the following differential-difference equations can be derived for the solution $f(n, m)$ of (10.14):

$$p\frac{\partial f}{\partial p} = \tfrac{1}{2}n\left(\underset{\sim}{f} - \widetilde{f}\right), \tag{10.15a}$$

$$q\frac{\partial f}{\partial q} = \tfrac{1}{2}m\left(\underset{\frown}{f} - \widehat{f}\right). \tag{10.15b}$$

These equations are, by construction, compatible with the lattice equation (10.13), which can be verified also by direct computation.

Observe that we have used the continuous dependence on the parameters p, q to derive (10.15), so they are PDEs in p, q. This is the dual to the discrete equation (10.13), in which n, m were the independent variables of the PΔE. We will use this duality to get similarity constraints.

Remark 10.3.1 Note that we can use equations (10.15) and (10.13) together also to derive an equation that only contains partial derivatives in p, q without any shifts. First we derive the following coupled system in terms of f and $g := \widehat{f} - \widetilde{f}$:

$$\left[\tfrac{1}{2}\partial_p\partial_q + (n\partial_q q - m\partial_p p)\frac{1}{p^2 - q^2}\right] g + \frac{p-q}{p+q}\,\partial_p\partial_q f = \frac{1}{(p+q)^2}\,(p\partial_p - q\partial_q) f, \tag{10.16a}$$

$$\left[\tfrac{1}{2}\partial_p\partial_q - \frac{1}{p^2 - q^2}(nq\partial_q - mp\partial_p\right] f = \frac{nm}{p^2 - q^2} g, \tag{10.16b}$$

with the lattice parameters p and q as independent variables. After eliminating g one arrives at

$$\partial_p\partial_q\,(p^2 - q^2)\partial_p\partial_q\, f = 4(n\partial_q - m\partial_p)\frac{1}{p^2 - q^2}\,(nq^2\partial_q - mp^2\partial_p)\, f. \tag{10.17}$$

Here the lattice variables n and m now play the role of parameters of the equation. Equation (10.17) was given in Tongas (2007) and in Lobb and Nijhoff (2009), and is an example of

[3] Note that the measure may contain additional discrete variables so that multidimensional consistency can be imposed, by setting more generally:

$$d\mu(k) = \left[\prod_\nu \left(\frac{p_\nu + k}{p_\nu - k}\right)^{n_\nu}\right] d\mu^0(k),$$

involving an arbitrary number of discrete variables n_ν with associated lattice parameters p_ν. Note also that (10.13) is satisfied by the plane-wave factors $\rho(k)$ of (5.4).

what we will later call a *generating PDE*. The main feature of this equation, which also follows by construction, is that this PDE exhibits a continuous analogue of the property of multidimensional consistency: one can simultaneously impose the same PDE but in different pairs of independent variables (changing only the corresponding parameters).

Now we consider the issue of similarity reductions of the equations (10.13) and (10.17) (cf. Tongas, 2007). In order to find the discrete similarity constraint we impose the requirement that the equation should be invariant under the action of the similarity operator $\boldsymbol{X} = p\partial_p + q\partial_q$ (in the terminology of symmetry analysis it is said that the vector field $\boldsymbol{X}$ is the *generator of the scaling symmetry*). If we let $\boldsymbol{X}$ act on the quadrilateral expression $Q := (p+q)(\widehat{f} - \widetilde{f}) - (p-q)(\widehat{\widetilde{f}} - f)$ (cf. (10.13)) we get

$$\begin{aligned} \boldsymbol{X}Q &= Q + (p+q)\boldsymbol{X}(\widehat{f} - \widetilde{f}) - (p-q)\boldsymbol{X}(\widehat{\widetilde{f}} - f) \\ &= Q + (p+q)(\widehat{v} - \widetilde{v}) - (p-q)(\widehat{\widetilde{v}} - v), \end{aligned} \tag{10.18}$$

where $v := (p\partial_p + q\partial_q)f$. We see that we get the symmetry $\boldsymbol{X}Q = Q$ if we set

$$\widehat{v} = \widetilde{v}, \quad \widehat{\widetilde{v}} = v. \tag{10.19}$$

A solution to these equations is given by[4]

$$v = \mu + \lambda(-1)^{n+m}, \tag{10.20}$$

where μ, λ are constant in n, m. Furthermore, $\boldsymbol{X}$ must also annihilate equations (10.15) and from this it follows that $\partial_p v = \partial_q v$ and therefore μ, λ are constants in p, q as well.

Recalling the definition of v we get from (10.20)

$$(p\partial_p + q\partial_q)f = \mu + \lambda(-1)^{n+m} \tag{10.21}$$

then from (10.15)

$$\frac{n}{2}(\widetilde{f} - \underset{\sim}{f}) + \frac{m}{2}(\widehat{f} - \underset{\frown}{f}) + \mu + \lambda(-1)^{n+m} = 0. \tag{10.22}$$

This is the discrete similarity constraint, derived from the assumption $\boldsymbol{X}Q = Q$, which is compatible with the lattice equation $Q = 0$. (More generally we could have $\boldsymbol{X}Q = F(Q)$ as long as $F(0) = 0$.) Thus we have the compatible pair (10.22) relating values on a star-shaped configuration and (10.13) which is a quad equation; see Figure 10.1. These allow local initial value problems defined on configurations such as Figure 10.2 to be iterated through the lattice leading to a single-valued function $f_{n,m}$ of the lattice variables.

[4] There can be more general solutions for v than (10.20) such as adding a term of the form σf to v but we will not consider that possibility here.

Discrete reduction

We now proceed with the actual symmetry reduction, which is to derive from the system comprising (10.13) together with the similarity constraint (10.25) an OΔE in terms of only one single shift-operation, say the tilde-shift. Then we would need to eliminate the hat-shift by using these two equations in conjunction. (Alternatively we could aim to eliminate the tilde-shift.)

The OΔE we are looking for will be in terms of the reduced variable

$$x := \widehat{\widetilde{f}} - f,$$

and its shifts in the tilde direction. Setting for convenience $t := (p-q)/(p+q)$, we find the following equations from (10.13) and its backward shift in m-direction:

$$\begin{aligned} \widehat{f} - \widetilde{f} &= t(\widehat{\widetilde{f}} - f), \\ \widetilde{f} - \underaccent{\hat}{f} &= t^{-1}(f - \underaccent{\hat}{\widetilde{f}}). \end{aligned}$$

The computations become simpler if we use the intermediate variables $\mathfrak{a} := \widetilde{f} - \underaccent{\tilde}{f}$ and $\mathfrak{b} := \widehat{f} - \underaccent{\hat}{f}$. Then and by adding the above two relations we find:

$$\mathfrak{b} - t^{-1}\widetilde{\mathfrak{b}} = (t - t^{-1})\,x. \tag{10.23}$$

Next for $\widetilde{f} - \underaccent{\tilde}{f}$ we have, using again (10.13) and the definition of x

$$\mathfrak{a} = \widetilde{f} - \underaccent{\tilde}{f} = \widehat{f} - \underaccent{\tilde}{f} - (\widehat{f} - \widetilde{f}) = \underaccent{\tilde}{x} - t\,x. \tag{10.24}$$

Note also that equation (10.22) can be written as

$$\frac{n}{2}\mathfrak{a} + \frac{m}{2}\mathfrak{b} + \mu + \lambda(-1)^{n+m} = 0. \tag{10.25}$$

Now solving $\mathfrak{b}$ from (10.25) and $\mathfrak{a}$ from (10.24), equation (10.23) becomes

$$\mu(t^{-1} - 1) - \lambda(-1)^{n+m}(1 + t^{-1}) + \tfrac{n}{2}(t\,x - \underaccent{\tilde}{x}) - \tfrac{n+1}{2t}(t\,\widetilde{x} - x) = \tfrac{m}{2}(t - t^{-1})x. \tag{10.26}$$

This is a second-order linear nonautonomous OΔE equation in the independent variable n, obtained as a reduction of PΔE for f (10.13). The equation has four parameters t, μ, λ, m of which one can be scaled away. We could have equally well derived a similar equation with m as the independent variable in which n would be a parameter.

We can apply the same method on the equations (10.15) and obtain

$$\begin{aligned} p\partial_p\mathfrak{a} = p\partial_p(\widetilde{f} - \underaccent{\tilde}{f}) &= -\frac{n+1}{2}\widetilde{\mathfrak{a}} + \frac{n-1}{2}\underaccent{\tilde}{\mathfrak{a}} \\ &= \frac{m}{2}(\widetilde{\mathfrak{b}} - \underaccent{\tilde}{\mathfrak{b}}) = \frac{m}{2}t\left[\mathfrak{b} - (t - t^{-1})x\right] - \frac{m}{2}\left[t^{-1}\mathfrak{b} + (t - t^{-1})\,\underaccent{\tilde}{x}\right] \\ &= (t - t^{-1})\left[-\frac{n}{2}\mathfrak{a} - \mu - \lambda(-1)^{n+m}\right] - \frac{m}{2}(t^2 - 1)(2x + t^{-1}\mathfrak{a}), \end{aligned}$$

and similarly

$$p\partial_p(tx) = \frac{n}{2}\left(2t\mathfrak{a} + x(t^2-1)\right) + \frac{m}{2}(t^2-1)x + (t-1)\mu + (t+1)\lambda(-1)^{n+m}.$$

Noting again that $g = tx$ this yields a coupled system of two first-order equations in the derivatives with respect to p, namely:

$$\left(p\partial_p + N\tau\right)\mathfrak{a} = -(mg+\nu)\tau\ , \tag{10.27a}$$

$$\left(p\partial_p - N\tau\right)g = nt\mathfrak{a} + t\nu - \widetilde{\nu}, \tag{10.27b}$$

in which we have abbreviated: $N := (n+m)/2$, $\tau := t - t^{-1}$, and $\nu := \mu + \lambda(-1)^{n+m}$. This is a system of ODEs in the independent variable p, with parameters n, m, μ, λ of which one can be scaled away. A similar system of equations can be derived for q as independent variable. From this system we can derive in a straightforward manner a second-order ODE for either g or a.

In conclusion, we were able to derive the OΔE (10.26) and the associated differential-difference equations (10.27). The dual nature of these results, in which the lattice parameters p, q and lattice variables n, m interchangeably play the role of independent variables and parameters of the equations, demonstrates the interplay between the continuous and the discrete in these reductions.

10.4 Similarity constraints for the lattice KdV family

We will now investigate similarity reductions for the nonlinear case of the KdV family of equations. We will pursue this by analogy with the linear case, keeping in mind that the nonlinear case is much more elaborate and will involve three different equations (possessing the same linearized limit). The three equation we consider are the lattice potential KdV (3.6), the lattice modified KdV (3.11) and the lattice Schwarzian KdV (3.13). These form a family on the basis of the Miura transforms connecting them (3.12), (3.14). These equations are known to have (among others) the following symmetries, respectively

$$\frac{\partial w}{\partial p} = -\frac{2np}{\underset{\sim}{w} - \widetilde{w}}\ , \qquad \frac{\partial w}{\partial q} = -\frac{2mq}{\underset{\wedge}{w} - \widehat{w}}\ , \tag{10.28a}$$

$$\frac{\partial v}{\partial p} = -v\frac{n}{p}\frac{\widetilde{v} - \underset{\sim}{v}}{\widetilde{v} + \underset{\sim}{v}}\ , \qquad \frac{\partial v}{\partial q} = -v\frac{m}{q}\frac{\widehat{v} - \underset{\wedge}{v}}{\widehat{v} + \underset{\wedge}{v}}\ , \tag{10.28b}$$

$$\frac{\partial z}{\partial p} = -\frac{2n}{p}\frac{(\widetilde{z} - z)(z - \underset{\sim}{z})}{\widetilde{z} - \underset{\sim}{z}}\ , \qquad \frac{\partial z}{\partial q} = -\frac{2m}{q}\frac{(\widehat{z} - z)(z - \underset{\wedge}{z})}{\widehat{z} - \underset{\wedge}{z}}\ . \tag{10.28c}$$

As symmetries these equations are compatible with the corresponding KdV equation, i.e. after computing p-derivative of one of the KdV equations and then using the above to eliminate the derivatives we get an expression that vanishes by virtue of the equation (and its shifts). Furthermore, comparing with the skew semi-continuum limits (5.26, 5.27, 5.28) we find that the operator ∂_p has the same effect as $n\partial_\tau$. (The reason why this replacement could work can be understood by studying the derivatives of the plane-wave factor.)

10.4.1 The discrete similarity constraint

We will now impose the scaling invariance $\boldsymbol{X}Q = 0$, where as before, the vector field is given by $\boldsymbol{X} = p\partial_p + q\partial_q$, on solutions of the lattice mKdV equation

$$Q := pv\widehat{v} + q\widehat{v}\widehat{\widetilde{v}} - qv\widetilde{v} - p\widetilde{v}\widehat{\widetilde{v}} = 0. \tag{10.29}$$

This yields

$$\begin{aligned} \boldsymbol{X}Q = Q &+ p(\boldsymbol{X}v)\widehat{v} + pv(\boldsymbol{X}\widehat{v}) + q(\boldsymbol{X}\widehat{v})\widehat{\widetilde{v}} + q\widehat{v}(\boldsymbol{X}\widehat{\widetilde{v}}) \\ &- q(\boldsymbol{X}v)\widetilde{v} - qv(\boldsymbol{X}\widetilde{v}) - p(\boldsymbol{X}\widetilde{v})\widehat{\widetilde{v}} - p\widetilde{v}(\boldsymbol{X}\widehat{\widetilde{v}}). \end{aligned}$$

Setting $\boldsymbol{X}v = \nu v$ for some ν to be determined, we get

$$\boldsymbol{X}Q = (1 + \widehat{\widetilde{\nu}} + \widehat{\nu})Q + (\nu - \widehat{\widetilde{\nu}})v(p\widehat{v} - q\widetilde{v}) + (\widehat{\nu} - \widetilde{\nu})\widetilde{v}(qv + p\widehat{\widetilde{v}}).$$

For this to vanish on all solutions of $Q = 0$ we must once again have $\nu = \widehat{\widetilde{\nu}}$ and $\widehat{\nu} = \widetilde{\nu}$, which have the solution (10.20), where μ, λ are independent of n, m.

To show that ν (and therefore μ, λ) is independent of p, we operate by $\boldsymbol{X}$ on the p-equation of (10.28b) and find

$$\boldsymbol{X}(\partial_p v) = (\partial_p \nu)v + (\nu - 1)\partial_p v = -(\nu - 1)v\frac{n}{p}\frac{\widetilde{v} - \underset{\sim}{v}}{\widetilde{v} + \underset{\sim}{v}}.$$

Thus we conclude that $\partial_p \nu = 0$. A similar argument applied to the q-counterpart leads to $\partial_q \nu = 0$.

Now we have established that

$$\boldsymbol{X}v = (p\partial_p + q\partial_q)v = (\mu + \lambda(-1)^{n+m})v \tag{10.30}$$

can be used as a similarity constraint for the lattice mKdV equation (3.11) and its differential-difference companion (10.28b). The discrete version of the above constraint can be obtained immediately using (10.28b) and its q-counterpart, which yields

$$n\frac{\widetilde{v} - \underset{\sim}{v}}{\widetilde{v} + \underset{\sim}{v}} + m\frac{\widehat{v} - \underset{\wedge}{v}}{\widehat{v} + \underset{\wedge}{v}} + \mu + \lambda(-1)^{n+m} = 0. \tag{10.31}$$

Note that in contrast to the continuous form of the constraint (10.10) this is nonlinear. Furthermore the equation (10.31) is in the form of (10.12) and it can be shown to be compatible with the lattice equation (10.29) according to the diagram of Figure 10.3.

So far we have constructed the continuous (10.30) and discrete (10.31) similarity constraint for the lattice mKdV. The same can be done for lattice potential KdV and for the lattice SKdV. In summary, for these three equations we have the continuous similarity constraints

$$p\frac{\partial w}{\partial p}+q\frac{\partial w}{\partial q}=(1+2\lambda(-1)^{n+m})w, \tag{10.32a}$$

$$p\frac{\partial v}{\partial p}+q\frac{\partial v}{\partial q}=(\mu+\lambda(-1)^{n+m})v, \tag{10.32b}$$

$$p\frac{\partial z}{\partial p}+q\frac{\partial z}{\partial q}=-(1-2\mu)z, \tag{10.32c}$$

and their discrete nonlinear forms

$$\frac{np^2}{2p+\underset{\sim}{w}-\widetilde{w}}+\frac{mq^2}{2q+\underset{\wedge}{w}-\widehat{w}}+(\lambda(-1)^{n+m}+\tfrac{1}{2})(w-np-mq)=0, \tag{10.33a}$$

$$n\frac{\widetilde{v}-\underset{\sim}{v}}{\widetilde{v}+\underset{\sim}{v}}+m\frac{\widehat{v}-\underset{\wedge}{v}}{\widehat{v}+\underset{\wedge}{v}}+\mu+\lambda(-1)^{n+m}=0, \tag{10.33b}$$

$$2n\frac{(\widetilde{z}-z)(z-\underset{\sim}{z})}{\widetilde{z}-\underset{\sim}{z}}+2m\frac{(\widehat{z}-z)(z-\underset{\wedge}{z})}{\widehat{z}-\underset{\wedge}{z}}-(1-2\mu)z=0. \tag{10.33c}$$

The λ,μ have been chosen so that they respect the Miura transforms connecting the equations (see Exercise 10.2). Note that whereas the similarity constraint for the mKdV equation contains two parameters, that of the KdV and SKdV equations allow for only one parameter each.

10.4.2 Implementing the similarity constraint

In this section we show how to use the similarity constraint to derive a reduction of the original equation. To illustrate this process we focus on the mKdV equation. This proceeds very much in analogy to the linear case. In the mKdV case the appropriate reduced variables are

$$x:=\frac{v}{\widehat{\widetilde{v}}}\,,\quad y:=\frac{\widetilde{v}}{\widehat{v}}\,. \tag{10.34}$$

In terms of these variables, the mKdV equation (10.29) can be rewritten as a fractional linear relation:

$$x=\frac{py-q}{p-qy}\quad\Leftrightarrow\quad y=\frac{px+q}{p+qx}\,. \tag{10.35}$$

Also in analogy with the linear case we define auxiliary variables

$$\mathfrak{a}:=\frac{\widetilde{v}-\underset{\sim}{v}}{\widetilde{v}+\underset{\sim}{v}}\,,\quad \mathfrak{b}:=\frac{\widehat{v}-\underset{\wedge}{v}}{\widehat{v}+\underset{\wedge}{v}}\,, \tag{10.36}$$

in terms of which the differential-difference equation (10.28b) and its q-counterpart can be written as

$$p\frac{\partial}{\partial p}\log v=-n\mathfrak{a}\ ,\quad q\frac{\partial}{\partial q}\log v=-m\mathfrak{b}. \tag{10.37}$$

From this we also get the conservation law

$$m\,p\,\partial_p\mathfrak{b}=n\,q\,\partial_q\mathfrak{a}. \tag{10.38}$$

Having defined the relevant variables we can proceed using the similarity constraint (10.31), which for convenience we write as

$$n\mathfrak{a}+m\mathfrak{b}+\mu+\lambda(-1)^{n+m}=0, \tag{10.39}$$

to derive separately a reduction on the one hand to an ordinary difference equation (OΔE), and on the other hand to ordinary differential equation (ODE).

In order to derive the OΔE, consider the lattice mKdV and its hat-downshift

$$\begin{aligned}pv\widehat{v}+q\widehat{\widetilde{v}}&=qv\widetilde{v}+p\widetilde{v}\,\widehat{\widetilde{v}},\\ p\underset{\frown}{v}v+qv\widetilde{v}&=q\underset{\frown}{v}\underset{\frown}{\widetilde{v}}+p\underset{\frown}{\widetilde{v}}\,\widetilde{v}.\end{aligned}$$

Adding these relations and dividing by the combinations $\widehat{v}+\underset{\frown}{v}$ and $\widehat{\widetilde{v}}+\underset{\frown}{\widetilde{v}}$ we can write the resulting expression in terms of x, y, $\mathfrak{b}$ as follows:

$$(px+q)\widetilde{\mathfrak{b}}+px=(py-q)\mathfrak{b}+py. \tag{10.40}$$

In the same way one can derive

$$(qx+p)\widehat{\mathfrak{a}}+qx=(q/y-p)\mathfrak{a}+q/y. \tag{10.41}$$

Furthermore, we want to express $\mathfrak{a}$ in terms of the reduced variables, which can be done by noting that $y/\underset{\sim}{x}=\widetilde{v}/\underset{\sim}{v}$, leading to

$$\mathfrak{a}=\frac{y-\underset{\sim}{x}}{y+\underset{\sim}{x}}=\frac{p(x-\underset{\sim}{x})+q(1-x\underset{\sim}{x})}{p(x+\underset{\sim}{x})+q(1+x\underset{\sim}{x})}. \tag{10.42}$$

We can now solve $\mathfrak{b}$ from (10.39) and substitute into (10.40) and then eliminate $\mathfrak{a}$ by (10.42) and y by (10.35) and obtain a closed-form OΔE, involving only x and its tilde shifts:

$$\begin{aligned}&(n+1)(q+px)(p+qx)\frac{p(\widetilde{x}-x)+q(1-x\widetilde{x})}{p(\widetilde{x}+x)+q(1+x\widetilde{x})}\\ &\quad-n(p^2-q^2)x\frac{p(x-\underset{\sim}{x})+q(1-x\underset{\sim}{x})}{p(x+\underset{\sim}{x})+q(1+x\underset{\sim}{x})}\\ &=-\mu q(p+2qx+px^2)+p\lambda(-1)^{n+m}(q+2px+qx^2)-mpq(1-x^2).\end{aligned} \tag{10.43}$$

Equation (10.43) is a second-order nonautonomous OΔE with four free parameters q/p, m, μ and λ.

This is a discrete Painlevé equation (dP) (Nijhoff et al., 2001), which contains other discrete Painlevé equations as limits. For example, if we take the limit $q \to \infty, m/q \to \sigma$, it is easy to see that (10.43) reduces to

$$\frac{n+1}{x_n x_{n+1}+1} + \frac{n}{x_n x_{n-1}+1} = n + \frac{1}{2} + \mu + \frac{1}{2}\sigma\left(x_n - \frac{1}{x_n}\right), \tag{10.44}$$

which in the literature is referred to as the alternate $\mathrm{P_{II}}$ equation; see also (11.15).

Another coalescence limit is obtained by writing $q/p = 1 + \delta$ and taking the limit $\delta \to 0$, while setting $n = n' - m$, where $n, m \to \infty$ such that n' is fixed (in the limit) and $\delta m \to \eta$ finite. In that limit we also take $(1 - x_n)/(1 + x_n) \to a_{n+1}$ and then (10.43) reduces to the following equation

$$\frac{1}{2}\eta(a_{n+1} + a_{n-1}) = \frac{-\mu + \lambda(-1)^n + n a_n}{1 - a_n^2} \tag{10.45}$$

(where we have repressed the primes). Equation (10.45) is the so-called discrete (asymmetric) Painlevé II equation ($\mathrm{dP_{II}}$).

Thus, we see that the dP equation (10.43) contains various significant other dPs as subcases for special choices of or limits on the parameters. However, this equation is by no means the most general Painlevé difference equation as we shall see in Chapter 11.

10.4.3 Continuous similarity reduction and P_{VI}

We now proceed to derive the reduction to an ODE, starting from the same set of variables (10.34) and (10.36) and using the same system of equations, i.e. the lattice equation in the form (10.35), the differential-difference equations (10.37) (and the fact that they are compatible with the lattice equation) and the similarity constraint (10.39) and their consequences (10.40) and (10.41). However, in the continuous reduction procedure we will implement a different elimination process, namely like in the linear case by trying to get rid of the lattice shifts in favor of differentiations in one or the other of the lattice parameters.

Since the process is somewhat involved we describe it in detail. First from (10.36) we get

$$\frac{\widetilde{v}}{\underset{\sim}{v}} = \frac{1 + \mathfrak{a}}{1 - \mathfrak{a}}, \tag{10.46}$$

and from (10.34)

$$\underset{\sim}{x} = \frac{\underset{\sim}{v}}{\widehat{v}} = \frac{1 + \mathfrak{a}}{1 - \mathfrak{a}}\, y. \tag{10.47}$$

Then we have

$$\begin{aligned}
\frac{2}{(1-\mathfrak{a})^2}\, q\,\partial_q \mathfrak{a} &= q\partial_q \log\left(\frac{1+\mathfrak{a}}{1-\mathfrak{a}}\right) = q\partial_q \log\left(\frac{\widetilde{v}}{\underset{\sim}{v}}\right) && \text{from} \quad (10.46)\\
&= m(\underset{\sim}{\mathfrak{b}} - \widetilde{\mathfrak{b}}) && \text{from} \quad (10.37)\\
&= m\left[\frac{(p\underset{\sim}{x}+q)(1+\mathfrak{b})-2q}{p\underset{\sim}{y}-q} - \frac{(py-q)(1+\mathfrak{b})+2q}{px+q}\right]. && \text{from} \quad (10.40)
\end{aligned}$$

At this point we use equations (10.35) and (10.46) to obtain

$$\underset{\sim}{y} = \frac{p(1+\mathfrak{a})y + q(1-\mathfrak{a})}{q(1+\mathfrak{a})y + p(1-\mathfrak{a})}$$

and also to eliminate $\underset{\sim}{x}$, x, after which the result can be written as

$$mp\frac{\partial \mathfrak{b}}{\partial p} = nq\frac{\partial \mathfrak{a}}{\partial q} = \frac{mnpq}{p^2-q^2}\left[(1-\mathfrak{a})(1+\mathfrak{b})y - (1+\mathfrak{a})(1-\mathfrak{b})\frac{1}{y}\right], \tag{10.48}$$

where the first equality is from (10.38). Thus, the relation (10.48) gives us the p-derivative of $\mathfrak{b}$ in terms of $\mathfrak{a}$, $\mathfrak{b}$ and y.

For the other equation we start with (10.34) and using (10.37) obtain

$$\begin{aligned}
p\partial_p \log y &= p\partial_p \log(\widetilde{v}/\widehat{v}) = n\widehat{\mathfrak{a}} - (n+1)\widetilde{\mathfrak{a}} && \text{from} \quad (10.34, 10.37)\\
&= n\widehat{\mathfrak{a}} + m\widetilde{\mathfrak{b}} + \mu - \lambda(-1)^{n+m} && \text{from} \quad (10.39)\\
&= \mu - \lambda(-1)^{n+m} + \frac{(qy^{-1}-p)\mathfrak{a} + q(y^{-1}-x)}{qx+p} \\
&\quad + m\,\frac{(py-q)\mathfrak{b} + p(y-x)}{px+q}. && \text{from} \quad (10.40, 10.41)
\end{aligned}$$

After eliminating x using (10.35) we get

$$\begin{aligned}
p\partial_p \log y =&\mu - \lambda(-1)^{n+m} + \frac{(py-q)(p-qy)}{(p^2-q^2)y}(m\mathfrak{b} - n\mathfrak{a})\\
&- \frac{pq(n+m)}{p^2-q^2}\left(y - \frac{1}{y}\right).
\end{aligned} \tag{10.49}$$

Finally we eliminate $\mathfrak{a}$ from (10.48, 10.49) using (10.39), solve $\mathfrak{b}$ from (10.49) and substitute into (10.48), which then yields

$$\begin{aligned}
&p(p^2-q^2)^2 y(qy-p)(py-q)\frac{\partial^2 y}{\partial p^2}\\
&=\frac{1}{2}p(p^2-q^2)^2\left[pq(3y^2+1)-2(p^2+q^2)y\right]\left(\frac{\partial y}{\partial p}\right)^2\\
&+(q^2-p^2)\left[2p^2y(py-q)(qy-p)+(q^2-p^2)^2y^2\right]\frac{\partial y}{\partial p}\\
&+\frac{1}{2}q\left[(\alpha y^2-\beta)(py-q)^2(qy-p)^2\right.\\
&\left.+(p^2-q^2)y^2\left((\gamma-1)(qy-p)^2-(\delta-1)(py-q)^2\right)\right], \qquad (10.50)
\end{aligned}$$

where we have introduced the following parameters

$$\begin{aligned}
\alpha &:= (\mu+\nu-m+n)^2, & \beta &:= (\mu+\nu+m-n)^2,\\
\gamma &:= (\mu-\nu+m+n+1)^2, & \delta &:= (\mu-\nu-m-n-1)^2, \qquad (10.51)
\end{aligned}$$

and abbreviated $\nu := \lambda(-1)^{n+m}$.

The complicated-looking ODE (10.50) is actually quite a well-known equation and can be identified as such if we make the simple change of variables

$$t := (p/q)^2, \quad w(t) := (p/q)y(p), \qquad (10.52)$$

after which it gets the standard form

$$\begin{aligned}
\frac{d^2w}{dt^2} &= \frac{1}{2}\left(\frac{1}{w}+\frac{1}{w-1}+\frac{1}{w-t}\right)\left(\frac{dw}{dt}\right)^2-\left(\frac{1}{t}+\frac{1}{t-1}+\frac{1}{w-t}\right)\frac{dw}{dt}\\
&+\frac{w(w-1)(w-t)}{8t^2(t-1)^2}\left(\alpha-\beta\frac{t}{w^2}+(\gamma-1)\frac{t-1}{(w-1)^2}-(\delta-1)\frac{t(t-1)}{(w-t)^2}\right). \qquad (10.53)
\end{aligned}$$

This is nothing less than the famous Painlevé VI equation with four parameters.

The above results provide a continuous/discrete hybrid system, which has P_{VI} in it as well as discrete Painlevé. That means we have a common solution space with functions of the form $P(t,s;n,m)$ which obey P_{VI} in both the variables $t=p^2$ and another P_{VI} in the variable $s=q^2$, while the same functions, now considered as a function of the discrete variables n,m obey (10.43) in terms of n and another copy of the same equation in terms of m.

In the case of the lattice Schwarzian KdV equation, the reduction can be obtained, but the derivation exhibits a remarkable twist in that it requires *two* similarity constraints rather than a single one. In fact, in addition to (10.33c) a second similarity constraint for the quantity z can be obtained by considering the similarity constraint for the lattice mKdV equation, i.e. equation (10.33b) and using the discrete Cole–Hopf transformations (3.14) to cast it in the form:

$$\mu+\lambda(-1)^{n+m}+n\frac{z_{n+1,m}+z_{n-1,m}-2z_{n,m}}{z_{n+1,m}-z_{n-1,m}}+m\frac{z_{n,m+1}+z_{n,m-1}-2z_{n,m}}{z_{n,m+1}-z_{n,m-1}}=0, \tag{10.54}$$

which looks distinctly different from (10.33c). Obviously these two constraints must be compatible when one takes into account the lattice SKdV equation itself, but the way to show that is highly nontrivial. Implementing both constraints simultaneously the following third-order OΔE can be derived

$$\begin{aligned}
&(t^2-1)(z_{n+1}-z_n)^2 \\
&=\left[2t^2\frac{\mu z_{n+1}(z_{n+2}-z_n)-(n+1)(z_{n+2}-z_{n+1})(z_{n+1}-z_n)}{(m+\mu-\lambda(-1)^{n+m})(z_{n+2}-z_n)+(n+1)(z_{n+2}-2z_{n+1}+z_n)}+z_{n+1}-z_n\right] \\
&\quad\times\left[2t^2\frac{\mu z_n(z_{n+1}-z_{n-1})-n(z_{n+1}-z_n)(z_n-z_{n-1})}{(m+\mu+\lambda(-1)^{n+m})(z_{n+1}-z_{n-1})+n(z_{n+1}-2z_n+z_{n-1})}+z_n-z_{n+1}\right],
\end{aligned} \tag{10.55}$$

for the quantity $z =: z_n$ (suppressing the dependence on the variable m which now becomes a parameter of the reduced equation. Equation (10.55) which was derived in Nijhoff et al. (2000b) is the discrete analogue of the continuous Schwarzian dP equation (10.6).

The explicit similarity reduction of the lattice potential KdV to an OΔE, using the constraint (10.33a), remains an open problem.

Remark 10.4.1 The sixth Painlevé equation has a natural higher-order extension called the Garnier system (Garnier, 1912). As we have obtained the sixth Painlevé equation and its contiguity relation (discrete P_{VI}) (10.43) from the lattice structure, it is no surprise that in the present approach we can also derive higher-order extensions of this continguity equation, which is related to the Garnier system (see Appendix C). This relies on the multidimensional consistency of quadrilateral lattices well as on an extension of the similarity constraint of the form

$$n\frac{\widetilde{v}-\underset{\sim}{v}}{\widetilde{v}+\underset{\sim}{v}}+m\frac{\widehat{v}-\underset{\frown}{v}}{\widehat{v}+\underset{\frown}{v}}+h\frac{\bar{v}-\underline{v}}{\bar{v}+\underline{v}}+\mu+\lambda(-1)^{n+m+h}=0\,, \tag{10.56}$$

which contains contributions from all three lattice directions. This gives a reduction to coupled systems of OΔEs (Nijhoff and Walker, 2001), which form the higher-order analogue of equation (10.43).

10.4.4 Generating PDEs

So far, we have two systems of equations that are interlinked, namely PΔEs and the differential-difference equations where the continuous independent variable is the lattice parameter. To complete the picture, we now present PDEs with respect to the lattice parameters:

$$-\partial_p\partial_q \log v = \frac{1}{p^2-q^2}\left[(n+p\partial_p \log v)(m-\partial_q \log v)y \right. \\ \left. -(n-\partial_p \log v)(m+\partial_q \log v)\frac{1}{y}\right], \tag{10.57a}$$

$$pq\partial_p\partial_q \log y = q\partial_q\left[(p-qy)\frac{p(py-q)\partial_p \log v + n(q+py)}{(p^2-q^2)y}\right] \\ - p\partial_p\left[(p-qy)\frac{-q(py-q)\partial_q \log v + m(q+py)}{(p^2-q^2)y}\right]. \tag{10.57b}$$

This coupled system of nonlinear PDEs is the nonlinear analogue of the coupled system of linear PDEs (10.16). This system of PDEs is *multidimensionally consistent* in a similar way to the lattice equations of Chapter 3: assuming that the dependent variables v and y depend not only on the independent variables p and q (with associated parameters n and m) but on an additional variable r as well, with an associated additional parameter (or, lattice variable) h associated with it:

$$v = v(p,q,r;n,m,h) = v_{n,m,h}(p,q,r).$$

Not only can we impose lattice equations in all pairs of the three lattice variables n, m and h, but we can also impose different coupled systems of the form (10.57) in every pair of variables p and q, p and r and q and r with the parameters n, m and h replaced accordingly.

The claim is that all these systems of coupled PDEs are compatible, which means that they can be imposed simultaneously on the solutions v. One technical detail is that the second dependent variable y changes as well from $y=\widetilde{v}/\widehat{v}$ to $Y=\widetilde{v}/\overline{v}$ to $Z=\widehat{v}/\overline{v}$.

In principle, as in the linear case of (10.17), we could obtain a higher-order scalar PDE in terms of a single dependent variable, v say, from the coupled system (10.57). However, the resulting equation is complicated by the fact that the expression for y in terms of v is quadratic for y. In the lattice Schwarzian KdV case, a derivation of such an equation has been done and the result is:

$$z_{sstt} = z_{sst}\left(\frac{z_{st}}{z_s}+\frac{z_{tt}}{z_t}\right) + z_{stt}\left(\frac{z_{st}}{z_t}+\frac{z_{ss}}{z_s}\right) - z_{st}\left(\frac{z_{st}z_{ss}}{z_s^2}+\frac{z_{st}z_{tt}}{z_t^2}+\frac{z_{ss}z_{tt}}{z_s z_t}\right) \\ + \frac{1}{s-t}\left[\frac{s}{t}\left(z_{sst}-\frac{z_{st}z_{ss}}{z_s}-\frac{1}{2}\frac{z_{st}^2}{z_t}\right) - \frac{t}{s}\left(z_{stt}-\frac{z_{st}z_{tt}}{z_t}-\frac{1}{2}\frac{z_{st}^2}{z_s}\right)\right] \\ - \frac{1}{(s-t)^2}\left[n^2\frac{s^2}{t^2}\frac{z_s}{z_t}\left(z_{st}-\frac{z_s z_{tt}}{z_t}\right) + m^2\frac{t^2}{s^2}\frac{z_t}{z_s}\left(z_{st}-\frac{z_t z_{ss}}{z_s}\right)\right] \\ - \frac{1}{2}\frac{1}{(s-t)^3}\left[n^2\frac{s}{t}z_s\left(1+\frac{(3s-4t)}{t^2}\frac{z_s}{z_t}\right) - m^2\frac{t}{s}z_t\left(1+\frac{(3t-4s)}{s^2}\frac{z_t}{z_s}\right)\right], \tag{10.58}$$

where the independent variables are $t := p^2$ and $s := q^2$. Equation (10.58) was introduced in Nijhoff et al. (2000a) where it was shown that apart from multidimensional consistency (as explained above) it had several other redeeming properties: i) a simple Lagrangian structure (see Chapter 12); ii) Möbius invariance (in fact, it is invariant under the group $PSL_2(\mathbb{C})$), iii) it has a non-isospectral Lax pair; iv) it allows for scaling-invariant solutions leading to the full (i.e. for generic parameters) Painlevé VI equation; v) it encodes in a single equation the entire hierarchy of Schwarzian KdV equations. Furthermore, as was shown in Tongas et al. (2001b), (10.58) can be viewed as a generalization of the Ernst-Weyl equation of general relativity.

10.4.5 Isomonodromic deformation problems

In this section we show how to apply the similarity reduction procedure to obtain the Lax pairs of the reduced equations found earlier in this chapter. This leads to *isomonodromic deformation problems* associated with those equations. The example considered here leads to a Lax pair that consists of a differential system and a discrete system. We note that other reductions have led to fully discrete Lax pairs for q-discrete equations (Hay et al., 2007; Ormerod et al., 2013). Thus the structure of the Lax pair depends on the similarity reduction.

In Chapter 3 we showed how to derive the Lax pair for the quadrilateral lattice equations from the multidimensional consistency property. In the case of the lattice mKdV system this was done by considering an additional direction in the lattice associated with a lattice parameter k which was reinterpreted as a spectral parameter, by linearizing the equation using the shifted variable $\overline{v} = f/g$. Thus, we obtained a discrete zero-curvature Lax pair for the lattice equation (in this case the lattice mKdV equation) of the form (3.21),

$$\widetilde{\boldsymbol{\phi}} = \boldsymbol{L}\boldsymbol{\phi} = \begin{pmatrix} -p & k\widetilde{v} \\ k/v & -p\widetilde{v}/v \end{pmatrix}\boldsymbol{\phi}\,, \quad \widehat{\boldsymbol{\phi}} = \boldsymbol{M}\boldsymbol{\phi} = \begin{pmatrix} -q & k\widehat{v} \\ k/v & -q\widehat{v}/v \end{pmatrix}\boldsymbol{\phi}, \tag{10.59}$$

with $\boldsymbol{\phi} = (f, g)^T$. As was shown, the discrete zero-curvature condition $\widehat{\boldsymbol{L}}\boldsymbol{M} = \widetilde{\boldsymbol{M}}\boldsymbol{L}$ is satisfied iff the variable v obeys the lattice mKdV equation (3.11).

We next proceed to deriving the Lax pair for the system of differential-difference equations (10.37). We start by applying the shift in the third direction on (10.28b):

$$-p\frac{\partial}{\partial p}\log v = n\frac{\widetilde{v}-\underset{\sim}{v}}{\widetilde{v}+\underset{\sim}{v}} \quad \Rightarrow \quad -p\frac{\partial}{\partial p}\log\overline{v} = n\frac{\overline{\widetilde{v}}-\overline{\underset{\sim}{v}}}{\overline{\widetilde{v}}+\overline{\underset{\sim}{v}}}\,. \tag{10.60}$$

We now eliminate the shifted variables $\overline{\widetilde{v}}$, $\overline{\underset{\sim}{v}}$ by using the lattice equation and its bar-shifted version:

$$\overline{\widetilde{v}} = v\frac{p\overline{v}-k\widetilde{v}}{p\widetilde{v}-k\overline{v}}, \quad \overline{\underset{\sim}{v}} = v\frac{k\underset{\sim}{v}+p\overline{v}}{p\underset{\sim}{v}+k\overline{v}}$$

and inserting this into (10.60) we get:

$$-\,p\frac{\partial}{\partial p}\log\overline{v}=n\frac{2pk(\overline{v}^2-\widetilde{v}\underaccent{\tilde}{v})-(p^2+k^2)(\widetilde{v}-\underaccent{\tilde}{v})\overline{v}}{(p^2-k^2)(\widetilde{v}+\underaccent{\tilde}{v})\overline{v}}\,, \tag{10.61}$$

which is a Riccati equation for $\overline{v}$. As before, we use $\overline{v}=f/g$ and splitting the result into two linear equations we obtain $p\partial_p\boldsymbol{\phi}=\boldsymbol{N}\boldsymbol{\phi}$, where $\boldsymbol{\phi}:=(f,g)^T$ and

$$\boldsymbol{N}:=\frac{n}{(p^2-k^2)(\widetilde{v}+\underaccent{\tilde}{v})}\begin{pmatrix}(p^2+k^2)\widetilde{v}+(p^2-k^2)\underaccent{\tilde}{v} & 2pk\widetilde{v}\underaccent{\tilde}{v}\\ 2pk & 2p^2\underaccent{\tilde}{v}\end{pmatrix}. \tag{10.62}$$

Let us now consider the following compatibility conditions:

$$p\frac{\partial}{\partial p}\boldsymbol{L}=\widetilde{\boldsymbol{N}}\boldsymbol{L}-\boldsymbol{L}\boldsymbol{N}, \tag{10.63a}$$

$$p\frac{\partial}{\partial p}\boldsymbol{M}=\widehat{\boldsymbol{N}}\boldsymbol{M}-\boldsymbol{M}\boldsymbol{N}, \tag{10.63b}$$

with the matrix $\boldsymbol{L}$, $\boldsymbol{M}$ given as above. The Lax equation (10.63a) gives us the first equation in (10.37), and (10.63b) gives in addition (10.41) (when expressed in terms of v using (10.34) and (10.36)). Similarly, we have a linear equation of the form $q\partial_q\boldsymbol{\phi}=\boldsymbol{K}\boldsymbol{\phi}$, where

$$\boldsymbol{K}:=\frac{m}{(q^2-k^2)(\widehat{v}+\underaccent{\hat}{v})}\begin{pmatrix}(q^2+k^2)\widehat{v}+(q^2-k^2)\underaccent{\hat}{v} & 2qk\widehat{v}\underaccent{\hat}{v}\\ 2qk & 2q^2\underaccent{\hat}{v}\end{pmatrix}, \tag{10.64}$$

with the consistency relations

$$q\frac{\partial}{\partial q}\boldsymbol{M}=\widehat{\boldsymbol{K}}\boldsymbol{M}-\boldsymbol{M}\boldsymbol{K}, \tag{10.65a}$$

$$q\frac{\partial}{\partial q}\boldsymbol{L}=\widetilde{\boldsymbol{K}}\boldsymbol{L}-\boldsymbol{L}\boldsymbol{K}, \tag{10.65b}$$

yielding the second member of (10.37) and (10.40). Furthermore from $p\partial_p\boldsymbol{\phi}=\boldsymbol{N}\boldsymbol{\phi}$ and $q\partial_q\boldsymbol{\phi}=\boldsymbol{K}\boldsymbol{\phi}$, we have a continuous Lax equation (similar to the zero curvature condition (3.24))

$$q\partial_q\boldsymbol{N}-p\partial_p\boldsymbol{K}+[\boldsymbol{N},\boldsymbol{K}]=0, \tag{10.66}$$

which yields (10.57).

It turns out that the similarity constraint for the vector $\boldsymbol{\phi}$ can be chosen in the following form

$$(p\partial_p+q\partial_q+k\partial_k)\boldsymbol{\phi}=\boldsymbol{\Lambda}\,\boldsymbol{\phi},\quad\text{where}\quad\boldsymbol{\Lambda}=\begin{pmatrix}\Lambda_1(n,m) & 0\\ 0 & \Lambda_2(n,m)\end{pmatrix}, \tag{10.67}$$

in analogy with (10.30), generalized in the sense of multidimensional consistency. After shifting this in the tilde direction and using the first member of the discrete Lax pair (10.59) we get the compatibility condition

$$k\partial_k \boldsymbol{L} + p\partial_p \boldsymbol{L} + q\partial_q \boldsymbol{L} = \widetilde{\boldsymbol{\Lambda}}\boldsymbol{L} - \boldsymbol{L}\boldsymbol{\Lambda}.$$

This contains the constraint (10.39), provided that we choose $\Lambda_1(n,m) = n + m + \gamma$ and $\Lambda_2(n,m) = n + m + \gamma - \mu - \lambda(-1)^{n+m}$ where γ is a constant. Inserting the expressions obtained for $\partial_p \boldsymbol{\phi}$ and $\partial_q \boldsymbol{\phi}$ from (10.62) and (10.64) we obtain a differential equation in the spectral parameter k of the form (Nijhoff et al., 2001)

$$k\partial_k \boldsymbol{\phi} = (\boldsymbol{\Lambda} - \boldsymbol{N} - \boldsymbol{K})\boldsymbol{\phi}. \tag{10.68}$$

The compatibility of this equation with (10.59) yields the equations (10.39) and both (10.40) and (10.41), which are the equations that lead to the reduction (10.43). Thus, the system of equations (10.68) and (10.59) constitute an isomonodromic deformation problem for (10.43). At the same the compatibility of (10.68) with either $p\partial_p \boldsymbol{\phi} = \boldsymbol{N}\boldsymbol{\phi}$ or $q\partial_q \boldsymbol{\phi} = \boldsymbol{K}\boldsymbol{\phi}$, gives the continuous P_{VI} in the form (10.50).

10.5 Notes

It was Sophus Lie who introduced the notion of continuous symmetry groups (i.e. groups of transformations depending continuously on parameters, or "Lie groups") primarily with the aim to investigate and describe solutions of (ordinary) differential equations. This has evolved into a well-established branch of mathematics, cf. e.g. the monographs (Bluman and Cole, 1974; Stephani, 1989; Olver, 2000; Gaeta, 2012). The extension of Lie group methods to the study of difference equations has been developed in the last two decades, cf. e.g. Shabat and Yamilov (1990); Levi and Yamilov (1997); Levi and Winternitz (2006); Hydon (2014) where the formal Lie theory for difference equations was developed.

However, for the integrable class of P∆Es more direct approaches, like the ones described in this chapter, were first developed in their own right, and subsequently given symmetry-theoretic underpinning in Orfanidis (1978); Tongas et al. (2005); Tongas (2007); Tongas et al. (2007); Levi and Yamilov (2011); Rasin and Hydon (2007b); Levi et al. (2007); Levi and Petrera (2007). More recently, the language of difference algebra was used to develop symmetries and conservation laws for quadrilateral lattices, (Mikhailov et al., 2011a,b). As was pointed out in Section 2.5, the existence of an infinite sequence of conservation laws is a hallmark of integrability and goes hand-in-hand with the existence of infinite higher-order symmetries. For integrable P∆Es such infinite sets of conservation laws and symmetries were only developed relatively recently (Rasin and Hydon, 2005, 2007a; Maruno and Quispel, 2006; Rasin and Schiff, 2009; Rasin, 2010; Xenitidis and Nijhoff, 2012; Xenitidis, 2011; Zhang et al., 2013).

The notion of generating PDEs was introduced in Nijhoff et al. (2000a), and those associated with the KdV and SKdV equation were identified in Tongas et al. (2001a) as generalizations of the Ernst-Weyl equation. Similar generating PDEs were constructed for the Boussinesq class in Tongas and Nijhoff (2005b) (which lead to a system that generalizes the Einstein–Maxwell–Weyl equations), and for all equations in the ABS list in Tsoubelis and Xenitidis (2009). While the generating PDE for the KdV hierarchy reduces to the full $\mathrm{P_{VI}}$ equation (Nijhoff et al., 2000a), the generating PDE for the Boussinesq class leads to a higher-order reduction (Tongas and Nijhoff, 2005b). On the discrete level the reduction of the Boussinesq lattice (Tongas and Nijhoff, 2006) leads to a Garnier-type discrete equation. Alternatively, a discrete version of the Garnier system was obtained from the multidimensional reduction of mKdV lattice following the method of this chapter, cf. Nijhoff and Walker (2001); Walker (2001).

We have not discussed in this chapter similarity reductions of q-difference type. This can be developed in a similar way, using q-difference analogues of the similarity constraints, and using (q-difference type) nonautonomous version of the lattice equation, cf. Field et al. (2008). Alternatively, q-difference Painlevé equations were deduced as reductions of a q-KP hierarchy (Kajiwara et al., 2002). Furthermore, periodic similarity reductions of nonautonomous integrable lattice equations were studied by several authors (Hay et al., 2007; Grammaticos et al., 2005; Ormerod, 2012; Ormerod et al., 2013). See also Joshi et al. (2014, 2015) for reductions based on affine Weyl groups. q-difference reductions to Garnier systems are discussed in Ormerod and Rains (2016).

We mention also that the reduction to $\mathrm{P_{VI}}$ is intimately related to the issues associated with discrete holomorphic functions, in particular discrete Z^γ (Bobenko and Pinkall, 1996a; Bobenko, 1999; Agafonov and Bobenko, 2000; Agafonov, 2003; Agafonov and Bobenko, 2003; Agafonov, 2005; Ando et al., 2014; Hay et al., 2011). The theory of discrete holomorphic functions forms an interesting development, closely related to the topic of this book. The linear theory discussed in Section 10.3 is linked to the linear theory of discrete holomorphic function (Bobenko et al., 2005; Mercat, 2001, 2008; Smirnov, 2010).

Exercises

10.1 Note that when we have an additional constraint equation we have in fact two equations for one function and then the two equations must be compatible in the sense that cross differentiation does not produce new independent equations. Show by direct computation that (10.10) is compatible with (10.4b).

10.2 Derive the similarity constraint (10.33b) from the constraint (10.33c) by using the discrete Cole–Hopf transformation (3.14). Use the same transformation once again to establish (10.54).

10.3 Perform the computation of the coalescence limit from (10.43) to the $\mathrm{dP_{II}}$ equation (10.45).

Hint: First derive the following intermediate equation:

$$(n'+1)(1-x_n^2)-(\mu+(-\lambda)^{n'})(1+x_n)^2=2\eta x_n\left(\frac{1-x_{n+1}}{1+x_{n+1}}+\frac{1-x_{n-1}}{1+x_{n-1}}\right),$$

where we consider x_n to be a function of n' rather than n.

10.4 Perform the similarity reduction for the lattice potential KdV equation by using the similarity constraint

$$u+xu_x+3tu_t=f(t),$$

where we have an arbitrary function $f(t)$ on the right-hand side. Thus, show that the quantity $U=u_x+\frac{x}{6t}$ obeys the P_{34} equation given by:

$$UU_{xx}-\frac{1}{2}U_x^2+2U^3-\frac{x}{3t}U^2+\frac{\bar{\mu}^2}{8t^2}=0.$$

Next, by using the standard Miura transformation (2.75) between the mKdV and KdV equation, derive the following Miura-type transformations relating P_{II} and P_{34}:

$$\bar{v}=\frac{\frac{\bar{\mu}}{2t}-U_x}{2U}\Leftrightarrow U=\bar{v}_x-\bar{v}^2+\frac{x}{6t}.$$

10.5 (a) Show that using (3.14) the similarity constraint (10.33c) for the variable $z_{n,m}$ can be cast into the form:

$$\mu-\frac{1}{2}+(n+1)\frac{v_{n+2,m}}{v_{n+2,m}+v_{n,m}}-n\frac{v_{n-1,m}}{v_{n+1,m}+v_{n-1,m}}$$
$$+m\frac{p}{q}\left(\frac{v_{n+1,m+1}v_{n+1,m-1}}{v_{n,m}(v_{n+1,m+1}+v_{n+1,m-1})}-\frac{v_{n,m+1}v_{n,m-1}}{v_{n+1,m}(v_{n,m+1}+v_{n,m-1})}\right)=0,$$

which by using next (3.11) to transform the last two terms can be cast into the form

$$\mu+\frac{1}{2}(n+1)\frac{v_{n+2,m}-v_{n,m}}{v_{n+2,m}+v_{n,m}}+\frac{1}{2}n\frac{v_{n,m}-v_{n-1,m}}{v_{n+1,m}+v_{n-1,m}}$$
$$+m\frac{v_{n+1,m+1}v_{n,m+1}-v_{n+1,m-1}v_{n,m-1}}{(v_{n+1,m+1}+v_{n+1,m-1})(v_{n,m+1}+v_{n,m-1})}=0.$$

By observing that the right-hand side of the latter relation is actually the sum of two terms plus their n-shifted counterparts, by performing a discrete integration show that one can obtain the similarity constraint (10.31) as a consequence.

(b) Use the discrete Cole–Hopf relations (3.14) to show that from the similarity constraint (10.33a) for the lattice potential KdV equation one can derive the following *third* form of the similarity constraint for the lattice Schwarzian KdV equation:

$$\left(1+2\lambda(-1)^{n+m}\right)z_{n,m}=2n\frac{z_{n,m}^2-z_{n+1,m}z_{n-1,m}}{z_{n+1,m}-z_{n-1,m}}+2m\frac{z_{n,m}^2-z_{n,m+1}z_{n,m-1}}{z_{n,m+1}-z_{n,m-1}}, \tag{10.69}$$

and show that all three constraints (10.33c), (10.54) and (10.69) are compatible subject to the lattice Schwarzian KdV equation.

10.6 Derive the generating PDE system for the lattice potential KdV system, by using the system of equations

$$\frac{\partial w}{\partial p}=\frac{2np}{\widetilde{w}-\underset{\sim}{w}},\quad \frac{\partial w}{\partial q}=\frac{2mq}{\widehat{w}-\underset{\frown}{w}},$$

together with the lattice equation

$$(\widehat{w}-\widetilde{w})(\widehat{\widetilde{w}}-w)=p^2-q^2.$$

Thus, show that one can derive the coupled system of equations:

$$(p^2-q^2)\frac{\partial_p\partial_q w}{(\partial_p w)(\partial_q w)}=\widehat{w}+\underset{\frown}{w}-\widetilde{w}-\underset{\sim}{w}=2W-\frac{2np}{\partial_p w}-\frac{2mq}{\partial_q w},$$

$$\partial_p\partial_q W=-\frac{\partial}{\partial q}\left(\frac{(2np+W\partial_p w)W}{p^2-q^2}\right)+\frac{\partial}{\partial p}\left(\frac{(2mq-W\partial_q w)W}{p^2-q^2}\right),$$

where $W:=\widehat{w}-\widetilde{w}$. By eliminating W derive from this system a higher-order PDE for w alone.

10.7 Work out the compatibility relation between the two continuous members of the Lax pair (10.62) and (10.64), to show that it gives rise to the generating PDE system (10.57) when the enties are expressed in terms of the quantities $\mathfrak{a}$, $\mathfrak{b}$ and x of (10.36).

11

Discrete Painlevé equations

In earlier chapters we explained how to derive discrete Painlevé equations through different perspectives, relying on symmetries, deautonomization through singularities, and geometry. Discrete Painlevé equations were found as recurrence relations of the Painlevé equations in Section 2.3 of Chapter 2, identified through the singularity confinement property in Section 7.1, found as similarity reductions of lattice equations in Chapter 10, and described through a unifying geometric point of view (Sakai, 2001) as mappings on rational surfaces in Section 7.3 in Chapter 7.

Discrete Painlevé equations arise in various mathematical and physical contexts, for example in random matrix theory, in models of quantum gravity, and as correlation functions in statistical mechanics. They also appear as nonlinear equations satisfied by coefficients of linear three-term recurrences governing orthogonal polynomials with semi-classical weights. They are pushing the boundary of what we define to be nonlinear special functions because they are regarded as defining equations for an entirely new class of special functions.

The focus of this chapter lies on the intrinsic properties of discrete Painlevé equations as integrable systems. After starting with a brief history of their discovery, we describe and deduce transformations between solutions, such as Bäcklund or Miura transformations. Then we focus on their associated linear problems and describe the different types that arise. The theory developed by Birkhoff and his school for solving linear difference equations is outlined and we suggest ways in which compatible linear problems could be used to solve nonlinear discrete Painlevé equations. Finally, we describe special solutions of the discrete Painlevé equations along with methods developed for deducing such solutions.

In the literature, there are three different types of discrete Painlevé equations, distinguished by the curves along which they are iterated, i.e. a straight line, a spiral or an elliptic curve in the complex plane, leading to what are respectively called additive-type, multiplicative-type or elliptic-type discrete Painlevé equations. If the equation is iterated by a shift in the independent variable n, say, then these cases are distinguished by the types of coefficient functions $z(n)$ appearing in the equation. In an additive-type equation, z is a linear function of n,[1] in a multiplicative equation, z is an exponential function of n, while

[1] Additive-type examples such as (2.68) with (2.67) also contain $(-1)^n$, but this is included by changing to a system of equations for odd and even iterates.

in an elliptic-type equation, z is an elliptic function of n. We saw examples of additive discrete Painlevé equations in Section 2.3 and multiplicative-type discrete Painlevé equations in Section 7.1.2. In this chapter, we will also describe Sakai's elliptic discrete Painlevé equation (see Section 11.8), which is regarded as the master equation.

Note that in this chapter we use the bar-shift notation, i.e. $\overline{w} = w(n+1)$, $w = w(n)$, $\underline{w} = w(n-1)$.

11.1 Early discoveries of discrete Painlevé equations

Many discrete Painlevé equations were first discovered as scalar second-order difference[2] equations. Two early discoveries were

$$\mathrm{dP_I}: \quad w\,(\overline{w} + w + \underline{w}) = a\,n + b + c\,w \tag{11.1}$$

$$\mathrm{dP_{II}}: \quad \overline{w} + \underline{w} = \frac{(a\,n+b)\,w + c}{1 - w^2} \tag{11.2}$$

where a, b and c are constants. Equation (11.1) was first described by Shohat (1939), as a condition satisfied by coefficients of the recurrence relation of a family of orthogonal polynomials. More than four decades later, it was rediscovered in models of gauge field theory (Bessis et al., 1980) and quantum gravity (Brezin and Kazakov, 1990). Fokas, Its and Zhou identified the equation as a discrete Painlevé equation (Fokas et al., 1992a), and in Its et al. (1990) it was shown how to solve it through a linear problem. Equation (11.2) was first discovered by Periwal and Shevitz (1990) and independently by Nijhoff and Papageorgiou (1991) who provided a linear problem for this equation and called it a "discrete second Painlevé equation" due its continuum limit to the second Painlevé equation.

The next sequence of discoveries came from the application of the singularity confinement method. Ramani et al. (1991) identified the following equations as discrete Painlevé equations:

$$\mathrm{qP_{III}}: \quad \overline{w}\,\underline{w} = cd\,\frac{(w - a\,q^n)(w - b\,q^n)}{(w-c)(w-d)} \tag{11.3}$$

$$\mathrm{dP_{IV}}: \quad (\overline{w} + w)\,(w + \underline{w}) = \frac{(w^2 - a^2)\,(w^2 - b^2)}{(w - (a\,n+b))^2 - c^2} \tag{11.4}$$

$$\mathrm{qP_V}: \quad (\overline{w}w - 1)\,(w\underline{w} - 1) = cd\,q^{2n}\,\frac{(w-a)\,(w-1/a)\,(w-b)\,(w-1/b)}{(w - c\,q^n)(w - d\,q^n)} \tag{11.5}$$

where a, b, c, d and q are constants, with $q \neq 0, 1$. Note that equations (11.3) and (11.5) are examples of q-difference equations, in which the iteration $n \mapsto n+1$ moves the non-autonomous coefficients in the equation on a spiral, while equations (11.1), (11.2) and (11.4) are difference equations, in which the iteration moves on a straight line. Once again,

[2] We use the term difference and discrete interchangeably in this chapter, but the word "difference" is preferred where we require analytic solutions in a nonempty domain. See Section 1.1.3 for precise definitions of these terms.

the nomenclature of discrete third, fourth and fifth Painlevé equations for each of equations (11.3)–(11.5) reflects their continuum limits to the respective Painlevé equations.

The discrete sixth Painlevé equation was discovered by Jimbo and Sakai (1996) as a system of equations for the dependent variables f and g

$$\mathrm{qP_{VI}}\ : \left\{ \begin{aligned} \frac{\overline{f}\, f}{b_3\, b_4} &= \frac{(g - t\, a_1)\,(g - t\, a_2)}{(g - a_3)\,(g - a_4)} \\ \frac{\overline{g}\, g}{a_3\, a_4} &= \frac{(\overline{f} - t\, b_1)\,(\overline{f} - t\, b_2)}{(\overline{f} - b_3)\,(\overline{f} - b_4)} \end{aligned} \right. \tag{11.6}$$

where a_i, b_i, $1 \le i \le 4$ are constants, which satisfy $a_3 a_4 b_1 b_2 = q a_1 a_2 b_3 b_4$, and $t = t_0\, q^n$.

It is interesting to note that $\mathrm{qP_{III}}$ is a reduction of $\mathrm{qP_{VI}}$. That is, if we take $f(t) = w(t/\lambda)$, $g(t) = w(t)$, with $\overline{f} = w(\lambda t)$ and $\overline{g} = w(\lambda^2 t)$, $\lambda = \sqrt{q}$ and $b_i = \sqrt{q}\, a_i$ for $i = 1, 2$, $b_j = a_j$ for $j = 3, 4$, then $\mathrm{qP_{VI}}$ becomes $\mathrm{qP_{III}}$ with q replaced by λ.

Equations (11.1)–(11.5) are called *symmetric* versions of discrete Painlevé equations. (A symmetric form of $\mathrm{qP_{VI}}$ was given in Grammaticos and Ramani (1999).) This terminology arises from their autonomous forms, which are examples of QRT mappings (6.31). As noted in Section 6.3, QRT mappings are called symmetric when the matrices $\mathbf{A}_i$, $i = 0, 1$ (used to define equations (6.32)) are symmetric and *asymmetric* otherwise. The same adjective is applied to the corresponding discrete Painlevé equations.

To see how the symmetric versions given above are related to asymmetric ones, consider $\mathrm{dP_I}$ with an extra term in the nonautonomous coefficient

$$w\,(\overline{w} + w + \underline{w}) = \big(\alpha\, n + \beta + \gamma\,(-1)^n\big) + \sigma\, w \tag{11.7}$$

where α, β, γ and σ are constants. Equation (11.1) given above as $\mathrm{dP_I}$ is the reduced case $\gamma = 0$ of this equation. The odd and even iterates $f_k = w_{2k}$, $g_k = w_{2k+1}$ of equation (11.7) satisfy the system of equations

$$\left\{ \begin{aligned} f_k\,(g_k + f_k + g_{k-1}) &= \big(z_k + \gamma\big) + \sigma\, f_k \\ g_k\,(f_{k+1} + g_k + f_k) &= \big(z_k + \alpha - \gamma\big) + \sigma\, g_k \end{aligned} \right. \tag{11.8}$$

where $z_k = 2\,\alpha\, k + \beta$. This system or equivalently equation (11.7) is called the asymmetric $\mathrm{dP_I}$ equation. It is interesting to note that the continuum limit of this system, obtained by taking $f = 1 + \epsilon\, r + \epsilon\, s$, $g = 1 - \epsilon\, r + \epsilon^2\, s$, $z = 1 - \epsilon^2\, t$, $\alpha = 2$, $\gamma = -\epsilon^3\, c/4$, leads to the second Painlevé equation $r_{tt} = 2\, r^3 + 2\, t\, r + c$, with $s = (r^2 - r_t + t)/4$.

By the convention adopted in the early studies, the asymmetric $\mathrm{dP_I}$, i.e. equation (11.7), could be renamed as an alternative $\mathrm{dP_{II}}$. In other words, equation (11.8), $\mathrm{dP_{II}}$ and $\mathrm{qP_{II}}$ provide three different integrable discrete versions of $\mathrm{P_{II}}$. The reduction of $\mathrm{qP_{VI}}$ to $\mathrm{qP_{III}}$ described above shows that $\mathrm{qP_{III}}$ is a symmetric version of $\mathrm{qP_{VI}}$.

The existence of more than one integrable discrete analogue for each continuous Painlevé equation was well known by the end of the 1990s. (At one stage, as many as

16 discrete versions of P_I had been identified.) Many discrete Painlevé equations were discovered via approaches explained in Chapter 7. The study of degeneracies or coalescence limits of these equations (Ramani and Grammaticos, 1996) (as explained in the next section) led to several more discrete equations. To prove that they are not transformable into each other is a nontrivial exercise.

11.2 Discrete Painlevé equations from Sakai's classification

The number of integrable discrete versions of Painlevé equations continued to increase until Sakai (2001), which provided a method of classifying and unifying the plethora of equations. Sakai showed how discrete Painlevé equations arise as dynamical systems on rational surfaces obtained as a nine-point blow-up of $\mathbb{CP}^2$ (or equivalently an eight-point blow-up of $\mathbb{CP}^1 \times \mathbb{CP}^1$).

The following is a list of asymmetric discrete Painlevé equations, provided in Sakai (2001). The labels listed below provide the type of rational surface associated with each system. Note that equation (11.6) is associated with the rational surface $A_3^{(1)}$.

$$A_4^{(1)} - \mathrm{qP_V} : \begin{cases} \overline{f}\, f = \dfrac{a_1 a_2}{a_3} \dfrac{1-g}{(1+a_2 g)(1-a_1 a_2 g)} \\ \overline{g}\, g = \dfrac{a_0\, a_4}{a_2} \dfrac{(1-a_3 \overline{f})(1-\overline{f})}{\overline{f}\,(a_4 - q\overline{f})} \\ \overline{a}_0 = q\, a_0, \overline{a}_4 = a_4/q \end{cases} \tag{11.9}$$

where a_1, a_2, a_3 are constants,

$$A_5^{(1)} - \mathrm{qP_{IV}} : \begin{cases} \overline{f}\, f = -\left(a_2\, b_0 - \dfrac{a_0 b_1}{g}\right)(1-g) \\ a_1\, b_0\, \overline{g}\, g = -\dfrac{(\overline{f} - a_2\, b_0)}{(\overline{f} + a_2)} \\ \overline{b}_0 = q\, b_0, \overline{b}_1 = b_1/q \end{cases} \tag{11.10}$$

where $q = a_0 a_1 a_2 = b_0 b_1$, with a_0, a_1, a_2 being constants,

$$A_5^{(1)} - \mathrm{qP_{III}} : \begin{cases} g\overline{f}\, f = -a_1 \dfrac{(g - b_1)}{(a_0 g - b_1)} \\ \overline{g}\, g\, \overline{f} = b_1 \dfrac{(\overline{f}\, g + b_1)}{(a_2 a_0 \overline{f}\, g + b_1)} \\ \overline{a}_2 = b_2/q, \overline{a}_0 = q\, a_0 \end{cases} \tag{11.11}$$

with a_1, b_1, b_0 being constants.

Note that the systems (11.10) and (11.11) above share the same rational surface. To understand how different equations can arise on the same rational surface, we need to understand its symmetries. Each rational surface S is represented in terms of an extended affine Weyl group, which is generated by its simple roots (see Section 11.5). But each set of simple roots gives rise to an accompanying set of orthogonal roots in the Picard group.[3] The span of these orthogonal roots provides another affine Weyl group whose elements leave S invariant, and therefore provide its symmetries. For $A_5^{(1)}$, the symmetry group is given by $(A_2 + A_1)^{(1)}$ (Sakai, 2001) and equations (11.10) and (11.11) are realized by different translations or Cremona isometries in this group. Degenerations of the $A_5^{(1)}-\mathrm{qP_{III}}$ equation lead to q-discrete second and first Painlevé systems associated with rational surfaces $A_6^{(1)}$ and $A_7^{(1)}$ respectively (see Section 11.3).

The systems classified by Sakai also include additive-type equations:

$$D_4^{(1)} - \mathrm{dP_V} : \begin{cases} \overline{f} + f = a_3 + \dfrac{a_1}{g+1} + \dfrac{a_0}{sg+1} \\ \overline{g}\, g = \dfrac{(\overline{f} - \lambda + a_2)(\overline{f} - \lambda + a_2 + a_4)}{s\overline{f}(\overline{f} - a_3)} \\ \overline{a}_0 = a_0 + \lambda, \overline{a}_1 = a_1 + \lambda, \overline{a}_2 = a_2 - \lambda, \end{cases} \tag{11.12}$$

where $\lambda = a_0 + a_1 + a_2 + a_3 + a_4$, with a_3, a_4, s being constants. Equation (11.12) is a recurrence relation obtained from Bäcklund transformations of $\mathrm{P_{VI}}$.

$$D_5^{(1)} - \mathrm{dP_{IV}} : \begin{cases} \overline{f}\, f = \dfrac{s\overline{g}}{(\overline{g} - a_3 + \lambda)(\overline{g} + a_0 + \lambda)} \\ \overline{g} + g = \dfrac{s}{\overline{f}} + \dfrac{a_0 + a_1}{1 - f} - \lambda + a_3 - a_0 \\ \overline{a}_0 = a_0 + \lambda, \overline{a}_3 = a_3 - \lambda, \end{cases} \tag{11.13}$$

where $\lambda = a_0 + a_1 + a_2 + a_3$, with a_1, a_2, s being constants. This equation arises from Bäcklund transformations of $\mathrm{P_V}$ as a recurrence relation.

$$D_5^{(1)} - \mathrm{dP_{III}} : \begin{cases} \overline{f} + f = 1 + \dfrac{a_0 - \lambda}{s - \overline{g}} + \dfrac{a_2 - \lambda}{\overline{g}} \\ \overline{g} + g = s - \dfrac{a_1 + a_0}{1 - f} + \dfrac{a_3 + a_0}{f} \\ \overline{a}_0 = a_0 - \lambda, \overline{a}_2 = a_2 + \lambda, \end{cases} \tag{11.14}$$

where $\lambda = a_0 + a_1 + a_2 + a_3$, with a_1, a_3, s being constants. Equation (11.14) can be obtained from Bäcklund transformations of $\mathrm{P_V}$ as a recurrence relation.

$$D_6^{(1)} - \mathrm{alt\ dP_{II}} : \begin{cases} f + \overline{f} = b_0 - a_0 - g - \frac{s}{g} \\ g\,\overline{g} = s - \dfrac{b_0 s}{\overline{f}} \\ \overline{b}_0 = b_0 + \lambda, \overline{b}_1 = b_1 - \lambda, \end{cases} \tag{11.15}$$

[3] The Picard group $\mathrm{Pic}(S)$ equipped with the intersection form can be described as a 10-dimensional lattice (Duistermaat, 2010), spanned by the two coordinate axes and the eight exceptional lines that arise from blowing up base points.

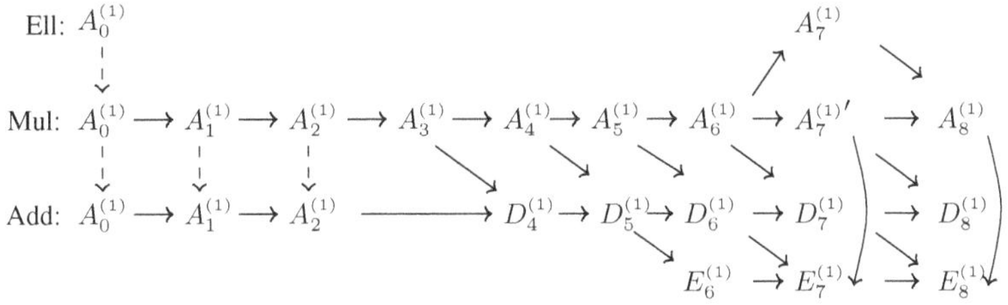

Figure 11.1 Sakai's diagram of rational surfaces of discrete Painlevé equations. Note that the abbreviations *Ell, Mul, Add* refer to the three different types of discrete Painlevé equations as explained at the beginning of the chapter.

where $\lambda = a_0 + a_1 = b_1 + b_0$, with a_1, a_2, s being constants. This equation arises from Bäcklund transformations of $\mathrm{P_{III}}$ as a recurrence relation.

$$E_6^{(1)} - \mathrm{dP_{II}}\ : \quad \begin{cases} \overline{f} = s - f - g + \dfrac{a_0}{g} \\ \overline{g} = s - g - \overline{f} - \dfrac{a_2 - \lambda}{\overline{f}} \\ \overline{a}_0 = a_0 + \lambda, \overline{a}_2 = a_2 - \lambda, \end{cases} \tag{11.16}$$

where $\lambda = a_0 + a_1 + a_2$, with a_1, s being constants. This equation arises from Bäcklund transformations of $\mathrm{P_{IV}}$ as a recurrence relation.

$$E_7^{(1)} - \mathrm{alt\ dP_{I}}\ : \quad \begin{cases} \overline{f} = -f - \dfrac{a_0}{g} \\ \overline{g} = s + f^2 - g \\ \overline{a}_0 = a_0 + \lambda, \overline{a}_1 = a_1 - \lambda, \end{cases} \tag{11.17}$$

where $\lambda = a_0 + a_1$, with s being a constant. Equation (11.17) arises from Bäcklund transformations of $\mathrm{P_{II}}$ as a recurrence relation.

The list above is not exhaustive. There are more q-discrete and d-discrete equations in Sakai's classification, which have not been included due to space constraints. Figure 11.1 provides a summary of the classification, with solid lines denoting inclusion in the group at the end of each respective line and dashed lines indicating different parameterizations of translations. The master equation at the top of the classification corresponds to the rational surface $A_0^{(1)}$ and is reproduced in Section 11.8.

The continuum limits of many systems such as the system given by equations (11.76) are not known and, therefore, the corresponding system is not labeled as a discrete version of a specific Painlevé equation. Nevertheless they are members of the collection of equations known as "discrete Painlevé equations" because of their common geometric properties.

In Sections 11.6.3, 11.7 and 11.8, we will see explicit ways of constructing special solutions of several discrete Painlevé equations. It has been proved that the general solutions of the $A_7^{(1)}$ q-discrete first Painlevé equation (11.19) (see Nishioka, 2010) cannot be expressed in terms of finite combinations of earlier known functions (including through solutions of linear difference equations with such functions as coefficients). Although such a proof is not yet available for other discrete Painlevé equations, their general solutions are believed to provide new transcendental functions, like the solutions of the continuous Painlevé equations.

Their transcendentality makes it difficult to provide explicit formulas that describe general solutions of discrete Painlevé equations, except in asymptotic limits. Asymptotic behaviors have been studied for two cases when the independent variable approaches infinity, i.e. for solutions of equations (11.1) and (11.21). For the additive-type equation (11.1), it is known that the scaled solutions $w_n = \sqrt{n} u_n$ (Joshi, 1997; Vereschagin, 1996; Joshi and Lustri, 2015) are either asymptotic to elliptic-functions or have power series expansions (with hidden exponentially small terms) in the limit $n \to \infty$. For the multiplicative-type equation (11.21), only solutions asymptotic to power series expansions have been studied (Joshi, 2015).

11.3 Coalescences and degeneracies of the discrete Painlevé equations

For the continuous Painlevé equations, it is well known that there exists a sequence of limits, which leads from P_{VI} to P_V, then to P_{IV} or P_{III}, then to P_{II} and finally to P_I. This sequence of limits is called a "coalescence cascade", because of the merger of types of movable singularities of each equation at each step. As more discrete Painlevé equations were discovered, a natural question arose as to whether they possessed limiting forms. The answer is yes (Ramani and Grammaticos, 1996), and in this section, we provide examples of such limits.

We focus first on qP_{III}, i.e. equation (11.3), as a prototypical example. Note that this equation can be rewritten as

$$\overline{w}\,\underline{w} = -\frac{\gamma\, w^2 + \zeta\, q^n\, w + \mu\, q^{2n}}{\alpha\, w^2 + \beta\, w + \gamma} \tag{11.18}$$

after scaling $w \to \sqrt{\alpha} w$, and renaming the parameters in the equation as $\gamma = -c\,d$, $\zeta = c\,d\,(a+b)/\alpha$, $\mu = -c\,d\,a\,b/\alpha^2$ and $\beta = \alpha\,(c+d)$.

The limit $\gamma \to 0$ leads to

$$qP_{II}: \quad \overline{w}\,\underline{w} = -\frac{\zeta\, q^n\, w + \mu\, q^{2n}}{w(\alpha\, w + \beta)}, \tag{11.19}$$

which has a continuum limit to P_{II} (after taking $\beta = -2 + \epsilon^3\,\theta$, $\zeta = -2$, $\mu = 1$, with $q = 1 + \epsilon^3/2$ and $\epsilon \to 0$, where θ is a constant parameter). We consider its transformations in Section 11.4 and its linear problem in Section 11.6.

Taking the limit $\beta \to 0$, we find

$$\overline{w}\,\underline{w} = -\frac{\zeta_0\, q^n}{w} + \frac{\mu_0\, q^{2n}}{w^2}, \tag{11.20}$$

where $\zeta_0 = \zeta/\alpha$ and $\mu_0 = \mu/\alpha$. The transformation $w = x\, q^{n/2}$ maps this to

$$\mathrm{qP_I}: \quad \overline{x}\,\underline{x} = -\frac{\zeta_0\, q^{-n/2}}{x} + \frac{\mu_0}{x^2}. \tag{11.21}$$

The name arises from the fact that this equation has a continuum limit to $\mathrm{P_I}$. Note that each pair of zeroes of the numerator and denominator on the right side of $\mathrm{qP_{III}}$ have merged (or coalesced) to provide only one zero in each of the numerator and denominator of $\mathrm{qP_I}$.

Other types of limits are possible. Suppose we start with the autonomous form of equation $\mathrm{qP_{III}}$, in which the numerator and denominator of the right side have a common zero. Then applying singularity confinement leads to an equation of the form

$$\overline{w}\,\underline{w} = \frac{a_0 q^n\, w + b}{c\, w + d_0\, q^n}. \tag{11.22}$$

If $c \neq 0$, then we can scale w to obtain $c = 1$ without loss of generality. Taking the limit $d_0 \to 0$, then provides

$$\overline{w}\,\underline{w} = a_0 q^n + \frac{1}{w}, \tag{11.23}$$

which is another q-discrete form of $\mathrm{P_I}$. To see this, apply the continuum limit $w = w_0(1 + \epsilon^2\, y)$, with $w_0^3 = -1/2$, $a_0 = 3\, w_0^2$, $q = 1 - \epsilon^5/3$, which in the limit $\epsilon \to 0$ lead to the equation $y_{tt} + 3\, y^2 + t = 0$, which is equivalent to $\mathrm{P_I}$. If, on the other hand, we had taken the limit $a_0 \to 0$ in equation (11.22), we would get an equation which is equivalent to equation (11.23).

Additive discrete equations also arise as degeneracies of q-discrete equations. To see this, consider $\mathrm{qP_{VI}}$, i.e. equation (11.6). Assume that $a_3 = c$, $a_4 = 1/c$, $b_3 = r$, $b_4 = 1/r$. Taking new variables $g = 1 + \delta\, y$, and parameters $b_1/q = c(1 + \delta\beta)$, $b_2/q = (1 + \delta\alpha)/c$, $a_1 = 1 + \delta\mu$, $a_2 = 1 + \delta\, \nu$, $r = 1 + \delta\, \rho$, $q = 1 + \delta\lambda$ and assuming the new independent variable to be $\zeta = n\, \lambda$, we find an additive discrete system in the limit $\delta \to 0$

$$\begin{aligned} y + \underline{y} &= -\frac{\zeta + \beta}{f/c - 1} - \frac{\zeta + \alpha}{c\, f - 1} \\ \overline{f}\, f &= \frac{(y - \zeta - \lambda/2 - \mu)\,(y - \zeta - \lambda/2 - \nu)}{(y - \rho)\,(y + \rho)}. \end{aligned} \tag{11.24}$$

This equation is often called a discrete form of $\mathrm{P_V}$ and is equivalent to (10.43). We refer the reader to details of the continuum limit in Grammaticos et al. (1998).

11.4 Bäcklund and other transformations of discrete Painlevé equations

Knowledge of the affine Weyl symmetry group of a discrete Painlevé equation provides transformations between solutions of the same equation, with different parameter values. In this section, we explain a method of directly constructing explicit transformations of scalar symmetric versions of discrete Painlevé equations. Further information about the theory of affine Weyl groups related to such transformations can be found in the next section.

The direct method starts with the singularity structure of a given discrete Painlevé equation. Ramani et al. (1995) showed that it is possible to construct τ-functions which are analytic at these singularities, analogous to the τ-functions constructed by Hirota for PDEs and partial difference equations (see Chapter 8). Such functions turn out to satisfy multilinear equations, whose structure leads us to a sequence of transformations (Joshi et al., 1998) between the original discrete Painlevé equation and another such equation.

11.4.1 An additive discrete second Painlevé equation

Here we show how to use the singularity structure of $\mathrm{dP_{II}}$ to find an invertible Miura transformation to another discrete Painlevé equation. (This method is similar to the one used to bilinearize equations in Section 8.5.1.)

The singularity pattern of $\mathrm{dP_{II}}$ (equation (11.2)) is given by $\ldots, \pm 1, \infty, \mp 1, \ldots$. (See Section 7.1.2 for the definition of singularity patterns.) We search for τ-functions by expressing w as a ratio of functions which vanish at each step in a singularity pattern. Since there are two singularity patterns, we need two τ-functions, which we call F and G. Requiring that F vanishes where $\underline{w} = -1$ and that G vanishes where $\underline{w} = +1$, Ramani et al. (1995) showed that we get the following transformations

$$w = 1 - \frac{\underline{F}\,\overline{G}}{F\,G} \tag{11.25a}$$

$$= -1 + \frac{\overline{F}\,\underline{G}}{F\,G}\,. \tag{11.25b}$$

Here, we focus on the discrete logarithmic derivative of one of these τ-functions by taking $\overline{v} = G/\overline{G}$. Then equations (11.25) become

$$w = 1 - \frac{\underline{F}}{\overline{v}\,F} \tag{11.26a}$$

$$= -1 + \frac{v\,\overline{F}}{F}\,. \tag{11.26b}$$

Now we take iterates of equation (11.26b) to obtain expressions for w, $\overline{w}$, $\underline{w}$ and $\underline{\underline{w}}$. Concurrently, we solve for $\underline{F}$, $\underline{\underline{F}}$ and $\overline{\overline{F}}$ from equation (11.26a) and its iterates and use these in $\mathrm{dP_{II}}$ and its backward iterate $\underline{\mathrm{dP_{II}}}$. The former equation is quadratic in $\overline{F}$ and solving this quadratic leads to

$$\overline{F} = \begin{cases} \dfrac{2F}{v} \\[2mm] \dfrac{(z+b-c)\,\overline{v}\,\overline{\overline{v}} + 2\,v\,\overline{v}}{b\,v\,\overline{v}\,\overline{\overline{v}} + v^2\,\overline{v} + v\,\underline{v}\,\overline{\overline{v}} + v\,\overline{v}\,\overline{\overline{v}}\,z}\,F\,. \end{cases} \tag{11.27}$$

Substituting the first case into $\underline{\mathrm{dP_{II}}}$ leads to an identity. However, the second case is not at all trivial, in fact the result is highly nonlinear, but it factorizes into three parts each of which is homogeneous in v, with degree 1, 3 and 5 respectively. Because of homogeneity, we change variables to $\overline{v} = r\,\underline{v}$ and find three equations for r. The first two factors lead to first-order equations for r, and the last one leads to

$$\begin{aligned}
&\overline{r}\,\underline{r}\Big(r^3\big((a-b+c-z)(b+c+z)\big) + r^2\big((b+z)^2 - (4+a)(b+z)+2a\big)\\
&\quad + r\big(2(b+z)-a-4\big)+1\Big) \qquad (11.28)\\
&\quad + \overline{r}\,r\big(r(b+z)+1\big) + r\underline{r}\,\big(r(-a+b+z)+1\big) + r^2 = 0.
\end{aligned}$$

This equation can be transformed to

$$\mathrm{dP_{34}}: \quad (\overline{s}+s)(s+\underline{s}) = -\frac{4\,s^2-(a+2\,c)^2/4}{(s-z/2+a/4-b/2)} \qquad (11.29)$$

by $r = 1/(s-z/2+a/4-b/2)$. Equation (11.29) is a discrete version of an equation often called the “thirty-fourth” Painlevé equation or $\mathrm{P_{34}}$ (due to its 34th place in the classification of second-order ODEs by Painlevé and his colleagues). It is well known (Ince, 1956) that the ODEs $\mathrm{P_{II}}$ and $\mathrm{P_{34}}$ are related to each other through a Miura transform. So it is not surprising to find such a relationship between their integrable discrete versions.

Retracing the above transformations, and eliminating $1/r = \underline{v}/\overline{v}$ by using equations (11.26), we find that $\mathrm{dP_{II}}$ and $\mathrm{dP_{34}}$ are related through the transformation

$$s = \frac{z}{2} - \frac{a}{4} + \frac{b}{2} + (1+\overline{w})\,(1-w) \qquad (11.30)$$

where the second equation is obtained by eliminating $\underline{v}/\overline{v}$ by using equations (11.26). By using $\mathrm{dP_{II}}$, we can also invert this transformation to find

$$w = \frac{a-2\,c-2\,s+2\,\underline{s}}{2(a+s+\underline{s})}. \qquad (11.31)$$

11.4.2 A multiplicative discrete second Painlevé equation

In this subsection, we find transformations of a $\mathrm{qP_{II}}$ equation given by

$$\mathrm{qP_{II}}: \quad \overline{w}\,\underline{w} = a\,z\,\frac{w+z}{w\,(w-1)}, \qquad (11.32)$$

where $z = z_0\,q^n$ and $\overline{w} = w(q\,z)$, $\underline{w} = w(z/q)$. This equation has singularity patterns $\{1,\infty,0\}$ and $\{0,\infty,1\}$ which suggests that there are two τ-functions F and G, given by

$$w = 1 + \frac{F\,\overline{G}}{\underline{F}\,G} \tag{11.33a}$$

$$= \frac{\overline{F}\,\underline{G}}{F\,G}\,. \tag{11.33b}$$

We introduce $u = F/\overline{F}$ and, as in the previous section, upshifting and downshifting the collection of equations $\mathrm{qP_{II}}$ and (11.33) to eliminate G, we find a homogeneous equation for u alone.

Introducing $r = \underline{u}/\overline{u}$ and eliminating G, we find the equation

$$(r\,\underline{r} - a\,z^2)\,(r\,\overline{r} - a\,q^2\,z^2) = -q\,a\,z\,(r - a\,z)\,(r - q\,z), \tag{11.34}$$

which turns out to be a q-discrete version of $\mathrm{P_{34}}$, more recognizable in the standard form

$$(y\,\underline{y} - 1)\,(y\,\overline{y} - 1) = -\frac{1}{z}\,(y - c)\,(y - 1/c), \tag{11.35}$$

where $r = \sqrt{a\,q}\,z\,y$ and $c = \sqrt{a/q}$. Notice that from equations (11.33), we can eliminate $\overline{G}/G$ between the expressions for w and $\overline{w}$, which leads to the transformation

$$r = \frac{\underline{u}}{\overline{u}} = \overline{w}\,(w - 1). \tag{11.36}$$

Conversely, by using equation (11.32), we find that w is given in terms of r by

$$w = \frac{r\underline{r} - az^2}{az - r}. \tag{11.37}$$

The pair of equations (11.36), (11.37) constitutes an (invertible) Miura transformation between equations (11.32) and (11.34).

We can repeat the same process, this time eliminating F rather than G, after introducing $v = \underline{G}/G$. This alternative leads to an equation for $\rho = \underline{v}/\overline{v}$, whose canonical form is

$$(Y\,\underline{Y} - 1)\,(Y\,\overline{Y} - 1) = -\frac{1}{z}\,(Y - \gamma)\,(Y - 1/\gamma), \tag{11.38}$$

where $\rho = z\,Y\sqrt{a/q}$ and $\gamma = \sqrt{aq} = c\,q$. Moreover, the analogous process that led to equation (11.36) shows

$$\rho = \underline{w}\,(w - 1). \tag{11.39}$$

Notice that y and Y satisfy the same equation but with apparently different parameters. But, we can consider equations (11.35) and (11.38) to be two instances of a sequence of equations with parameter values c mapped to $\widetilde{c} = \gamma$, where $\sqrt{\widetilde{a}/q} = \widetilde{c} = \gamma = \sqrt{aq}$. This implies that we have a mapping between a solution w of equation (11.32) with parameter

a and a solution $\widetilde{w}$ of a copy of the same equation but this time with parameter $\widetilde{a}$, where $\widetilde{a} = q^2 a$.

Rewriting equation (11.32) as an equation for $\widetilde{w}$, with parameter $\widetilde{a}$, as a starting point for the first elimination process that led to r and y, we have

$$r = \sqrt{\widetilde{a}\, q}\, z\, y = \sqrt{\widetilde{a}\, q}\, z\, Y = q^2 \rho. \tag{11.40}$$

Now equations (11.37) and (11.40) give

$$\begin{aligned}\widetilde{w} &= \frac{r\underline{r} - \widetilde{a}z^2}{\widetilde{a}z - r} \\ &= \frac{q^4\, \rho\underline{\rho} - \widetilde{a}z^2}{\widetilde{a}z - q^2\, \rho} \\ &= \frac{q^2\, \underline{w}(w-1)\, \underline{\underline{w}}(\underline{w}-1) - az^2}{az - \underline{w}(w-1)}\end{aligned} \tag{11.41}$$

and using equation (11.32) again, we obtain the Bäcklund transformation

$$\widetilde{w} = \frac{az\big(q\, \underline{w}(w-1) - z\big)}{w\big(az - \underline{w}(w-1)\big)} = \frac{z\big(qa\,(w+z) - w\overline{w}\big)}{w\big(w(\overline{w}-1) - z\big)} \tag{11.42}$$

where the two expressions on the right are equivalent because w satisfies equation (11.32). In an analogous way, we can also find

$$\underset{\sim}{w} = \frac{az\big(q\, \overline{w}(w-1)/q - z\big)}{w\big(az - \overline{w}(w-1)\big)} = \frac{z\big(a\,(w+z)/q - w\underline{w}\big)}{w\big(w(\underline{w}-1) - z\big)}, \tag{11.43}$$

where $\underset{\sim}{w}$ is a solution of equation (11.32) with parameter $\underset{\sim}{a} = a/q^2$.

11.5 Affine Weyl groups

Weyl groups are reflection groups. Affine Weyl groups incorporate integer translations of each reflection, and, thereby, create a lattice. The transformations we found in the previous section are elements of extended affine Weyl groups associated with each discrete Painlevé equation. In this section, we explain the basic elements of this theory and illustrate it for the qP_{IV} equation. Readers who already know the defining properties of affine and extended affine Weyl groups may wish to go straight to Section 11.5.3.

11.5.1 A two-dimensional example

In this section, we describe the construction of an affine Weyl group in two dimensions, starting with a particular choice of two linearly independent vectors.

Consider the simple example provided by the vectors labeled α and β, drawn in Figure 11.2. Note that all vectors in the plane can be described as linear combinations of these (i.e. they generate the two-dimensional real plane). The dotted lines s_1, s_2 in this figure are perpendicular to α, β respectively. If we take a reflection of β across s_1, we get a new vector $\alpha + \beta$, which is labeled γ in Figure 11.3.

Similarly, we can generate $-\alpha$ by reflecting α in the vertical line. Applying successive reflections, we see that only a finite set of vectors is obtained. These are the six vectors shown in Figure 11.4. Note that the only multiples of α, β that occur here are given by $\pm\alpha$, $\pm\beta$. Moreover, because the angle $\theta_{\alpha\beta}$ between α and β is equal to $2\pi/3$, we have $\cos(\theta_{\alpha\beta}) = -1/2$. Note also that $\theta_{s_1 s_2} = \pi/3$ and we have $\theta_{\alpha\beta} = \pi - \theta_{s_1 s_2}$.

We could also have chosen to include vectors that are translations of the reflections of these vectors. For example, consider the collection of vectors that are the translation of the vectors in Figure 11.3 to the left by α or 2α. This yields the additional lines in Figure 11.5. If we translate the vectors in Figure 11.4 by an integer multiple of the so-called lowest root $-\alpha - \beta$ and then take the collection of all such vectors, the result is called an affine Weyl group and denoted $A_2^{(1)}$ (for the notation and further details, see Humphreys, 1992).

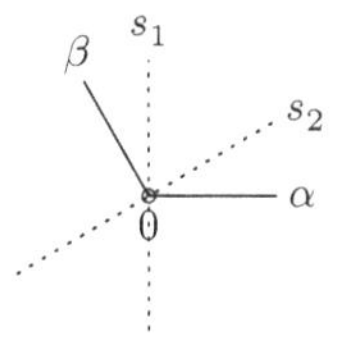

Figure 11.2 Two vectors α and β making angle $2\pi/3$. The lines s_1, s_2 are perpendicular to α, β respectively.

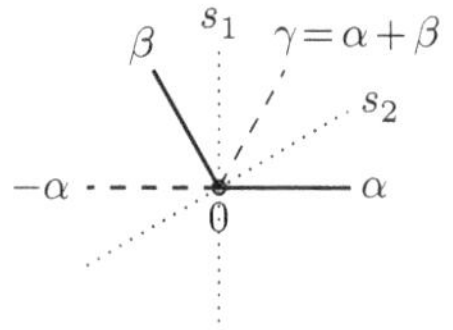

Figure 11.3 The vector $\gamma = \alpha + \beta$ denotes the reflection of β in the line s_1.

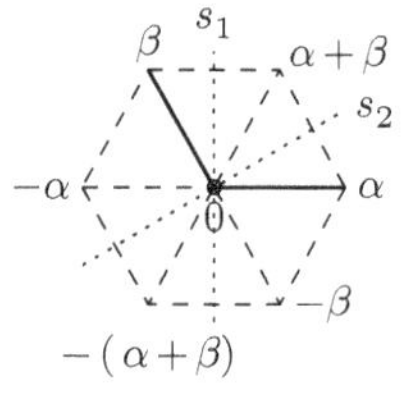

Figure 11.4 Vectors rotated by multiples of angle $\pi/3$.

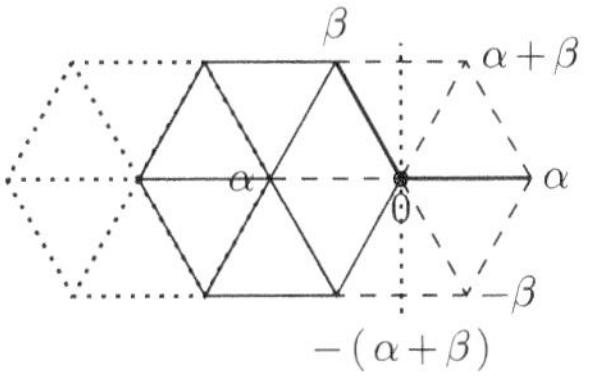

Figure 11.5 An affine translation of A_2.

Note that the latter is now an infinite-dimensional group because the integer parameters in the linear combination are free. The affine Weyl group yields a triangular lattice of the plane.

11.5.2 General affine Weyl groups

In this section, we describe how the ideas above generalize naturally to higher dimensions (and to other two-dimensional vectors satisfying a property called the crystallographic property).

Inner product

In any finite-dimensional real Euclidean space V, there is a natural inner product (or dot product) of vectors $x = (x_1, \ldots, x_n)$ and $y = (y_1, \ldots, y_n)$ defined by

$$(x, y) = \sum_{i=1}^{n} x_i\, y_i.$$

Recall that this product satisfies $(x, x) > 0$ for all $x \neq 0$ and that the length of the vector x is denoted $|x| = \sqrt{(x, x)}$. We also have $(x, y) = |x||y|\cos(\theta_{xy})$, where θ_{xy} is the angle between the vectors x and y.

Reflections

Given a vector y, we define

$$w_y(x) = x - 2\,\frac{(x, y)}{(y, y)}\,y \tag{11.44a}$$

$$= x - (x, y^{\vee})y \tag{11.44b}$$

for all $x \in V$, where

$$y^{\vee} := \frac{2y}{(y, y)}. \tag{11.45}$$

The function $w_y(x)$ has the following properties

$$\text{(i)} \quad w_y(y) = -y \tag{11.46a}$$

$$\text{(ii)} \quad w_y(x) = x \quad \Rightarrow (x, y) = 0 \tag{11.46b}$$

$$\text{(iii)} \quad (w_y(x), w_y(z)) = (x, z) \tag{11.46c}$$

which show that w_y is an isometry of V (from (iii)) given by a reflection across a hyperplane perpendicular to y (from (i) and (ii)).

Notice that $w_y(x)$ shifts the vector x by a multiple of $-y$, where the multiple is given by a projection

$$A_{xy} = \frac{2(x, y)}{(y, y)}. \tag{11.47}$$

This implies

$$A_{xy}A_{yx} = \frac{2(x, y)}{(y, y)}\frac{2(y, x)}{(x, x)} = 4\cos^2(\theta_{xy}).$$

The crystallographic property $A_{\alpha\beta} \in \mathbb{Z}$ places a strong constraint on the possible angles between roots in Φ. The property $4\cos^2(\theta_{\alpha\beta}) = n$, for $n \in \mathbb{N}$, implies $0 \leq |n| \leq 4$ and so the only possibilities are $\cos(\theta_{\alpha\beta}) = 0, \pm 1/2, \pm 1/\sqrt{2}, \pm\sqrt{3}/2$, or ± 1. The simplest choices are taken to be $\theta_{\alpha\beta} = \pi/2, 2\pi/3, 3\pi/4, 5\pi/6$. Noting that $\theta_{\alpha\beta} = \pi - \theta_{s_1 s_2}$, we define $\theta_{s_1 s_2} = \pi/m$, for each of these choices. That is, we have $m = 2, 3, 4, 6$, respectively.

Root systems

A *root system* Φ is a finite set of vectors satisfying the following properties:

(i) The group generated by Φ is the whole vector space: $\langle\Phi\rangle = V$.
(ii) For all $\alpha, \beta \in \Phi$, $w_\alpha(\beta) \in \Phi$.
(iii) The only scalar multiples of an element $\alpha \in \Phi$ that also belong to Φ are α itself and $-\alpha$.
(iv) For all $\alpha, \beta \in \Phi$, we have the *crystallographic property* $A_{\alpha\beta} \in \mathbb{Z}$.

The elements of Φ are called "roots". Corresponding to each root α, there is a coroot $\check{\alpha}$ defined by equation (11.45). The set of coroots also forms its own root system, called the dual root system. Given a root system Φ, a Weyl group W is defined as the group generated by the reflections

$$W = \langle w_\alpha | \alpha \in \Phi\rangle.$$

There are many choices of root systems possible. The subset Π of the root system Φ, which obeys the following properties is called the set of "simple roots":

(i) Π is a basis of V.
(ii) W is the span $\langle w_\alpha | \alpha \in \Pi\rangle$.
(iii) Each $\alpha \in \Phi$ can be written as a linear combination of Π where the coefficients of the linear combination are all either positive, negative (or zero) integers.

It is easy to check from Figure 11.4 that $\Pi = \{\alpha, \beta\}$ forms a simple root system.

Figure 11.6 Dynkin diagram of A_2.

Another way to encode the values of $A_{\alpha\beta}A_{\beta\alpha}$ is through a Dynkin diagram, which

(i) converts each simple root to one node;
(ii) places a single line joining two nodes if $A_{\alpha\beta}A_{\beta\alpha} = 1$, i.e. $\theta_{\alpha\beta} = 2\pi/3$;
(iii) places a double line joining two nodes if $A_{\alpha\beta}A_{\beta\alpha} = 2$, i.e. $\theta_{\alpha\beta} = 3\pi/4$;
(iv) places a triple line joining two nodes if $A_{\alpha\beta}A_{\beta\alpha} = 3$, i.e. $\theta_{\alpha\beta} = 5\pi/6$.

The root system is called *irreducible* if the corresponding Dynkin diagram is connected.

Figure 11.6 illustrates the Dynkin diagram of A_2, with simple roots shown in Figure 11.2. It is clear that this diagram and hence the corresponding root system is irreducible.

Cartan matrices

The Cartan matrix corresponding to the root system is the matrix A with entries given by $a_{ij} = A_{\alpha_i\alpha_j}$, where $A_{\alpha_i\alpha_j}$ is defined in equation (11.47) with α_i being elements of the root system. Notice that this means

(i) $a_{ii} = 2$
(ii) $a_{ij} < 0$ for $i \neq j$
(iii) $a_{ij} = 0$ if and only if $a_{ji} = 0$, for $i \neq j$
(iv) A can be written as the product DS, where D is a diagonal matrix and S is a symmetric matrix.

Here the entries of D are given by $\delta_{ij}/(\alpha_i, \alpha_j)$ and those of S by $2(\alpha_i, \alpha_j)$. The *affine Cartan matrix* includes an extra row and column, which includes entries $A_{\alpha_i\alpha_j}$ for the root system together with the lowest root.

So the Cartan matrix of A_2 is

$$A = \begin{pmatrix} 2 & -1 \\ -1 & 2 \end{pmatrix}$$

and the affine Cartan matrix corresponding to $A_2^{(1)}$ is

$$A = \begin{pmatrix} 2 & -1 & -1 \\ -1 & 2 & -1 \\ -1 & -1 & 2 \end{pmatrix},$$

where the lowest root is $-\alpha - \beta$.

Affine Weyl groups

Given a root system Φ and an integer k, an affine reflection is given by

$$w_{y,k}(x) = x - 2\frac{(x, y)}{(y, y)}y - k\,y^{\vee},$$

and an *affine Weyl group* $W^{(1)}$ is defined to be the group generated by all such affine reflections

$$W^{(1)} = \langle w_{r,k} | r \in \Phi,\ k \in \mathbb{Z} \rangle .$$

Moreover, an *extended affine Weyl group* $\widetilde{W}$ is the affine Weyl group with the added reflection $\alpha \leftrightarrow \beta$.

Another way to realize the extended affine Weyl group is to consider it as a semi-direct product of the affine Weyl group with the co-weight lattice. The weight lattice of a root system Φ consists of all $x \in \Phi$ such that $2(x, y)/(y, y)$ is an integer for all $y \in \Phi$. For A_2, with simple roots α, β, the weight lattice is generated by $(2\alpha + \beta)/3$ and $(\alpha + 2\beta)/3$.

When the root system is irreducible, the affine Weyl group is said to have a *nice generating set* composed of the reflections across hyperplanes corresponding to the simple roots and one additional affine reflection, namely $w_{y,1}$, where y is the unique "lowest" root, i.e. which has minimal sum of coefficients when written as a linear combination of the simple roots.

Coxeter relations

The nice generating set of the affine Weyl group, described above, is also a Coxeter generating set. That is the relations satisfied by the generators all follow from equations of the form $s^2 = 1$ and $(s\,t)^m = 1$, where s, t are in the generating set and m is an integer larger than 1. The number m is the same number in $\theta_{s_1 s_2} = \pi/m$ defined above to satisfy the crystallography property.

In the example of A_2, we have $m = 3$. In other words, we have Coxeter generators s_1, s_2 satisfying the relations

$$s_1^2 = 1,\, s_2^2 = 1,\, (s_1 s_2)^3 = 1.$$

The affine Weyl group $A_2^{(1)}$ has a nice generating set consisting of $w_{\alpha,0}$, $w_{\beta,0}$ and $w_{-\alpha-\beta,1}$, often called s_1, s_2 and s_0 respectively. The Coxeter relations are

$$s_i^2 = 1,\, (s_i s_{i+1})^3 = 1,\, i \in \mathbb{Z}/3\,\mathbb{Z}.$$

11.5.3 $q\mathrm{P}_{\mathrm{IV}}$ and its symmetries

In this subsection, we will describe the affine Weyl group of transformations of the $q\mathrm{P}_{\mathrm{IV}}$ equation:

$$\overline{f}_0 = a_0 a_1\, f_1 \frac{1 + a_2 f_2 + a_2 a_0 f_0 f_2}{1 + a_0 f_0 + a_0 a_1 f_0 f_1} \tag{11.48a}$$

$$\overline{f}_1 = a_1 a_2\, f_2 \frac{1 + a_0 f_0 + a_0 a_1 f_0 f_1}{1 + a_1 f_1 + a_1 a_2 f_1 f_2} \tag{11.48b}$$

$$\overline{f}_2 = a_2 a_0\, f_0 \frac{1 + a_1 f_1 + a_1 a_2 f_1 f_2}{1 + a_2 f_2 + a_2 a_0 f_2 f_0} \tag{11.48c}$$

where a_j, $j = 0, 1, 2$ are constant parameters with $a_0 a_1 a_2 = q$, and $f_0 f_1 f_2 = q\,t^2$, where $\overline{t} = q\,t$. The variables f_j, $j = 0, 1, 2$ are taken to be dependent on t and $\overline{f}_j = f_j(q\,t)$. The system of equations (11.48) was derived by Kajiwara et al. (2001) and these authors showed that it admits a family of transformations generated by s_i, $i = 0, 1, 2$, and by π, which satisfy

$$s_i^2 = 1, (s_i\, s_{i+1})^3 = 1, \pi^3 = 1, \pi\, s_i = s_{i+1}\, \pi,\ i = 0, 1, 2 \bmod 3.$$

Here π refers to the permutation of simple roots and is also referred to as a diagram automorphism, meaning an automorphism of the Coxeter–Dynkin diagram. These fundamental transformations act on the parameters as follows:

$$s_0(a_0) = a_0^{-1}, \qquad s_0(a_1) = a_1\, a_0, \qquad s_0(a_2) = a_2\, a_0 \tag{11.49a}$$
$$s_1(a_0) = a_0 a_1, \qquad s_1(a_1) = a_1^{-1}, \qquad s_1(a_2) = a_2\, a_1 \tag{11.49b}$$
$$s_2(a_0) = a_0 a_2, \qquad s_2(a_1) = a_1\, a_2, \qquad s_2(a_2) = a_2^{-1} \tag{11.49c}$$
$$\pi(a_0) = a_1, \qquad \pi(a_1) = a_2, \qquad \pi(a_2) = a_0. \tag{11.49d}$$

The actions of s_i and π on the dependent variables are given by the following equations:

$$s_0(f_0) = f_0, \qquad s_0(f_1) = f_1\,\frac{a_0 + f_0}{1 + a_0 f_0}, \qquad s_0(f_2) = f_2\,\frac{1 + a_0 f_0}{a_0 + f_0} \tag{11.50a}$$
$$s_1(f_0) = f_0\,\frac{1 + a_1 f_1}{a_1 + f_1}, \qquad s_1(f_1) = f_1, \qquad s_1(f_2) = f_2\,\frac{a_1 + f_1}{1 + a_1 f_1} \tag{11.50b}$$
$$s_2(f_0) = f_0\,\frac{a_2 + f_2}{1 + a_2 f_2}, \qquad s_2(f_1) = f_1\,\frac{1 + a_2 f_2}{a_2 + f_2}, \qquad s_2(f_2) = f_2 \tag{11.50c}$$
$$\pi(f_0) = f_1, \qquad \pi(f_1) = f_2, \qquad \pi(f_2) = f_0. \tag{11.50d}$$

Notice that the first nine transformations can be summarized as

$$s_i(a_j) = a_j\, a_i^{-a_{ij}}, s_i(f_j) = f_j \left(\frac{a_i + f_i}{1 + a_i f_i}\right)^{u_{ij}}, \quad i, j = 0, 1, 2, \tag{11.51}$$

where $A = (a_{ij})_{i,j=0}^2$, $U = (u_{ij})_{i,j=0}^2$ are matrices given by

$$A = \begin{pmatrix} 2 & -1 & -1 \\ -1 & 2 & -1 \\ -1 & -1 & 2 \end{pmatrix}, \quad U = \begin{pmatrix} 0 & -1 & -1 \\ -1 & 0 & -1 \\ -1 & -1 & 0 \end{pmatrix}.$$

Notice that A is the extended Cartan matrix of type $A_2^{(1)}$. We can verify that these transformations commute with the time evolution of the equation (11.48).

11.6 Linear problems

In Chapter 2, we saw that discrete Painlevé equations, such as dP_I arise as recurrence relations after eliminating derivatives from the Bäcklund transformations of continuous Painlevé equations. This relationship also provides transformations of the linear problems or Lax pairs of the latter equations, often called "Schlesinger transformations". In Chapter 10, Lax pairs were also obtained by similarity reductions. In this way, we can obtain a pair of linear problems for which the compatibility condition turns out to be additive discrete Painlevé equations. Linear problems for multiplicative- or elliptic-type discrete Painlevé equations also possess associated linear problems for which they act as compatibility conditions, although these had to be found in other ways. We focus on linear problems for q-discrete Painlevé equations in this section and explain the concepts through examples.

11.6.1 Compatibility of linear q-difference equations

A pair of linear problems

$$\psi(q\,x, t) = A(x, t)\,\psi(x, t) \tag{11.52a}$$

$$\psi(x, \lambda\,t) = B(x, t)\,\psi(x, t) \tag{11.52b}$$

where A and B are non-singular $N \times N$ matrices, ψ is an N-component (column) vector and q and λ are given complex numbers neither of which are equal to 0 or 1, define $\psi(q\,x, \lambda\,t)$ in two different ways: one by taking $t \mapsto \lambda\,t$ in (11.52a) and the second by taking $x \mapsto q\,x$ in (11.52b). The first example of an isomonodromic deformation problem of q-difference type (11.52) was given in Papageorgiou et al. (1992) for the discrete P_{III} and for discrete P_{VI} a Lax pair of this type was given in Jimbo and Sakai (1996).

Imposing the compatibility condition:

$$A(x, \lambda\,t)B(x, t) = B(q\,x, t)\,A(x, t), \tag{11.53}$$

ensures that the two systems are consistent.

qP_{II}

Consider the following matrices (Joshi and Shi, 2011), which arose from a study of reductions of the linear problem for lattice mKdV (Hay et al., 2007)

$$\begin{aligned} A(x, t) = \begin{pmatrix} e_1 & 0 \\ 0 & e_2 \end{pmatrix} + \begin{pmatrix} 0 & m_1(t) \\ m_2(t) & 0 \end{pmatrix} x \\ + \begin{pmatrix} n_1(t) & 0 \\ 0 & n_2(t) \end{pmatrix} x^2 + \begin{pmatrix} 0 & f_1(t) \\ f_2(t) & 0 \end{pmatrix} x^3, \end{aligned} \tag{11.54a}$$

$$B(x,t) = \begin{pmatrix} \dfrac{\sigma t}{\lambda g} & \dfrac{t\,x}{\sigma\,\lambda^2} \\ \dfrac{t\,x}{\sigma} & \dfrac{\lambda \underline{g}}{\sigma t} \end{pmatrix}, \tag{11.54b}$$

where $\sigma^2 = -\,e_1/e_2$ and the functions of t are given by

$$m_1(t) = \frac{e_2 g\left(\frac{t}{\lambda}\right)\left(e_2\lambda^2 g(t)\left(g\left(\frac{t}{\lambda}\right)-1\right)-e_1t^2\right)}{e_1 t^2 \lambda} \tag{11.55a}$$

$$m_2(t) = -\frac{t^2\left(g\left(\frac{t}{\lambda}\right)\left(e_1\lambda^2 - e_2 g(t)\right)+e_1t^2\right)}{\lambda g(t) g\left(\frac{t}{\lambda}\right)^2} \tag{11.55b}$$

$$n_1(t) = \frac{e_2\left(e_1t^2 + e_2 g(t)\left(g\left(\frac{t}{\lambda}\right)-1\right)g\left(\frac{t}{\lambda}\right)\right)}{e_1 g\left(\frac{t}{\lambda}\right)} \tag{11.55c}$$

$$n_2(t) = -\frac{e_2\left(-e_1t^2\lambda^2 g\left(\frac{t}{\lambda}\right) - e_1t^4 + e_2\lambda^4 g(t) g\left(\frac{t}{\lambda}\right)^2\right)}{e_1\lambda^2 g(t) g\left(\frac{t}{\lambda}\right)} \tag{11.55d}$$

$$f_1(t) = -\frac{e_2^2 t^2}{e_1\lambda} \tag{11.55e}$$

$$f_2(t) = -\frac{e_2^2 t^2 \lambda}{e_1} \tag{11.55f}$$

where $g = g(t)$, with $\overline{g} = g(\lambda\, t)$ and $\lambda^2 = q$. Substituting into equation (11.53), we find a compatibility condition for $g(t)$ which turns out to be a forward iteration of a scaled version of $\mathrm{qP_{II}}$, i.e. equation (11.32)

$$\overline{g}\,\underline{g} = \alpha\, t^2\, \frac{(g+t^2)}{g\,(g-1)}, \tag{11.56}$$

with $\alpha = e_1/e_2$.

11.6.2 Early results on linear q-difference equations

Birkhoff, and his students Adams, Carmichael, Le Caine and Guenther, developed a comprehensive theory of linear q-difference equations in the early 1900s. In this section, we will focus on major results provided by Carmichael (1912). He showed that (under certain conditions) these equations have solutions given in terms of convergent series in neighborhoods of the origin and infinity and that there is a connection matrix that relates the corresponding fundamental solutions in the overlapping neighborhood. These results are described more precisely in the following theorem, where the solution around infinity is denoted $Y^{(\infty)}$, that around the origin is denoted $Y^{(0)}$ and the connection matrix that relates them is given by $C(x)$, where $Y^{(\infty)}(x) = Y^{(0)}C(x)$.

Theorem 11.6.1 (Carmichael 1912, p. 159) *Given $q \in \mathbb{C}$, $q \neq 0, 1$, consider the linear matrix system*

$$Y(q\,x) = q^{\rho}\, x^{\rho}\, A^{(0)}(x)Y(x) = q^{\sigma}\, x^{\sigma}\, A^{(\infty)}(x)Y(x) \tag{11.57}$$

where ρ, σ are constants, $|q| \neq 1$, and $A^{(0)}(x)$, $A^{(\infty)}(x)$ are $N \times N$ matrix-valued functions that are analytic in a neighborhood of 0 and ∞ respectively with expansions

$$A^{(0)}(x) = \sum_{j=1}^{\infty} A_j^{(0)} x^j, \; |x| \leq r$$

$$A^{(\infty)}(x) = \sum_{j=1}^{\infty} \frac{A_j^{(\infty)}}{x^j}, \; |x| \geq R$$

for some nonzero r and R, such that the eigenvalues of each of $A_0^{(0)}$ and $A_0^{(\infty)}$ are finite, nonzero and whose ratios do not equal any integer powers of q. Then there exist fundamental matrix solutions with entries

1. $Y_{ij}^{(\infty)}(x) = \exp\Big(\rho \log\big(q(\eta^2 + \eta)\big)\Big) V_{ij}^{(\infty)}(x)x^{\mu_j}$
2. $Y_{ij}^{(0)}(x) = \exp\Big(\sigma \log\big(q(\eta^2 + \eta)\big)\Big) V_{ij}^{(0)}(x)x^{\nu_j}$

for some constants μ_j, ν_j, $j = 1, \ldots, N$ where $\eta = \ln x/\ln q$ and $V^{(\infty)}(x)$, $V^{(0)}(x)$ are $N \times N$ matrix-valued analytic functions, around $x = \infty$, $x = 0$ respectively. Moreover, there exists an $N \times N$ connection matrix $C(x)$ such that $Y^{(\infty)}(x) = Y^{(0)}(x)C(x)$, which satisfies $C(q\,x) = C(x)$.

Consider the case of equation (11.54a). The eigenvalues of the leading-order matrix around the origin are e_1 and e_2, while those near infinity are $\pm e_2\, q^5\, t^2/\alpha$. So the ratio of eigenvalues near 0 is $e_1/e_2 = \alpha$, while that near ∞ is -1. Carmichael's hypothesis imposes the condition that α cannot be an integer power of q or equal to 0. (In the next section, we show that nevertheless, explicit solutions can be found in the cases that do not necessarily satisfy these conditions.)

Suppose Carmichael's condition is satisfied for the general equation (11.52a), which is compatible with equation (11.52b). Writing the fundamental matrix in a neighborhood of $x = 0$ as $Y^{(0)}(x, t)$ and that in a neighborhood of $x = \infty$ as $Y^{(\infty)}(x, t)$, we have

$$Y^{(\infty)}(x, t) = Y^{(0)}(x, t)C(x, t), \tag{11.58}$$

where $C(qx, t) = C(x, t)$. Substituting this into equation (11.52b), we find

$$\begin{aligned} Y^{(\infty)}(x, \lambda\, t) &= B(x, t)Y^{(\infty)}(x, t) \\ \Rightarrow\; Y^{(0)}(x, \lambda\, t)C(x, \lambda\, t) &= B(x, t)Y^{(0)}(x, t)C(x, t) \\ \Rightarrow\; B(x, t)Y^{(0)}(x, t)C(x, \lambda\, t) &= B(x, t)Y^{(0)}(x, t)C(x, t)\,. \end{aligned} \tag{11.59}$$

Therefore, in regions where both B or $Y^{(0)}$ are non-singular, we get the result $C(x, \lambda t) = C(x, t)$; that is, C is also conserved in the deformation parameter t. This is the discrete analogue of the *isomonodromy* property, well known for linear problems associated with continuous Painlevé equations (Flaschka and Newell, 1980).

11.6.3 Special solutions and linear problems

Many discrete Painlevé equations have special solutions that can be expressed in terms of rational functions or previously known special functions. In this section, we consider special solutions of equation (11.56), which occur when the parameter α in the equation is given by $\alpha = \alpha_k := 1/q^{4k}$, for $k \in \mathbb{N}$. Note that this case violates Carmichael's condition for equation (11.54a).

First consider the case $k = 0$. Then the corresponding solution $g_0(t)$ satisfies

$$\overline{g}_0\, \underline{g}_0 = t^2 \frac{(g_0 + t^2)}{g_0\,(g_0 - 1)}.$$

This equation admits a solution of the form $g_0(t) = a_0\, t$ if the constant a_0 satisfies

$$a_0^3\, t(a_0 t - 1) = t\,(a_0 + t) \;\Rightarrow\; a_0^4 = 1 \text{ and } -a_0^3 = a_0 \;\Rightarrow\; a_0 = \pm i.$$

Since the equation is invariant under $t \mapsto -t$, these two cases of a_0 are equivalent and we focus on the case $a_0 = -i$. Substituting this into equation (11.54a), and setting also $e_1 = e_2 = 1$ we find

$$A(x, t) = \begin{pmatrix} 1 + x^2\, t(i + i\lambda - t/\lambda) & x(1 + it/\lambda + it/\lambda^2 - x^2t^2/\lambda) \\ x(\lambda^2 + i\lambda t + it - \lambda x^2\, t^2) & 1 + i(1 + \lambda)x^2 t - x^2t^2/\lambda \end{pmatrix}. \tag{11.60}$$

Observe that if we take a gauge transformation from ψ to Y given by

$$\psi(x, t) = \frac{1}{2}\begin{pmatrix} 1/\lambda & 1/\lambda \\ 1 & -1 \end{pmatrix} Y(x, t),$$

then equation (11.52a) can be written as

$$Y(q\, x, t) = \begin{pmatrix} 1 + \lambda\, x & 0 \\ 0 & 1 - \lambda\, x \end{pmatrix} \begin{pmatrix} 1 + i x\, t & 0 \\ 0 & 1 - i x\, t \end{pmatrix} \begin{pmatrix} 1 + i\frac{x\,t}{\lambda} & 0 \\ 0 & 1 - i\frac{x\,t}{\lambda} \end{pmatrix} Y(x, t). \tag{11.61}$$

Because the coefficient matrix is a product of diagonal matrices, we can solve for each component explicitly. The solution for each component Y_1, Y_2 of Y is

$$\begin{aligned} Y_1(x, t) &= \Gamma_{\lambda^2}\,(1 + \lambda\, t)\, \Gamma_{\lambda^2}\,(1 + i x\, t)\, \Gamma_{\lambda^2}\,(1 + i x\, t/\lambda) \\ Y_2(x, t) &= \Gamma_{\lambda^2}\,(1 - \lambda\, t)\, \Gamma_{\lambda^2}\,(1 - i x\, t)\, \Gamma_{\lambda^2}\,(1 - i x\, t/\lambda) \end{aligned} \tag{11.62}$$

given in terms of q-Gamma functions $\Gamma_q(1-z)$ which are defined as solutions of

$$y(q\,z) = (1-z)y(z)$$

given by

$$\Gamma_q(1-z) = \frac{1}{(z;q)_\infty}$$

where $(z;q)_\infty = (1-z)(1-q\,z)(1-q^2\,z)\ldots$.

From the results found for Bäcklund transformation of qP$_{\rm II}$ in Section 11.4.2, we have a mapping from a solution $g_k(t)$ of qP$_{\rm II}$ with parameter α_k to a solution $g_{k+1}(t)$ of qP$_{\rm II}$ with parameter α_{k+1}:

$$g_{k+1} = \frac{\left(\alpha_k\,t^2 - q^2\underline{g_k}g_k + \alpha_k g_k\right)t^2}{q^2\,g_k\left(\underline{g_k}g_k - g_k - t^2\right)}\,. \tag{11.63}$$

Remark 11.6.2 To avoid confusion with the notation of Section 11.4.2 recall that z in equation (11.32) has been replaced by t^2 in equation (11.56) (to avoid square roots of q in its linear problem). This implies that the multiplicative parameter q here is the square root of the parameter assumed in equation (11.32).

In particular, applying this transformation to $g_0(t) = -i\,t$, we get

$$g_1(t) = \frac{\left((q+1)t - i\right)t}{iq\left((q+1)t - iq\right)}\,.$$

In this way, we find an infinite sequence of rational solutions of qP$_{\rm II}$.

If we had started instead with k being a *half*-integer, we would also have found explicit solutions, but this time in terms of q-hypergeometric functions. We refer the reader to Joshi and Shi (2012) for details.

11.7 Linearization of discrete Painlevé equations

In the previous Section 11.6, we used the linear problem associated with a discrete Painlevé equation to find the latter's exact solutions. In this section, we show how to deduce such solutions directly from the nonlinear equation without using its linear problem.

We focus on special-function solutions of the Painlevé equation in this section and use an approach that is based on the following observation. The discrete Riccati equation

$$\overline{F} = \frac{a\,F + b}{c\,F + d} \tag{11.64}$$

where $a\,d - b\,c \neq 0$ can be linearized by taking $F = u/v$, where

$$\begin{cases} \overline{u} = K\,(a\,u + b\,v) \\ \overline{v} = K\,(c\,u + d\,v) \end{cases}$$

for some appropriate constant K. This linear system can be explicitly solved in terms of known special functions for a certain class of coefficients $a, \ldots, d$ (see Section 1.2.2).

To see how this observation helps to solve a Painlevé system, consider the example of qP_{VI} equation (11.6), under the assumption (Jimbo and Sakai, 1996)

$$\overline{f} = b_4\,\frac{g - a_2 t}{g - a_4}. \tag{11.65}$$

Substituting this into the first equation in (11.6), we find after canceling of common factors that

$$f = b_3\,\frac{g - a_1 t}{g - a_3}. \tag{11.66}$$

At the same time, equation (11.65) can be inverted to show that

$$g = a_4\,\frac{\overline{f} - a_2\,b_4\,t/a_4}{\overline{f} - b_4} \tag{11.67}$$

while equation (11.66) can be inverted and iterated to give

$$\overline{g} = a_3\,\frac{\overline{f} - q\,a_1\,b_3\,t/a_3}{\overline{f} - b_3}. \tag{11.68}$$

Multiplying these two equations and using the result to replace $\overline{g}\,g$ in the second equation of equation (11.6), we find an identity for $\overline{f}$, which can only be satisfied for arbitrary $\overline{f}$ if the following two equations are satisfied:

$$\frac{b_1}{b_3} = q\,\frac{a_1}{a_3}, \quad \frac{b_2}{b_4} = \frac{a_2}{a_4}.$$

Under these conditions on the parameters, let

$$t = \frac{b_3}{b_1}\,s, \; a = \frac{a_3}{a_4}, \; b = \frac{a_2\,b_4}{a_4\,b_1}, \; c = \frac{a_3\,b_4}{a_4\,b_3}.$$

Then equations (11.65)–(11.66) can be linearized by taking

$$g = a_3\,\frac{u}{v}, \; \overline{f} = b_4\,\frac{u - (b\,s/c)\,v}{u - v/a} \tag{11.69}$$

under which equation (11.68) gives

$$\frac{\overline{u}}{\overline{v}} = \frac{\left(1-\frac{as}{c}\right)u + \frac{1-b}{c}sv}{\left(1-\frac{a}{c}\right)u + \frac{1-bs}{c}v}. \tag{11.70}$$

Take the linear system of equations

$$\begin{cases} \overline{u} &= \frac{1}{1-(ab/c)s}\left((1-\frac{as}{c})u + \frac{1-b}{c}sv\right), \\ \overline{v} &= \frac{1}{1-(ab/c)s}\left((1-\frac{a}{c})u + \frac{1-bs}{c}v\right). \end{cases} \tag{11.71}$$

This system is solved by the following q-hypergeometric functions (which are also called basic hypergeometric functions) (Gasper and Rahman, 2004)

$$u = {}_2\phi_1\left(\begin{matrix} a\ b \\ c \end{matrix}; q, s\right), \quad v = \frac{c-a}{c-1}\,{}_2\phi_1\left(\begin{matrix} a\ qb \\ qc \end{matrix}; q, s\right). \tag{11.72}$$

11.8 Sakai's elliptic discrete Painlevé equation

The aim of this section is to reproduce Sakai's elliptic discrete Painlevé equation which was provided first in Sakai (2001). We present the version written down in Murata et al. (2003), which has a slightly more concise form. The statement of the equation requires some preliminary notation:

$$w = \begin{pmatrix} a & b \\ c & d \end{pmatrix} z \quad \Leftrightarrow \quad w = \frac{az+b}{cz+d} \tag{11.73}$$

$$M(h,\kappa_1,\kappa_2,s) = M_0(h,\kappa_1,\kappa_2,s)\,M_1(h,\kappa_1,\kappa_2,s)\,M_2(h,\kappa_1,\kappa_2,s), \tag{11.74}$$

where

$$M_0(h,\kappa_1,\kappa_2,s) = \begin{pmatrix} -\wp\left(2s - \frac{(-\kappa_1+\kappa_2)}{2}\right) & \wp\left(2s - \frac{(\kappa_1-\kappa_2)}{2}\right) \\ 1 & 1 \end{pmatrix} \tag{11.75a}$$

$$\begin{aligned} &M_1(h,\kappa_1,\kappa_2,s) = \\ &\operatorname{diag}\Big(\big(h-\wp(\kappa_2)\big)\big(\wp(2s)-\wp(2s-\kappa_2)\big) \\ &\qquad\qquad \times \big(\wp(2s-(\kappa_1+\kappa_2)/2) - \wp(2s-(\kappa_1-\kappa_2)/2)\big), \\ &\quad \big(h-\wp(\kappa_1)\big)\big(\wp(2s)-\wp(2s-\kappa_1)\big) \\ &\qquad\qquad \times \big(\wp(2s-(\kappa_1+\kappa_2)/2) - \wp(2s-(-\kappa_1+\kappa_2)/2)\big)\Big), \end{aligned} \tag{11.75b}$$

$$M_2(h,\kappa_1,\kappa_2,s) = \begin{pmatrix} 1 & -\wp(2s-\kappa_1) \\ 1 & -\wp(2s-\kappa_2) \end{pmatrix}. \tag{11.75c}$$

The equation involves 8 constant parameters b_i, $i = 1,\ldots,8$ and an independent variable t. We let $\lambda = \frac{1}{2}\sum_{i=1}^{8} b_i$, and also define $c_i = b_i + t$, $d_i = t - b_i$.

Sakai's elliptic difference Painlevé equation is then the following system, governing $(f(t), g(t))$:

$$\overline{g} = M\left(f, c_7, c_8, t - \sum_{i=1}^{6} c_i/4\right) M\left(f, c_5, c_6, t - \sum_{i=1}^{4} c_i/4\right) \times M\left(f, c_3, c_4, t - \sum_{i=1}^{2} c_i/4\right) M\left(f, c_1, c_2, t\right) g \tag{11.76a}$$

$$\underline{f} = M\left(g, d_7, d_8, t - \sum_{i=1}^{6} d_i/4\right) M\left(g, d_5, d_6, t - \sum_{i=1}^{4} d_i/4\right) \times M\left(g, d_3, d_4, t - \sum_{i=1}^{2} d_i/4\right) M\left(g, d_1, d_2, t\right) f \tag{11.76b}$$

where $\overline{g} = g(t+\lambda)$, $\underline{f} = f(t-\lambda)$. Recall that $M(g,a,b,c)f = M_0(g,a,b,c)\, M_1(g,a,b,c)\, M_2(g,a,b,c)f$ where M_j are defined in equations (11.75) and multiplication by a matrix actually means carrying out the linear fractional action defined in equation (11.73).

It is interesting to note that the space of initial values is constructed by blowing up $\mathbb{P}^1 \times \mathbb{P}^1$ at eight points in general position, which are given by

$$p_i: \quad \big(\wp(t+b_i), \wp(t-b_i)\big), \; i = 1, \ldots, 8. \tag{11.77}$$

Only one blow-up is needed at each p_i to regularize the space. Each corresponding exceptional line e_i (replacing p_i) is a curve of self-intersection number -1. Moreover, the bi-degree (2, 2)-curves (i.e. curves defined by polynomials that are quadratic in f and in g) passing through $\{p_i\}_{i=1}^{8}$ are not degenerate. This means the surface is a rational surface of type $A_0^{(1)}$.

11.9 Notes

The early history of discrete Painlevé equations was already mentioned in Section 11.1. Some extensive reviews of this relatively new subject were given in Grammaticos et al. (1999), and Grammaticos and Ramani (2004) for their general properties and in Tamizhmani et al. (2004) for the special solutions for those discrete Painlevé equations. This has been a thriving subject to which many researchers have contributed.

Starting with the examples of $\mathrm{dP_I}$ and $\mathrm{dP_{II}}$, which emerged from the Hermitian and unitary matrix ensembles, random matrix theory has been a rich source of results for discrete Painlevé equations; see e.g. Forrester (2003); Forrester and Witte (2006); Ormerod et al. (2011); Forrester (2010). The corresponding class of special functions, which arise as coefficients in the recurrence relations for semi-classical orthogonal polynomials lead to important classes of Painlevé functions (Shohat, 1939; Freud, 1976; Nevai, 1986; Magnus, 1995; van Assche and Foupouagnigni, 2003; Forrester and Witte, 2004; Ormerod et al., 2011; Witte, 2012). It is interesting to note that the appearance of discrete Painlevé equations historically in orthogonal polynomial theory consisted of additive-type equations, but recent discoveries show that multiplicative-type equations are also possible (Boelen et al., 2010; Boelen and van Assche, 2010).

For the continuous Painlevé equations, the investigation of the isomonodromic deformation problems, designed to study the transcendental solutions and their asymptotic properties (connecting them to Riemann–Hilbert problems), has been well-developed (Fokas et al., 2006). The corresponding isomonodromic theory for the discrete Painlevé equations is still in its early stages. The starting point is the Lax pairs for the dPs, which for some of them have been found from early on, cf. Nijhoff and Papageorgiou (1991); Levi et al. (1992); Joshi et al. (1992); Papageorgiou et al. (1992). The actual analysis has so far only been performed only in a handful of cases: the differential monodromy cases of dP_{I} (Fokas et al., 1992b) and dP_{II} (Muğan and Santini, 2011), and a few instances of difference (Borodin, 2004), and q-difference systems (Joshi and Shi, 2011, 2012; Ormerod, 2011). Thus, the study of the transcendental solutions of q-Painlevé equations and most notably of the Sakai's elliptic difference equation still remains largely open. A Lax pair for Sakai's elliptic difference equation has been found (Arinkin and Borodin, 2006; Rains, 2008; Yamada, 2009; Noumi et al., 2013; Rains, 2011).

Isomonodromic deformation problems of q-type Painlevé equations (Murata, 2009) can be obtained from the one for qP_{VI} given by Jimbo and Sakai (1996). For the remaining q-Painlevé equations they can be deduced from higher-order discrete systems as (order) reductions (Sakai, 2005). However, the isomonodromic analysis (Mano, 2010) of such equations remains to be carried out and a similar analysis of higher-order discrete equations and proof of their irreducibility, cf. e.g. Nishioka (2012, 2010) remain interesting open problems for the future.

In his studies of the continuous Painlevé equation K. Okamoto stressed the importance of the affine Weyl Group (Okamoto, 1979, 1981, 1986, 1999). This is also the point of view taken in the book Noumi (2004). Sakai developed the Okamoto point of view in the seminal paper (Sakai, 2001), where both continuous and discrete Painlevé equations were united in a single framework. In Kajiwara et al. (2001, 2003, 2004, 2005a,b), this geometric framework was used to deduce special function solutions of discrete Painlevé equations. A universal description (Kajiwara et al., 2006) using point configurations provides a further explanation of Sakai's elliptic Painlevé equation. Furthermore Ohta et al. (2001) provide an alternative construction of an elliptic discrete equation with affine Weyl symmetry group $E_8^{(1)}$. The review of Kajiwara et al. (2006) gives a comprehensive overview of the geometric approach to discrete Painlevé equations.

We finally mention in this context also the higher-order and higher-rank Painlevé-type equations. The Garnier system (Garnier, 1912), which is given in Appendix C, constitutes what one could call a P_{VI} hierarchy. The construction of hierarchies of Painlevé equations goes back to Flaschka and Newell (1981). For discrete Painlevé equations such hierarchies were constructed in Cresswell and Joshi (1999a,b), Clarkson et al. (2003). At the same time discrete analogues of the Garnier system were given in Nijhoff and Walker (2001), Tongas and Nijhoff (2006) for the differential case, while q-discrete version of the Garnier system was constructed by Sakai (2005).

Exercises

11.1 As described in Section 11.1, each of the scalar discrete Painlevé equations possesses an *asymmetric* version. Consider the following examples and find the system of equations satisfied by the odd and even iterates $u_k = w_{2k}$, $v_k = w_{2k+1}$ in each case.

(a)

$$\mathrm{dP_{II}}: \quad \overline{w} + \underline{w} = \frac{z_n\, w + \sigma}{1 - w^2},$$

where $z_n = \alpha\, n + \beta + \gamma\, (-1)^n$.

(b)

$$\mathrm{qP_{III}}: \quad \overline{w}\, \underline{w} = cd\, \frac{(w - a_n\, q^n)(w - b_n\, q^n)}{(w - c_n)(w - d_n)},$$

where

$$a_n = \begin{cases} a_e, & n = 2k \\ a_o, & n = 2k+1 \end{cases}, \qquad b_n = \begin{cases} b_e, & n = 2k \\ b_o, & n = 2k+1 \end{cases}$$

$$c_n = \begin{cases} c_e, & n = 2k \\ c_o, & n = 2k+1 \end{cases}, \qquad d_n = \begin{cases} d_e, & n = 2k \\ d_o, & n = 2k+1 \end{cases}$$

with $q\, a_e\, b_e\, c_o\, d_o = a_o\, b_o\, c_e\, d_e$. Furthermore, show that the system you find is the same as $\mathrm{qP_{VI}}$, i.e. equation (11.6).

11.2 The symmetric $\mathrm{qP_V}$ can be written as (Ramani et al., 1991, 1995)

$$(\overline{w}w - 1)\,(w\underline{w} - 1) = z^2\, \frac{\gamma(w^4+1) + \kappa\, w(w^2+1) + \mu\, w^2}{\alpha w^2 + \beta z w + \gamma z^2}, \tag{11.78}$$

where $\alpha, \beta, \gamma, \kappa$ and μ are constant, while $z = q^n$.

(a) Show that equation (11.78) can be mapped to

$$(2x-1)\overline{x}\underline{x} - x(\overline{x} + \underline{x}) = \frac{(c_0 - a_0\, z^2)x^3 + \left(e_0 + \frac{1}{2}(c_0 + a_0\, z^2 - b_0 z)\right)x^2 - 2\, d_0 x + d_0}{2\, a_0\, z^2\, x^2 + (b_0 z - 2\, a_0 z^2)x + \frac{1}{2}(c_0 + a_0\, z^2 - b_0 z)}$$

under the transformation $x = (1+w)/2$, by finding relations between the parameters $\alpha, \beta, \gamma, \kappa, \mu$ of $\mathrm{qP_V}$ and a_0, b_0, c_0, d_0, e_0. Find the continuum limit of this equation after taking $c_0 = \mathcal{O}(1/\delta^2)$, $e_0 = \mathcal{O}(1/\delta^2)$, as $\delta \to 0$ with the remaining parameters being held finite. Simultaneously, in this limit, we assume $q = 1 + 2\,p\,\delta$, $n = t/\delta$, which implies that q^n becomes $\exp(2\,p\,t)$. The constant $2p$ can be assumed to be unity without loss of generality by rescaling t. Show that this limit leads to P_{V} with independent variable transformed to e^t.

(b) Show that the case $\gamma = 0$ approaches P_{IV} in the continuum limit. Hint: take $w = 1 + \epsilon y$, with $q = 1 - \epsilon^2$ and find scalings of the remaining parameters that lead to $y_{tt} = y_t^2/(2y) + 3\,y^3/2 + 4ty^2 + 2y(t^2 + a) + b/w$ as $\epsilon \to 0$ for appropriate a and b.

(c) Show that the case $\alpha = 0$ with $\gamma = 0$ approaches P_{34} in a continuum limit. Hint: in this case, you will need to consider $w = 1 + \epsilon^2 y$ and $q = 1 - \epsilon^3/2$.

(d) Show that the case $\gamma = 0$ with $\kappa = 0$ approaches P_{II} in a continuum limit. Hint: in this case, you will need to consider $w = i + \epsilon y$ and $q = 1 - \epsilon^3/2$.

(e) Show that the case $\alpha = \kappa = 0$ with $\gamma = 0$ approaches P_{I} in a continuum limit. Hint: in this case, you will need to consider $w = w_0(1 + \epsilon^2 y)$, with $w_0^2 = -1/3$ and $q = 1 - \epsilon^5/4$.

11.3 Consider the system of equations

$$y_n = (1 + x_n)\,(1 - x_{n+1}) - \frac{1}{2}\,(z_n + z_{n+1}) \tag{11.79a}$$

$$x_n = \frac{m + y_n - y_{n-1}}{y_n + y_{n-1}}\,. \tag{11.79b}$$

By eliminating either x_n or y_n and their respective iterates, find the second-order difference equations satisfied by each variable and verify that these turn out to be dP_{34} or dP_{II}.

11.4 Find an exact solution of the asymmetric dP_I, i.e. equations (11.8) by assuming $g = -f + \sigma$ and following the steps below.

(a) Show that

$$f = -g + \sigma$$
$$\underline{g} = \frac{z + \gamma}{f}$$
$$\overline{f} = \frac{z + \alpha - \gamma}{g}\,.$$

(b) Assuming that $\overline{z} - z = \alpha/2$, show that these results are consistent only if $\alpha = 4\,\gamma$.

(c) Using the above results, show that g must satisfy the discrete Riccati equation

$$g = \sigma - \frac{z + \gamma}{\underline{g}}$$

and linearize this equation by using $g = u/v$ to get

$$\overline{v} - \sigma\, v + (z + \gamma)\underline{v} = 0.$$

This is a discrete Airy equation, which is equivalent to a recurrence relation of the parabolic cylinder equation.

11.5 Show that Sakai's elliptic discrete Painlevé equation (11.76) has the following one-parameter solution

$$f = \wp(2t^2/\lambda + t + h_0), \;\; g = \wp(-2t^2/\lambda + t - h_0),$$

where h_0 is a free parameter determined by initial conditions. Hint: you will first need to prove that

$$\wp(2t - (c_1 + c_2)/2 - c) = M(\wp(c), c_1, c_2, t)\wp(2t - c),$$

for arbitrary c. This relies on the identity (A.13) in Appendix A, relating the difference between two Weierstrass $\wp$-functions to a ratio of sigma functions. Sakai's elliptic discrete Painlevé equation then leads to the following equations

$$\wp(t + \lambda - \overline{h}) = \wp(-3t - \lambda - h)$$
$$\wp(t + \lambda - \underline{h}) = \wp(-3t - \lambda + h)$$

whose compatibility condition gives $\overline{h} = h + 4t + 2\lambda$, so that we get $h = 2t^2/\lambda + h_0$.

12

Lagrangian multiform theory

In this chapter we give a description of the Lagrangian theory of systems that are integrable in the sense of multidimensional consistency. We have already encountered Lagrangians in the context of DIS, namely in Chapter 6 as generating functions for the discrete-time systems and integrable maps. In principle, such Lagrangian descriptions can also hold in nonintegrable dynamical systems. However, as we shall see, the notion of multidimensional consistency will shed new light on the nature of the variational formalism.

Many integrable PDEs admit a Lagrangian description, i.e. they can be derived by variational approach from an action functional based on the *least-action principle*. This principle has a long history going back to Leibniz, Maupertuis and Euler at the beginning of the eighteenth century, and it was incorporated in the theories of dynamics by Lagrange and Hamilton (see Gelfand and Fomin, 2000). The variational description of the equations we study, including the integrable class, is important because it enables a systematic approach to the search for symmetries and conservation laws, e.g. through Noether's famous theorem (Noether, 1918). Furthermore, it is a starting point for the transition to quantum theory: quantization of physical systems through *the Feynman path integral* can be thought of as the quantum analogue of the Lagrangian approach to classical systems.

Figure 12.1 Joseph-Louis Lagrange (1736–1813).

Most integrable systems allow a Lagrangian description. The dynamical equations arise from applying the least-action principle to an action functional S, obtained from a Lagrangian function through an integral or a sum over the independent variables. This allows us to derive one equation for each dependent variable from a single Lagrangian.

What was still missing from this standard Lagrangian approach was the description of *multidimensional consistency*, which is a key feature of integrable systems (see Chapter 3). This means that one function can solve many consistent equations simultaneously and it is then natural to expect that this property should be visible in the Lagrangian framework.

The question of how multidimensional consistency can be made manifest in the variational approach was answered in Lobb and Nijhoff (2009), Lobb (2010), which pointed out that the Lagrangians of the system should be reinterpreted as a part of an extended object, that is, as a *Lagrangian multiform*.

The development of this approach is the main aim of this chapter. We will start with the conventional Lagrangian approach (continuous and discrete) and then describe Lagrangian multiforms. Two key new observations are: first, the replacement of the traditional Lagrangian function by a differential or difference p-form which is *closed on solutions of the equations of the motion*, and second, that *the Lagrangians themselves are solutions of the constitutive equations arising from the variational principle.*

12.1 Conventional Lagrange theory and its discrete analogue

We recall the standard theory of Lagrangians in this section. In Section 12.2, we will describe a new perspective on the variational principle, which arises naturally in integrable systems. In order to distinguish and clarify the difference in perspective, we summarize the conventional Lagrange formulation, in the discrete as well as the continuous case and in one and several dimensions, and give some applications to integrable cases.

12.1.1 The continuous Lagrange formalism

Let us consider a system of functions $\boldsymbol{x}(t) = (x_1(t), \ldots, x_N(t))$, which may be thought of as describing particle positions depending on a time-variable t. The least-action, or variational principle, forms a universal way of encoding the equations of motion for such a system of particles. The main quantity here is the so-called action functional, i.e. a function of the functions $\boldsymbol{x}(t)$, which trace out a "path" in configuration space. This action functional takes the form:

$$S[\boldsymbol{x}(t)] = \int_{t_0}^{t_1} \mathcal{L}(\boldsymbol{x}(t), \dot{\boldsymbol{x}}(t))\, dt \ , \quad \dot{\boldsymbol{x}} := \frac{d\boldsymbol{x}}{dt} \ , \tag{12.1}$$

where the function $\mathcal{L}$, depending on the positions $\boldsymbol{x}(t)$ and the velocities $\dot{\boldsymbol{x}}(t)$ is called the *Lagrange function* or more succinctly the *Lagrangian*. The choice of the Lagrangian will determine the dynamical model that we want to consider, and usually it is in this choice that the specific physical application of the formalism is represented. However, without relying on any particular form of this Lagrangian, the variational principle will allow us to derive a set of universal equations for it, which then in turn will provide the equations of motion of the physical system under consideration.

The variational (least-action) principle states the following: *As a function of paths* $\{\boldsymbol{x}(t)\, ,\ t_0 \le t \le t_1\}$ *the "classical path" is the one for which the action S acquires an extremum.* In the same way as the search for the local minima or critical points of a function $f(x)$ relies on varying the independent variable x, $x \to x + \delta x$ and looking for points at which the first-order deviations from the value of f vanish, i.e.

$$f(x+\delta x) = f(x) + f'(x)\,\delta x + \mathcal{O}(\delta x^2) \quad \Rightarrow f'(x) = 0\,,$$

we can obtain the critical *functions* $\boldsymbol{x}(t)$ of the action functional by considering arbitrary infinitesimal *local* variations $\boldsymbol{x}(t) \to \boldsymbol{x}(t) + \delta\boldsymbol{x}(t)$, and setting the first-order contributions of those variations to the action functional to zero, i.e. $\delta S = 0$. This will then lead to a system of conditions for the *critical paths* which define the desired classical trajectories of the corresponding dynamical system. Those infinitesimal local variations can be illustrated by the following diagram:

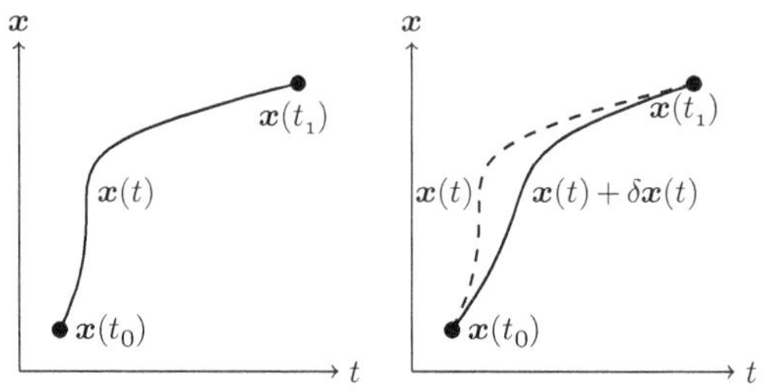

Figure 12.2 Infinitesimal local deformation of the curve Γ.

Under those variations we obtain, up to first order, the following variation of the action functional

$$\begin{aligned}
S + \delta S &\approx \int_{t_0}^{t_1} \mathcal{L}(\boldsymbol{x}(t) + \delta\boldsymbol{x}(t), \dot{\boldsymbol{x}}(t) + \delta\dot{\boldsymbol{x}}(t))\,dt \\
&\approx \int_{t_0}^{t_1} \left[\mathcal{L}(\boldsymbol{x}(t), \dot{\boldsymbol{x}}(t)) + \frac{\partial\mathcal{L}}{\partial\boldsymbol{x}} \cdot \delta\boldsymbol{x}(t) + \frac{\partial\mathcal{L}}{\partial\dot{\boldsymbol{x}}} \cdot \delta\dot{\boldsymbol{x}}(t)\right] dt \\
&\approx S + \int_{t_0}^{t_1} \left[\frac{\partial\mathcal{L}}{\partial\boldsymbol{x}} - \frac{d}{dt}\left(\frac{\partial\mathcal{L}}{\partial\dot{\boldsymbol{x}}}\right)\right] \cdot \delta\boldsymbol{x}(t)\,dt + \left.\frac{\partial\mathcal{L}}{\partial\dot{\boldsymbol{x}}} \cdot \delta\boldsymbol{x}(t)\right|_{t_0}^{t_1} \\
\delta S &\approx \int_{t_0}^{t_1} \left[\frac{\partial\mathcal{L}}{\partial\boldsymbol{x}} - \frac{d}{dt}\left(\frac{\partial\mathcal{L}}{\partial\dot{\boldsymbol{x}}}\right)\right] \cdot \delta\boldsymbol{x}(t)\,dt
\end{aligned}$$

while fixing the end points of the paths $\delta\boldsymbol{x}(t_0) = \delta\boldsymbol{x}(t_1) = 0$. (Here we have used integration by parts in the third step.) Assuming that those first-order contributions vanish, as required by the least-action principle, and taking into account that the local variations $\delta\boldsymbol{x}(t)$ are arbitrary, we obtain the following set of equations

$$\frac{\partial\mathcal{L}}{\partial x_i} - \frac{d}{dt}\left(\frac{\partial\mathcal{L}}{\partial\dot{x}_i}\right) = 0\,, \quad i = 1, \ldots, N \tag{12.2}$$

which are the well-known *Euler–Lagrange (EL) equations*. Given any specific choice of Lagrangian, these EL equations determines the classical path under given initial conditions.

12.1.2 The discrete-time Lagrange formalism

A discrete version of the Lagrange formalism, where time t is discrete, was first studied by Cadzow (1970); Logan (1973); Maeda (1981). Replacing $t \in \mathbb{R}$ by $n \in \mathbb{Z}$, and functions $\boldsymbol{x}(t)$ of continuous time by functions $\boldsymbol{x}_n := (x_1(n), \ldots, x_N(n))$ of discrete time n, and the action functional becomes a sum:

$$S[\boldsymbol{x}_n] = \sum_{n=n_0}^{n_1-1} \mathcal{L}(\boldsymbol{x}_n, \boldsymbol{x}_{n+1}) , \tag{12.3}$$

in which $\mathcal{L}$ is now a discrete Lagrange function, i.e. it depends on the coordinates $\boldsymbol{x}_n$ and their shift $\boldsymbol{x}_{n+1}$, which replace the derivative with respect to t. In this case, the variation of the action under local infinitesimal variations $\boldsymbol{x}_n \to \boldsymbol{x}_n + \delta\boldsymbol{x}_n$ yields:

$$\begin{aligned} S + \delta S &\approx \sum_{n=n_0}^{n_1-1} \mathcal{L}(\boldsymbol{x}_n + \delta\boldsymbol{x}_n, \boldsymbol{x}_{n+1} + \delta\boldsymbol{x}_{n+1}) \\ &\approx \sum_{n=n_0}^{n_1-1} \left[\mathcal{L}(\boldsymbol{x}_n, \boldsymbol{x}_{n+1}) + \frac{\partial\mathcal{L}}{\partial\boldsymbol{x}_n} \cdot \delta\boldsymbol{x}_n + \frac{\partial\mathcal{L}}{\partial\boldsymbol{x}_{n+1}} \cdot \delta\boldsymbol{x}_{n+1} \right] \\ &\approx S + \sum_{n=n_0}^{n_1-1} \left[\frac{\partial\mathcal{L}}{\partial\boldsymbol{x}_n} + T^{-1}\left(\frac{\partial\mathcal{L}}{\partial\boldsymbol{x}_{n+1}} \right) \right] \cdot \delta\boldsymbol{x}_n \end{aligned}$$

where T denotes the shift operator $T\boldsymbol{x}_n = \boldsymbol{x}_{n+1}$ and its inverse is $T^{-1}\boldsymbol{x}_n = \boldsymbol{x}_{n-1}$. Note that in the third line, instead of an integration by parts, we do a simple resummation of the discrete variable n in order to bring out a common factor $\delta\boldsymbol{x}_n$, and we ignore contributions from the boundary since we want to fix $\delta\boldsymbol{x}_{n_0} = \delta\boldsymbol{x}_{n_1} = 0$. Thus, because all the variations $\delta\boldsymbol{x}_n$ are arbitrary, the condition that the first-order terms vanish leads to the following relations:

$$\frac{\partial\mathcal{L}}{\partial x_i(n)} + T^{-1}\left(\frac{\partial\mathcal{L}}{\partial x_i(n+1)} \right) = 0 , \quad i = 1, \ldots, N \tag{12.4}$$

which are the *discrete Euler–Lagrange equations*. They determine the classical equations of the motion in discrete time. (Note the seeming sign difference between (12.4) and (12.2): it is due to the fact that in the discrete case we use a backward shift in (12.4).)

12.1.3 Higher-dimensional Lagrange formalism

Continuous case

So far we have considered the Lagrange formalism for functions of one independent variable. However, if we consider functions of more than one variable (referred to as a "field" by physicists), say in the simplest case $u(x, t)$ with independent variables x and t, an action functional takes the form

$$S[u(x,t)] = \iint_\Omega dx\, dt\, \mathcal{L}(u, u_x, u_t, u_{xx}, u_{xt}, u_{tt}, \dots)\,, \tag{12.5}$$

i.e. an integral over a domain Ω in the space of the independent variables x and t of a *Lagrangian density* $\mathcal{L}$ depending on the field variable u and on its partial derivatives $u_x, u_t, \dots$ up to a certain order. The least-action principle now states that the relevant field equations, i.e. a system of PDEs, are obtained by requiring the action functional, viewed as a function of all possible field configurations, has a critical point under infinitesimal local variations $u(x,t) \to u(x,t) + \delta u(x,t)$. Thus, computing the effect on the functional S of such variations up to first order in the variations $\delta u(x,t)$, we get:

$$\begin{aligned}\delta S &= \iint_\Omega dx\, dt \delta\mathcal{L}(u, u_x, u_t, \dots)\\ &= \iint_\Omega dx\, dt \left[\frac{\partial \mathcal{L}}{\partial u}\delta u + \frac{\partial \mathcal{L}}{\partial u_x}\delta u_x + \frac{\partial \mathcal{L}}{\partial u_t}\delta u_t + \cdots\right]\\ &= \iint_\Omega dx\, dt \left[\frac{\partial \mathcal{L}}{\partial u} - \frac{\partial}{\partial x}\left(\frac{\partial \mathcal{L}}{\partial u_x}\right) - \frac{\partial}{\partial t}\left(\frac{\partial \mathcal{L}}{\partial u_t}\right) + \cdots\right]\delta u\end{aligned}$$

where we used the fact that $\delta u_x = (\delta u)_x, \delta u_t = (\delta u)_t, \dots$, and where integrations by parts has been successively been performed, ignoring boundary terms, by imposing that at the boundary of the integration region Ω we have $\delta u|_{\partial\Omega} = 0$, and the same for the derivatives of the variations. Now, since we want to set $\delta S = 0$ for arbitrary δu in the inside of Ω, we must have that its coefficients within the integral vanish, leading to the *field-theoretic Euler–Lagrange equations*:

$$\frac{\partial \mathcal{L}}{\partial u} - \frac{\partial}{\partial x}\left(\frac{\partial \mathcal{L}}{\partial u_x}\right) - \frac{\partial}{\partial t}\left(\frac{\partial \mathcal{L}}{\partial u_t}\right) + \cdots = 0. \tag{12.6}$$

The terms included in these EL equations depend on how many partial derivatives are included in the Lagrange density.

Example 12.1.1 Consider the following action

$$S[w(x,t)] = \iint_\Omega dx\, dt\, \mathcal{L}(w, w_x, w_t, w_{xx}, \dots) \quad \text{with} \quad \mathcal{L} = \tfrac{1}{2}w_x w_t - \tfrac{1}{2}w_{xx}^2 + w_x^3\,. \tag{12.7}$$

The relevant EL equation in this case is;

$$\frac{\partial \mathcal{L}}{\partial w} - \frac{\partial}{\partial x}\left(\frac{\partial \mathcal{L}}{\partial w_x}\right) - \frac{\partial}{\partial t}\left(\frac{\partial \mathcal{L}}{\partial w_t}\right) + \frac{\partial^2}{\partial x^2}\left(\frac{\mathcal{L}}{\partial w_{xx}}\right) = 0 \tag{12.8}$$

and this is easily seen to lead to the KdV equation in the form

$$w_{xt} + w_{xxxx} + 3(w_x^2)_x = 0 \,,$$

i.e. the KdV equation for $u := w_x$.

Discrete case

In the discrete case we can also have higher-dimensional variational principles, namely when the number of independent discrete variables is larger than one. For simplicity, let us consider discrete fields $u(n_1, n_2)$ depending on two independent discrete variables n_1, n_2. A corresponding action functional would take the form:

$$S[u(n_1, n_2)] = \sum_{n_1, n_2 \in \mathbb{Z}} \mathcal{L}(u, T_1 u, T_2 u; p_1, p_2) \,,$$

where the Lagrangian "density" now is a function of the field u and its elementary shifts $T_1 u = u(n_1 + 1, n_2)$, $T_2 u = u(n_1, n_2 + 1)$ on the lattice of the independent discrete variables. Once again we apply the least-action principle, stating that the relevant system of PΔEs is obtained by requiring the action to be a critical point under infinitesimal "local" variations $u(n_1, n_2) \to u(n_1, n_2) + \delta u(n_1, n_2)$. A similar computation as in Section 12.1.2, following similar steps to compute the variation δS up to first order, yields:

$$\begin{aligned}\delta S = \sum_{n_1, n_2 \in \mathbb{Z}} \Big\{ &\frac{\partial}{\partial u} \mathcal{L}(u, T_1 u, T_2 u; p_1, p_2) \delta u + \frac{\partial}{\partial T_1 u} \mathcal{L}(u, T_1 u, T_2 u; p_1, p_2) \\ &\delta(T_1 u) + \frac{\partial}{\partial T_2 u} \mathcal{L}(u, T_1 u, T_2 u; p_1, p_2) \delta(T_2 u) \Big\} = 0 \,.\end{aligned}$$

Setting $\delta(T_i u) = T_i \delta u$, and resumming each of the terms we get:

$$\begin{aligned}0 = \sum_{n_1, n_2 \in \mathbb{Z}} \Big\{ &\frac{\partial}{\partial u} \mathcal{L}(u, T_1 u, T_2 u; p_1, p_2) + \frac{\partial}{\partial u} \mathcal{L}(T_1^{-1} u, u, T_1^{-1} T_2 u; p_1, p_2) \\ &+ \frac{\partial}{\partial u} \mathcal{L}(T_2^{-1} u, T_1 T_2^{-1} u, u; p_1, p_2) \Big\} \delta u \,,\end{aligned}$$

ignoring boundary terms. Since δu is arbitrary, the discrete field-theoretic EL equations are given by:

$$\begin{aligned}\frac{\partial}{\partial u} \Big(&\mathcal{L}(u, T_1 u, T_2 u; p_1, p_2) + \mathcal{L}(T_1^{-1} u, u, T_1^{-1} T_2 u; p_1, p_2) \\ &+ \mathcal{L}(T_2^{-1} u, T_1 T_2^{-1} u, u; p_1, p_2) \Big) = 0 \,.\end{aligned} \tag{12.9}$$

The latter forms a system of PΔEs which provide the relevant equations under consideration.

Example 12.1.2 Lattice potential KdV equation: Consider an action functional with a Lagrangian given by

$$\mathcal{L}(u, T_1 u, T_2 u; p_1, p_2) = u(T_1 - T_2)u + (p_1^2 - p_2^2)\ln(T_1 u - T_2 u)\,. \tag{12.10}$$

The discrete Euler–Lagrange equations (12.9) lead to the following PΔE for the function $u(n_1, n_2)$:

$$T_1 u - T_2^{-1} u + \frac{p_1^2 - p_2^2}{u - T_1 T_2^{-1} u} + T_1^{-1} u - T_2 u + \frac{p_1^2 - p_2^2}{u - T_1^{-1} T_2 u} = 0\,. \tag{12.11}$$

This equation can be seen as a "gluing" of two copies of a quadrilateral equation, namely the lattice potential KdV equation (H1). In fact, if u obeys the quadrilateral equation

$$Q(u, T_1 u, T_2 u, T_1 T_2 u; p_1, p_2) := (u - T_1 T_2 u)(T_1 u - T_2 u) + p_1^2 - p_2^2 = 0\,, \tag{12.12}$$

which can be written in the following **3-leg form**:

$$(u + T_1 u) - (u + T_2 u) + \frac{p_1^2 - p_2^2}{u - T_1 T_2 u} = 0\,, \tag{12.13}$$

then (12.11) is automatically satisfied (the reverse is not necessarily true!). The latter equation is, in fact, nothing but the lattice potential KdV equation introduced in Section 3.8.1. Pictorially, the variational equation can be represented in the following form:

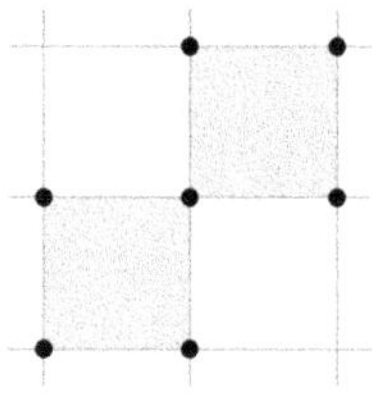

Figure 12.3 Stencil of the variational equation (12.11).

The fact that we get a "weaker" equation than the actual quad-equation (12.12) for the variable u, with respect to which we vary the action, is not surprising: it is analogous to what happens in the example of the continuous KdV equation of the previous subsection. Like evolution equation of the form $u_t = f(u, u_x, u_{xx}, \ldots)$ which are only indirectly variational, the quad-equations are also only indirectly variational with the variational equation being a derived equation from the quadrilateral one.[1]

[1] We note that the latter seems to be a particular feature of quadrilateral lattice equations, while it is not present in higher-order equations, such as the lattice BSQ equation (3.71), which is defined on the 9-point stencil of Figure 3.9. In Exercise 12.7 it is shown that the latter, and not a derived form, arises as the EL equations of a 4-point Lagrangian.

12.1.4 Actions for quadrilateral lattice equations

The Lagrange formalisms described above in various settings were not restricted to the integrable situation alone, and could hold for any system, integrable or nonintegrable, so long as it admits a description in terms of a least-action principle. We will now look at the specific situation of integrable PΔEs and, in particular, the case of quadrilateral lattice equations as described in Chapter 3.

3-leg forms and 3-point Lagrangians

We have seen in Example 12.1.2 that the lattice potential KdV equation (12.12) can be written in a 3-leg form (12.13), which arises from the variational formulation (12.11) of the lattice potential KdV equation. In fact, it turns out that all the quad-equations in the ABS list (Adler et al., 2003) allow a 3-leg form, which can be written in one of the two following ways:

$$\text{additive:} \quad \psi(u, u_1; p_1) - \psi(u, u_2; p_2) = \phi(u, u_{1,2}; p_1, p_2), \tag{12.14a}$$

$$\text{multiplicative:} \quad \frac{\psi(u, u_1; p_1)}{\psi(u, u_2; p_2)} = \phi(u, u_{1,2}; p_1, p_2)\,, \tag{12.14b}$$

in which ϕ and ψ are referred to as "long leg" and "short leg" respectively. Pictorially, the 3-leg forms can be represented in the following way:

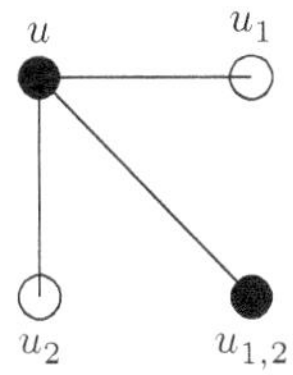

Figure 12.4 Diagram of the 3-leg form of quadrilateral lattice equations.

In equation (12.13), we have an additive form with the identifications $\psi(u, v; p) = u+v$ and $\phi(u, v; p, q) = (p^2 - q^2)/(u - v)$. In the case of Q4 (3.46) we have a multiplicative form with the identifications:

$$\psi(u, v; p) = \frac{\sigma(\xi + \eta + \alpha)\,\sigma(\xi - \eta + \alpha)}{\sigma(\xi + \eta - \alpha)\,\sigma(\xi + \eta - \alpha)}\,,$$

$$\phi(u, v; p, q) = \frac{\sigma(\xi + \eta + \alpha - \beta)\,\sigma(\xi - \eta + \alpha - \beta)}{\sigma(\xi + \eta - \alpha + \beta)\,\sigma(\xi + \eta - \alpha + \beta)}\,,$$

in which we have introduced uniformizing variables ξ, η and parameters α, β related to u, v and p, q respectively, by the relations $u = \wp(\xi)$, $v = \wp(\eta)$, $p = \wp(\alpha)$, $q = \wp(\beta)$. Here $\wp$ and σ are Weierstrass elliptic functions; see Appendix B.3.

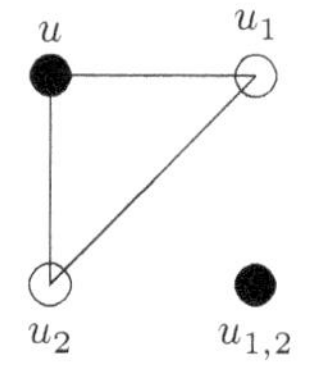

Figure 12.5 Graph of the configuration of the 3-point Lagrangians.

Note that the above 3-leg forms (12.14) involve four points on the lattice as shown in Figure 12.4. From these one can deduce actions which involve contributions from all four points on each elementary quadrilateral, as presented in Adler et al. (2003).

We will next see that Lagrangians of the form $\mathcal{L}(u, u_1, u_2; p_1, p_2)$ depending only on three points of each elementary quadrilateral, as shown in Figure 12.5, are possible. This choice turns out to be essential for developing the Lagrangian multiform theory describing multidimensionally consistent systems.

It is possible to find such Lagrangians, depending on three points, for all equations in the ABS list. We now give such a "universal" description of the Lagrange structure for scalar affine-linear quadrilateral lattice equations of the form

$$Q(u, u_i, u_j, u_{ij}; p_i, p_j) = 0, \tag{12.15}$$

for a lattice with given directions $\boldsymbol{e}_i, \boldsymbol{e}_j$ and associated lattice parameters p_i, p_j.

For simplicity, we will denote shifts on the lattice by indices, i.e. for a dependent variable $u = u(\boldsymbol{n})$ we write:

$$u_i = T_i u := u(\boldsymbol{n} + \boldsymbol{e}_i)\ , \quad u_{ij} = T_i T_j u := u(\boldsymbol{n} + \boldsymbol{e}_i + \boldsymbol{e}_j) \tag{12.16}$$

where $\boldsymbol{n} = \sum_j n_j \boldsymbol{e}_j$, with $n_j \in \mathbb{Z}$, and $\boldsymbol{e}_j$ denote a system of elementary vectors.

The function Q is assumed to be affine-linear and to possess Kleinian symmetry (see Section 3.1.2). If the relevant Lagrangian functions have 3-point form, then the action takes the form

$$S[u(\boldsymbol{n})] = \sum_{\boldsymbol{n} \in \boldsymbol{\Lambda}} \mathcal{L}_{p_i, p_j}(u(\boldsymbol{n}), T_i u(\boldsymbol{n}), T_j u(\boldsymbol{n})) \tag{12.17}$$

and it follows that the EL equations become (cf. (12.9))

$$\left(\frac{\partial}{\partial u} + T_i^{-1} \frac{\partial}{\partial u_i} + T_j^{-1} \frac{\partial}{\partial u_j} \right) \mathcal{L}_{p_i, p_j}(u, u_i, u_j) = 0. \tag{12.18}$$

A result of Adler et al. (2003) is that the discriminants of the affine-linear function $Q = Q(u, u_i, u_j, u_{ij}; p_i, p_j)$ take the form (cf. (3.51))

$$Q_{u_j} Q_{u_{ij}} - Q\, Q_{u_j u_{ij}} =: k(p_i, p_j) h_{p_i}(u, u_i) \tag{12.19a}$$

$$Q_{u_i} Q_{u_{ij}} - Q\, Q_{u_i u_{ij}} =: k(p_j, p_i) h_{p_j}(u, u_j) \tag{12.19b}$$

$$Q_u Q_{u_{ij}} - Q\, Q_{u u_{ij}} =: -k(p_i, p_j) h_{p_{ij}}(u_i, u_j) \tag{12.19c}$$

where $h_p(u, v)$ is a symmetric biquadratic function of u and v depending on a single parameter p, with $k(p, q) = -k(q, p)$ being a function of the lattice parameters only and where p_{ij} depends only on the lattice parameters p_i and p_j. The biquadratic functions h_p are essential ingredients in the context of the Lagrangian theory.

The main result of this section is the following. There exists, up to a multiplicative constant, a unique three-point Lagrangian $\mathcal{L}(u, u_i, u_j)$, whose Euler–Lagrange equations vanish on solutions of the quad-equation $Q(u, u_i, u_j, u_{ij}; p_i . p_j) = 0$, which is given by

$$\begin{aligned}\mathcal{L}(u, u_i, u_j) = &\int_{u^0}^{u}\int_{u_i^0}^{u_i} \frac{dx\,dy}{h_p(x, y)} - \int_{u^0}^{u}\int_{u_j^0}^{u_j} \frac{dx\,dy}{h_q(x, y)} - \int_{u_i^0}^{u_i}\int_{u_j^0}^{u_j} \frac{dx\,dy}{h_r(x, y)} \\ &+ \int_{u_i^0}^{u_i} dx \int_{u_j^0}^{Y(u^0,x,u_{ij}^0)} \frac{dy}{h_r(x, y)} + \int_{u_j^0}^{u_j} dy \int_{u_i^0}^{X(u^0,y,u_{ij}^0)} \frac{dx}{h_r(x, y)}\end{aligned} \tag{12.20}$$

where the functions X and Y are solutions of the respective equations

$$Q(u^0, x, Y, u_{ij}^0; p, q) = 0, \quad Q(u^0, X, y, u_{ij}^0; p, q) = 0\,.$$

We refer to Xenitidis et al. (2011) for details of the proof. The main idea of the proof is to use implicit differentiation of Q with respect to two of the four points on a quadrilateral, while keeping the other two fixed. Above we assumed that the lower limits u^0 were kept fixed while varying u, but it is equally valid to consider (12.20) as a Lagrangian under variations of u^0 while keeping u fixed.

12.2 Lagrangian 2-form structure

In this section, we describe the idea of **Lagrangian multiforms** (Lobb and Nijhoff, 2009), which offers a new perspective on variational principles.

The property of multidimensional consistency is an important theme in the development integrability in this book. The Lagrangian structure provides a fundamental description of dynamics in the sense that the Lagrangian encodes basic features of the equations it represents: the equations of the motion themselves, their symmetries and in some sense also the global properties of solutions. The conventional Lagrangian framework applies equally well to integrable and nonintegrable equations and does not distinguish between them. However, if the Lagrangian is to encode all the essential features of the system, it is essential for integrable systems that multidimensional consistency is incorporated into the Lagrangian structure.

The standard Lagrangian approach provides a single equation for each dependent variable. However, due to multidimensional consistency, there is an arbitrary number of (in fact, infinitely many) equations imposed on one and the same dependent variable. This creates a conundrum. The solution to this problem is obvious in hindsight: to incorporate many self-consistent equations for one and the same dependent variable, *the Lagrangian must be an extended object, namely a differential or difference form.*[2]

The next immediate question is: where is multidimensional consistency hidden? The answer comes from the requirement that the *Lagrangian multiform is closed on the solutions of the system.* This property of closure guarantees that the action functional *is independent of the surface* on which it is evaluated. This surface-independence allows us to deform the action integral in any direction we choose, and subsequently the EL equations derived on those surfaces yield the multitude of compatible equations forming the multidimensionally consistent system. In the subsections below, we set up a program to realize these ideas.

12.2.1 Discrete variational principle for integrable lattice systems

Let us consider the 3-point Lagrangians

$$\mathcal{L}_{ij}(\boldsymbol{n}) = \mathcal{L}(u(\boldsymbol{n}), u(\boldsymbol{n}+\boldsymbol{e}_i), u(\boldsymbol{n}+\boldsymbol{e}_j); p_i, p_j) \tag{12.21}$$

in terms of fields $\boldsymbol{u}$ (where by "field" we now mean the collection of all dependent variables u at each lattice location) defined on elementary plaquettes in a multidimensional lattice defined by a vector $\boldsymbol{n}$; see (12.16). These plaquettes are characterized by ordered triplets $\sigma_{ij}(\boldsymbol{n}) = (\boldsymbol{n}, \boldsymbol{n}+\boldsymbol{e}_i, \boldsymbol{n}+\boldsymbol{e}_j)$; see Figure 12.6. We will impose the condition that it be anti-symmetric under interchange of the direction labels i and j

$$\mathcal{L}_{ij} = -\mathcal{L}_{ji},$$

and consider them as functions with an orientation associated with the plaquette on which they are defined. Therefore, we consider them as discrete analogues of 2-forms (Flanders, 1963; Von Westenholz, 2009).

Embedded in a multidimensional lattice, we can now choose a surface σ, consisting of a connected configuration of such elementary plaquettes $\sigma_{ij}(\boldsymbol{n})$; see Figure 12.7. We choose a corresponding action:

$$S[u(\boldsymbol{n});\sigma] = \sum_{\sigma_{ij}(\boldsymbol{n})\in\sigma} \mathcal{L}_{ij}(\boldsymbol{n}), \tag{12.22}$$

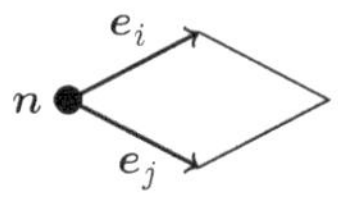

Figure 12.6 Elementary plaquette in the lattice representing a lattice 2-form.

[2] Note that by a Lagrangian multiform, we do not mean that the Lagrangian is a volume form; that is, an object whose degree as a form is the same as the dimension of the space in which it is given.

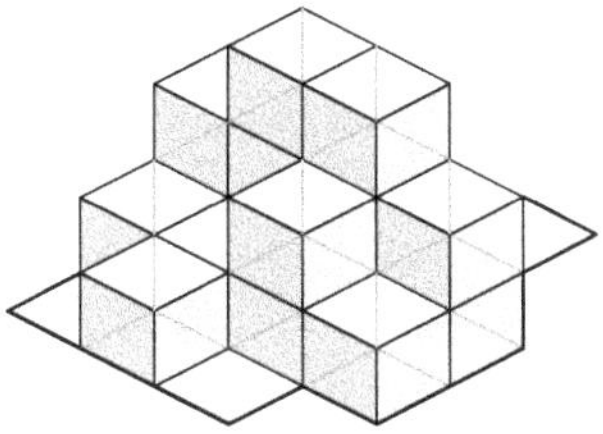

Figure 12.7 A discrete surface embedded in a multidimensional lattice.

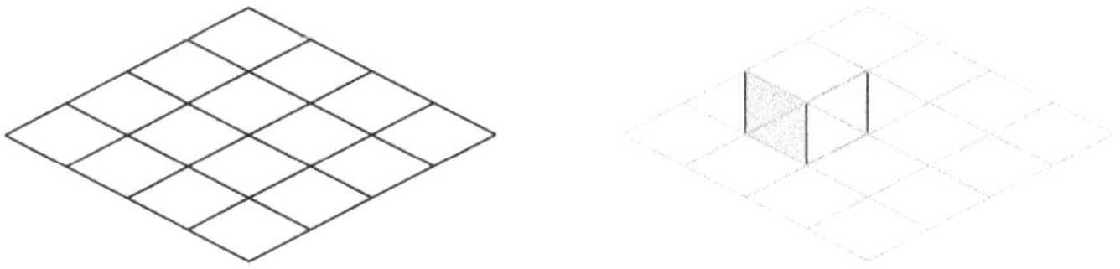

Figure 12.8 Elementary deformation of a flat surface to a non-flat one.

which is the sum over all contributions from the Lagrangian components (12.21) on all the plaquettes in the surface, taking into account their orientation. This action is now a functional not only of the functions u but also of the surface σ. In this description, the geometry of σ in the space of the independent (discrete) variables is part of the variational calculus. Consequently, we will need to consider variations in σ as well as variations in the field u.

The simplest case of an elementary variation is that of a flat piece of a discrete surface, which corresponds to the popping out of a small cube as illustrated in Figure 12.8.[3] This implies a change in the action from S to S' given by

$$\begin{aligned} S' = S &- \mathcal{L}_{ij}(u, T_i u, T_j u) + \mathcal{L}_{ij}(T_k u, T_k T_i u, T_k T_j u) + \mathcal{L}_{jk}(T_i u, T_i T_j u, T_i T_k u) \\ &+ \mathcal{L}_{ki}(T_j u, T_j T_k u, T_i T_j u) - \mathcal{L}_{jk}(u, T_j u, T_k u) - \mathcal{L}_{ki}(u, T_k u, T_i u). \end{aligned}$$

This incorporates changes in each of six plaquettes: one plaquette is removed and five others introduced, taking into account orientations. The variation principle requires that the action be stationary under such changes, that is $S' = S$. This requirement is

$$\begin{aligned} &\mathcal{L}_{jk}(T_i u, T_i T_j u, T_i T_k u) - \mathcal{L}_{jk}(u, T_j u, T_k u) \\ &+ \mathcal{L}_{ki}(T_j u, T_j T_k u, T_i T_j u) - \mathcal{L}_{ki}(u, T_k u, T_i u) \\ &+ \mathcal{L}_{ij}(T_k u, T_k T_i u, T_k T_j u) - \mathcal{L}_{ij}(u, T_i u, T_j u) = 0, \end{aligned}$$

which can be written as the **closure relation**

$$\Delta_i \mathcal{L}_{jk} + \Delta_j \mathcal{L}_{ki} + \Delta_k \mathcal{L}_{ij} = 0, \qquad (12.23)$$

[3] Another elementary variation of a discrete surface such as in Figure 12.7 corresponds to flipping of three adjacent plaquettes to another such three adjacent plaquettes across an elementary cube.

where Δ_i is the difference operator in the direction given by the label i, i.e. $\Delta_i = T_i - \mathrm{id}$. The closure relation (12.23) means that the Lagrangian 2-form is closed in an analogous sense to that of differential forms.

The closure relation may be identically satisfied, for example if $\mathcal{L}_{ij} = u_i - u_j$ (or in general, if the Lagrangian is an "exact 2-form") as can be easily verified, but then it fails to give us any further information about the fields. In this case, a discrete analogue of the Stokes' theorem shows that we can integrate the Lagrangian to reduce the action to a discrete line integral along the boundary of the surface and so the system is independent of the surface σ. It is, therefore, important to require that the closure relation is satisfied for only those fields that satisfy the Euler–Lagrange equations.

The function(s) $\boldsymbol{u}(\boldsymbol{n})$ solving a multidimensionally consistent system are those for which an action S

(a) is invariant under local deformations of the surface on which S is defined, and
(b) attains an extremum under infinitesimal local deformations of the dependent variable(s) $\boldsymbol{u}(\boldsymbol{n})$.

At this point we have two fundamental types of equations that may derived from a Lagrangian, first the usual Euler–Lagrange equations on a fixed surface and second, the closure relations. Since the closure relations are not assumed to be satisfied identically, we have an over-determined system of equations for the field $\boldsymbol{u}$. This would lead to trivial field configurations, unless the equations are all compatible. This now imposes a condition on the Lagrangians, which implies that the Lagrangians themselves are determined by the equations. Such Lagrangians will be called *admissible Lagrangians*. We will see that the natural class of systems that arise from this approach are the integrable ones.

The set of fundamental EL equations follow from a basic set of configurations on the lattice involving a set of basic elementary surfaces consisting of a minimal number of plaquettes, on each of which we write down the corresponding EL equations around a central vertex. These configurations are provided in Figures 12.9–12.10. Figure 12.9 represents the "planar" EL equations, while Figure 12.10 represents corner elements.

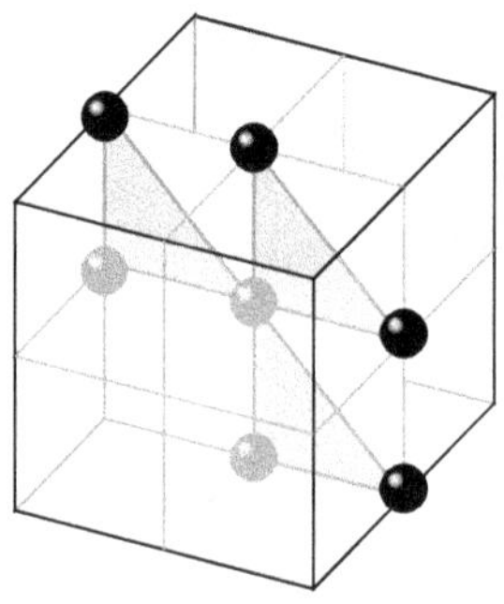

Figure 12.9 Flat 2D space.

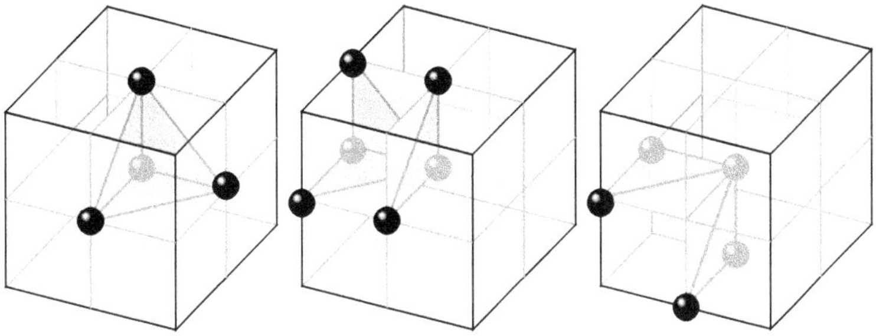

Figure 12.10 Configurations when embedded in 3D.

These elementary lattice configurations correspond to the following basic set of EL equations:

$$\frac{\partial}{\partial u}\Big(\mathcal{L}(T_i^{-1}u, u, T_i^{-1}T_ju; p_i, p_j) + \mathcal{L}(u, T_iu, T_ju; p_i, p_j) + \mathcal{L}(T_j^{-1}u, T_iT_j^{-1}u, u; p_i, p_j)\Big) = 0, \tag{12.24a}$$

$$\frac{\partial}{\partial u}\big(\mathcal{L}(u, T_iu, T_ju; p_i, p_j) + \mathcal{L}(u, T_ju, T_ku; p_j, p_k) + \mathcal{L}(u, T_ku, T_iu; p_k, p_i)\big) = 0, \tag{12.24b}$$

$$\frac{\partial}{\partial u}\Big(\mathcal{L}(T_i^{-1}u, u, T_i^{-1}T_ju; p_i, p_j) - \mathcal{L}(u, T_ju, T_ku; p_j, p_k) + \mathcal{L}(T_i^{-1}u, T_i^{-1}T_ku, u; p_k, p_i)\Big) = 0, \tag{12.24c}$$

$$\frac{\partial}{\partial u}\Big(\mathcal{L}(T_j^{-1}(u), u, T_j^{-1}T_ku; p_j, p_k) + \mathcal{L}(T_i^{-1}u, T_i^{-1}T_ku, u; p_k, p_i)\Big) = 0, \tag{12.24d}$$

complemented by the closure relation (12.23).

In the continuous case, the analogous situation is given by an action of the form (Lobb and Nijhoff, 2009; Xenitidis et al., 2011):

$$S[u(\boldsymbol{p}); \sigma] = \int_\sigma \sum_{i<j} \mathcal{L}_{ij} \mathrm{d}p_i \wedge \mathrm{d}p_j \tag{12.25}$$

which is a functional of the field configuration $u(\boldsymbol{p})$, where $\boldsymbol{p} = (p_i)$ denotes a set of continuous variables p_i (taking real or complex values) and of a smooth surface σ. Here the Lagrangians $\mathcal{L}_{ij}$ are components of a differential 2-form, which is not identically or trivially a closed form, but is closed when evaluated for field configurations, which are solutions of the EL equations on the surface σ. The closure relation in this case is

$$\partial_{p_i}\mathcal{L}_{jk} + \partial_{p_j}\mathcal{L}_{ki} + \partial_{p_k}\mathcal{L}_{ij} = 0. \tag{12.26}$$

It is equivalent to the closure of the Lagrangian 2-form written in terms of coordinates. This condition guarantees that the action integral is independent under local deformations of the surface $\sigma \to \sigma + \delta\sigma$, but only when the Lagrangians $\mathfrak{L}_{ij}$ are evaluated for field configurations corresponding to solutions of the equations of the motion.

12.2.2 Linear case

Consider the following linear quad-equation:

$$\begin{aligned} &Q_{p_i,p_j}(w, T_i w, T_j w, T_i T_j w) \\ &\quad = (p_i + p_j)(T_i w - T_j w) - (p_i - p_j)(w - T_i T_j w) = 0 \end{aligned} \tag{12.27}$$

employing the same notation as before. A skew-symmetric Lagrangian for this equation is given by

$$\mathfrak{L}_{ij} = w(T_i w - T_j w) - \frac{1}{2}\left(\frac{p_i + p_j}{p_i - p_j}\right)\left(T_i w - T_j w\right)^2 . \tag{12.28}$$

The EL equations yield

$$T_i w + T_i^{-1} w - T_j w - T_j^{-1} w - \frac{p_i + p_j}{p_i - p_j}\left(2w - T_i^{-1} T_j w - T_j^{-1} T_i w\right) = 0 ,$$

which vanish whenever w obeys (12.27). Note that the Lagrangian $\mathfrak{L}_{ij}$ obeys a closure relation (12.23) on solutions of the lattice equation. (See Exercise 12.8 for details of the computation.)

It is this closure relation, plus the fact that the Lagrangian is skew symmetric with regard to the interchanging of shifts and parameters, that allows us an interpretation of the Lagrangian as a component of a *difference form*, which is closed on solutions of the system (here the quadrilateral equation). Thus, this is a first example of a Lagrangian multiform structure.

Parallel to the discrete case, we also have the continuous equation given by the non-autonomous PDE (10.17) of Section 10.3, which we rewrite for the present purpose as

$$\partial_{p_i}\partial_{p_j}(p_i^2 - p_j^2)\partial_{p_i}\partial_{p_j} w = 4(n_j\partial_{p_i} - n_i\partial_{p_j})\frac{1}{p_i^2 - p_j^2}(n_j p_i^2 \partial_{p_i} - n_i p_j^2 \partial_{p_j}) w . \tag{12.29}$$

This equation admits a Lagrangian:

$$\begin{aligned} \mathfrak{L}_{ij} = \frac{1}{n_j n_i} &\left\{ \frac{1}{2}(p_i^2 - p_j^2)(\partial_{p_i}\partial_{p_j} w)^2 + n_j^2(\partial_{p_i} w)^2 - n_i^2(\partial_{p_j} w)^2 \right. \\ &\left. + \frac{p_i^2 + p_j^2}{p_i^2 - p_j^2}(n_j \partial_{p_i} w - n_i \partial_{p_j} w)^2 \right\} . \end{aligned} \tag{12.30}$$

The continuous Lagrangian $\mathcal{L}_{ij}$ obeys the closure relation:

$$\partial_{p_i}\mathcal{L}_{jk} + \partial_{p_j}\mathcal{L}_{ki} + \partial_{p_k}\mathcal{L}_{ij} = 0 \tag{12.31}$$

on solutions of the continuous equation. But, since the PDE is of higher order, to establish the closure relation (12.31) requires an additional relation, namely:

$$n_k(p_i^2 - p_j^2)\partial_{p_i}\partial_{p_j}w + n_i(p_j^2 - p_k^2)\partial_{p_j}\partial_{p_k}w + n_j(p_k^2 - p_i^2)\partial_{p_i}\partial_{p_k}w = 0\,, \tag{12.32}$$

which can be viewed as a linear and continuous analogue of the tetrahedral-type relation arising from the 3-leg form. Alternatively, we can consider the system (10.16), rewritten as

$$\left[\tfrac{1}{2}\partial_{p_i}\partial_{p_j} + (n + i\partial_{p_j}p_j - n_j\partial_{p_i}p_i)\frac{1}{p_i^2 - p_j^2}\right]g + \frac{p_i - p_j}{p_i + p_j}\partial_{p_i}\partial_{p_j}w$$
$$= \frac{1}{(p_i + p_j)^2}(p_i\partial_{p_i} - p_j\partial_{p_j})w\,, \tag{12.33a}$$
$$\left[\tfrac{1}{2}\partial_{p_i}\partial_{p_j} - \frac{1}{p_i^2 - p_j^2}(n_i p_j\partial_{p_j} - n_j p_i\partial_{p_i}\right]w = \frac{n_i n_j}{p_i^2 - n_j^2}v\,, \tag{12.33b}$$

in terms of which we also have a Lagrangian

$$\mathcal{L}_{ij} = \tfrac{1}{2}\frac{p_i - p_j}{p_i + p_j}(\partial_{p_i}w)\partial_{p_j}w - \tfrac{1}{2}v\partial_{p_i}\partial_{p_j}w$$
$$+\frac{1}{p_i^2 - p_j^2}\left[\tfrac{1}{2}n_i n_j v^2 + v(n_i p_j\partial_{p_j} - n_j p_i\partial_{p_i})w\right]. \tag{12.34}$$

Remark 12.2.1 Note that inbetween the linear PΔE (12.27) and the linear PDE (12.29) there reside intermediate equations, which are the differential difference equation (10.28a), rewritten in the present context as:

$$2p_i\frac{\partial}{\partial p_i}w = n_i(T_i^{-1}w - T_i w)\,. \tag{12.35}$$

Note that although these equations are "partial" in the sense that they involve two independent variables (one discrete and one continuous), they are also in some sense one-dimensional because the lattice parameter p_i (playing the role as a continuous variable) and the discrete variable n_i represent the same direction in the multidimensional space. Nonetheless, also these semi-discrete equations admit a Lagrangian structure with the following semi-discrete Lagrangian:

$$\mathcal{L}_i = n_i w\frac{\partial}{\partial p_i}T_i w - p_i\left(\frac{\partial}{\partial p_i}w\right)^2. \tag{12.36}$$

It is an interesting open question whether these semi-discrete Lagrangians also admit a closure relation of some sort, and what role these would play in the geometrical picture associated with the multiform structure.

12.2.3 Nonlinear case

We now consider the lattice potential KdV equation (H1) given by

$$(u - T_i T_j u)(T_j u - T_i u) = p_i^2 - p_j^2 \,, \tag{12.37}$$

which, as we have seen, admits the Lagrangian

$$\mathcal{L}_{ij}(u, T_i u, T_j u) = u(T_i u - T_j u) + (p_i^2 - p_j^2)\ln(T_i u - T_j u) \,. \tag{12.38}$$

By direct computation, using the explicit form of the Lagrangians, we have

$$\begin{aligned}
&\Delta_1 \mathcal{L}(u, u_2, u_3; p_2, p_3) + \Delta_2 \mathcal{L}(u, u_3, u_1; p_3, p_1) + \Delta_3 \mathcal{L}(u, u_1, u_2; p_1, p_2) \\
&= (u_{1,2} - u_{1,3})u_1 + (p_2^2 - p_3^2)\ln(u_{1,2} - u_{1,3}) - (u_2 - u_3)u - (p_2^2 - p_3^2)\ln(u_2 - u_3) \\
&+(u_{2,3} - u_{1,2})u_2 + (p_3^2 - p_1^2)\ln(u_{2,3} - u_{1,2}) - (u_3 - u_1)u - (p_3^2 - p_1^2)\ln(u_3 - u_1) \\
&+(u_{1,3} - u_{2,3})u_3 + (p_1^2 - p_2^2)\ln(u_{1,3} - u_{2,3}) - (u_1 - u_2)u - (p_1^2 - p_2^2)\ln(u_1 - u_2).
\end{aligned}$$

We now use equation (12.37) to compute the double-shifted difference:

$$\begin{aligned}
u_{1,2} - u_{1,3} &= \frac{(p_2^2 - p_3^2)u_1 + (p_3^2 - p_1^2)u_2 + (p_1^2 - p_2^2)u_3}{(u_1 - u_2)(u_2 - u_3)(u_3 - u_1)}(u_2 - u_3) \\
&=: A_{1,2,3}(u_2 - u_3)
\end{aligned}$$

where the expression $A_{1,2,3}$ is invariant under permutations of the indices. Inserting the latter result, and the ones obtained by cyclically permuting indices, the left-hand side of the closure relation reduces to

$$\begin{aligned}
A_{1,2,3}(u_2 - u_3)u_1 &+ (p_2^2 - p_3^2)\ln\big(A_{1,2,3}(u_2 - u_3)\big) \\
&-(u_2 - u_3)u - (p_2^2 - p_3^2)\ln(u_2 - u_3) \\
+A_{1,2,3}(u_3 - u_1)u_2 &+ (p_3^2 - p_1^2)\ln\big(A_{1,2,3}(u_3 - u_1)\big) \\
&-(u_3 - u_1)u - (p_3^2 - p_1^2)\ln(u_3 - u_1) \\
+A_{1,2,3}(u_1 - u_2)u_3 &+ (p_1^2 - p_2^2)\ln\big(A_{1,2,3}(u_1 - u_2)\big) \\
&-(u_1 - u_2)u - (p_1^2 - p_2^2)\ln(u_1 - u_2) = 0
\end{aligned}$$

and hence all terms contributing to the closure relation cancel.

Also in the nonlinear case, we have a semi-discrete equation, as we have already encountered in Chapter 10, namely (10.28). Rewritten in the present notations it reads:

$$\frac{\partial u}{\partial p_i} = \frac{2n_i p_i}{T_i u - T_i^{-1} u}, \tag{12.39}$$

and as we pointed out earlier, it can be shown that this equation is compatible with H1. Also here we have a Lagrangian, namely

$$\mathcal{L}_i = u \frac{\partial}{\partial p_i} T_i u + 2n_i p_i \log\left(\frac{\partial}{\partial p_i} T_i u\right). \tag{12.40}$$

The following coupled system of PDEs follows from the DΔE and the PΔE:

$$\frac{\partial^2 u}{\partial p_i\, \partial p_j} = \frac{1}{p_i^2 - p_j^2}\left[2n_i p_i \frac{\partial u}{\partial p_j} - 2n_j p_j \frac{\partial u}{\partial p_i} - 2v_{ij} \frac{\partial u}{\partial p_i}\frac{\partial u}{\partial p_j}\right], \tag{12.41a}$$

$$\frac{\partial^2 v_{ij}}{\partial p_i\, \partial p_j} = \frac{\partial}{\partial p_i}\left[\frac{(\partial u/\partial p_j) v_{ij}^2 + 2n_j p_j v_{ij}}{p_i^2 - p_j^2}\right] + \frac{\partial}{\partial p_j}\left[\frac{(\partial u/\partial p_i) v_{ij}^2 - 2n_i p_i v_{ij}}{p_i^2 - p_j^2}\right], \tag{12.41b}$$

where $v_{ij} := (T_i - T_j)u$. The PDE (12.41) derives as EL equations from the following Lagrangian:

$$\mathcal{L}_{ij} = v_{ij}\left(\frac{\partial^2 u}{\partial p_i\, \partial p_j} - \frac{2n_i p_i (\partial u/\partial p_j) - 2n_j p_j (\partial u/\partial p_i)}{p_i^2 - p_j^2}\right) + \frac{1}{p_i^2 - p_j^2} v_{ij}^2 \frac{\partial u}{\partial p_i}\frac{\partial u}{\partial p_j}. \tag{12.42}$$

Here the variational equations are given by

$$\frac{\delta \mathcal{L}_{ij}}{\delta u} = 0, \quad \frac{\delta \mathcal{L}_{ij}}{\delta v_{ij}} = 0.$$

The latter Lagrangian can be shown to obey the continuous closure relation (12.26), but again since the closure is of first-order while the equations are of higher-order we need an additional relation which is verified on the solutions of the system, namely the following relation for the "form field" v_{ij}:

$$\frac{\partial v_{ij}}{\partial p_k} = \frac{v_{ik}^2 (\partial u/\partial p_k) + 2n_k p_k v_{ik}}{p_i^2 - p_k^2} - \frac{v_{jk}^2 (\partial u/\partial p_k) + 2n_k p_k v_{jk}}{p_j^2 - p_k^2},$$

as well as the condition $v_{ij} + v_{jk} + v_{ki} = 0$, which follows from the definition of v_{ij}.

An alternative description of the continuous PDE is to eliminate the form field v_{ij} and write a closed-form higher-order equation for u alone. That equation derives from a higher-degree Lagrangian of the form

$$\mathcal{L}_{ij} = \frac{1}{4}(p_i^2 - p_j^2)\frac{(\partial_{p_i}\partial_{p_j}u)^2}{(\partial_{p_i}u)\partial_{p_j}u} + \frac{1}{p_i^2 - p_j^2}\left(n_i^2 p_i^2 \frac{\partial_{p_j}u}{\partial_{p_i}u} + n_j^2 p_j^2 \frac{\partial_{p_i}u}{\partial_{p_j}u}\right), \tag{12.43}$$

which contains only u and its derivatives. The relevant Euler–Lagrange equation:

$$\frac{\partial}{\partial p_i}\frac{\partial}{\partial p_j}\left(\frac{\partial \mathcal{L}_{ij}}{\partial(\partial_{p_i}\partial_{p_j}u)}\right) - \frac{\partial}{\partial p_i}\left(\frac{\partial \mathcal{L}_{ij}}{\partial(\partial_{p_i}u)}\right) - \frac{\partial}{\partial p_j}\left(\frac{\partial \mathcal{L}_{ij}}{\partial(\partial_{p_j}u)}\right) = 0\,,$$

gives rise to a higher-order PDE, which is multidimensionally consistent in the sense that it is compatible with similar PDEs with different choices of the label i and j.

12.3 Lagrangian 1-form structure

In this section, we show how to construct a Lagrangian formulation of systems in which there are multiple times, which form commuting flows. In the above sections, we constructed a Lagrangian 2-form for multidimensionally consistent systems. Here, multidimensional consistency means commutativity in multiple time flows of a one-dimensional system, which in the continuous case means

$$\frac{\partial \boldsymbol{x}}{\partial t_j} = F_j(\boldsymbol{x}), \tag{12.44}$$

where F_j are some functions of $\boldsymbol{x}$ describing the flow. The commutativity of these flows means that on solutions of the system of equations we have the identity

$$\frac{\partial}{\partial t_j}\left(\frac{\partial \boldsymbol{x}}{\partial t_k}\right) = \frac{\partial}{\partial t_k}\left(\frac{\partial \boldsymbol{x}}{\partial t_j}\right). \tag{12.45}$$

In the discrete case, the analogous system is

$$T_j\boldsymbol{x} = \mathfrak{F}_j(\boldsymbol{x}), \tag{12.46}$$

where now the $\mathfrak{F}_j$ are functions of $\boldsymbol{x}$ describing the discrete-time evolution, and where the commutativity means

$$\mathfrak{F}_j\big(\mathfrak{F}_k(\boldsymbol{x})\big) = \mathfrak{F}_k\big(\mathfrak{F}_j(\boldsymbol{x})\big). \tag{12.47}$$

We now describe how to set up a Lagrangian 1-form structure for systems of the type (12.46) subject to (12.47).

12.3.1 Variational principle for continuous Lagrangian 1-forms

For simplicity of the formulae, let us illustrate the method in the case of a two-time system, i.e. with a two-dimensional space of independent variables, while for now we will keep the numbers of components N of the dependent variable unspecified. The variational approach starts from an action functional of the form

$$
\begin{aligned}
S[\boldsymbol{x}(t_1,t_2);\Gamma] &= \int_\Gamma (\mathcal{L}_1 dt_1 + \mathcal{L}_2 dt_2) \\
&= \int_{s_0}^{s_1} \left(\mathcal{L}_1(t_1(s),t_2(s))\frac{dt_1}{ds} + \mathcal{L}_2(t_1(s),t_2(s))\frac{dt_2}{ds} \right) ds\,,
\end{aligned}
\tag{12.48}
$$

where the *Lagrangian 1-form* is given by $\mathbf{L} := \mathcal{L}_1 dt_1 + \mathcal{L}_2 dt_2$, with Lagrangian components

$$
\mathcal{L}_i = \mathcal{L}_i(\boldsymbol{x}(t_1,t_2), \boldsymbol{x}_{t_1}(t_1,t_2), \boldsymbol{x}_{t_2}(t_1,t_2), \dots)\,, \quad i = 1,2. \tag{12.49}
$$

Furthermore, the action integral is over an arbitrary (smooth) parameterized curve Γ in the space of independent variables $\boldsymbol{t} = (t_1, t_2)$, parameterized as follows:

$$
\Gamma : \ (t_1(s), t_2(s))\,, \quad s \in [s_0, s_1]\,.
$$

Thus, the action generally depends not only on the field configuration, i.e. the choice of functions $\boldsymbol{x}(\boldsymbol{t})$, but also of the curve Γ. The main point of the variational principle of the Lagrangian multiform theory is that we need to require *that for solutions of the relevant EL equations the action is also at a critical point for the variations associated with the deformations of the curve*, which implies then that for those particular field configurations (i.e. the solutions of the equations of the motion) the action actually is independent of the curve (keeping the end points fixed), and hence the curve can be (locally at least) chosen in any direction. Along those directions we can then derive the usual EL equations on a fixed (but arbitrary) curve, whose solutions in turn imply that the deformations are allowed in the first place. This sounds very convoluted and circular as a procedure, but it can be made explicit, through setting up a system of fundamental equations for the Lagrangian.

This can be seen as a two-step procedure containing two types of variations:

1. variations $\boldsymbol{t}(s) \to \boldsymbol{t}(s) + \delta\boldsymbol{t}(s)$ of the parameterized curve, i.e. $\Gamma \to \Gamma'$ (as indicated in Figure 12.11(a));
2. variations $\boldsymbol{x}(\boldsymbol{t}) \to \boldsymbol{x} + \delta\boldsymbol{x}(\boldsymbol{t})$ of the dependent variables on a fixed, arbitrary curve Γ (as indicated in Figure 12.11(b)).

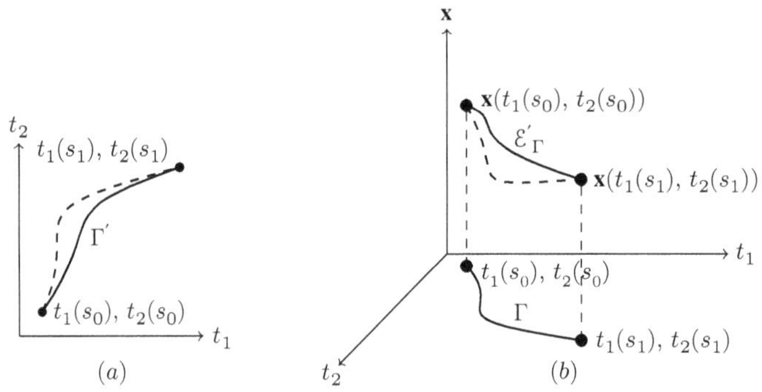

Figure 12.11 (a) Deformation of the discrete curve Γ. (b) Variations on the fixed curve Γ.

Applying the variations of the curve, we regard the functions $[\boldsymbol{x}(t_1, t_2); \Gamma] = \mathcal{S}[\boldsymbol{t}(s)]$, i.e. as a functional of the independent variables as functions of the parameter s, with effectively the scalar Lagrangian

$$L(\boldsymbol{t}(s)) := \mathcal{L}_1(t_1(s), t_2(s))\frac{dt_1}{ds} + \mathcal{L}_2(t_1(s), t_2(s))\frac{dt_2}{ds} ,$$

and apply the usual variational formalism (as reviewed earlier). This leads to the closure relation

$$\frac{\partial \mathcal{L}_2}{\partial t_1} = \frac{\partial \mathcal{L}_1}{\partial t_2} . \tag{12.50a}$$

In the variations of the evaluation curve $\mathcal{E}_\Gamma$ (for a fixed curve Γ in the space of the independent variables), we have to take into account variations of the derivatives of $\delta\boldsymbol{x}$ along the curve and in transversal directions to the curve. Thus, we need to split the variations $\delta\boldsymbol{x}_{t_i}$ $(i = 1, 2)$ into components $\delta(\boldsymbol{x}_{t_i})_\parallel + \delta(\boldsymbol{x}_{t_i})_\perp$. As a consequence, the variation of the action will lead to two types of equations, both of which have to hold simultaneously: on the one hand we have the contributions from $\delta(\boldsymbol{x}_{t_i})_\parallel$ yielding the system of generalized EL equations:

$$\frac{\partial \mathcal{L}_1}{\partial \boldsymbol{x}}\frac{dt_1}{ds} + \frac{\partial \mathcal{L}_2}{\partial \boldsymbol{x}}\frac{dt_2}{ds} - \frac{d}{ds}\left\{\frac{1}{\|d\boldsymbol{t}/ds\|^2}\times\right. \tag{12.50b}$$
$$\left.\left[\left(\frac{dt_1}{ds}\right)^2\frac{\partial \mathcal{L}_1}{\partial \boldsymbol{x}_{t_1}} + \left(\frac{dt_1}{ds}\right)\left(\frac{dt_2}{ds}\right)\left(\frac{\partial \mathcal{L}_1}{\partial \boldsymbol{x}_{t_2}} + \frac{\partial \mathcal{L}_2}{\partial \boldsymbol{x}_{t_1}}\right) + \left(\frac{dt_2}{ds}\right)^2\frac{\partial \mathcal{L}_2}{\partial \boldsymbol{x}_{t_2}}\right]\right\} = 0 ,$$

while the contributions from $\delta(\boldsymbol{x}_{t_i})_\perp$ yields a set of additional constraints of the form:

$$\frac{\partial \mathcal{L}_2}{\partial \boldsymbol{x}_{t_1}}\left(\frac{dt_2}{ds}\right)^2 + \left(\frac{\partial \mathcal{L}_1}{\partial \boldsymbol{x}_{t_1}} - \frac{\partial \mathcal{L}_2}{\partial \boldsymbol{x}_{t_2}}\right)\frac{dt_1}{ds}\frac{dt_2}{ds} - \frac{\partial \mathcal{L}_1}{\partial \boldsymbol{x}_{t_2}}\left(\frac{dt_1}{ds}\right)^2 = 0 . \tag{12.50c}$$

The system (12.50) constitutes the extended EL system for two-time Lagrangian 1-forms. It was derived (in a slightly different form) in Yoo-Kong et al. (2011); Yoo-Kong (2011) and generalized in Suris (2012).

The first example of a Lagrangian 1-form structure, involving two-time variables is given by the Calogero–Moser model. The form of corresponding Lagrangian components were obtained in Yoo-Kong et al. (2011) from systematic continuum limits on the corresponding discrete-time Lagrangians (see the next subsection). They were of the form:

$$\mathcal{L}_1 = \sum_{i=1}^{N} \frac{1}{2}\left(\frac{\partial x_i}{\partial t_1}\right)^2 + \sum_{i\neq j}^{N} V(x_i - x_j)\,, \tag{12.51a}$$

$$\mathcal{L}_2 = \sum_{i=1}^{N}\left[\left(\frac{\partial x_i}{\partial t_1}\right)\frac{\partial x_i}{\partial t_2} + \alpha\left(\frac{\partial x_i}{\partial t_1}\right)^3\right] + \sum_{\substack{i,j=1\\ i\neq j}}^{N} \frac{\partial x_i}{\partial t_1} W(x_i - x_j)\,. \tag{12.51b}$$

The implementation of the extended EL system (12.50) leads to conditions on the potentials $V(x)$ and $W(x)$, from which it follows that $W(x) = -6\alpha V(x) + \text{constant}$, and furthermore, that $W(x)$ has to obey the functional relation

$$W'(x)\left[W(y) - W(x+y)\right] - W'(y)\left[W(x) - W(x+y)\right] = W'(x+y)\left[W(x) - W(y)\right]\,.$$

It is easily recognized (see Appendix B) that this leads to the general solution $W(x) = \beta\wp(x) + \gamma$ with β and γ constant. Thus, we obtain the full elliptic CM system as a solution of the 1-form Lagrangian system.

A second example is given by the Ruijsenaars–Schneider (RS) model (Ruijsenaars and Schneider, 1986), i.e. the relativistic variant of the CM model. The Lagrangians in this case can be chosen of the form:

$$\mathcal{L}_1 = \sum_{i=1}^{N} \frac{\partial x_i}{\partial t_1} \ln\left(\frac{\partial x_i}{\partial t_1}\right) + \sum_{i\neq j}^{N} \frac{\partial x_j}{\partial t_1} V(x_i - x_j)\,, \tag{12.52a}$$

$$\begin{aligned}\mathcal{L}_2 = &\sum_{i=1}^{N}\left(\frac{\partial x_i}{\partial t_2}\ln\left(\frac{\partial x_i}{\partial t_1}\right) + \alpha\left(\frac{\partial x_i}{\partial t_1}\right)^2 + \beta\frac{\partial x_i}{\partial t_2}\right)\\ &+ \sum_{i\neq j}^{N}\left[\frac{\partial x_j}{\partial t_2} W(x_i - x_j) + \frac{\partial x_i}{\partial t_1}\frac{\partial x_j}{\partial t_1} U(x_i - x_j)\right].\end{aligned} \tag{12.52b}$$

These forms of the Lagrangians follow once again from those of the discrete-time version of the Lagrangians for the RS class (Yoo-Kong and Nijhoff, 2011), by systematic continuum limits. The implementation of the extended EL system[4] leads to conditions on the

[4] These results and the parallel results for the Calogero-Moser system were obtained in collaboration with one of the authors (FWN) and S. Yoo-Kong (paper in preparation).

three potentials $V(x)$, $W(x)$ and $U(x)$, which we can take to be even functions of x. Then we have that $V(x) = W(x) + \text{const.} = \ln(U(x) - \alpha) + \text{const.}$, with $V(x)$ obeying the functional relation

$$e^{2V(x)}\left(V'(y) + V'(z)\right) - e^{2V(y)}\left(V'(x) + V'(z)\right) = e^{2V(z)}\left(V'(x) - V'(y)\right) ,$$

with $z = x+y$. A solution of this functional equation is given by $V(x) = -\ln(a + b\wp(x))$ for a, b constant, and this leads to the full elliptic case of the RS system.

12.3.2 Discrete Lagrangian 1-form structure

As in the continuous case, consider a system described by N dependent variables $\boldsymbol{x}(\boldsymbol{n}) = (x_1(\boldsymbol{n}), \ldots, x_N(\boldsymbol{n}))$, which we can consider as particle positions depending on set of discrete multi-time variables $\boldsymbol{n} = (\ldots, n_i, \ldots, n_j, \ldots)$, where for simplicity we assume $n_j \in \mathbb{Z}$. The action functional is

$$S(\boldsymbol{x}(\boldsymbol{n}); \Gamma) = \sum_{\boldsymbol{\gamma}(\boldsymbol{n})\in\Gamma} \mathcal{L}_i(\boldsymbol{x}(\boldsymbol{n}), \boldsymbol{x}(\boldsymbol{n}+\boldsymbol{e}_i)), \tag{12.53}$$

where the sum is evaluated over an arbitrary discrete curve Γ (with or without boundary points) in the space of independent discrete variables $\boldsymbol{n}$. By a *discrete curve* we mean a collection of oriented connected links $\boldsymbol{\gamma}(\boldsymbol{n})$, composed of elementary links $\gamma_i(\boldsymbol{n}) = (\boldsymbol{n}, \boldsymbol{n}+\boldsymbol{e}_i)$ on the edges of the lattice, between adjacent sites on a regular lattice, as indicated in Figure 12.12.

The action (12.53) is evaluated by summing the contributions $\mathcal{L}_i$ from each of the oriented links γ_i along the entire curve, where the corresponding Lagrangian functions $\mathcal{L}_i(\gamma_i(\boldsymbol{n}))$ defined on these links can be interpreted as components of a lattice 1-form, given by

$$\mathcal{L}_i(\gamma_i(\boldsymbol{n})) = \mathcal{L}_i(\boldsymbol{x}(\boldsymbol{n}), \boldsymbol{x}(\boldsymbol{n}+\boldsymbol{e}_i)) , \tag{12.54}$$

where these components may also depend on parameters p_i associated with the link.

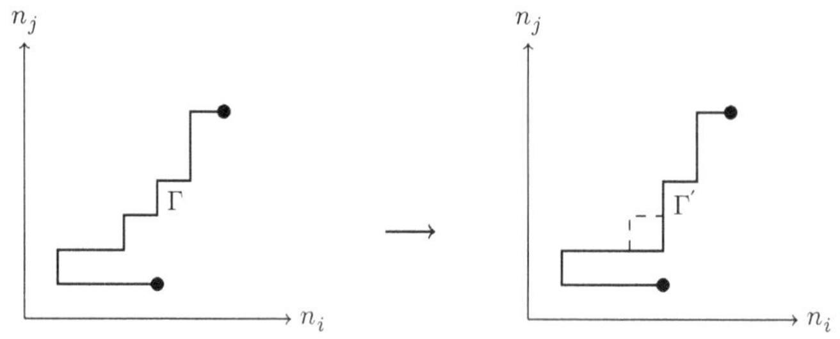

Figure 12.12 Example of a discrete curve Γ and its elementary deformation.

We will consider the action (12.53) as a functional not only with respect to the field configurations, i.e. the values of the dependent variables $\boldsymbol{x}(\boldsymbol{n})$ along a given discrete curve Γ, but also as a functional of the discrete curves. In other words, we can vary the curve by making elementary deformations (as indicated in Figure 12.12) of the curve, while keeping the end points fixed. Requiring the action to be stationary under such elementary moves, it is not hard to see that the Lagrangian has to obey a *1-form discrete closure relation* of the form:

$$\Delta_i \mathcal{L}_j = \Delta_j \mathcal{L}_i \ , \tag{12.55}$$

which is required to hold only on solutions $\boldsymbol{x}(\boldsymbol{n})$ of the relevant EL equations, as discussed before.

Specializing to the case of a system with two discrete times $\boldsymbol{n}_1, \boldsymbol{n}_2$, the appropriate elementary components are given in the diagrams 12.13.

The actions (12.53) associated with the elementary curves in Figure 12.13 provide us with the basic relations constituting the discrete multi-time Euler–Lagrange (EL) equations, which read:

1 From the action for the discrete curve in this case

$$S[\boldsymbol{x}; \Gamma_{(a)}] = \mathcal{L}_i(\boldsymbol{x}, T_i\boldsymbol{x}) + (T_i\mathcal{L}_j)(T_i\boldsymbol{x}, T_iT_j\boldsymbol{x})$$
$$\Rightarrow \quad \frac{\partial \mathcal{L}_i(\boldsymbol{x}, T_i\boldsymbol{x})}{\partial T_i\boldsymbol{x}} + \frac{\partial (T_i\mathcal{L}_j)(T_i\boldsymbol{x}, T_iT_j\boldsymbol{x})}{\partial T_i\boldsymbol{x}} = 0 \ . \tag{12.56a}$$

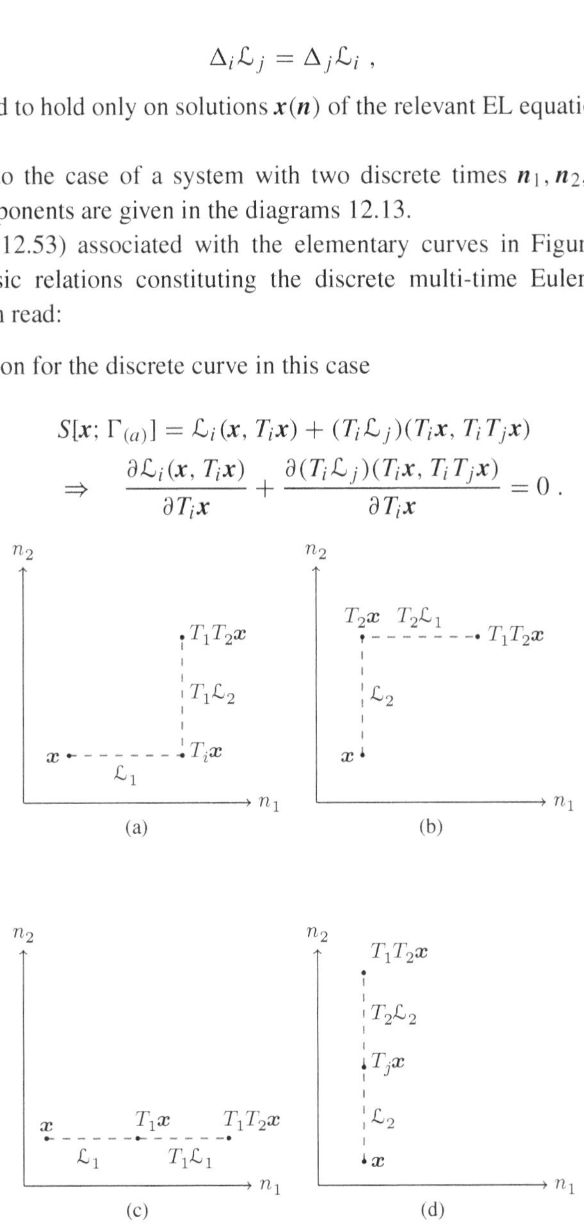

Figure 12.13 Elementary discrete curves for n_1 and n_2 variables.

2 Here the action reads

$$S[\boldsymbol{x};\Gamma_{(b)}]=\mathcal{L}_j(\boldsymbol{x},T_j\boldsymbol{x})+(T_j\mathcal{L}_i)(T_j\boldsymbol{x},T_iT_j\boldsymbol{x})$$
$$\Rightarrow\quad \frac{\partial\mathcal{L}_j(\boldsymbol{x},T_j\boldsymbol{x})}{\partial T_j\boldsymbol{x}}+\frac{\partial(T_j\mathcal{L}_i)(T_j\boldsymbol{x},T_iT_j\boldsymbol{x})}{\partial T_j\boldsymbol{x}}=0\,. \tag{12.56b}$$

3 The action for a discrete curve in case (c) is

$$S[\boldsymbol{x};\Gamma_{(c)}]=\mathcal{L}_i(\boldsymbol{x},T_i\boldsymbol{x})+(T_i\mathcal{L}_i)(T_i\boldsymbol{x},T_iT_j\boldsymbol{x})$$
$$\Rightarrow\quad \frac{\partial\mathcal{L}_i(\boldsymbol{x},T_i\boldsymbol{x})}{\partial T_i\boldsymbol{x}}+\frac{\partial(T_i\mathcal{L}_i)(T_i\boldsymbol{x},T_iT_j\boldsymbol{x})}{\partial T_i\boldsymbol{x}}=0\,. \tag{12.56c}$$

4 The action in this case is

$$S[\boldsymbol{x};\Gamma_{(d)}]=\mathcal{L}_j(\boldsymbol{x},T_j\boldsymbol{x})+(T_j\mathcal{L}_j)(T_j\boldsymbol{x},T_iT_j\boldsymbol{x})$$
$$\Rightarrow\quad \frac{\partial\mathcal{L}_j(\boldsymbol{x},T_j\boldsymbol{x})}{\partial T_j\boldsymbol{x}}+\frac{\partial(T_j\mathcal{L}_j)(T_j\boldsymbol{x},T_iT_j\boldsymbol{x})}{\partial T_j\boldsymbol{x}}=0\,. \tag{12.56d}$$

The system of equations (12.56) together with the closure relation (12.55) form the basic system of equations for the discrete Lagrangian 1-form structure. They were first given in Yoo-Kong (2011), cf. also Yoo-Kong et al. (2011); Yoo-Kong and Nijhoff (2011).

The first examples for which the Lagrangian 1-form structure was elaborated, cf. Yoo-Kong et al. (2011), were the discrete-time Calogero–Moser systems, for which the Lagrangian 1-form components are given by:

$$\mathcal{L}_i(\boldsymbol{x},T_i\boldsymbol{x};p_i)=\sum_{k,l=1}^{N}f(x_k-T_ix_l)-\frac{1}{2}\sum_{\substack{k,l=1\\ l\neq k}}^{N}[f(x_k-x_l)+f(T_ix_k-T_ix_l)]-p_i(\Xi-T_i\Xi)\,, \tag{12.57}$$

where $\Xi=\sum_{k=1}^{N}x_k$. It turns out that the integrable cases obeying the entire system of EL equations (12.56) as well as the closure relation (12.55) of the function f adopts either one of the forms:

$$f(x)=\log(x),\quad f(x)=\log\sinh(x),\quad f(x)=\log\sigma(x), \tag{12.58}$$

associated with the rational discrete-time rational, hyperbolic and elliptic discrete-time CM system respectively.

A second class of examples are the discrete-time Ruijsenaars–Schneider systems (relativistic Calogero–Moser systems), cf. Yoo-Kong and Nijhoff (2011). In the relativistic case we have a parameter-deformation of the previous case, where the Lagrangian components take the forms:

$$\mathcal{L}_i(\boldsymbol{x}, T_i\boldsymbol{x}; p_i) = \sum_{k,l=1}^{N} [F(x_k - T_i x_l) - F(x_k - T_i x_l - \lambda)] - \frac{1}{2}\sum_{\substack{k,l=1\\ l\neq k}}^{N} [F(x_k - x_l + \lambda) + F(T_i x_k - T_i x_l + \lambda)] - \ln|p_i|\,(\Xi - T_i\,\Xi), \tag{12.59}$$

where the function F adopts one of the following forms:

$$F(x) = x(\log x - 1), \quad F(x) = \int^x \log\sinh(x)\,\mathrm{d}x, \quad F(x) = \int^x \log\sigma(x)\,\mathrm{d}x \tag{12.60}$$

corresponding to the rational, hyperbolic and elliptic discrete-time Ruijsenaars model (Yoo-Kong and Nijhoff, 2011).

The point of view we want to press here is that the above explicit formulae for the Lagrangians are to be considered as particular solutions of the fundamental variational system comprising the elementary EL equations together with the closure relation. The questions as to what other solutions of that system can be found, or a forteriori whether we can find the general solution of that system, is an intriguing open one.

12.4 Notes

The Lagrangian formalism forms one of the foundations of both classical as well as quantum physics. It is, therefore, somewhat surprising that in the modern theory of integrable systems its role has been somewhat underplayed, in contrast to the Hamilton formalism (where bi-Hamiltonian structure have played a major role in the development). Nonetheless, there are many advantages in the Lagrange formalism, for example often symmetries are more easily made manifest in the Lagrangian. Furthermore, through Noether's theorem (Noether, 1918) a direct connection can be made between symmetries and conservation laws. There have been valuable attempts in the past to incorporate the Lagrangian into the theory of integrable systems, cf. e.g. Zakharov and Mikhailov (1980); Nijhoff (1987); Dickey (1990), but integrability itself was not a manifest aspect of the Lagrangian structure, even in cases where entire hierarchies were described by appropriate multiple Lagrangians, cf. e.g. Bustamante and Hojman (2003).

The variational principle that incorporated integrability was first proposed in Lobb and Nijhoff (2009) and in this chapter we have described the basic ideas underlying this new formalism, called *Lagrangian multiform theory*. We have seen that integrability, in the form of multidimensional consistency, is encoded in the closure relation, which holds on solutions of the equations of the motion. This property was first discovered for cases of the ABS list of quad-lattice equations, but was soon extended to other systems in a series of papers, e.g. to the multi-component systems (Lobb and Nijhoff, 2010; Atkinson et al., 2012), to higher-dimensional lattice systems, (Lobb et al., 2009), to the continuous integrable systems such as the multidimensionally consistent generating PDEs (Xenitidis et al., 2011), as well as to one-dimensional systems such as the finite hierarchies of commuting flows of integrable many-body systems (Yoo-Kong et al., 2011; Yoo-Kong and Nijhoff, 2011).

The ideas of the multiform theory were taken up and expanded in work by the Berlin group in a number of papers (Bobenko and Suris, 2010; Boll et al., 2013a,b; Bobenko and Suris, 2015; Boll et al., 2015), in which the subject was renamed *pluri-Lagrangian structures*, because of its relation with the notion of pluri-harmonic functions in complex analysis, while we prefer the name *Lagrangian multiform*, introduced in Lobb and Nijhoff (2009), to emphasize the connection to multidimensional consistency.

The principles for obtaining the variational equations from the Lagrangian 1-form action were set out in Yoo-Kong et al. (2011); Yoo-Kong and Nijhoff (2011); Yoo-Kong (2011), as described in Section 12.3.1, and for discrete 2-forms in Xenitidis and Nijhoff (2012); Lobb and Nijhoff (2013) and independently in Boll et al. (2013a, 2016). For continuous 2-forms the detailed derivation of the corresponding EL equations was given recently in Vermeeren (2014), cf. also Suris (2013). What distinguishes the multiform theory is the point of view that the Lagrangian components of the Lagrangian p-form are to be considered as solutions of the (extended) system of EL equations, where the closure represents the variational principle applied to the relevant geometry in the space of independent variables.

Interestingly, the Lagrangians for the Q4 equation appears also in relation to the "master solution" of the star-triangle relations of statistical mechanics, cf. Bazhanov and Sergeev (2012a,b).

Exercises

12.1 Consider the Lagrangian 1-form action (12.48) with $\mathcal{L}_1$ and $\mathcal{L}_2$ given by (12.51) in the two-time case. Deduce that the generalized EL-equations along the t_1 and t_2 directions are given by

$$\frac{\partial \mathcal{L}_1}{\partial \boldsymbol{x}} - \frac{\partial}{\partial t_1}\left(\frac{\partial \mathcal{L}_1}{\partial \boldsymbol{x}_{t_1}}\right) = 0, \quad \frac{\partial \mathcal{L}_2}{\partial \boldsymbol{x}} - \frac{\partial}{\partial t_2}\left(\frac{\partial \mathcal{L}_2}{\partial \boldsymbol{x}_{t_2}}\right) = 0, \quad \frac{\partial \mathcal{L}_2}{\partial \boldsymbol{x}_{t_1}} = 0, \quad (12.61)$$

together with the closure relations (12.50). Derive the condition $W'(x) + 6\alpha V'(x) = 0$ and the functional relation for $W(x)$.

12.2 Show that discrete EL equations for the following discrete Lagrangian

$$\mathcal{L}(\psi, T_1\psi, T_2\psi) = \psi(T_1\psi - T_2\psi) + \ln\left(1 + e^{T_1\psi - T_2\psi + \log(\gamma)}\right) \quad (12.62)$$

give rise to the lattice mKdV equation (3.90), with $W = e^{T_1\psi - T_2\psi}$.

12.3 In Adler and Suris (2004) it was pointed out that the 3-leg form makes the multi-dimensional consistency of the equation manifest. Show that by adding a third direction to the lattice, the following *tetrahedral lattice equation* can be derived from the multiplicative the 3-leg form (12.14b)

$$\frac{\phi(u, u_{1,2}; p_1, p_2)}{\phi(u, u_{1,3}; p_1, p_3)} = \phi(u, u_{2,3}; p_3, p_2).$$

12.4 Consider the most general case, namely the Q4 equation, in Jacobi form, given by:

$$\begin{aligned} Q_{\mathfrak{p}_i,\mathfrak{p}_j} &= p_i(u\,u_i + u_j u_{ij}) - p_j(u\,u_j + u_i u_{ij}) \\ &- p_{ij}(u\,u_{ij} + u_i u_j) + p_i p_j p_{ij}(1 + u\,u_i u_j u_{ij}), \end{aligned} \tag{12.63}$$

where the parameters can be parameterized in terms of Jacobi elliptic functions:

$$p_i = \sqrt{k}\,\mathrm{sn}(\alpha_i; k)\,, \quad p_j = \sqrt{k}\,\mathrm{sn}(\alpha_j; k)\,, \quad p_{ij} = \sqrt{k}\,\mathrm{sn}(\alpha_{ij}; k)$$

with $\alpha_{ij} = \alpha_i - \alpha_j$. Computing the biquadratics from the discriminants (12.19), we obtain:

$$h_{\mathfrak{p}}(x, y) = p(1 + x^2 y^2) - \frac{1}{p}(x^2 + y^2) + 2\frac{P}{p}xy, \quad K_{\mathfrak{p}_i,\mathfrak{p}_j} = -p_i p_j p_{ij}$$

where in this example $\mathfrak{p} = (p, P)$ is a point on the elliptic curve given by $P^2 = p^4 - (k + 1/k)p^2 + 1$ and in which k is the modulus of the Jacobi elliptic function. In fact, the double integral in the Lagrangian can be evaluated as:

$$\int_{x_0}^{x_1}\int_{y_0}^{y_1} \frac{dx\,dy}{h_{\mathfrak{p}}(x, y)} = -2\int_{\eta_0}^{\eta_1} d\eta\, \log\left(\frac{\mathrm{sn}(\xi_1) - \mathrm{sn}(\eta + \alpha)}{\mathrm{sn}(\xi_1) - \mathrm{sn}(\eta - \alpha)}\,\frac{\mathrm{sn}(\xi_0) - \mathrm{sn}(\eta - \alpha)}{\mathrm{sn}(\xi_0) - \mathrm{sn}(\eta + \alpha)}\right),$$

with $x_i = \sqrt{k}\,\mathrm{sn}(\xi_i; k)$, $y_i = \sqrt{k}\,\mathrm{sn}(\eta_i; k)$ and $p = \sqrt{k}\,\mathrm{sn}(\alpha; k)$.

12.5 Establish the 3-leg form for the Q4 equation in the Weierstrass form. Similarly show the same for the Jacobi form using the ingredients:

$$\phi(u, u_i) = \psi(u, u_i) = \left(\frac{\mathrm{sn}(\xi_i) - \mathrm{sn}(\xi + \alpha_i)}{\mathrm{sn}(x_i) - \mathrm{sn}(\xi - \alpha_i)}\right)\frac{\Theta(\xi + \alpha_i)}{\Theta(\xi - \alpha_i)},$$

where in the latter: $u = \sqrt{k}\,\mathrm{sn}(\xi)$ and $p_i = \sqrt{k}\,\mathrm{sn}(\alpha_i)$.

12.6 Use the formula (12.20) to show that the Lagrangian associated with the quadrilateral form for the lattice potential KdV equation (12.37) leads to (12.38), and for the lattice potential mKdV equation (3.11) yields (12.62). Starting with (12.20) derive also the Lagrangian for the lattice Schwarzian KdV equation (3.13).

12.7 (a) Show that the following is a Lagrangian

$$\mathcal{L} = \frac{p^2+pq+q^2}{p-q}(\rho^2 - \widetilde{\rho}\widehat{\rho}) - (p+2q)\rho\widetilde{\rho} + (q+2p)\rho\widehat{\rho} - (p-q)\rho\widehat{\widetilde{\rho}}\,, \quad (12.64)$$

for the linearized BSQ equation (5.49).

(b) Show that the following 4-point Lagrangian gives rise to the lattice BSQ equation (3.71):

$$\begin{aligned}\mathcal{L} = (p^3 - q^3)\ln(p - q + \widehat{u} - \widetilde{u}) + qu\widehat{u} - pu\widetilde{u}\\ -(p+q+u)(p+q-\widehat{\widetilde{u}})(p-q+\widehat{u}-\widetilde{u})\,.\end{aligned} \quad (12.65)$$

12.8 This exercise is about proving the closure relation (12.23) for the linear case on solutions of the lattice equation (12.27). By substituting the Lagrangian into the left-hand side of the closure formula derive

$$\begin{aligned}&\Delta_i\mathcal{L}_{jk} + \Delta_j\mathcal{L}_{ki} + \Delta_k\mathcal{L}_{ij} =\\ &= (T_k)(T_kT_iw - T_kT_jw) - w(T_iw - T_jw)\\ &-\tfrac{1}{2}\left[(T_kT_iw - T_kT_jw)^2 - (T_iw - T_jw)^2\right] + \text{cycl.}(i,j,k)\end{aligned}$$

where cycl.(i, j, k) denotes the addition of similar terms obtained by cyclic permutation of the indices i, j, k of the given ones. As an intermediate step use (12.27) to evaluate

$$T_kT_iw - T_kT_jw = \frac{2p_k(p_j - p_i)}{(p_i - p_k)(p_j - p_k)}T_kw - \frac{p_i + p_k}{p_i - p_k}T_iw + \frac{p_j + p_k}{p_j - p_k}T_jw,$$

and inserting this into the above expression show that the closure relation is satisfied.

Appendix A

Elementary difference calculus and difference equations

In this appendix we will give a brief account of some of the basic tools to deal with (ordinary) difference equations. We will start with difference operators and their properties, define discrete integration, and then provide several methods for solving linear ordinary difference equations. Further techniques for solving linear higher-order equations are also provided.

A.1 Difference operators and discrete integration

Difference equations can arise in several contexts. In this section we consider difference equations as discrete versions of differential equations. Many properties carry over from the differential to the discrete setting. However, there are important distinctions between them, due to the absence of the chain rule and the Leibniz rule.

A.1.1 Difference versus differential operators

We have already introduced the difference operator Δ_δ in (1.9): it depends on a stepsize parameter δ, and its action on a function $y(x)$, of a single independent variable x, is given by the formula

$$\Delta_\delta y(x) := \frac{1}{\delta}\,(T_\delta - \mathrm{id})\, y(x), \tag{A.1}$$

where we can see that it essentially builds on the *shift operator*

$$T_\delta y(x) := y(x+\delta)\,. \tag{A.2}$$

It is tempting to expect that one could build a theory based on the difference operator (A.1) in a fairly similar way as for the differential operator d/dx, but there are important distinctions. Let us recall some of the main properties of the differential operator and compare them with the difference operator:

- **Linearity:** For any two functions $y(x)$, $w(x)$ and constants α, β we have:

$$\frac{d}{dx}\left(\alpha y(x)+\beta w(x)\right)=\alpha\,\frac{dy}{dx}+\beta\,\frac{dw}{dx}.$$

This can be easily seen to work the same way for differentials:

$$\Delta(\alpha x_n+\beta y_n)=\alpha\Delta x_n+\beta\Delta y_n\,, \tag{A.3}$$

for two functions $x, y:\mathbb{Z}\to\mathbb{R}$, and arbitrary constants $\alpha,\ \beta$.

- **Product rule:** For the product of any two functions $y(x)$, $w(x)$ we have for derivatives the Leibniz rule:

$$\frac{d}{dx}(y(x)w(x))=\frac{dy(x)}{dx}w(x)+y(x)\frac{dw(x)}{dx}.$$

For the difference operator the product rule takes a subtly different form: by direct computation we get

$$\begin{aligned}\Delta_\delta(y(x)w(x)) &= \frac{1}{\delta}\left[y(x+\delta)w(x+\delta)-y(x)w(x)\right]\\ &= \frac{1}{\delta}y(x+\delta)[w(x+\delta)-w(x)]+\frac{1}{\delta}[y(x+\delta)-y(x)]w(x)\\ &= y(x+\delta)(\Delta_\delta w(x))+(\Delta_\delta y(x))w(x),\end{aligned}$$

in other words

$$\Delta_\delta(yw)=(T_\delta y)(\Delta_\delta w)+(\Delta_\delta y)w. \tag{A.4}$$

This looks similar to the differential product rule, but has one crucial extra element: it involves necessarily the shift operator T_δ as well. Obviously we could also write (A.4) in the form

$$\Delta_\delta(yw)=y\,(\Delta_\delta w)+(\Delta_\delta y)(T_\delta w), \tag{A.5}$$

but in either way the shift operator T_δ cannot be avoided. In fact, both (A.4) and (A.5) are reformulations of the more fundamental product property of the shift operator:

$$T_\delta(yw)=(T_\delta y)(T_\delta w). \tag{A.6}$$

Note that if f is a smooth function of x then the property (A.6) can be seen as a consequence of the shift operator acting as an infinite-order differential operator:

$$T_\delta f(x) = \exp(\delta\, d/dx)\, f(x) = f(x) + \frac{\delta}{1!} f'(x) + \frac{\delta^2}{2!} f''(x) + \frac{\delta^3}{3!} f'''(x) + \cdots, \quad \text{(A.7)}$$

i.e. as a Taylor series.

- **Chain rule:** For two functions $z(y)$ and $y(x)$ forming the composed function $Z(x) = z \circ y(x) = z(y(x))$ by substitution, we have the chain rule

$$\frac{dZ}{dx} = z'(y(x))y'(x)$$

where $z'(y) = dz/dy$, $y'(x) = dy/dx$.

For the difference operator this does not even make sense, because we cannot usually compose functions. If both the domain and range were the same, say $\mathbb{Z}$ (the "ultra-discrete case"), the composition would be natural. However, that is what we usually do not want to do, and most commonly we want the functions y_n to take values in a continuous set, e.g. the reals $\mathbb{R}$ or the complex numbers $\mathbb{C}$.

A.1.2 The finite-difference operator

We will next concentrate on functions depending on a single discrete variable n, taking values in the integers $n \in \mathbb{Z}$, which we can denote by y_n. This setting can be retrieved from the analytic difference case, where we have a function $y(x)$, by the identification (1.24). Then the shift operator can be represented by a map: $y_n \mapsto y_{n+1}$. Here the discrete variable n will be considered to be the **independent variable**, while the function value y_n is the **dependent variable**. The calculus we develop in this section will concentrate on this particular situation.

We shall from now on work with the (forward) **finite-difference** and **finite-shift** operators:

$$(\Delta y)_n = y_{n+1} - y_n =: \Delta y_n, \quad (Ty)_n = y_{n+1} =: Ty_n, \quad \text{(A.8)}$$

and we set $T^{-1} y_n = y_{n-1}$ as being the inverse of the shift operator T. Thus, we have $\Delta = T - \text{id}$. If the shift is always the same, then we can rescale the variables so that the shift is by one unit:

$$y(x_0 + n\delta) = y(\delta(z_0 + n)) = w(z_0 + n) = w_n,$$

and this is often assumed.

Apart from Δ we could also introduce other types of difference operators, for instance:

$$\nabla y_n = y_n - y_{n-1} = (\text{id} - T^{-1})y_n, \quad \Box y_n = y_{n+1} - y_{n-1} = (T - T^{-1})y_n,$$

i.e. the backward difference operator and the symmetric difference operator, respectively, but we will make little use of them.[1] Although difference equations can now be written in terms of these operators, it is often more useful to write linear difference equations in terms of the shift operator T. In fact, the higher-order difference operators can be expressed in terms of T as follows:

$$\begin{aligned}
\Delta y_n &= y_{n+1} - y_n = (T - \mathrm{id})\, y_n, \\
\Delta^2 y_n &= y_{n+2} - 2y_{n+1} + y_n = \left(T^2 - 2T + \mathrm{id}\right) y_n, \\
\dots & \quad \dots \\
\Delta^k y_n &= \sum_{j=0}^{k} \binom{k}{j} (-1)^j\, y_{n+k-j} = (-1)^k \sum_{j=0}^{k} \binom{k}{j} (-T)^{k-j}\, y_n.
\end{aligned}$$

The difference operator Δ has similar properties as Δ_δ of the previous subsection, namely linearity

$$\Delta(\alpha y_n + \beta w_n) = \alpha \Delta y_n + \beta \Delta w_n, \tag{A.9a}$$

for two functions $y, w : \mathbb{Z} \to \mathbb{R}$, and arbitrary constants α, β. But once again, the product (Leibniz) rule has to be modified from the one for derivatives, namely:

$$\Delta(y_n w_n) = y_{n+1} w_{n+1} - y_n w_n = (T y_n)\Delta w_n + (\Delta y_n) w_n = (\Delta y_n) T w_n + y_n \Delta w_n, \tag{A.9b}$$

noting as we did earlier that one cannot write the product rule in terms of difference operators Δ alone, but one needs to involve the shift operator T as well.

To reiterate, we have encountered two essential differences between continuous and discrete analysis; first, the difference version of the Leibnitz rule cannot be solely expressed in terms of the difference operator, and second, there is no natural analogue to the chain rule.

A.1.3 Difference analogues for some familiar functions

The difference operator acts in some ways similarly to the derivative (e.g. linearity) but due to the change in the Leibniz rule many things are also different. Nevertheless we would like to have analogues to familiar functions, so that Δ would act on them in a similar way as d/dx.

In particular we would like to have a function on which Δ would operate as d/dx operates for x^k, and replace the rule

$$\frac{d}{dx} x^k = k\, x^{k-1},$$

[1] The notation varies in the literature and sometimes ∇ is used for the average: $\nabla y_n = y_{n+1} + y_n$.

by something analogous. If we think of x being replaced by the discrete variable n, such a function should obey the rule:

$$\Delta f(n;k) = k\, f(n;k-1)\,. \tag{A.10}$$

It turns out that the corresponding function can be given in terms of the so-called *falling factorial* or *discrete power function*[2] defined by

$$a^{(k)} = a\,(a-1)\,(a-2)\,\ldots(a-k+1), \quad \text{for } k>0, \quad a^{(0)} \equiv 1\,, \tag{A.12}$$

in which k is a nonnegative integer, but where a could still take on any (real) value. The analogue of negative powers is defined by Milne-Thomson (2000, p. 25)

$$a^{(-k)} = \frac{1}{(a+1)\,(a+2)\,\ldots(a+k)}, \quad \text{for } k>0. \tag{A.13}$$

The following computation

$$\begin{aligned}
\Delta n^{(k)} &= (n+1)^{(k)} - n^{(k)} \\
&= [(n+1)n(n-1)\cdots(n-k+2)] - [n(n-1)\cdots(n-k+2)(n-k+1)] \\
&= n(n-1)\cdots(n-k+2)[(n+1)-(n-k+1)] \\
&= k[n(n-1)\cdots(n-k+2)] \\
&= k\,n^{(k-1)},
\end{aligned}$$

shows that $f(n;k) = n^{(k)}$ is a solution of the difference equation (A.10) establishing the analogy with the power function.

Since Δ is a linear operator, we can get an analogue to any function defined by a power series solution, simply by considering linear combinations of power functions of different degree (the degree being the variable k).

[2] We adopt the notation used by Boole (1860, p. 6). In the literature there appears also the *raising factorial* or *Pochhammer symbol*, defined by products of increasing factors instead of (A.12). In this book we use notation $(a)_k$ for it

$$(a)_k = a(a+1)\cdots(a+k-1). \tag{A.11}$$

Note that k does not need to be an integer for this definition to make sense. In fact, one can extend the definition of $(a)_k$ to noninteger values of k by setting

$$(a)_k = \frac{\Gamma(a+k)}{\Gamma(a)},$$

where Γ denotes the standard Gamma-function (see Abramowitz and Stegun, 1970, Section 6; Olver et al., 2010; DLMF, 2010, Chapter 5). At the moment it is sufficient to know that $\Gamma(z)$ is a complex-valued function, generalizing the factorial, and obeying the difference equation: $\Gamma(z+1) = z\,\Gamma(z)$. In the case of q-difference equations, see Section 1.1.7, the rising factorial is preferred as the standard notation. However, this has the disadvantage that it is a less natural generalization of the power function.

Example A.1.1 As the analogue of the exponential function $\exp(ax)$, we can take

$$e(n;a) = \sum_{k=0}^{\infty} \frac{a^k\, n^{(k)}}{k!} = (1+a)^n. \tag{A.14}$$

We can then show that

1 the discrete exponential function $e(n;a)$ obeys the difference equation

$$\Delta e(n;a) = a\, e(n;a);$$

2 as a consequence of the second equality in (A.14) we have the relation

$$e(n;a)e(m;a) = e(n+m;a).$$

Example A.1.2 Many analogues of ordinary differential results exist also for the q-difference case. First of all we have for powers of z

$$D_q z^n = [n]_q z^{n-1}, \quad \text{where} \quad [n]_q = \frac{1-q^n}{1-q},$$

where D_q is the derivative operator (1.27). As a consequence the q-exponential function defined by

$$\exp_q(x) \equiv e_q^z := \sum_{k=0}^{\infty} \frac{z^k}{[k]_q!}, \quad \text{where} \quad [k]_q! := [1]_q[2]_q \cdots [k]_q,$$

satisfies the natural property

$$D_q\, e_q^z = e_q^z.$$

To make a comparison between the analogues of exponential function in the two examples above, it is useful to write e_q in terms of the *q-shifted factorial*

$$(a;q)_k = \frac{(a;q)_\infty}{(q^k a;q)_\infty} \quad \text{where} \; (a;q)_\infty = \prod_{j=0}^{\infty}(1-aq^j), \quad |q|<1,$$

which is also called q-Pochhammer symbol. In terms of these symbols the q-exponential also has other definitions, since we have the identity[3]:

$$\frac{1}{(x;q)_\infty} = \sum_{k=0}^{\infty} \frac{x^k}{(q;q)_k} =: e_q(x) = \exp_q\left(\tfrac{x}{1-q}\right).$$

[3] In the literature there are various different notations for the q-exponential function, in particular the notation $e_q(x)$ is standard in the theory of q-hypergeometric functions (Koekoek et al., 2010).

A.1.4 Discrete integration: the antidifference operator

In the same way as the notion of the **integral** arises as the inverse of the operation of the derivative, i.e. the **antiderivative** of the function $f(x)$:

$$f(x) = \frac{dF(x)}{dx} \qquad \Rightarrow \qquad F(x) = \int^x f(x')\,dx',$$

where the indefinite integral denotes a primitive of the function $f(x)$ (fixing a lower limit of integration amounts to specifying an integration constant). The inverse operation of the difference operator Δ, i.e. its **antidifference**, is given by the summation:

$$f_n = \Delta F_n \qquad \Rightarrow \qquad F_n = \Delta^{-1} f_n := \sum_j^{n-1} f_j.$$

Note that the upper index in the summation is $n-1$, since Δ is defined by the *forward* differentiation. In both cases the inverse is not unique, but is determined up to a constant: an **integration constant** with respect to the variable x in the derivative case, and a constant with respect to the variable n in the difference case. In the indefinite integration/summation the constant is left unspecified, while in the definite integration/summation this constant can be fixed by specifying the lower limit in the integral/sum, e.g.

$$F(x) = \int_0^x f(x)\,dx \qquad \text{and} \qquad F_n = \sum_{j=0}^{n-1} f_j.$$

This lower limit is purely a matter of choice and the choice depends on the problem at hand (e.g. initial values when we deal with differential or difference equations). We can also define the summation so that n is as the lower index:

$$G_n := \sum_{j=n}^{M} f_j, \qquad \Delta G_n = -f_n.$$

Example A.1.3 In the continuous case the antiderivative (up to an integration constant) of the function $f(x) = x^k$ is $F(x) = x^{k+1}/(k+1)$ and the antiderivative of the exponential function $\exp(ax)$ is $\exp(ax)/a$. The antidifference (up to an integration constant) of the discrete power function $f(n;k) = n^{(k)}$ is given by $F(n;k) = n^{(k+1)}/(k+1)$, and the antidifference of the exponential function is $\Delta^{-1} e(n;a) = e(n;a)/a$.

The usual rules of linearity apply to the antidifferential:

$$\sum(\alpha x_n + \beta y_n) = \alpha \sum x_n + \beta \sum y_n. \tag{A.15}$$

There are also rules corresponding to integration by parts:

$$\sum_{j=m}^{n-1} f_j \Delta g_j = f_n g_n - f_m g_m - \sum_{j=m}^{n-1} \Delta f_j \, T g_j$$

The summation symbol can also be used to define a **generating function**. Let

$$F(z) := \sum_{k=0}^{\infty} f_k \, z^k, \quad \text{or} \quad \overline{F}(z) := \sum_{k=0}^{\infty} f_k \, \frac{z^k}{k!}, \tag{A.16}$$

then f_n can be recovered as follows:

$$f_n = \frac{1}{n!} \left(\frac{d}{dz} \right)^n F(z) \bigg|_{z=0} = \left(\frac{d}{dz} \right)^n \overline{F}(z) \bigg|_{z=0}.$$

If the summation (A.16) can be performed explicitly the resulting generating function $F(z)$ or $\bar{F}(z)$ readily encodes the whole list of quantities f_n.

So far we have dealt with the inverse of the usual difference operator. In an analogous way, the inverse of the q-difference operator D_q (1.27) is given by the (definite) **Jackson integral**

$$\int_0^x f(x') d_q x' := (1-q)x \sum_{k=0}^{\infty} q^k f(q^k x), \tag{A.17}$$

where the "integral" is in fact a sum. (In difference calculus we often use the term "integrate" for performing a summation.) One can then show that the fundamental theorem of integration holds with respect to the q-difference operator: i.e.

$$D_q \int_0^x f(x') d_q x' = f(x) = \int_0^x (D_q f)(x') d_q x'.$$

A.2 Linear difference equations

We note first, that by defining above the antidifference operators, we have already "solved" (i.e. integrated) some difference equations, namely equations given in the simple form:

$$\Delta y_n = f_n, \quad D_q y(x) = f(x) \quad \text{with} \quad f_n, f(x) \text{ given.}$$

This would correspond to the simplest linear *in* homogeneous case.

The next class of difference equations that we consider is the class of general *linear* difference equations. Linear differential equations are far easier to treat, in general, than nonlinear differential equations, because the property of linearity allows the superposition of solutions (at least in the linear *homogeneous* case). Linear difference equations have the same property and are expected to be simpler than nonlinear equations. Thus, we would expect that some general theory could be developed along similar lines as for differential equations.

For both linear differential as well as linear difference equations, one has to further distinguish between

i) homogeneous versus non-homogeneous equations;
ii) constant-coefficients versus variable-coefficients equations.

With regard to i), we will later see that once one has a general solution to the homogeneous part one can construct a solution to the non-homogeneous equation using the method of *variation of constants*. With regard to ii), for the constant-coefficient case elementary methods exist to solve them even in explicit form. For the non-constant-coefficient case, the situation is much more complicated, and often solutions can only be given through power series, and they are subject to subtle considerations of the singular points where the coefficients vanish or blow up. Here we will mostly restrict ourselves to considering the constant-coefficient case and some simple examples of the nonconstant situation.

A.2.1 Linear constant-coefficient difference equations

The general case of *first-order*, constant-coefficient, linear finite-difference equations is not very exciting. Such equations can be written in the form

$$ay_{n+1} + by_n = 0, \ (a, b \text{ constant}),$$

in the homogeneous case, leading to the solution $y_n = (-b/a)^n y_0$, and in the inhomogeneous case (with given function f_n) we can follow the procedure of the example below.

Example A.2.1 We now solve the *inhomogeneous* first-order difference equation

$$ay_{n+1} + by_n = f_n, \quad a, b \text{ constants}, \quad f_n \text{ a given function of } n.$$

The solution of the homogeneous equation (replacing f_n by 0) is given above. By a principle which is called **variation of constants**, we set $y_n = (-b/a)^n w_n$ (i.e. replacing the constant y_0 by a function w_n of n yet to be determined). Inserting this into the equation yields:

$$\left(-\frac{b}{a}\right)^n \left[a\left(-\frac{b}{a}\right) w_{n+1} + bw_n\right] = f_n \quad \Rightarrow \quad w_{n+1} - w_n = -\frac{1}{b}\left(-\frac{a}{b}\right)^n f_n ,$$

and the latter equation can be "integrated" by a summation, yielding

$$w_n = w_0 - \frac{1}{b}\sum_{j=0}^{n-1}\left(-\frac{a}{b}\right)^j f_j, \quad w_0 \ \text{(constant) initial value.}$$

Thus, returning to the variable y_n, the general solution is given by:

$$y_n = \left(-\frac{b}{a}\right)^n y_0 - \frac{1}{b}\sum_{j=0}^{n-1}\left(-\frac{b}{a}\right)^{n-j} f_j\,,$$

in which initial value y_0 plays the role of an "integration constant".

Let us next consider *second-order* constant-coefficient, linear finite-difference equations. The general form of the equation in the homogeneous case is:

$$a y_{n+1} + b y_n + c y_{n-1} = 0\,. \tag{A.18}$$

In a procedure similar to the standard method for solving constant-coefficient differential equations, we have first to find the **characteristic equation** associated with the equation. Recall that in the differential case this was done by using the exponential trial solution $y(x) = e^{\lambda x}$, which yields a polynomial equation for λ. In the difference case we use the ansatz $y_n = \lambda^n$, where again λ is to be determined. Inserting this into (A.18) we get the characteristic equation:

$$a\lambda^{n+1} + b\lambda^n + c\lambda^{n-1} = 0 \quad \Rightarrow \quad a\lambda^2 + b\lambda + c = 0\,.$$

Let the roots of this quadratic equation be λ_1, λ_2, then we obtain as possible solutions: λ_1^n and λ_2^n and if these roots are distinct the general solution is given by a linear superposition of these solutions. If the roots are not distinct, i.e. if $\lambda_1 = \lambda_2$, then we have λ_1^n and $n\lambda_1^n$ as the two independent solutions.[4] Thus, there are two possibilities:

distinct roots, $\lambda_1 \neq \lambda_2$: the general solution in this case is $y_n = c_1\lambda_1^n + c_2\lambda_2^n$;
coinciding roots, $\lambda_1 = \lambda_2$: the general solution is given by $y_n = (c_1 + c_2 n)\lambda_1^n$,

where c_1, c_2 are arbitrary parameters (integration constants) of the solution. They can be fixed by specifying initial data, such as the values of y_0, y_1.

Example A.2.2 The Fibonacci numbers are given by the recursion formula

$$F_{n+1} = F_n + F_{n-1}, \quad \text{or} \quad (T^2 - T - 1)F_n = 0,$$

[4] In fact, this situation can be seen to arise from a limiting procedure: by taking as independent solutions λ_1^n and $\frac{\lambda_2^n - \lambda_1^n}{\lambda_2 - \lambda_1}$, where the latter is a linear combination of the two solutions given above, and taking now the limit that the two roots coincide, i.e. $\lambda_2 = \lambda_1 + \epsilon$, with $\epsilon \to 0$, the second solution goes in this limit over into $n\lambda_1^{n-1}$.

with initial values $F_0 = F_1 = 1$. What is the general solution? This is a three-point second-order equation and therefore we need two exponents. Substituting $F_n = \lambda^n$ yields

$$\lambda^{n-1}(\lambda^2 - \lambda - 1) = 0,$$

with roots $\lambda = \frac{1}{2}(1 \pm \sqrt{5})$. The general solution is therefore

$$F_n = a\left[\tfrac{1}{2}(1+\sqrt{5})\right]^n + b\left[\tfrac{1}{2}(1-\sqrt{5})\right]^n,$$

and the above-mentioned initial values imply $a + b = 1$, $a - b = 1/\sqrt{5}$, yielding

$$F_n = \tfrac{1}{\sqrt{5}}\left(\left[\tfrac{1}{2}(1+\sqrt{5})\right]^{n+1} - \left[\tfrac{1}{2}(1-\sqrt{5})\right]^{n+1}\right).$$

The generating function for Fibonacci numbers is then

$$F(z) = \sum_{n=0}^{\infty} F_n z^n = \frac{1}{z\sqrt{5}} \sum_{n=0}^{\infty}\left(\left[\tfrac{1}{2}(1+\sqrt{5})z\right]^{n+1} - \left[\tfrac{1}{2}(1-\sqrt{5})z\right]^{n+1}\right) = \frac{1}{1-z-z^2}.$$

The case of *in*homogeneous second-order constant-coefficient difference equation

$$a y_{n+1} + b y_n + c y_{n-1} = f_n\,, \tag{A.19}$$

can be treated once again by the method of variation of constants mentioned on page 371. Alternatively, it is easy to check by direct computation that a *particular solution* of the inhomogeneous equation can be constructed as follows (in the case of distinct roots λ_1, λ_2 of the characteristic equation)

$$y_n^{\text{part}} = \frac{1}{a}\sum_{j=0}^{n-1} \frac{\lambda_2^{n-j} - \lambda_1^{n-j}}{\lambda_2 - \lambda_1} f_j\,, \tag{A.20}$$

after which the general solution of the equation is obtained by adding to this particular solution the general solution of the homogeneous equation. In the case of coinciding roots the analogous formula is obtained by taking the limit $\lambda_2 \to \lambda_1$ in (A.20)

$$y_n^{\text{part}} = \frac{1}{a}\sum_{j=0}^{n-1} (n-j)\lambda_1^{n-j-1} f_j\,. \tag{A.21}$$

The general solution of the inhomogeneous equation is now given by the combination $y_n = y_n^{\text{hom}} + y_n^{\text{part}}$ of the general solution of the homogeneous equation and the particular solution of the inhomogeneous equation. The solution depends on integration constants c_1,

c_2, which can subsequently be determined (but only once we have the full solution of the inhomogeneous problem) by imposing initial conditions, e.g. by giving y_0 and y_1.

A.2.2 General linear first-order difference equations

Let us now consider a simple case of *nonconstant* coefficient first-order linear difference equation, which can be treated to by a technique analogous to the method of integrating factors in differential equations. Let us consider a difference equation of the type:

$$y_{n+1} - g_n y_n = h_n, \tag{A.22}$$

where $g_n (\neq 0)$, h_n are some given functions. In the same way as for differential equations, the complete solution of (A.22) is a sum of the general solution of the homogeneous part and a particular solution of the inhomogeneous equation.

It is easy to see that the homogeneous version of (A.22), namely

$$w_{n+1} - g_n \, w_n = 0, \tag{A.23}$$

can be solved by

$$w_n := \prod_{j=j_0}^{n-1} g_j = w_0 \prod_{j=0}^{n-1} g_j. \tag{A.24}$$

Now let

$$y_n = w_n z_n, \tag{A.25}$$

where z is the new unknown dependent variable, then (A.22) becomes

$$z_{n+1} - z_n = h_n / w_{n+1}. \tag{A.26}$$

Here the left-hand side is Δz_n and the equation can be integrated with solution

$$z_n := \sum_{k=k_0}^{n-1} h_k / w_{k+1}. \tag{A.27}$$

Thus the general solution of (A.22) is

$$y_n := \left(C + \sum_{k=k_0}^{n-1} \frac{h_k}{\prod_{j=j_0}^{k} g_j} \right) \prod_{j=j_0}^{n-1} g_j. \tag{A.28}$$

Here C is the one necessary "summation constant", associated with the homogeneous solution, its value being determined by an initial condition. This is a formal solution, and in practice we would like to express the sums and products in closed form, but there are no general methods for that. Sometimes it is useful to note that $\prod g_j = e^{\sum \log g_j}$.

Example A.2.3 In order to solve

$$y_{n+1} - n y_n = n,$$

we use (A.28) with $g_n = n,\ h_n = n$. For convenience we take $j_0 = k_0 = 1$. Then we get

$$y_n = (n-1)!\left(C + \sum_{k=1}^{n-1} \frac{1}{(k-1)!}\right).$$

A.2.3 General linear higher-order difference equations

Most of the above considerations can be extended to the higher-order case as well. The general linear equation is of the form

$$a_n^{(N)} y_{n+N} + a_n^{(N-1)} y_{n+N-1} + \cdots + a_n^{(1)} y_{n+1} + a_n^{(0)} y_n = f_n \tag{A.29}$$

where $a_n^{(j)}$, f_n are some given functions and $a_n^{(N)} a_n^{(0)} \neq 0$. In the case of constant coefficients (i.e. when all $a_n^{(j)} = a_0^{(j)}$ do not depend on the independent variable n) the homogeneous problem reduces to solving the corresponding characteristic equation:

$$a_0^{(N)} \lambda^N + a_0^{(N-1)} \lambda^{N-1} + \cdots a_0^{(1)} \lambda + a_0^{(0)} = 0\,,$$

leading to roots $\lambda_1, \ldots, \lambda_N$ (some of which may coincide). If all roots are mutually distinct, the solution would take the form

$$y_n = \sum_{j=1}^{N} c_j \lambda_j^n, \quad c_j \text{ constant coefficients}$$

but if roots coincide we will have to supply different independent solution. In fact, if a root λ_j has multiplicity μ_j (i.e. μ_j is the power of the factor $(\lambda - \lambda_j)$ in the factorization of the polynomial given by the characteristic equation), then the corresponding linearly independent solutions are $\lambda_j^n,\ n\lambda_j^n,\ \ldots,\ n^{\mu_j - 1}\lambda_j^n$. The general solution of (A.29) in the inhomogeneous case is given by a linear combination of N linearly independent general solutions of the homogeneous part plus a particular solution of the inhomogeneous equation.

Alternatively, the Nth-order equation (A.29) can be converted into a $N \times N$ matrix first-order linear difference equation and can be treated on that level (see the Problems section).

Example A.2.4 We will later need the solution of the difference equation

$$y_{n+3} - y_{n+2} - y_{n+1} + y_n = 0. \tag{A.30}$$

The characteristic equation is $\lambda^3 - \lambda^2 - \lambda + 1 = 0$ and its roots are $-1, 1, 1$. The root 1 corresponds to a constant solution, since $1^n = 1$, but since this is a double root the other root corresponds to n. The general solution is therefore (verify this!)

$$y_n = \alpha + \beta n + \gamma(-1)^n. \tag{A.31}$$

In the case with nonconstant coefficients (A.29) the situation can be markedly more complicated, and to find basic solutions one needs to resort to developing power series in the same way as is done in the Fuchsian theory of differential equations. Such power series solutions often depend crucially on the occurrence of singularities of the coefficients. We will refrain from treating this theory further here; the only issue we want to mention at this point is the nontrivial matter of *linear independency* of solutions of the difference equation. In the continuous case of linear differential equations the linear independence of solutions is expressed through the non-vanishing of a particular determinant associated with the differential equation, the so-called **Wronski determinant** or **Wronskian**. In the case of linear difference equations the role of the Wronskian is replaced by the so-called **Casorati determinant** or **Casoratian**. This determinant is constructed as follows. Suppose that we have N discrete functions $y_n^{(j)}$, $j = 1, \ldots, N$, then the Casoratian of these functions is given by

$$C(y_n^{(1)}, y_n^{(2)}, \ldots, y_n^{(N)}) = \begin{vmatrix} y_n^{(1)} & y_n^{(2)} & \cdots & y_n^{(N)} \\ y_{n+1}^{(1)} & y_{n+1}^{(2)} & \cdots & y_{n+1}^{(N)} \\ \vdots & \vdots & \ddots & \vdots \\ y_{n+N-1}^{(1)} & y_{n+N-1}^{(2)} & \cdots & y_{n+N-1}^{(N)} \end{vmatrix},$$

and the functions are independent iff their Casoratian is nonzero. (It is easy to see that the continuous limit of the Casoratian is proportional to the Wronskian: by subtracting the first row from the second we get a discrete derivative in the second row (up to a common factor), and this can be continued to all rows.)

Furthermore, setting $f_n \equiv 0$, the n-dependence of the Casoratian is given by:

$$C(y_{n+1}^{(1)}, y_{n+1}^{(2)}, \ldots, y_{n+1}^{(N)}) = (-1)^N \frac{a_n^{(0)}}{a_n^{(N)}} C(y_n^{(1)}, y_n^{(2)}, \ldots, y_n^{(N)})$$

which, provided $a_n^{(0)}/a_n^{(N)} \neq 0$, guarantees that independent solutions remain so as functions of n.

A.2.4 Variation of constants – matrix form

Let us now consider the $N \times N$ matrix difference equation:

$$\boldsymbol{y}_{n+1} = \boldsymbol{A}_n \boldsymbol{y}_n + \boldsymbol{f}_n \,, \tag{A.32}$$

for the N-component vector $\boldsymbol{y}_n$, with $\boldsymbol{A}_n$ given matrix coefficient (with entries functions of n), and $\boldsymbol{f}_n$ given vector function of n. Let us assume we know a fundamental matrix solution of the *homogeneous* equation

$$\boldsymbol{Y}_{n+1} = \boldsymbol{A}_n \boldsymbol{Y}_n, \qquad \text{such that} \;\; \det(\boldsymbol{Y}_n) \neq 0. \tag{A.33}$$

Formally the solution can be given by the ordered product

$$\boldsymbol{Y}_n = \boldsymbol{A}_{n-1}\boldsymbol{A}_{n-2}\cdots\boldsymbol{A}_1\boldsymbol{A}_0\boldsymbol{Y}_0.$$

Whether or not this solution can be given in reasonable closed form depends on the particular nature of the coefficients in the equation.

Note that any matrix $\widetilde{\boldsymbol{Y}}_n$ obtained from $\boldsymbol{Y}_n$ by matrix multiplication from the right by an *invertible* constant matrix $\boldsymbol{C}$, i.e.

$$\boldsymbol{Y}_n \;\rightarrow\; \widetilde{\boldsymbol{Y}}_n = \boldsymbol{Y}_n \boldsymbol{C},$$

is again a fundamental matrix solution of (A.33).

The variation of constants approach now amounts to finding a solution of the inhomogeneous vector equation (A.32) of the form $\boldsymbol{y}_n = \boldsymbol{Y}_n \boldsymbol{c}_n$ where the N-component vector $\boldsymbol{c}_n$ remains to be determined as a function of n. Plugging this form of $\boldsymbol{y}_n$ into the inhomogeneous equation and performing the simple computation

$$\boldsymbol{Y}_{n+1}\boldsymbol{c}_{n+1} = \boldsymbol{A}_n\boldsymbol{Y}_n\boldsymbol{c}_{n+1} = \boldsymbol{A}_n\boldsymbol{Y}_n\boldsymbol{c}_n + \boldsymbol{f}_n$$

$$\Rightarrow \quad \boldsymbol{c}_{n+1} - \boldsymbol{c}_n = \boldsymbol{Y}_{n+1}^{-1}\boldsymbol{f}_n \quad \Rightarrow \quad \boldsymbol{c}_n = \boldsymbol{c}_0 + \sum_{j=0}^{n-1} \boldsymbol{Y}_{j+1}^{-1}\boldsymbol{f}_j$$

we obtain the solution in the form

$$\boldsymbol{y}_n = \boldsymbol{Y}_n\boldsymbol{c}_n = \boldsymbol{Y}_n\boldsymbol{c}_0 + \sum_{j=0}^{n-1} \boldsymbol{Y}_n\boldsymbol{Y}_j^{-1}\boldsymbol{A}_j^{-1}\boldsymbol{f}_j \,. \tag{A.34}$$

Thus, once $\boldsymbol{Y}_n$ is known, the last formula provides the solution of the vector difference equation (A.32), and this is also the general solution, because the first term contains an arbitrary constant N-component vector $\boldsymbol{c}_0$ which corresponds to the N integration constants of the equation.

As an application let us now reconsider the Nth-order equation (A.29) from the matrix perspective. After introducing new variables

$$y_n^{[j]} := y_{n+j-1}, \quad j = 1, \ldots, N,$$

the equation can be written as the first-order system

$$y_{n+1}^{[j]} = y_n^{[j+1]}, \quad j = 1, \ldots, N-1, \tag{A.35a}$$

$$y_{n+1}^{[N]} = \frac{1}{a_n^{(N)}}\left(f_n - a_n^{(N-1)} y_n^{[N]} - \cdots - a_n^{(1)} y_n^{[2]} - a_n^{(0)} y_n^{[1]}\right). \tag{A.35b}$$

Thus by introducing the N-component vector

$$\boldsymbol{y}_n = \left(y_n^{[1]}, y_n^{[2]}, \ldots, y_n^{[N]}\right)^T = (y_n, y_{n+1}, \ldots, y_{n+N-1})^T$$

we can write the system of equations (A.35), in vector form as follows:

$$\boldsymbol{y}_{n+1} = \boldsymbol{A}_n \boldsymbol{y}_n + \boldsymbol{f}_n\,, \tag{A.36}$$

with $\boldsymbol{A}_n$ a matrix of coefficient, and $\boldsymbol{f}_n$ a vector given by:

$$\boldsymbol{f}_n = \begin{pmatrix} 0 \\ \vdots \\ \vdots \\ \vdots \\ 0 \\ \frac{f_n}{a_n^{(N)}} \end{pmatrix}, \quad \boldsymbol{A}_n = \begin{pmatrix} 0 & 1 & 0 & \cdots & \cdots & 0 \\ \vdots & 0 & 1 & 0 & \cdots & \vdots \\ \vdots & & \ddots & \ddots & \ddots & \vdots \\ \vdots & & & 0 & 1 & 0 \\ 0 & 0 & \cdots & \cdots & 0 & 1 \\ -\frac{a_n^{(0)}}{a_n^{(N)}} & -\frac{a_n^{(1)}}{a_n^{(N)}} & \cdots & \cdots & -\frac{a_n^{(N-2)}}{a_n^{(N)}} & \frac{a_n^{(N-1)}}{a_n^{(N)}} \end{pmatrix}.$$

Suppose now that we know N independent solutions $y_n^{(1)}, \ldots, y_n^{(N)}$ of the scalar equation (A.29), then with each one of them we can associate a vector $\boldsymbol{y}_n^{(j)}$, and form the matrix $\boldsymbol{Y}_n = \left(\boldsymbol{y}^{(1)}, \boldsymbol{y}^{(2)}, \ldots, \boldsymbol{y}^{(N)}\right)$ with these vectors as its columns:

$$\boldsymbol{Y}_n = \begin{pmatrix} y_n^{(1)} & y_n^{(2)} & \cdots & y_n^{(N)} \\ y_{n+1}^{(1)} & y_{n+1}^{(2)} & \cdots & y_{n+1}^{(N)} \\ \vdots & \vdots & & \vdots \\ y_{n+N-1}^{(1)} & y_{n+N-1}^{(2)} & \cdots & y_{n+N-1}^{(N)} \end{pmatrix}.$$

This can be identified as the fundamental matrix solution of (A.36), for which the determinant $\det(\boldsymbol{Y}_n)$ can be identified with the Casoratian of the N independent solutions. i.e. $\det(\boldsymbol{Y}_n) = C(y^{(1)}, y^{(2)}, \dots, y^{(N)})$. Therefore the independence of the solutions guarantees the invertibility of the matrix $\boldsymbol{Y}_n$, and the solution is given by (A.34).

Example A.2.5 Consider the second-order inhomogeneous linear difference equations:

$$a_n y_{n+1} + b_n y_n + c_n y_{n-1} = f_n, \quad a_n c_n \neq 0 ,$$

with coefficients a_n, b_n, c_n and inhomogeneous term f_n given. Let $y_n^{(1)}$, $y_n^{(2)}$ be two independent solutions of the corresponding *homogeneous* equation.

i) Construct the Casorati *matrix*

$$\boldsymbol{Y}_n =: \begin{pmatrix} y_n^{(1)} & y_n^{(2)} \\ y_{n+1}^{(1)} & y_{n+1}^{(2)} \end{pmatrix}$$

which obeys the equation

$$\boldsymbol{Y}_n = \boldsymbol{A}_n \boldsymbol{Y}_{n-1} \quad \text{where} \quad \boldsymbol{A}_n = \begin{pmatrix} 0 & 1 \\ -c_n/a_n & -b_n/a_n \end{pmatrix}.$$

ii) A particular solution of the *in*homogeneous equation can be found in the form $y_n^{\text{part}} = a_n y_n^{(1)} + b_n y_n^{(2)}$ with coefficients a_n, b_n given by

$$a_n = a_0 - \sum_{j=0}^{n-1} \frac{f_j y_j^{(2)}}{a_j C_j}, \qquad b_n = b_0 + \sum_{j=0}^{n-1} \frac{f_j y_j^{(1)}}{a_j C_j},$$

with a_0, b_0 arbitrary and $C_j := \det \boldsymbol{Y}_j$.

iii) Hence the general solution of the inhomogeneous difference equation (with two arbitrary integration constants) can be written as:

$$y_n = a_n y_n^{(1)} + b_n y_n^{(2)}.$$

Thus once we know the solution of the homogeneous problem $\boldsymbol{y}_{n+1} = \boldsymbol{A}_n \boldsymbol{y}_n$ we can solve the inhomogeneous equation by this method. This solution can always be given as an ordered product, but obtaining a nice closed-form expression is a difficult problem. This is even more involved in the case of analytic difference equations.

A.2.5 Factorization method

It is sometime useful to write the equation in terms of the shift operator as follows

$$\mathcal{T}y_n = f_n, \text{ where } \mathcal{T} := a_n^{(N)}T^N + a_n^{(N-1)}T^{N-1} + \cdots a_n^{(1)}T + a_n^{(0)}\,\mathrm{id}. \tag{A.37}$$

Recalling now that polynomials in one variables can be factorized in an algebraically closed field (such as $\mathbb{C}$), we may expect to do the same on the operator $\mathcal{T}$, i.e. write it as an ordered product

$$\mathcal{T} = a_n^{(N)}(T - b_n^{(N)}\mathrm{id})(T - b_n^{(N-1)}\mathrm{id})\cdots(T - b_n^{(1)}\mathrm{id}). \tag{A.38}$$

Note that here the order of the factors is important, because they do not necessarily commute. (In what follows we often omit writing the identity operator id.) Finding such a factorization may be highly complicated and itself requires solving a system of difference equations.[5] However, sometimes a factorization can be found in explicit form and we will here only consider examples of such cases.

Once we have a factorization we can reduce the order of the equation as follows. Let

$$w_n := (T - b_n^{(N-1)})\cdots(T - b_n^{(1)})y_n \tag{A.39}$$

then equation (A.37) becomes the first-order equation

$$a_n^{(N)}(T - b_n^{(N)})w_n = f_n \tag{A.40}$$

for w_n. After solving this, and inserting w_n into (A.39), the equation for y_n becomes

$$(T - b_n^{(N-1)})\cdots(T - b_n^{(1)})y_n = w_n \tag{A.41}$$

where the inhomogeneous term is now given by w_n, the general solution of (A.40). This last equation is of order $N-1$, which means that we have reduced the order of the equation. This process can be repeated.

Example A.2.6 Let us consider the following difference equation

$$y_{n+1} + y_{n-1} = \frac{2n^2}{n^2-1}y_n \Leftrightarrow (T^2 - \tfrac{2n^2}{n^2-1}T + 1)y_{n-1} = 0.$$

The operator in the latter equation factorizes as

$$(T - \tfrac{n}{n-1})(T - \tfrac{n-1}{n})y_{n-1} = 0.$$

[5] Whether or not a factorization of the form (A.38) exists in the first place depends, in fact, on the field of function we are working with. Such problems are studied in the subjects of Difference Algebra (Levin, 2008) and Difference Galois Theory (van der Put and Singer, 1997).

Defining $w_{n-1} := (T - \frac{n-1}{n})y_{n-1}$, we obtain the system of equations

$$\begin{cases} (T - \frac{n-1}{n})y_{n-1} & = w_{n-1} \\ (T - \frac{n}{n-1})w_{n-1} & = 0 \end{cases}.$$

Solving the second equation gives $w_n = 3\,C\,n$, where C is a constant, and when this is substituted into the first equation it can be solved with the result

$$y_n = C(n^2 - 1) + D/n.$$

A.2.6 Reduction from inhomogeneous to homogeneous equation

If a particular solution of the inhomogeneous equation can be found it can be used to reduce the problem into an homogeneous one. Consider an equation of the type (A.37) and assume its particular solution is given by η_n, i.e. $\mathfrak{T}\eta_n = f_n$. We can then write $y_n = w_n + \eta_n$ and the equation for w_n becomes $\mathfrak{T}w_n = 0$.

Another method of reducing the problem to a homogeneous one is to operate on the equation with an operator that annihilates the inhomogeneous part. Then one solves the resulting (higher-order) homogeneous equation. The solution is then substituted into the inhomogeneous equation and the extra coefficients determined.

A.2.7 Reduction of order

With a known solution of the homogeneous equation the order of the equation can be reduced, using the variation of constants method, discussed before for matrices.

Example A.2.7 Consider the second-order equation

$$a_n^{(2)} y_{n+2} + a_n^{(1)} y_{n+1} + a_n^{(0)} y_n = f_n, \quad \forall n.$$

Suppose η_n is a solutions of the homogeneous part. Then substituting $y_n = \eta_n z_n$ and using the homogeneous equation for η_n yields

$$a_n^{(2)} \eta_{n+2}(z_{n+2} - z_{n+1}) - a_n^{(0)} \eta_n (z_{n+1} - z_n) = f_n.$$

which is first-order in $w_n := z_{n+1} - z_n$.

Example A.2.8 We use the previous example to illustrate the method. The difference equation

$$y_{n+1} - 2\xi y_n + 2n y_{n-1} = 0, \tag{A.42}$$

has a particular solution given by Hermite polynomials $y_n = H_n(\xi)$. (For Hermite polynomials the equation (A.42) is a recurrence relation.) In order to get the general solution, we set $y_n = H_n(\xi)z_n(\xi)$. Inserting this into the equation, we obtain for $w_n =: z_{n+1} - z_n$:

$$H_{n+1}w_n - 2nH_{n-1}w_{n-1} = 0$$

which is solved by

$$w_n = 2^n n! \frac{H_0 H_1}{H_n H_{n+1}} w_0.$$

Next z_n is given by the sum

$$z_n = z_0 + \sum_{j=0}^{n-1} 2^j j! \frac{H_0 H_1}{H_j H_{j+1}} w_0,$$

and therefore the general solution is given by

$$y_n = y_0 \frac{H_n(\xi)}{H_0(\xi)} + H_n(\xi) \sum_{j=0}^{n-1} 2^j j! \frac{y_1 H_0(\xi) - y_0 H_1(\xi)}{H_j(\xi) H_{j+1}(\xi)}.$$

A.2.8 Generating functions – z-transforms

In treatment of the Fibonacci difference equation we have introduced the generating function of the Fibonacci numbers. This is a particular example of a z-transform for the difference equation which is an analogue of the Laplace transform for differential equations. By introducing suitable series expansions in terms of an auxiliary variable z, such as

$$Y(z) = \sum_{j=0}^{\infty} y_j z^j, \quad \text{or more generally} \quad Y(z) = \sum_{j=0}^{\infty} y_j \kappa_j z^j.$$

the linear difference equation can be often converted into an algebraic equation for $Y(z)$ (in the case of constant coefficients), or into a differential equation (for coefficients depending polynomially on the independent discrete variable n). Here the numerical factors κ_j depend on the case and must be carefully chosen. Note also that the choice of initial values is built in the form of the generating function.

Example A.2.9 Consider again the difference equation (A.42), now with initial values $y_{-1} = 0$, $y_0 = 1$. We look for a solution of the form:

$$Y(z) = \sum_{j=0}^{\infty} y_j \kappa_j z^j, \quad \kappa_j = \frac{1}{j!}.$$

Now an equation of $Y(z)$ can be obtained by manipulation the equation (A.42):

$$
\begin{aligned}
0 &= \sum_{j=0}^{\infty} \left[y_{j+1} - 2\xi y_j + 2j y_{j-1} \right] \frac{z^j}{j!} \\
&= \frac{d}{dz} \sum_{j=1}^{\infty} \frac{z^j}{j!} y_j - 2\xi \sum_{j=0}^{\infty} \frac{z^j}{j!} y_j + 2z \sum_{j=1}^{\infty} \frac{z^{j-1}}{(j-1)!} y_{j-1} \\
&= \frac{dY}{dz} - 2(\xi - z) Y(z).
\end{aligned}
$$

This can be solved with $Y(z) = e^{2\xi z - z^2} Y(0)$, where $y_0 = 1 \Rightarrow Y(0) = 1$. When $Y(z)$ is expanded in powers of z the coefficients of z^n are polynomials in the parameter ξ and in the discrete variable n, namely the Hermite polynomials $H_n(\xi)$.

Appendix B
Theta functions and elliptic functions

The theory of elliptic functions forms one of the most beautiful part of mathematics. In the last few decades it has become clear that many of the structures behind integrable systems share deep common features with the theory of elliptic functions and elliptic curves. We will see that in a sense elliptic functions form a microcosm, a paradigm, for the wider theory of integrable systems.

This appendix describes in a unified way the theory underlying elliptic functions and key properties that are needed in other parts of the book. The approach we follow is closest to the one in the book by Akhiezer (1990), which is "constructive" and highlights the interconnections between the many parts of this theory. From our point of view the *addition formulae* play a central role and provide a natural connection with the main subject of this book.

B.1 Definitions

In this section we will recall the basics of complex analytic function theory. For a more complete treatment we refer the reader to one of the standard books, e.g. Ablowitz and Fokas (2003); Ahlfors (1953).

B.1.1 Analytic functions

A function $f : \mathfrak{U} \to \mathbb{C}$ is said to be analytic at an interior point z_0 of an open domain $\mathfrak{U} \subset \mathbb{C}$ if the limit

$$\lim_{z \to z_0} \frac{f(z) - f(z_0)}{z - z_0} =: f'(z_0)\ , \tag{B.1}$$

exists. The function f is called *analytic* or *holomorphic* in $\mathfrak{U}$ if the limit exists for all $z_0 \in \mathfrak{U}$. The condition of complex differentiability on a function f in the complex plane is much more restrictive than in the real-valued case. Note that because z is complex, we can approach z_0 from different directions in the complex plane but the limit should be the same. An example of a function $f(z)$ which is not differentiable at 0 is $f(z) = |z|$. The differential quotient in (B.1) is then $|z|/z = e^{-i\theta}$ where θ is the argument of z. So the limit depends on the direction of approach to 0.

Decomposing the function $f(z)$ into its real and imaginary parts, we can view f as a function of two real variables $x = \Re z$, $y = \Im z$

$$f(z) = u(x, y) + iv(x, y), \quad z = x + iy \,.$$

The derivative f' (if it exists) can be separated into its real and imaginary parts after expressing it in terms of partial derivatives in x, y. This implies the following conditions on the functions u and v

$$\frac{\partial u}{\partial x} = \frac{\partial v}{\partial y}, \quad \frac{\partial u}{\partial y} = -\frac{\partial v}{\partial x}, \tag{B.2}$$

which are called the *Cauchy–Riemann equations*. These, in turn, imply that both u and v obey Laplace's equation at the given point (x, y) corresponding to z, namely

$$\Delta u = \Delta v = 0, \quad \Delta = \frac{\partial^2}{\partial x^2} + \frac{\partial^2}{\partial y^2} \,. \tag{B.3}$$

A key result of complex analytic function theory is *Cauchy's theorem*, which states the following:

Theorem B.1.1 *If the function f is analytic in an open disk $\mathcal{D} \subset \mathbb{C}$, then on any closed curve $\Gamma \subset \mathcal{D}$ we have*

$$\int_\Gamma f(z)dz = 0 \,.$$

A corollary of Cauchy's theorem is *Cauchy's integral formula.*

Theorem B.1.2 *If f is analytic in an open disk $\mathcal{D} \subset \mathbb{C}$ and γ is a closed curve in $\mathcal{D}$ we have*

$$n(\gamma, z_0) f(z_0) = \frac{1}{2\pi i} \int_\Gamma \frac{f(z)}{z - z_0} dz \,, \tag{B.4}$$

where $n(\gamma, z_0)$ is the winding number of γ around z_0. (If we go around the curve only once in the counter-clockwise direction, then $n(\gamma, z_0) = 1$.)

Cauchy's integral formula has several powerful consequences. Firstly, all values of the analytic function f for points interior to the curve can be found from the values of f *on* the curve. Secondly, formula (B.4) can be differentiated arbitrarily many times to give the nth derivative $f^{(n)}(z_0)$ of f at z_0, for arbitrary integer n. Thirdly, it shows that f has a *convergent* Taylor series expansion around z_0 in $\mathcal{D}$:

$$f(z) = f(z_0) + \frac{f'(z_0)}{1!}(z - z_0) + \frac{f''(z_0)}{2!}(z - z_0)^2 + \cdots, \quad z_0 \in \mathcal{D}. \tag{B.5}$$

B.1.2 Singularities

If a function $f(z)$ is analytic in a disk around a except at the point a then it is said to be singular at a and a is called an isolated singularity. If we can define $f(a)$ so that f becomes analytic at a then that point is a removable singularity. If $\lim_{z\to a} = \infty$ then the singularity is a pole. If a function is analytic everywhere in $\mathbb{C}$ then we call the function an *entire function*.

If $f(z)$ has a pole at $z = a$, the smallest integer n for which $\lim_{z\to a}(z-a)^n f(z)$ is finite is called the order of the pole, and if $n = 1$ then a is called a simple pole. For example, the function $1/(z(z-1)^2)$ has a simple pole at $z = 0$ and a pole of order 2 at $z = 1$. The functions e^z and $\sin(z)$ are examples of entire functions.

We expect all analytic functions to be single valued, but we will also have to work with multivalued functions such as $\sqrt{z}$. In that case we will restrict ourselves to regions in which they are single valued and analytic. To describe this region we use the idea of branch cuts. For example, consider $f(z) = z^{1/2}$. Let us take $z = e^{i\theta}$ and change θ continuously from 0 to 2π, then $z \to z$ but $f \to -f$. To prevent this multivaluedness we remove a line from 0 to ∞ which is a branch cut.

If f has an isolated singularity at z_0 then it has a *Laurent series expansion*

$$f(z) = \sum_{n=-\infty}^{\infty} a_n (z - z_0)^n \,, \tag{B.6}$$

which converges uniformly on an annular region surrounding z_0 where f is analytic. The coefficients in (B.6) are given by

$$a_n = \frac{1}{2\pi i}\int_\Gamma \frac{f(\zeta)}{(\zeta - z_0)^{n+1}}\, d\zeta, \quad n = \ldots, -1, 0, 1, 2, \ldots \,.$$

If $a_n = 0$ for $n < -p$, z_0 is a pole of order p, and if $a_n = 0$ for $n < 0$ the function is analytic at z_0. The coefficient a_{-1} is called the *residue* of the function f at z_0. These values, denoted by $\mathrm{Res}_{z=z_0} f(z)$, play a major role in the residue calculus, which is a complex analytic technique (based on Cauchy's theorem) to evaluate definite integrals and infinite sums.

Functions which are analytic everywhere except for poles are called *meromorphic functions*.

B.1.3 Elliptic functions

In this section, we define elliptic functions as meromorphic, doubly periodic functions. We start by discussing the nuances of this definition. We will restrict ourselves from now on to meromorphic functions.

Let f be a complex function, then if for any regular point z we have

$$f(z + \Omega) = f(z), \quad \Omega \in \mathbb{C} \text{ fixed and nonzero}\,, \tag{B.7}$$

then f is called *periodic* with period Ω. Clearly, in that case we have $f(z+m\Omega) = f(z)$ for any integer m. If two functions f, g are periodic with the same period Ω, then pointwise sums and differences $f \pm g$, products fg and quotients f/g, and derivatives f', are periodic as well with the same period. Any integer multiple of a period Ω is also a period. If equation (B.7) holds and every period of the function f is an integer multiple of Ω, then Ω is called a *primitive period* of f.

A *multiply periodic function* is a function that has more than one primitive period, namely $\Omega_1, \Omega_2, \ldots, \Omega_n$, with

$$f(z + m_1\Omega_1 + \cdots + m_n\Omega_n) = f(z)\,,$$

with the Ω_i all independent (meaning that the only nontrivial integral linear combination of these periods adding up to zero is the one with all coefficients equal to zero). The set of all points of the form $m_1\Omega + \cdots + m_n\Omega_n$, with the m_i integer, is called the *period lattice*.

The exponential function $f(z) = e^{iz}$ is an example of a periodic function with period 2π. The possibilities of multiply periodic functions are restricted by Jacobi's theorem, which states (Akhiezer, 1990):

Theorem B.1.3 *1. There does not exist a nonconstant function with $n \geq 3$ primitive periods; 2. there exist nonconstant functions with $n = 2$ given primitive periods iff the ratio of these periods is not real.*

Definition B.1.1 *A complex function $f(z)$ is called an elliptic function if it is meromorphic and doubly periodic, i.e. it admits two independent primitive periods.*

At least one of the two primitive periods Ω_1, Ω_2 of an elliptic function should be complex since the ratio Ω_2/Ω_1 should be nonreal. Hence they form a parallellogram given in Figure B.1 and thus the complex plane can be tessellated by all the parallelograms formed by the period lattice obtained by translating this parallelogram over integer multiples of the two periods.

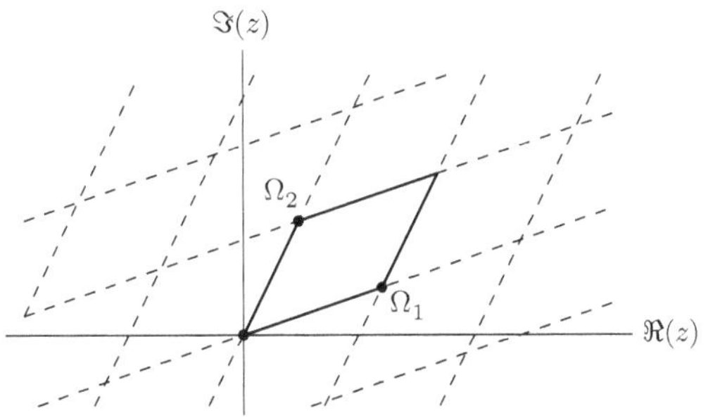

Figure B.1 Elliptic periodic parallelogram and tessellation of the complex plane.

The theory of elliptic functions provides a number of general results which we list here without proof:

i) there does not exist a nonconstant elliptic function that is regular in a period parallelogram, i.e. it must have at least one pole;
ii) the sum of the residues with respect to all the poles inside a single parallelogram of the period lattice is zero;
iii) the number of poles of an elliptic function in a period parallelogram counting multiplicity cannot be less than 2.

The elliptic functions are closely connected to a family of complex algebraic curves, called *elliptic curves*. These are curves which, in appropriate coordinates, can be cast in the form:

$$w^2 = \mathcal{R}(z) , \tag{B.8}$$

where $\mathcal{R}$ is a polynomial of order $p = 3$ or $p = 4$ in z, i.e.

$$\mathcal{R}(z) = \alpha z^3 + \beta z^2 + \gamma z + \delta \quad \text{or} \quad \mathcal{R}(z) = \alpha z^4 + \beta z^3 + \gamma z^2 + \delta z + \epsilon .$$

Although we can consider (B.8) for real variables, it is more useful to consider this equation for complex values of w and z. The elliptic functions arise as the functions in terms of which one can naturally parameterize such curves.

B.2 Theta functions

Elliptic functions can be expressed in terms of entire functions called theta functions. We will see that many properties of elliptic functions arise from the corresponding properties of the theta functions. For example, the zeros of theta functions correspond to poles of elliptic functions. The properties that we will derive include addition formulae, product formulae and transformation rules.

B.2.1 Definition of theta functions and relationships between them

We define the theta functions with characteristic $a, b = 0, 1$ and modulus τ as follows:

$$\theta_{ab}(x|\tau) = \sum_{n\in\mathbb{Z}} \exp\left[\pi i \tau \left(n + \frac{a}{2}\right)^2 + 2\pi i \left(n + \frac{a}{2}\right)\left(x + \frac{b}{2}\right)\right] . \tag{B.9}$$

The sum on the right-hand side being uniformly convergent for all $|x| \leq R$ and all $R > 0$ whenever $\Im\,\tau > 0$, (i.e. τ is in the Siegel half-plane $\tau \in \mathbb{H}$), the theta function is an entire function of $x \in \mathbb{C}$. It is doubly quasi-periodic with the periodicity relations being given by:

$$\theta_{ab}(x+1|\tau) = e^{\pi i a}\theta_{ab}(x|\tau), \quad \theta_{ab}(x+\tau|\tau) = e^{-\pi i(\tau+2x+b)}\theta_{ab}(x|\tau)\,. \tag{B.10}$$

The resummation $n \leftrightarrow -n$ in (B.9) leads to the relation

$$\theta_{ab}(-x|\tau) = \theta_{-a,-b}(x|\tau) = e^{\pi i(\tau a^2-2ax)}\theta_{ab}(x-a\tau|\tau) = e^{\pi i ab}\theta_{ab}(x|\tau)\,, \tag{B.11}$$

where the last two steps hold for a, b integer. Thus, it is immediately clear that $\theta_{00}, \theta_{01}, \theta_{10}$ are all even functions of its argument, but that θ_{11} is an odd function:

$$\begin{aligned}&\theta_{00}(-x|\tau) = \theta_{00}(x|\tau), \theta_{10}(-x|\tau) = \theta_{10}(x|\tau),\\ &\theta_{01}(-x|\tau) = \theta_{01}(x|\tau), \theta_{11}(-x|\tau) = -\theta_{11}(x|\tau).\end{aligned}$$

B.2.2 Fundamental addition formulae

We will now use the series (B.9) to obtain expressions for the products of two theta functions, namely

$$\begin{aligned}\theta_{ab}(x|\tau)\theta_{a'b'}(y|\tau) &= \sum_{n,m\in\mathbb{Z}} \exp\left\{\pi i\tau\left[(n+\frac{a}{2})^2 + (m+\frac{a'}{2})^2\right]\right.\\ &\qquad \left. + 2\pi i\left[(n+\frac{a}{2})(x+\frac{b}{2}) + (m+\frac{a'}{2})(y+\frac{b'}{2})\right]\right\}\\ &= \sum_{n,m\in\mathbb{Z}} \exp\left\{\frac{1}{2}\pi i\tau\left[(n+m+\frac{a+a'}{2})^2 + (n-m+\frac{a-a'}{2})^2\right]\right.\\ &\qquad + \pi i\left[(n+m+\frac{a+a'}{2})(x+y+\frac{b+b'}{2})\right.\\ &\qquad\quad \left.\left. +(n-m+\frac{a-a'}{2})(x-y+\frac{b-b'}{2})\right]\right\}.\end{aligned}$$

This invites the change of summation variables from n, m to $N = n+m$ resp. $M = n-m$. However, since $N+M = 2n \in 2\mathbb{Z}$ and $N-M = 2m \in 2\mathbb{Z}$ implying that either both N and M are even integers, or that they are both odd. Thus these replacements in the double sum lead to

$$\sum_{n,m\in\mathbb{Z}} \rightsquigarrow \sum_{\substack{N,M\in\mathbb{Z}\\ N,M\ \text{even}}} + \sum_{\substack{N,M\in\mathbb{Z}\\ N,M\ \text{odd}}}$$

Replacing N, M in the first sum by $2N, 2M$ and in the second sum by $2N+1, 2M+1$ we get

$$\begin{aligned}\dots = \sum_{N,M\in\mathbb{Z}} \exp\Bigg\{ & 2\pi i\tau\left[(N+\frac{a+a'}{4})^2+(M+\frac{a-a'}{4})^2\right] \\ & +2\pi i\left[(N+\frac{a+a'}{4})(x+y+\frac{b+b'}{2})\right. \\ & \left.+(M+\frac{a-a'}{4})(x-y+\frac{b-b'}{2})\right]\Bigg\} \end{aligned} \tag{B.12}$$

$$\begin{aligned} + \sum_{N,M\in\mathbb{Z}} \exp\Bigg\{ & 2\pi i\tau\left[(N+\frac{a+a'+2}{4})^2+(M+\frac{a-a'+2}{4})^2\right] \\ & +2\pi i\left[(N+\frac{a+a'+2}{4})(x+y+\frac{b+b'}{2})\right. \\ & \left.+(M+\frac{a-a'+2}{4})(x-y+\frac{b-b'}{2})\right]\Bigg\} \\ = \theta_{AB}(x+y|2\tau)\theta_{A'B'}(x-y|2\tau) & +\theta_{A+1,B}(x+y|2\tau)\theta_{A'+1,B'}(x-y|2\tau) \end{aligned} \tag{B.13}$$

with new characteristics:

$$A=\frac{a+a'}{2}\,,\quad B=b+b',$$
$$A'=\frac{a-a'}{2}\,,\quad B'=b-b'.$$

Clearly for $a, b, a', b' \in \mathbb{Z}_2$ to have integer characteristics we must take either $a = a' = 0$ or $a = a' = 1$, while b and b' can take on all values in $\mathbb{Z}_2$. This leads to the eight equations:

$$\begin{aligned}
\theta_{00}(x|\tau)\theta_{00}(y|\tau) &= \theta_{00}(x+y|2\tau)\theta_{00}(x-y|2\tau)+\theta_{10}(x+y|2\tau)\theta_{10}(x-y|2\tau)\\
\theta_{01}(x|\tau)\theta_{00}(y|\tau) &= \theta_{01}(x+y|2\tau)\theta_{01}(x-y|2\tau)+\theta_{11}(x+y|2\tau)\theta_{11}(x-y|2\tau)\\
\theta_{00}(x|\tau)\theta_{01}(y|\tau) &= \theta_{01}(x+y|2\tau)\theta_{01}(x-y|2\tau)-\theta_{11}(x+y|2\tau)\theta_{11}(x-y|2\tau)\\
\theta_{01}(x|\tau)\theta_{01}(y|\tau) &= \theta_{00}(x+y|2\tau)\theta_{00}(x-y|2\tau)-\theta_{10}(x+y|2\tau)\theta_{10}(x-y|2\tau)
\end{aligned}$$

as well as

$$\begin{aligned}
\theta_{10}(x|\tau)\theta_{10}(y|\tau) &= \theta_{10}(x+y|2\tau)\theta_{00}(x-y|2\tau)+\theta_{00}(x+y|2\tau)\theta_{10}(x-y|2\tau)\\
\theta_{11}(x|\tau)\theta_{10}(y|\tau) &= \theta_{11}(x+y|2\tau)\theta_{01}(x-y|2\tau)+\theta_{01}(x+y|2\tau)\theta_{11}(x-y|2\tau)\\
\theta_{10}(x|\tau)\theta_{11}(y|\tau) &= \theta_{11}(x+y|2\tau)\theta_{01}(x-y|2\tau)-\theta_{01}(x+y|2\tau)\theta_{11}(x-y|2\tau)\\
\theta_{11}(x|\tau)\theta_{11}(y|\tau) &= -\theta_{10}(x+y|2\tau)\theta_{00}(x-y|2\tau)+\theta_{00}(x+y|2\tau)\theta_{10}(x-y|2\tau)\,.
\end{aligned} \tag{B.14}$$

These identities will serve as the starting point from which all key identities for elliptic functions will be derived. Fortunately, these relations are not all independent, and neither are the four theta functions θ_{00}, θ_{10}, θ_{01}, θ_{11}. In fact, these functions are related through shifts over *half periods*, i.e. shifts over $\frac{1}{2}$ and $\frac{\tau}{2}$ in the argument. They are given by the relations

$$\theta_{a,b+1}(x|\tau) = \theta_{ab}(x + \frac{1}{2}|\tau), \quad \theta_{a+1,b}(x|\tau) = e^{\frac{1}{4}\pi i\tau + \pi i(x+\frac{b}{2})}\theta_{ab}(x + \frac{\tau}{2}|\tau). \tag{B.15}$$

Using these relations the fore-last equation in the list can be written as:

$$\begin{vmatrix} \theta_{11}(x|2\tau) & \theta_{11}(y|2\tau) \\ \theta_{01}(x|2\tau) & \theta_{01}(y|2\tau) \end{vmatrix} = \theta_{11}\left(\frac{x+y}{2} - \frac{1}{2} \mid \tau\right)\theta_{11}\left(\frac{x-y}{2} \mid \tau\right) \tag{B.16}$$

while the others can be deduced from (B.16) by appropriate shifts in the argument (shifts over τ and 1). Furthermore, taking $y = x + 1$ we obtain from (B.16) the relation

$$\theta_{11}(x|\tau) = 2e^{-\pi ix + \frac{1}{2}\pi i\tau}\frac{\theta_{11}(x|2\tau)\theta_{11}(x-\tau|2\tau)}{i\theta_{11}(-\frac{1}{2}|\tau)} \tag{B.17}$$

connecting elliptic functions of one set of period to the ones with one of the periods halved. This is related to the famous, *Landen transform*, cf. Whittaker and Watson (1902).

Below we present a general proof of the identities (B.16) and show that from (B.14) all well-known identities between elliptic functions of the same modulus (periods) can be derived, either in the Jacobian representation or in the Weierstrass representation. Also all five-term theta-function relations can be found, such as:

$$\begin{aligned} &2\theta_{00}(x)\theta_{00}(y)\theta_{00}(z)\theta(w) = \\ &\quad = \theta_{00}(X)\theta_{00}(Y)\theta_{00}(Z)\theta_{00}(W) + \theta_{01}(X)\theta_{01}(Y)\theta_{01}(Z)\theta_{01}(W) \\ &\quad + \theta_{10}(X)\theta_{10}(Y)\theta_{10}(Z)\theta_{10}(W) + \theta_{11}(X)\theta_{11}(Y)\theta_{11}(Z)\theta_{11}(W) \end{aligned} \tag{B.18}$$

in which all theta functions are of the same modulus τ and where we have abbreviated:

$$X = \frac{1}{2}(x+y+z+w)\ ,\quad Y = \frac{1}{2}(x+y-z-w)$$

$$Z = \frac{1}{2}(x-y+z-w)\ ,\quad W = \frac{1}{2}(x-y-z+w)$$

using the oddness of the θ_{11} one actually has a closed-form equation for this function, namely:

$$\begin{aligned} &\theta_{11}(x+y)\theta_{11}(x-y)\theta_{11}(z+w)\theta_{11}(z-w) \\ &\quad + \theta_{11}(x+z)\theta_{11}(x-z)\theta_{11}(w+y)\theta_{11}(w-y) \\ &\quad + \theta_{11}(x+w)\theta_{11}(x-w)\theta_{11}(y+z)\theta_{11}(y-z) = 0. \end{aligned} \tag{B.19}$$

The theta function relation (B.19) constitutes the key identity which we will use for the constitution of the Weierstrass family of elliptic functions in the next section.

B.2.3 Product formulae

One of the most important type of identities in combinatorics are the ones that relate infinite sums to infinite products, e.g. the famous Rogers–Ramanujan identities. Many of these identities have an origin in the representation theory of infinite-dimensional Lie algebras. One of the most fundamental of such relations is the *Jacobi triple product relation*, which can be most easily formulated as:

$$\sum_{n\in\mathbb{Z}} q^{n(n-1)/2}(-z)^n = \prod_{j=1}^{\infty}(1-q^{j-1}z)(1-q^j z^{-1})(1-q^j)\ , \quad |q|<1\,. \tag{B.20}$$

Proof The proof is elementary. If we call the right-hand side $\vartheta(z;q)$, it is easy to see that it converges uniformly for all z within a finite radius whenever $|q|<1$. By reordering the factors it is clear that the following q-difference equation holds:

$$\vartheta(z;q) = -z\vartheta(qz;q).$$

Thus, if we assume the expansion

$$\vartheta(z;q) = \sum_{n\in\mathbb{Z}} c_n z^n$$

we obtain the recursion relation $c_n = -q^{n-1}c_{n-1}$ for the coefficients, which leads to $c_n = (-1)^n q^{n(n-1)/2}c_0$ upon iteration. The coefficient $c_0 = c_0(q)$ can be determined by considering the series and product for $z = iq^{1/2}$, in which case we get

$$\prod_{j=1}^{\infty}(1+q^{2j-1})(1-q^j) = c_0(q)\sum_{k\in\mathbb{Z}}(-1)^k q^{2k^2}\ ,$$

while if we take $z = q^{1/2}$ we get

$$\prod_{j=1}^{\infty}(1-q^{j-1/2})^2(1-q^j) = c_0(q)\sum_{k\in\mathbb{Z}}(-1)^k q^{\frac{1}{2}k^2}\ ,$$

a comparison of which leads to the relation

$$\frac{c_0(q^4)}{c_0(q)} = \frac{\prod_{j=1}^{\infty}(1-q^{4j-2})^2(1-q^{4j})}{\prod_{j=1}^{\infty}(1+q^{2j-1})(1-q^j)} = 1$$

where the last equality is a consequence of the surprising relation

$$\prod_{j=1}^{\infty}(1-q^j) = \prod_{k=1}^{\infty}(1-q^{2k-1})(1-q^{2k}) = \prod_{k=1}^{\infty}(1-q^{2k-1})(1+q^k)(1-q^k)$$

$$\Rightarrow \prod_{k=1}^{\infty}(1-q^{2k-1})(1+q^k) = 1 \, .$$

Thus it follows that $c_0(q) = c_0(0) = 1$, since if $q = 0$ the left-hand side of (B.20) reduces to $1 - z$, while the right-hand side only gets contributions from $j = 0, 1$. Thus, we obtain the required identity. □

Equation (B.20) is a confirmation of the fact that $\vartheta(z; q)$ is an entire function of z with simple zeroes at $z = q^j$, $j \in \mathbb{Z}$. We mention that the product

$$\eta(q) = q^{1/24} \prod_{j=1}^{\infty}(1-q^j) \tag{B.21}$$

which enters in the coefficient of (B.20) plays an important role in combinatorics and is known as the *Dedekind η-function*. A classic result by Euler relates the infinite product in (B.21) to the infinite sum $\sum_{k\in\mathbb{Z}}^{(-1)^k} \exp((3k^2+k)/2)$, while its inverse serves as the generating function of partitions.

It is clear that by taking

$$z = e^{2\pi i x} \quad , \quad q = e^{2\pi i \tau}$$

the function $\vartheta(z; q)$ reduces to the θ_{11} function (apart from a factor). Thus, the triple product relation (B.20) directly yields the various expressions for the theta functions in terms of infinite products.

B.3 The Weierstrass family of elliptic functions

Here, we collect some useful formulae for the Weierstrass elliptic functions. The Weierstrass σ-function is basically the θ_{11} apart from a multiplicative factor and a scaling in its argument. It is the latter theta function for which we have a closed-form addition formula, namely (B.19), so understandably this function plays the main role in the theory.

Thus, we define the so-called Weierstrass σ-function to be given by

$$\sigma(z|2\omega_1, 2\omega_2) = 2\omega_1 \exp\left(\frac{\eta_1 z^2}{2\omega_1}\right) \frac{\theta_{11}(x|\tau)}{\theta'_{11}(0|\tau)} \quad , \quad \tau = \frac{\omega_2}{\omega_1} \quad , \quad z = 2\omega_1 x \, . \tag{B.22}$$

The exponential prefactor and scaling are motivated in order to make the function *automorphic* under the action of the group $SL(2,\mathbb{Z})$, i.e. invariant under linear transformations acting on the period lattice:

$$\begin{pmatrix} \omega_1 \\ \omega_2 \end{pmatrix} \mapsto \begin{pmatrix} \alpha & \beta \\ \gamma & \delta \end{pmatrix} \begin{pmatrix} \omega_1 \\ \omega_2 \end{pmatrix}. \tag{B.23}$$

The relations between the Weierstrass elliptic functions are given by

$$\zeta(z) = \frac{\sigma'(z)}{\sigma(z)} \ , \quad \wp(z) = -\zeta'(z) \ , \tag{B.24}$$

where $\sigma(z)$ and $\zeta(z)$ are odd functions and $\wp(z)$ is an even function of its argument.

We recall also that the $\sigma(z)$ is an entire function, and $\zeta(z)$ is a meromorphic function having simple poles at $\omega_{k,l}$, both being quasi-periodic, obeying

$$\zeta(x + 2\omega_{1,2}) = \zeta(x) + 2\eta_{1,2} \ , \quad \sigma(x + 2\omega_{1,2}) = -\sigma(x) e^{2\eta_{1,2}(x+\omega_{1,2})} \ , \tag{B.25}$$

in which

$$\eta_i \equiv \zeta(\omega_i) \ , \quad i = 1, 2 \ .$$

The relation between periods and the η_i is given by

$$\eta_1 \omega_2 - \eta_2 \omega_1 = \frac{\pi i}{2} \ .$$

The function $\wp(z)$ is doubly periodic, and it is only the $\wp$ function that is truly elliptic according to the definition of an elliptic function.

As a consequence of the product formula for the theta functions we obtain the following product formula for the Weierstrass σ-function:

$$\sigma(z) = z \prod_{(k,\ell)\neq(0,0)} \left(1 - \frac{z}{\omega_{k,\ell}}\right) \exp\left[\frac{z}{\omega_{k,\ell}} + \frac{1}{2}\left(\frac{z}{\omega_{k,\ell}}\right)^2\right], \tag{B.26a}$$

with $\omega_{kl} = 2k\omega_1 + 2\ell\omega_2$. Taking the logarithmic derivatives we obtain from (B.26a) the following double-sum expansions for the Weierstrass ζ- and $\wp$ functions:

$$\zeta(z; 2\omega_1, 2\omega_2) = \frac{1}{z} + \sum_{(k,\ell)\neq(0,0)} \left[\frac{1}{z + \omega_{k,\ell}} - \frac{1}{\omega_{k,\ell}} + \frac{z}{\omega_{k,\ell}^2}\right], \tag{B.26b}$$

$$\wp(z; 2\omega_1, 2\omega_2) = \frac{1}{z^2} + \sum_{(k,\ell)\neq(0,0)} \left[\frac{1}{(z + \omega_{k,\ell})^2} - \frac{1}{\omega_{k,\ell}^2}\right]. \tag{B.26c}$$

From a computational point of view, the most important property of these elliptic functions is the existence of a number of functional relations, the most fundamental being the three-term relation for the σ-function

$$\sigma(x+y)\sigma(x-y)\sigma(a+b)\sigma(a-b) = \tag{B.27}$$
$$\sigma(x+a)\sigma(x-a)\sigma(y+b)\sigma(y-b) - \sigma(x+b)\sigma(x-b)\sigma(y+a)\sigma(y-a),$$

which is a direct consequence of the similar relation for the θ_{11} functions (B.19). From this functional equation one can derive by suitable limits:

$$\zeta(x) + \zeta(y) + \zeta(z) - \zeta(x+y+z) = \frac{\sigma(x+y)\sigma(x+z)\sigma(y+z)}{\sigma(x)\sigma(y)\sigma(z)\sigma(x+y+z)}. \tag{B.28}$$

In fact, all identities for the Weierstrass functions as functions of their arguments can be derived from these functional relations.

Now it will be convenient to express the addition formulae in terms of a function $\Phi_\kappa(x)$ (with κ some complex number) given by

$$\Phi_\kappa(x) \equiv \frac{\sigma(x+\kappa)}{\sigma(x)\sigma(\kappa)}, \tag{B.29}$$

which is similar to the function Ψ used in the previous section.[1]

Equation (B.27) can also be cast into the following form

$$\Phi_\kappa(x)\Phi_\kappa(y) = \Phi_\kappa(x+y)\left[\zeta(\kappa) + \zeta(x) + \zeta(y) - \zeta(\kappa+x+y)\right]. \tag{B.30}$$

The following three-term relation for $\sigma(x)$ is a consequence of (B.27), and this equation can be cast into the following convenient form

$$\Phi_\kappa(x)\Phi_\lambda(y) = \Phi_\kappa(x-y)\Phi_{\kappa+\lambda}(y) + \Phi_{\kappa+\lambda}(x)\Phi_\lambda(y-x), \tag{B.31}$$

which is obtained from the elliptic analogue of the partial fraction expansion, i.e. equation (B.30).

From (B.28) various other addition formulae can be derived by simple manipulations, taking into account that these are functional identities valid for all values of the arguments (away from the singularities of the functions). Thus, by differentiation we can derive:

$$\zeta(x+y) - \zeta(x) - \zeta(y) = \frac{1}{2}\frac{\wp'(x) - \wp'(y)}{\wp(x) - \wp(y)}. \tag{B.32}$$

Furthermore, we have the relation

$$\wp(x) + \wp(y) + \wp(x+y) = \frac{1}{4}\left(\frac{\wp'(x) - \wp'(y)}{\wp(x) - \wp(y)}\right)^2. \tag{B.33}$$

[1] In the literature sometimes a different form for $\Phi_\kappa(x)$ is introduced, namely

$$\Phi(x;\kappa) := \frac{\sigma(x-\kappa)}{\sigma(x)\sigma(\kappa)}\, e^{\zeta(\kappa)x},$$

which is sometimes referred to as the *Lamé function*. We prefer to work with the form given in (B.29).

Equation (B.33) is the well-known addition formula for the Weierstrass $\wp$ function. Finally, we get the following remarkable addition formula:

$$\frac{\sigma(x-y)\sigma(x+y)}{\sigma^2(x)\sigma^2(y)} = \wp(y) - \wp(x)\,, \tag{B.34}$$

and the fundamental three-term relation form σ (B.27) can be easily seen to arise as the consistency condition for the functional relation (B.34), by rewriting the trivial identity

$$\begin{aligned}(\wp(x)-\wp(y))(\wp(a)-\wp(b)) &= (\wp(x)-\wp(a))(\wp(y)-\wp(b)) \\ &\quad - (\wp(x)-\wp(b))(\wp(y)-\wp(a))\end{aligned}$$

in terms of σ-functions. Thus, we have come full circle: from the three-term relation for σ we have derived a sequel of addition formulae at the end of which we recover the three-term relation itself!

A number of transformation properties of the Weierstrass functions relate functions of different periods together. The Landen transform (in terms of the Weierstrass functions) relates the functions of a given period to those of twice (or vice versa half) the period.[2] They are obtained from the fundamental relation (B.17) for the θ functions. Translating them to the Weierstrass family by using the correspondence (B.22) we obtain the following relations

$$\sigma(z) = e^{-\frac{1}{2}(e_1+\bar{e}_2)z^2+\bar{\eta}_2 z}\bar{\sigma}(z)\frac{\bar{\sigma}(z-2\omega_2)}{\bar{\sigma}(-2\omega_2)} \tag{B.35a}$$

$$\eta(z) = -(e_1+\bar{e}_2)z + \bar{\eta}_2 + \bar{\zeta}(z) + \bar{\zeta}(z-2\omega_2) \tag{B.35b}$$

$$\wp(z) = e_1 + \bar{e}_2 + \bar{\wp}(z) + \bar{\wp}(z-2\omega_2) \tag{B.35c}$$

in which $\sigma(z) = \sigma(z|2\omega_1, 2\omega_2)$, $\bar{\sigma}(z) = \sigma(z|2\omega_1, 4\omega_2)$, and $\zeta(z)$, $\wp(z)$ the corresponding Weierstrass functions with periods $2\omega_1, 2\omega_2$, respectively $\bar{\zeta}(z)$ and $\bar{\wp}(z)$ the corresponding Weierstrass functions with periods $2\omega_1, 4\omega_2$.

B.4 Half-period functions and the elliptic curve

Since the Weierstrass functions are periodic or quasi-periodic with primitive period, for convenience let us introduce as a third half-period $\omega_3 = -\omega_1 - \omega_2$, for which $\eta_3 \equiv \zeta(\omega_3) = -\eta_1 - \eta_2$. We then have also the relations:

[2] Usually the Landen transform is given in terms of the Jacobi elliptic functions, where the moduli change by a fractional linear transformation; see Whittaker and Watson (1902) and DLMF (2010), Section 22.7.

$$\eta_1\omega_2 - \eta_2\omega_1 = \frac{\pi i}{2} \quad \Rightarrow \quad \eta_2\omega_3 - \eta_3\omega_2 = \frac{\pi i}{2}, \quad \eta_3\omega_1 - \eta_1\omega_3 = \frac{\pi i}{2}.$$

Introducing now the three functions:

$$W_i(x) \equiv \Phi_{\omega_i}(x)e^{-\eta_i x}, \quad i = 1, 2, 3 \tag{B.36}$$

we can derive several relations from the addition formulae (B.30) and (B.31) for Φ as well using the periodicity conditions (B.25).

The following properties hold for the functions W_i:

i) The $W_i(x)$ are periodic with period $2\omega_i$, but picking up a minus sign with respect to the other periods:

$$W_i(x + 2\omega_i) = W_i(x), \quad W_i(x + 2\omega_j) = -W_i(x), \quad i, j, k = 1, 2, 3 \text{ cyclic}. \tag{B.37}$$

ii) The W_i are *odd* functions with respect to their argument:

$$W_i(-x) = -W_i(x). \tag{B.38}$$

iii) The addition formula:

$$W_i(x)W_j(z) + W_j(y)W_k(x) + W_k(z)W_i(y) = 0 \ , \ \ x + y + z = 0 \tag{B.39}$$

follows by applying (B.31) together with the periodicity relations (B.25).

iv) The following relations are a consequence of (B.34) together with the periodicity (B.25)

$$W_i^2(x) = \wp(x) - e_i, \quad e_i \equiv \wp(\omega_i), \tag{B.40}$$

which means that for fixed constants e_1, e_2, e_3 the (W_1, W_2, W_3) are coordinates on the intersection of three complex cylinders.

v) From (B.30) we obtain the relation

$$\frac{W_i(x)W_j(x)}{W_k(x)} = \eta_k + \zeta(x) - \zeta(x+\omega_k) = -\frac{1}{2}\frac{\wp'(x)}{\wp(x) - e_k}, \quad i, j, k = 1, 2, 3 \quad \text{cyclic}. \tag{B.41}$$

vi) Using the periodicity (B.25) we have

$$W_i(x)W_i(x + \omega_i) = -\frac{e^{\eta_i\omega_i}}{\sigma^2(\omega_i)}. \tag{B.42}$$

We are now in a position to derive the most important property, namely the equation for the elliptic curve. In fact, combining (B.40) and (B.41), we get

$$W_1(x)W_2(x)W_3(x) = -\frac{1}{2}\wp'(x), \tag{B.43}$$

and taking the square of this relation, using again (B.40), we obtain:

$$(\wp'(x))^2 = 4(\wp(x) - e_1)(\wp(x) - e_2)(\wp(x) - e_3) \,. \tag{B.44}$$

Using the fact that

$$e_1 + e_2 + e_3 = 0 \,,$$

and taking as (complex) coordinates $(z, w) = (\wp(x), \wp'(x))$, we can cast (B.44) in the form:

$$w^2 = \mathcal{R}(z) = 4z^3 - g_2 z - g_3 \,, \tag{B.45}$$

which is the standard form of the so-called *Weierstrass curve*, which is one of the standard forms of an elliptic curve. The constants g_2, g_3 are given in terms of the e_i through the formulae:

$$g_2 = -4(e_1e_2 + e_1e_3 + e_2e_3), \quad g_3 = 4e_1e_2e_3 \,, \tag{B.46}$$

and are called the *moduli of the curve*. The roots e_i, $i = 1, 2, 3$, of the cubic $\mathcal{R}(z)$ are called the *branch points* of the curve, and they are expressed through the formulae $e_i = \wp(\omega_i)$ in terms of the Weierstrass elliptic function.

We finish this account on Weierstrass functions by mentioning the relations:

$$\wp'(x) = -\frac{\sigma(2x)}{\sigma^4(x)}, \quad \frac{1}{2}\frac{\wp''(x)}{\wp'(x)} = \zeta(2x) - 2\zeta(x) \,, \tag{B.47}$$

as well as the so-called duplication formulae:

$$\sigma(2x) = 2\frac{\sigma(x)\,\sigma(x+\omega_1)\,\sigma(x+\omega_2)\,\sigma(x+\omega_3)}{\sigma(\omega_1)\,\sigma(\omega_2)\,\sigma(\omega_3)} \,, \tag{B.48a}$$

$$2\zeta(2x) = \zeta(x) + \zeta(x+\omega_1) + \zeta(x+\omega_2) + \zeta(x+\omega_3) \,, \tag{B.48b}$$

$$4\wp(2x) = \wp(x) + \wp(x+\omega_1) + \wp(x+\omega_2) + \wp(x+\omega_3) \,. \tag{B.48c}$$

Finally, we mention that by eliminating the derivatives from (B.33) using the relation of the elliptic curve (B.44), we obtain the purely "discrete" addition formula for the $\wp$ function, namely

$$\left(XY + XZ + YZ + \frac{g_2}{4}\right)^2 - (4XYZ - g_3)(X + Y + Z) = 0 \tag{B.49}$$

in which

$$X = \wp(x), \quad Y = \wp(y), \quad Z = \wp(z) \quad \text{such that} \quad x + y + z = 0 \,.$$

The *triquadratic* relation (B.49) can be viewed as a discretization of the formula for the elliptic curve: if we set $z = \varepsilon$ and take the limit $\varepsilon \to 0$, the dominant contribution in the relation (B.49) goes over into the defining equation for the Weierstrass curve in standard form.

B.5 Jacobi elliptic functions

The half-period functions of the latter chapter are up to a scaling equivalent to the other famous class of elliptic functions: the Jacobi class. In fact, by setting

$$u = \sqrt{e_1 - e_3}\, x, \quad k^2 \equiv \frac{e_2 - e_3}{e_1 - e_3}, \tag{B.50}$$

we can introduce the functions

$$\mathrm{sn}(u;k) = \frac{\sqrt{e_1 - e_3}}{W_3(x)}, \quad \mathrm{cn}(u;k) = \frac{W_1(x)}{W_3(x)}, \quad \mathrm{dn}(u;k) = \frac{W_2(x)}{W_3(x)}. \tag{B.51}$$

Albeit this was not the way in which these functions were introduced initially, in the context of the previous section it is a convenient way to make the connection with the Weierstrass class. In fact, we can rewrite (B.51) as

$$\mathrm{sn}(u;k) = \frac{\sqrt{e_1 - e_3}}{\sqrt{\wp(x) - e_3}}, \quad \mathrm{cn}(u;k) = \sqrt{\frac{\wp(x) - e_1}{\wp(x) - e_3}}, \quad \mathrm{dn}(u;k) = \sqrt{\frac{\wp(x) - e_2}{\wp(x) - e_3}}. \tag{B.52}$$

These functions are periodic with periods $4K$ and $4iK'$ in the argument u, where K and K' (the real and imaginary quarter periods) are given by

$$K = \sqrt{e_1 - e_3}\, \omega_1, \quad iK' = \sqrt{e_1 - e_3}\, \omega_2,$$

whereas the parameter k is called the modulus of the Jacobi elliptic functions. From the properties of the functions W_i, or in fact from (B.52), it follows that

$$\mathrm{sn}^2(u;k) + \mathrm{cn}^2(u;k) = 1, \quad k^2 \mathrm{sn}^2(u;k) + \mathrm{dn}(u;k)^2 = 1. \tag{B.53}$$

Various properties can be asserted for the Jacobi functions directly from the definitions above. In fact, it follows that sn is an *odd* function of u, while cn and dn are *even* functions:

$$\mathrm{sn}(u;k) = -\mathrm{sn}(u;k), \quad \mathrm{cn}(-u;k) = \mathrm{cn}(u;k), \quad \mathrm{dn}(-u;k) = \mathrm{dn}(u;k), \tag{B.54}$$

while

$$\mathrm{sn}(0;k) = 0, \quad \mathrm{cn}(0;k) = \mathrm{dn}(0;k) = 1, \tag{B.55}$$

and they allow the following series expansions:

$$\mathrm{sn}(u;k) = u - (1+k^2)\frac{u^3}{3!} + (1+14k^2+k^4)\frac{u^5}{5!} - \cdots \tag{B.56a}$$

$$\mathrm{cn}(u;k) = 1 - \frac{u^2}{2!} + (1+4k^2)\frac{u^4}{4!} - (1+44k^2+16k^4)\frac{u^6}{6!} + \cdots \tag{B.56b}$$

$$\mathrm{dn}(u;k) = 1 - k^2\frac{u^2}{2!} + k^2(4+k^2)\frac{u^4}{4!} - k^2(16+44k^2+k^4)\frac{u^6}{6!} + \cdots . \tag{B.56c}$$

Remark B.5.1 It is illustrative to compare the expansions (B.56) with the series expansions for the trigonometric functions $\sin(u)$, $\cos(u)$, (including the constant function 1) by setting $k = 0$.

From now on we shall consider the parameter k to be fixed and simply write $\mathrm{sn}(u)$, $\mathrm{cn}(u)$ and $\mathrm{dn}(u)$ for the Jacobi functions, suppressing the second argument.

The (quasi-)periodicity properties for these functions follow equally from the properties 1 and 5 of the $W_i(x)$ as listed in the previous section, namely

$$\mathrm{sn}(2K-u) = \mathrm{sn}(u), \quad \mathrm{cn}(2K-u) = -\mathrm{cn}(u), \quad \mathrm{dn}(2K-u) = \mathrm{dn}(u),$$
$$\mathrm{sn}(2iK'-u) = -\mathrm{sn}(u), \quad \mathrm{cn}(2iK'-u) = -\mathrm{cn}(u), \quad \mathrm{dn}(2iK'-u) = -\mathrm{dn}(u).$$

In fact, sn has periods $4K$ and $2iK'$, cn has periods $4K$ and $2K + 2iK'$, while dn has periods $2K$ and $4iK'$. The zeroes of sn are located at $u = 2mK + 2niK'$, of cn at $u = (2m+1)K + 2niK'$, while of dn they are at $u = (2m+1)K + 2n+1)iK'$, for $n, m \in \mathbb{Z}$. Poles of all three functions are located at $u = 2mK + (2n+1)iK'$, $n, m \in \mathbb{Z}$.

Differential relations follow from the differential relations for the $W_i(x)$. For instance, from

$$\frac{d}{dx}\ln W_3(x) = \zeta(x+\omega_3) - \zeta(x) - \eta_3 = -\frac{W_1(x)W_2(x)}{W_3(x)} \quad \Rightarrow$$
$$\Rightarrow \quad \frac{d}{du}\ln \mathrm{sn}(u) = \frac{1}{\sqrt{e_1-e_3}} W_3(x)\,\mathrm{cn}(u)\,\mathrm{dn}(u)$$

and hence we obtain the differential equation for sn, and similarly for the others, namely

$$\frac{d}{du}\mathrm{sn}(u) = \mathrm{cn}(u)\,\mathrm{dn}(u), \quad \frac{d}{du}\mathrm{cn}(u) = -\mathrm{sn}(u)\,\mathrm{dn}(u), \quad \frac{d}{du}\mathrm{dn}(u) = -k^2\,\mathrm{sn}(u)\,\mathrm{cn}(u)\,. \tag{B.57}$$

We note that from the first differential relations (B.57), eliminating cn and dn, the sn function obeys the Jacobi differential equation

$$\left(\frac{ds}{du}\right)^2 = \left(1-s^2\right)\left(1-k^2s^2\right), \tag{B.58}$$

which is related to the *Jacobi curve*:

$$w^2 = \mathcal{R}(z) = (1 - z^2)(1 - k^2 z^2) . \tag{B.59}$$

Thus, alternatively we could have introduced the sn function by defining it through the inversion of the *elliptic integral*

$$u = \int_0^{\mathrm{sn}(u)} \frac{ds}{\sqrt{(1 - s^2)(1 - k^2 s^2)}} , \tag{B.60}$$

which by the change of integration variables: $s = \sin\theta$, can also be written in the form:

$$u = \int_0^{\mathrm{am}(u)} \frac{d\theta}{\sqrt{1 - k^2 \sin^2\theta}} , \tag{B.61}$$

in which the upper integration limit is called the *amplitude function*. In terms of this function we have

$$\mathrm{sn}(u) = \sin(\mathrm{am}(u)) .$$

Clearly, if $k = 0$ we have $\mathrm{am}(u) = u$, and the sn function reduces to the usual sin function.

The quarter periods K and iK' can be obtained from the *complete elliptic integral of the first kind*:

$$K = \int_0^1 \left(1 - s^2\right)^{-1/2} \left(1 - k^2 s^2\right)^{-1/2} ds, \tag{B.62}$$

with K' obtained from (B.62) by replacing k by $k' \equiv \sqrt{1 - k^2}$. Furthermore, we have the *complete elliptic integrals of the second kind*:

$$E = \int_0^1 \left(1 - s^2\right)^{-1/2} \left(1 - k^2 s^2\right)^{1/2} ds, \tag{B.63}$$

and similarly E' obtained by replacing k by k'. The relation between these quantities is given by

$$EK' + E'K - KK' = \frac{1}{2}\pi ,$$

paralleling the relation between η, η', ω and ω' in the Weierstrass case.

Addition formulae follow from (B.39), from which we can infer

$$\mathrm{cn}(v)\,\mathrm{sn}(u + v) = \mathrm{sn}(u)\,\mathrm{dn}(v) + \mathrm{dn}(u)\,\mathrm{sn}(v)\,\mathrm{cn}(u + v) , \tag{B.64a}$$

$$\mathrm{dn}(u)\,\mathrm{sn}(u + v) = \mathrm{cn}(u)\,\mathrm{sn}(v) + \mathrm{sn}(u)\,\mathrm{cn}(v)\,\mathrm{dn}(u + v) , \tag{B.64b}$$

$$\mathrm{sn}(u)\,\mathrm{cn}(u + v) + \mathrm{sn}(v)\,\mathrm{dn}(u + v) = \mathrm{cn}(u)\,\mathrm{dn}(v)\,\mathrm{sn}(u + v) . \tag{B.64c}$$

By supplementing these relations with the ones with u and v interchanged, one obtains several linear systems, from which $\mathrm{sn}(u+v)$, $\mathrm{cn}(u+v)$ and $\mathrm{dn}(u+v)$ can be solved in terms of the Jacobi functions with single arguments u or v, leading after some manipulation to the following addition formulae:

$$\mathrm{sn}(u+v) = \frac{\mathrm{sn}(u)\,\mathrm{cn}(v)\,\mathrm{dn}(v) + \mathrm{sn}(v)\,\mathrm{cn}(u)\,\mathrm{dn}(u)}{1 - k^2\mathrm{sn}^2(u)\,\mathrm{sn}^2(v)}\,, \tag{B.65a}$$

$$\mathrm{cn}(u+v) = \frac{\mathrm{cn}(u)\,\mathrm{cn}(v) - \mathrm{sn}(u)\,\mathrm{dn}(u)\,\mathrm{sn}(v)\,\mathrm{dn}(v)}{1 - k^2\mathrm{sn}^2(u)\,\mathrm{sn}^2(v)}\,, \tag{B.65b}$$

$$\mathrm{dn}(u+v) = \frac{\mathrm{dn}(u)\,\mathrm{dn}(v) - k^2\mathrm{sn}(u)\,\mathrm{cn}(u)\,\mathrm{sn}(v)\,\mathrm{cn}(v)}{1 - k^2\mathrm{sn}^2(u)\,\mathrm{sn}^2(v)}\,. \tag{B.65c}$$

Proof We will follow Whittaker and Watson (1902, p. 495) in making use of the differential equation (B.58), from which we obtain immediately the following second-order differential equation

$$\frac{d^2 s}{du^2} = 2k^2 s^3 - (1+k^2)s\,. \tag{B.66}$$

Essentially, the proof of equation (B.65a) is due to Euler (1761), who considered a differential equation in the form:

$$\frac{dx}{\sqrt{X}} + \frac{dy}{\sqrt{Y}} = 0\,,$$

where X and Y are quartic polynomials of x, y respectively, leading to an integral of the form

$$\int \frac{dx}{\sqrt{X}} + \int \frac{dy}{\sqrt{Y}} = C\,,$$

with C constant. In the present context this corresponds to considering the arguments u and v to verify while keeping $u+v$ constant, say equal to c, i.e. implying that

$$\frac{dv}{du} = -1\,.$$

Now writing for the sake of brevity

$$s_1 := \mathrm{sn}(u), \quad s_2 := \mathrm{sn}(v)$$

and denoting by $\dot{}$ the derivative with respect to u

$$\dot{s}_1 := \frac{d}{du}\mathrm{sn}(u), \quad \dot{s}_2 = \frac{d}{du}\mathrm{sn}(v) = -\frac{d}{dv}\mathrm{sn}(v)\,,$$

we have from (B.66)

$$\ddot{s}_1\, s_2 - s_1\, \ddot{s}_2 = 2\,k^2\, s_1\, s_2\, \left(s_1^2 - s_2^2\right). \tag{B.67}$$

However, the right side can be re-expressed by using equation (B.58):

$$\begin{aligned}\dot{s}_1{}^2\, s_2^2 - s_1^2\, \dot{s}_2{}^2 &= s_2{}^2\left(1 - s_1^2\right)\left(1 - k^2\, s_1^2\right) - s_1{}^2\left(1 - s_2^2\right)\left(1 - k^2\, s_2^2\right) \\ &= s_2^2 - \left(1 + k^2\right) s_2^2\, s_1^2 + k^2\, s_2^2\, s_1^4 - s_1^2 + (1 + k)^2\, s_1^2\, s_2^2 - k^2\, s_1^2\, s_2^4 \\ &= \left(s_1^2 - s_2^2\right)\left(-1 + k^2\, s_1^2\, s_2^2\right).\end{aligned}$$

We use this result to replace $\left(s_1^2 - s_2^2\right)$ on the right side of equation (B.67). Notice that the left side of equation (B.67) is the derivative of $\dot{s}_1\, s_2 - s_1\, \dot{s}_2$. Putting these two results together, we can rewrite equation (B.67) as

$$\begin{aligned}\frac{\left(\dot{s}_1\, s_2 - s_1\, \dot{s}_2\right)^{\cdot}}{\dot{s}_1\, s_2 - s_1\, \dot{s}_2} &= \frac{2\,k^2\, s_1\, s_2\left(\dot{s}_1\, s_2 + s_1\, \dot{s}_2\right)}{k^2\, s_1^2\, s_2^2 - 1} \\ &= \frac{\left(1 - k^2\, s_1^2\, s_2^2\right)^{\cdot}}{1 - k^2\, s_1^2\, s_2^2}.\end{aligned}$$

Since both sides are derivatives with respect to u, we can integrate to get

$$\frac{\dot{s}_1\, s_2 - s_1\, \dot{s}_2}{1 - k^2\, s_1^2\, s_2^2} = -A \tag{B.68}$$

where A is a constant of integration. But this is a second integral of the first-order equation $du/dv = -1$. Therefore, its integration constant must be a function of the first constant c which we already obtained earlier; that is, $A = f(u + v)$. Letting $v \to 0$, we find from equation (B.68) that

$$f(u) = s_1 = \mathrm{sn}(u).$$

□

Appendix C

The continuous Painlevé equations and the Garnier system

C.1 The Painlevé equations

The problem of classifying ODEs by their singularity behavior was first proposed by L. Fuchs for the linear case and E. Picard for a general class of ODEs (Picard, 1887, 1890). A particularly important property used in this classification was to require that solutions be single-valued around all singularities whose locations depend on initial value(s). Such singularities are called *movable*. The requirement that all movable singularities be poles is now called the Painlevé property and has turned out to be crucial for integrable systems. Picard also solved the problem exhaustively for the first-order case (in which case it yields two classes: the Riccati equation and the first-order equation defining elliptic functions) and found some second-order results as well. However, the full problem of classifying second-order ODEs of the form $y'' = F(y, y', x)$, where F is analytic in x, rational in y and polynomial y', was investigated in the period 1890–1910 by Painlevé, Gambier and others (Painlevé 1902, 1906; Gambier 1906a,b, 1910; Ince 1956). Their work resulted in 50 equations (although the precise number depends on a choice for distinguishing different equations), of which six turned out to define new transcendental functions. They are as follows:

$$
\begin{aligned}
\mathrm{P_I}: &\quad y'' = 6y^2 + x \\
\mathrm{P_{II}}: &\quad y'' = 2y^3 + xy + \alpha \\
\mathrm{P_{III}}: &\quad y'' = \frac{y'^2}{y} - \frac{y}{x} + \frac{1}{x}(\alpha y^2 + \beta) + \gamma y^3 + \frac{\delta}{y} \\
\mathrm{P_{IV}}: &\quad y'' = \frac{y'^2}{2y} + \frac{3y^3}{2} + 4xy^2 + 2(x^2 - \alpha)y + \frac{\beta}{y} \\
\mathrm{P_V}: &\quad y'' = \left(\frac{1}{2y} + \frac{1}{y-1}\right)y'^2 - \frac{y'}{x} \\
&\qquad + \frac{(y-1)^2}{x^2 y}(\alpha y^2 + \beta) + \frac{\gamma y}{x} + \frac{\delta y(y+1)}{y-1} \\
\mathrm{P_{VI}}: &\quad y'' = \frac{1}{2}\left(\frac{1}{y} + \frac{1}{y-1} + \frac{1}{y-x}\right)y'^2 - \left(\frac{1}{x} + \frac{1}{x-1} + \frac{1}{y-x}\right)y' \\
&\qquad + \frac{y(y-1)(y-x)}{x^2(x-1)^2}\left(\alpha + \frac{\beta x}{y^2} + \frac{\gamma(x-1)}{(y-1)^2} + \frac{\delta x(x-1)}{(y-x)^2}\right)
\end{aligned}
$$

The Painlevé equations have become the subject of great interest after their reappearance in physics (in particular in statistical mechanics, quantum field theory and random matrix theory), and they have been intensively studied from the 1970s onward; see e.g. Conte and Musette (2008); Conte (1999).

C.2 The Garnier system

It is worth mentioning that the "top equation" in the Painlevé list, P_{VI}, was found first by R. Fuchs (1905, 1907), while solving a problem posed by his father L. Fuchs concerning a linear ODE. Garnier, in his 1912 paper, generalized the work by R. Fuchs, and considered an isomonodromic deformation problem with $n+3$ regular singularities, namely at 0, 1, ∞ and a moving singularity at any one of the variables t_i, $(i = 1, \dots, n)$, of the linear second-order differential equation:

$$\frac{d^2y}{dx^2} = p(x)y,$$

$$p(x) = \sum_{l=1}^{n}\left[\frac{c_l}{(x-t_l)^2} + \frac{\alpha_l}{x(x-1)(x-t_l)}\right] + \frac{c_{n+1}}{x^2} + \frac{c_{n+2}}{(x-1)^2} + \frac{c_{n+3}}{x(x-1)}$$
$$+\sum_{j=1}^{n}\left[\frac{3}{4(x-\lambda_j)^2} + \frac{\beta_j}{x(x-1)(x-\lambda_j)}\right].$$

The isomonodromic deformation can be given in terms of the additional differential relations:

$$\frac{\partial y}{\partial t_i} = A_i y + B_i \frac{\partial y}{\partial x}, \qquad (i = 1, \dots, n),$$

where the functions A_i, B_i can be computed from the consistency conditions.

The result is the following nonlinear system, called the **Garnier system** (Garnier, 1912), which involves N dependent variables λ_j $(j = 1, \dots, n$, depending on n independent variables t_i $(i = 1, \dots, n)$, which are the roots of the following polynomials:

$$\varphi(x) \equiv x(x-1)\prod_{l=1}^{n}(x-t_l), \qquad \psi(x) \equiv \prod_{j=1}^{n}(x-\lambda_j).$$

The Garnier system consists of n systems (one system for each independent variable t_i) of coupled ODEs of the form:

$$\frac{\partial^2\lambda_j}{\partial t_i^2} = \frac{1}{2}\left(\frac{\varphi'(\lambda_j)}{\varphi(\lambda_j)} - \frac{\psi''(\lambda_j)}{2\psi'(\lambda_j)}\right)\left(\frac{\partial\lambda_j}{\partial t_i}\right)^2 - \left(\frac{\varphi''(t_i)}{2\varphi'(t_i)} - \frac{\psi'(t_i)}{\psi(t_i)}\right)\frac{\partial\lambda_j}{\partial t_i}$$
$$+\frac{1}{2}\sum_{\substack{l=1\\ l\neq j}}^{n}\frac{\varphi(\lambda_j)\psi'(\lambda_l)(\lambda_l-t_i)^2}{\varphi(\lambda_l)\psi'(\lambda_j)(\lambda_j-t_i)^2(\lambda_j-\lambda_l)}\left(\frac{\partial\lambda_l}{\partial t_i}\right)^2$$

$$-\sum_{\substack{l=1\\ l\neq j}}^{n} \frac{\lambda_j - t_i}{(\lambda_l - t_i)(\lambda_l - \lambda_j)} \frac{\partial \lambda_j}{\partial t_i} \frac{\partial \lambda_l}{\partial t_i} + 2\frac{\psi^2(t_i)}{\varphi'^2(t_i)(\lambda_j - t_i)^2} \frac{\varphi(\lambda_j)}{\psi'(\lambda_j)}$$

$$\times \left[\sum_{k=1}^{n+3} \left(c_k + \frac{3}{4} \right) - 2 + \sum_{\substack{k=1\\ k\neq i}}^{n+2} \frac{\varphi'(t_k)}{\psi(t_k)} \frac{c_k + \frac{1}{4}}{\lambda_j - t_k} + \frac{\varphi'(t_i)}{\psi(t_i)} \frac{c_i}{\lambda_j - t_i} \right], \qquad j = 1, \ldots, n$$

where the remaining $t_k \neq t_i$ are parameters, and where the c_k, for $k = 1, \ldots, n+3$ are fixed constants, together with the system of linear PDEs:

$$\frac{\varphi'(t_i)(t_i - \lambda_j)}{\psi(t_i)} \frac{\partial \lambda_j}{\partial t_i} - \frac{\varphi'(t_k)(t_k - \lambda_j)}{\psi(t_k)} \frac{\partial \lambda_j}{\partial t_k}$$
$$= \frac{t_i - t_k}{(\lambda_j - t_i)(\lambda_j - t_k)} \frac{\varphi(\lambda_j)}{\psi'(\lambda_j)}, \qquad , i, j, k = 1, \ldots, n,$$

which guarantee the consistency between the n different systems of ODEs. The linear system for $n = 1$ reduces to the isomonodromic deformation problem for P_{VI} as was studied by R. Fuchs (1905, 1907).

We note that a q-difference analogue of the Garnier system was given in Sakai (2005).

Appendix D
Some determinantal identities

In Chapters 7, 8 and 9 solutions to integrable equations were given in terms of matrix determinants and proven by using various determinantal identities. In this appendix we mention some of them; for further literature we refer to the classics of Muir (1890); Scott (1904), for recent treatments to Brualdi and Schneider (1983); Vein and Dale (1999); Gelfand et al. (2008).

There are several identities attributed to Sylvester. The following is one (Akritas et al., 1996):

Theorem D.1 *Let $\boldsymbol{A}$ be an invertible $N \times N$ matrix, $\boldsymbol{B}$ an $N \times M$ matrix, $\boldsymbol{C}$ an $M \times N$ matrix, and $\boldsymbol{D}$ an invertible $M \times M$ matrix, then we have*

$$\det(\boldsymbol{A}) \det\left(\boldsymbol{D} - \boldsymbol{C}\boldsymbol{A}^{-1}\boldsymbol{B}\right) = \det(\boldsymbol{D}) \det\left(\boldsymbol{A} - \boldsymbol{B}\boldsymbol{D}^{-1}\boldsymbol{C}\right) \tag{D.1}$$

Proof Using LU decomposition we have

$$\begin{aligned}\det\begin{pmatrix} \boldsymbol{A} & \boldsymbol{B} \\ \boldsymbol{C} & \boldsymbol{D} \end{pmatrix} &= \det\left\{\begin{pmatrix} \boldsymbol{A} & \boldsymbol{0} \\ \boldsymbol{C} & \boldsymbol{1} \end{pmatrix}\begin{pmatrix} \boldsymbol{1} & \boldsymbol{A}^{-1}\boldsymbol{B} \\ \boldsymbol{0} & \boldsymbol{D} - \boldsymbol{C}\boldsymbol{A}^{-1}\boldsymbol{B} \end{pmatrix}\right\} \\ &= \det(\boldsymbol{A}) \det\left(\boldsymbol{D} - \boldsymbol{C}\boldsymbol{A}^{-1}\boldsymbol{B}\right),\end{aligned}$$

and by using UL decomposition on the same matrix we also have

$$\begin{aligned}\det\begin{pmatrix} \boldsymbol{A} & \boldsymbol{B} \\ \boldsymbol{C} & \boldsymbol{D} \end{pmatrix} &= \det\left\{\begin{pmatrix} \boldsymbol{1} & \boldsymbol{B} \\ \boldsymbol{0} & \boldsymbol{D} \end{pmatrix}\begin{pmatrix} \boldsymbol{A} - \boldsymbol{B}\boldsymbol{D}^{-1}\boldsymbol{C} & \boldsymbol{0} \\ \boldsymbol{D}^{-1}\boldsymbol{C} & \boldsymbol{1} \end{pmatrix}\right\} \\ &= \det(\boldsymbol{D}) \det\left(\boldsymbol{A} - \boldsymbol{B}\boldsymbol{D}^{-1}\boldsymbol{C}\right).\end{aligned}$$

□

A special case of (D.1), namely obtained by setting $\boldsymbol{A} = \boldsymbol{I}$ and $\boldsymbol{D} = \boldsymbol{I}$, is often useful.

Theorem D.2 *(Weinstein–Aronszajn formula) Let $\boldsymbol{B}$ be an $N \times M$ matrix and $\boldsymbol{C}$ an $M \times N$ matrix, then we have the following identity:*

$$\det_{N\times N}(\boldsymbol{I}_{N\times N}+\boldsymbol{BC})=\det_{M\times M}(\boldsymbol{I}_{M\times M}+\boldsymbol{CB}). \tag{D.2}$$

This formula is useful when we have matrices of the type $\boldsymbol{I}+$ finite rank. In particular if $M=1$ so that $\boldsymbol{B},\boldsymbol{C}$ can be interpreted as vectors $\boldsymbol{B}=\boldsymbol{b},\boldsymbol{C}=\boldsymbol{c}^T$ we have

$$\det_{N\times N}(\boldsymbol{I}_{i,j}+\boldsymbol{b}_i\boldsymbol{c}_j^T)=1+\boldsymbol{c}^T\cdot\boldsymbol{b}. \tag{D.3}$$

For the description and proof of the following identities (we follow mainly Gragg, 1972) we need to have notation for choosing rows and columns of matrices. For that purpose let us define

$$S_{M,K}=\{\alpha=(\alpha_1,\alpha_2,\ldots,\alpha_K):1\le\alpha_1<\alpha_2<\cdots<\alpha_K\le M\},$$

that is, $S_{M,K}$ is the set of ordered sets of K elements form $\{1,2,\ldots,M\}$. We also define the length of $\alpha\in S_{M,K}$ by

$$|\alpha|:=\alpha_1+\alpha_2+\cdots+\alpha_K.$$

Furthermore, if $\alpha\in S_{M,K}$ we define its complement $\alpha'\in S_{M,M-K}$ as the list that contains the remaining elements of the list $\{1,2,\ldots,M\}$ that were not included in α. Given $\alpha,\beta\in S_{M,K}$ the matrix $\boldsymbol{A}(\alpha,\beta)$ is a $K\times K$ matrix obtained from $\boldsymbol{A}=(a_{ij})$ by keeping those matrix elements a_{ij} for which $i\in\alpha,\ j\in\beta$. With the above definitions we can give several expansion results.

Theorem D.3 *(Cauchy–Binet) Let* $\boldsymbol{A}$ *be an* $N\times M$ *matrix,* $\boldsymbol{B}$ *an* $N\times K$ *matrix and* $\boldsymbol{C}$ *an* $K\times M$ *matrix, with* $\boldsymbol{A}=\boldsymbol{BC}$. *Choose* $\alpha\in S_{N,L}$, $\beta\in S_{N,L}$ *then* $\boldsymbol{A}(\alpha,\beta)$ *is an* $L\times L$ *matrix and*

$$\det\boldsymbol{A}(\alpha,\beta)=\sum_{\gamma\in S_{K,L}}\det\boldsymbol{B}(\alpha,\gamma)\boldsymbol{C}(\gamma,\beta). \tag{D.4}$$

Theorem D.4 *(Laplace expansion) Let* $\boldsymbol{A}$ *be an* $N\times N$ *matrix and fix some* $\alpha\in S_{N,K}$. *Then*

$$\det\boldsymbol{A}=\sum_{\beta\in S_{N,K}}(-1)^{|\alpha|+|\beta|}\det\boldsymbol{A}(\alpha,\beta)\det\boldsymbol{A}(\alpha',\beta'). \tag{D.5}$$

For example, an expansion along the first row of the matrix is obtained by choosing $K=1$, $\alpha=(1)$. Often we apply Laplace expansion to a $2N\times 2N$ matrix and choose $M=2N,\ K=N$.

The following theorem can be proven using Laplace expansion.

Theorem D.5 *(Extensional theorem) Let $\boldsymbol{a}, \boldsymbol{b}, \boldsymbol{c}, \boldsymbol{d}$ be $N \times 1$ column vectors and $\boldsymbol{M}$ an $N \times (N-2)$ matrix. Then*

$$\begin{aligned}\det(\boldsymbol{a}, \boldsymbol{b}, \boldsymbol{M}) \det(\boldsymbol{c}, \boldsymbol{d}, \boldsymbol{M}) &= \det(\boldsymbol{a}, \boldsymbol{c}, \boldsymbol{M}) \det(\boldsymbol{b}, \boldsymbol{d}, \boldsymbol{M}) \\ &\quad - \det(\boldsymbol{a}, \boldsymbol{d}, \boldsymbol{M}) \det(\boldsymbol{b}, \boldsymbol{c}, \boldsymbol{M}). \end{aligned} \tag{D.6}$$

In order to prove this we first take a matrix whose determinant is clearly zero and then use Laplace expansion. The determinant to consider is

$$\det \begin{vmatrix} \boldsymbol{M} & \boldsymbol{a} & \boldsymbol{b} & \boldsymbol{c} & \boldsymbol{d} & \boldsymbol{0} \\ \boldsymbol{0} & \boldsymbol{a} & \boldsymbol{b} & \boldsymbol{c} & \boldsymbol{d} & \boldsymbol{M} \end{vmatrix} = 0,$$

which can be shown to vanish by suitable row operations. Next apply Laplace expansion on the above $2N \times 2N$ determinant with $K = N$ and $\alpha = (1, 2, \ldots, N)$.

As a further specialization we have the identity variously attributed to Sylvester, Jacobi, Desnanot and Dogson:

Theorem D.6 *Let $\boldsymbol{A}$ be an $N \times N$ matrix and denote by $\boldsymbol{A}_{r;p}$ the matrix from which row r and column p have been removed, similarly by $\boldsymbol{A}_{rs;pq}$ the matrix where rows r, s and columns p, q have been removed. Then assuming $r < s$ and $p < q$ we have*

$$\det \boldsymbol{A} \ \det \boldsymbol{A}_{rs;pq} = \det \boldsymbol{A}_{r;p} \ \det \boldsymbol{A}_{s;q} - \det \boldsymbol{A}_{r;q} \ \det \boldsymbol{A}_{s;p}. \tag{D.7}$$

For $r = p = 1, s = q = N$ this result is obtained from (D.6) by choosing $a = (1, 0, \ldots, 0, 0)^T, b = (0, 0, \ldots, 0, 1)^T$ and $\boldsymbol{A} = (\boldsymbol{c}, \boldsymbol{M}, \boldsymbol{d})$ and by permuting the columns in the determinants in (D.6) to the form

$$\begin{aligned}\det(\boldsymbol{a}, \boldsymbol{M}, \boldsymbol{b}) \det(\boldsymbol{c}, \boldsymbol{M}, \boldsymbol{d}) &= -\det(\boldsymbol{a}, \boldsymbol{c}, \boldsymbol{M}) \det(\boldsymbol{M}, \boldsymbol{d}, \boldsymbol{b}) \\ &\quad + \det(\boldsymbol{a}, \boldsymbol{M}, \boldsymbol{d}) \det(\boldsymbol{c}, \boldsymbol{M}, \boldsymbol{b}). \end{aligned}$$

From this expression the result (D.7) can be read off. Other choices of $r < s$, $p < q$ can be obtained by row and column permutation.

The **Plücker relations**, which are relations between coordinates on a geometric object called a Grassmannian, are a very general class of identities for minors made from a collection of vectors. They are very easily deduced using the following lemma:

Lemma D.1 *Let $\boldsymbol{a}_1, \ldots, \boldsymbol{a}_{n+1}$ denote a collection of vectors in $\mathbb{R}^n$ or $\mathbb{C}^n$. Then we have the following relation*

$$\sum_{j=1}^{n+1} (-1)^{n+1-j} \left| \boldsymbol{a}_1, \boldsymbol{a}_2, \cdots, \not{\boldsymbol{a}_j}, \cdots, \boldsymbol{a}_{n+1} \right| \boldsymbol{a}_j = 0,$$

where $\not{a_j}$ means that the jth column is omitted. This formula expresses the fact that $n+1$ vector of dimension n must be linearly dependent. Considering now an additional set of vectors $\boldsymbol{a}_{n+2}, \dots, \boldsymbol{a}_{2n}$ in the same vector space, we get from the lemma the following quite general relation between minors:

Theorem D.7 *(Plücker relations) Let $\boldsymbol{a}_1, \boldsymbol{a}_2, \dots, \boldsymbol{a}_{2n}$ be an arbitrary set of $2n$ vectors in $\mathbb{R}^n$ or $\mathbb{C}^n$, then we have the following identity for the minors composed of these vectors:*

$$\sum_{j=1}^{n+1}(-1)^{n+1-j} \left|\boldsymbol{a}_1, \boldsymbol{a}_2, \cdots, \not{\boldsymbol{a}_j}, \cdots, \boldsymbol{a}_{n+1}\right| \left|\boldsymbol{a}_j, \boldsymbol{a}_{n+2}, \cdots, \boldsymbol{a}_{2n}\right| = 0 .$$

Although the derivation is almost trivial, the Plücker relations contain in a nutshell many well-known determinantal identities. In particular, if one chooses all but four of the vectors $\boldsymbol{a}_i$ to coincide, e.g. by setting $\boldsymbol{a}_{n+3} = \boldsymbol{a}_3, \dots, \boldsymbol{a}_{2n} = \boldsymbol{a}_n$, we recover the earlier identity Theorem D.5. For more on Plücker relations, see Chapter 8 of Miwa et al. (2000).

When computing with Casoratians we also need the following:

Theorem D.8 *(Column identity) Denote the determinant of an $N \times N$ matrix by giving the list of its columns: $|\mathbf{a}_1, \cdots, \mathbf{a}_N|$. Then we have the following identity*

$$\sum_{j=1}^{N} |\mathbf{a}_1, \cdots, \mathbf{a}_{j-1}, [\mathbf{ba}]_j, \mathbf{a}_{j+1}, \cdots, \mathbf{a}_N| = \left(\sum_{j=1}^{N} b_j\right) |\mathbf{a}_1, \cdots, \mathbf{a}_N|, \tag{D.8}$$

where $\mathbf{a}_j = (a_{1j}, \cdots, a_{Nj})^T$ is the j-th column and $\mathbf{b} = (b_1, \cdots, b_N)^T$ is some N-component column vector, and $[\mathbf{ba}]_j$ stands for the term-wise product $(b_1 a_{1j}, \cdots, b_N a_{Nj})^T$.

Let us also define some matrices and determinants that are often needed:

Definition D.1 *A **Hankel matrix** is a square matrix $A = (a_{i,j})$ with identical entries along each skew-diagonal, i.e. $a_{i,j} = a_{i-1,j+1}$.*

*A **Toeplitz matrix** is a square matrix $A = (a_{i,j})$ with identical entries on the diagonal, i.e. $a_{i,j} = a_{i+1,j+1}$.*

*A **Wronskian** is the determinant of a square matrix $\mathbf{W}$ with entries $W_{ij} = \partial_x^{j-1} f_i(x)$ where f_i are some given functions.*

*A **Casoratian** is the determinant of a square matrix $\boldsymbol{C}$ with entries $C_{ij} = f_i(s+j)$ where f_i are some given functions.*

References

Abarenkova, N., Anglès d'Auriac, J.-Ch., Boukraa, S., Hassani, S. and Maillard, J.-M. 1999. Topological entropy and Arnold complexity for two-dimensional mappings. *Phys. Lett. A*, **262**(1), 44–49.

Ablowitz, M. J. and Clarkson, P. A. 1991. *Solitons, Nonlinear Evolution Equations and Inverse Scattering*. London Mathematical Society Lecture Note Series, vol. 149. Cambridge, UK: Cambridge University Press.

Ablowitz, M. J. and Fokas, A. S. 2003. *Complex Variables: Introduction and Applications*. Cambridge, UK: Cambridge University Press.

Ablowitz, M. J. and Herbst, B. M. 1990. On homoclinic structure and numerically induced chaos for the nonlinear Schrödinger equation. *SIAM J. Appl. Math.*, **50**(2), 339–351.

Ablowitz, M. J. and Ladik, J. F. 1976. A nonlinear difference scheme and inverse scattering. *Stud. Appl. Math.*, **55**(3), 213–229.

Ablowitz, M. J. and Ladik, J. F. 1976/77. On the solution of a class of nonlinear partial difference equations. *Stud. Appl. Math.*, **57**(1), 1–12.

Ablowitz, M. J. and Segur, H. 1977. Exact linearization of a Painlevé transcendent. *Phys. Rev. Lett.*, **38**(20), 1103.

Ablowitz, M. J. and Segur, H. 1981. *Solitons and the Inverse Scattering Transform*. SIAM Studies in Applied Mathematics, vol. 4. Philadelphia, PA: Society for Industrial and Applied Mathematics (SIAM).

Ablowitz, M. J., Ramani, A. and Segur, H. 1978. Nonlinear evolution equations and ordinary differential equations of Painlevé type. *Lett. Nuovo Cimento*, **23**(9), 333–338.

Ablowitz, M. J., Ramani, A. and Segur, H. 1980a. A connection between nonlinear evolution equations and ordinary differential equations of P-type. I. *J. Math. Phys.*, **21**(4), 715–721.

Ablowitz, M. J., Ramani, A. and Segur, H. 1980b. A connection between nonlinear evolution equations and ordinary differential equations of P-type. II. *J. Math. Phys.*, **21**(5), 1006–1015.

Ablowitz, M. J., Halburd, R. and Herbst, B. 2000. On the extension of the Painlevé property to difference equations. *Nonlinearity*, **13**(3), 889–905.

Abramowitz, M. and Stegun, I. A. 1970. *Handbook of Mathematical Functions*. New York: Dover.

Adler, M. and van Moerbeke, P. 1999. Vertex operator solutions to the discrete KP-hierarchy. *Commun. Math. Phys.*, **203**(1), 185–210.

Adler, V. E. 1998. Bäcklund transformation for the Krichever-Novikov equation. *Int. Math. Res. Not.*, **1998**(1), 1–4.

Adler, V. E. 2001. Discrete equations on planar graphs. *J. Phys. A: Math. Gen.*, **34**(48), 10453–10460.

Adler, V. E. 2006. Some incidence theorems and integrable discrete equations. *Discrete Comput. Geom.*, **36**(3), 489–498.

Adler, V. E. and Suris, Yu. B. 2004. Q_4: integrable master equation related to an elliptic curve. *Int. Math. Res. Not.*, **2004**(47), 2523–2553.

Adler, V. E. and Veselov, A. P. 2004. Cauchy problem for integrable discrete equations on quad-graphs. *Acta Appl. Math.*, **84**(2), 237–262.

Adler, V. E., Marikhin, V. G. and Shabat, A. B. 2001. Lagrangian chains and canonical Bäcklund transformations. *Theor. Math. Phys.*, **129**(2), 1448–1465.

Adler, V. E., Bobenko, A. I. and Suris, Yu. B. 2003. Classification of integrable equations on quad-graphs. The consistency approach. *Commun. Math. Phys.*, **233**(3), 513–543.

Adler, V. E., Bobenko, A. I. and Suris, Yu. B. 2004. Geometry of Yang–Baxter maps: pencils of conics and quadrirational mappings. *Commun. Anal. Geom.*, **12**(5), 967–1007.

Adler, V. E., Bobenko, A. I. and Suris, Yu. B. 2009. Discrete nonlinear hyperbolic equations: classification of integrable cases. *Funktsional. Anal. i Prilozhen.*, **43**(1), 3–21.

Adler, V. E., Bobenko, A. I. and Suris, Yu. B. 2012. Classification of integrable discrete equations of octahedron type. *Int. Math. Res. Not.* **2012**(8), 1822–1889.

Agafonov, S. I. 2003. Imbedded circle patterns with the combinatorics of the square grid and discrete Painlevé equations. *Discrete Comput. Geom.*, **29**(2), 305–319.

Agafonov, S. I. 2005. Asymptotic behavior of discrete holomorphic maps z^c and $\log(z)$. *J. Nonlinear Math. Phys.*, **12**, Supp 2, 1–14.

Agafonov, S. I., and Bobenko, A. I. 2000. Discrete z^γ and Painlevé equations. *Int. Math. Res. Not.*, **2000**(4), 165–193.

Agafonov, S. I. and Bobenko, A. I. 2003. Hexagonal circle patterns with constant intersection angles and discrete Painlevé and Riccati equations. *J. Math. Phys.*, **44**(8), 3455–3469.

Ahlfors, L. V. 1953. *Complex Analysis, an Introduction to the Theory of Analytic Functions of One Complex Variable.* New York: McGraw-Hill Book Company.

Airault, H., McKean, H. P. and Moser, J. 1977. Rational and elliptic solutions of the Korteweg–de Vries equation and a related many-body problem. *Commun. Pure Appl. Math.*, **30**(1), 95–148.

Akhiezer, N. I. 1990. *Elements of the Theory of Elliptic Functions*. Translations of Mathematical Monographs, vol. 79. Providence, RI: American Mathematical Society. Translated from the second Russian edition by H. H. McFaden.

Akritas, A. G., Akritas, E. K. and Malaschonok, G. I. 1996. Various proofs of Sylvester's (determinant) identity. *Math. Comput. Simul.*, **42**(4), 585–593.

Ando, H., Hay, M., Kajiwara, K. and Masuda, T. 2014. An explicit formula for the discrete power function associated with circle patterns of Schramm type. *Funkcial. Ekvac.*, **57**(1), 1–41.

Andrews, G. E., Askey, R. and Roy, R. 1999. *Special Functions*. Encyclopedia of Mathematics and Its Applications. Cambridge, UK: Cambridge University Press.

Anglès d'Auriac, J.-Ch., Maillard, J.-M., and Viallet, C. M. 2006. On the complexity of some birational transformations. *J. Phys. A: Math. Gen.*, **39**(14), 3641.

Arinkin, D. and Borodin, A. 2006. Moduli spaces of d-connections and difference Painlevé equations. *Duke Math. J.*, **134**(3), 515–556.

Arnold, V. I. 1990. Dynamics of complexity of intersections. *Bol. Soc. Brasil. Mat. (N.S.)*, **21**(1), 1–10.

Arnold, V. I. 1997. *Mathematical Methods of Classical Mechanics*. Graduate Texts in Mathematics, vol. 60. New York: Springer.

Atkinson, J. 2008a. Bäcklund transformations for integrable lattice equations. *J. Phys. A: Math. Theor.*, **41**(13), 135202, 8pp.

Atkinson, J. 2008b. *Integrable Lattice Equations: Connection to the Möbius Group, Bäcklund Transformations and Solutions*. Ph.D. thesis, University of Leeds.

Atkinson, J. 2012. A multidimensionally consistent version of Hirota's discrete KdV equation. *J. Phys. A: Math. Theor.*, **45**(22), 222001.

Atkinson, J. and Joshi, N. 2013. Singular-boundary reductions of type-Q ABS equations. *Int. Math. Res. Not.*, **2013**(7), 1451–1481.

Atkinson, J. and Nieszporski, M. 2014. Multi-quadratic quad equations: integrable cases from a factorized-discriminant hypothesis. *Int. Math. Res. Not.*, **2014**(15), 4215–4240.

Atkinson, J. and Nijhoff, F. W. 2010. A constructive approach to the soliton solutions of integrable quadrilateral lattice equations. *Commun Math. Phys.*, **299**(2), 283–304.

Atkinson, J., Hietarinta, J. and Nijhoff, F. W. 2007. Seed and soliton solutions for Adler's lattice equation. *J. Phys. A: Math. Theor.*, **40**(1), F1–F8.

Atkinson, J., Hietarinta, J. and Nijhoff, F. W. 2008. Soliton solutions for Q3. *J. Phys. A: Math. Theor.*, **41**(14), 142001, 11pp.

Atkinson, J., Lobb, S. B. and Nijhoff, F. W. 2012. An integrable multicomponent quad-equation and its Lagrangian formulation. *Theor. Math. Phys.*, **173**(3), 1644–1653.

Babelon, O., Bernard, D. and Talon, M. 2003. *Introduction to Classical Integrable Systems*. Cambridge, UK: Cambridge University Press.

Babelon, O., Talon, M. and Peyranére, M. C. 2010. Kowalevski's analysis of the swinging Atwood's machine. *J. Phys. A: Math. Theor.*, **43**(8), 085207.

Bäcklund, A. V. 1883. Om ytor med konstant negative krökning, *Lund Univ. Årsskrift*, **19**, 1–48.

Baker, G. A. 1975. *Essentials of Padé Approximants*. New York: Academic Press.

Barnett, D. C., Halburd, R. G., Morgan, W. and Korhonen, R. J. 2007. Nevanlinna theory for the q-difference operator and meromorphic solutions of q-difference equations. *Proc. R. Soc. Edinburgh Sect. A*, **137**(3), 457–474.

Bauer, F. L. 1959. Sequential reduction to tridiagonal form. *J. Soc. Indust. Appl. Math.*, **7**, 107–113.

Baxter, R. J. 1982. *Exactly Solved Models in Statistical Mechanics*. London: Academic Press Inc.

Bazhanov, V. V. and Sergeev, S. M. 2012a. Elliptic gamma-function and multi-spin solutions of the Yang–Baxter equation. *Nucl. Phys. B*, **856**(2), 475–496.

Bazhanov, V. V. and Sergeev, S. M. 2012b. A master solution of the quantum Yang–Baxter equation and classical discrete integrable equations. *Adv. Theor. Math. Phys.*, **16**(1), 65–95.

Bellon, M. P. and Viallet, C.-M. 1999. Algebraic entropy. *Commun. Math. Phys.*, **204**(2), 425–437.

Bender, C. M. and Orszag, S. A. 1978. *Advanced Mathematical Methods for Scientists and Engineers*. New York: McGraw-Hill.

Bessis, D., Itzykson, C. and Zuber, J.-B. 1980. Quantum field theory techniques in graphical enumeration. *Adv. Appl. Math.*, **1**(2), 109–157.

Bianchi, L. 1899. *Vorlesungen über Differentialgeometrie*. Leipzig: B.G. Teubner.

Birkhoff, G. D. 1913. The generalized Riemann problem for linear differential equations and the allied problems for linear difference and q-difference equations. *Proc. Am. Acad. Arts Sci.*, **49**(9), 521–568.

Bluman, G. W. and Cole, J. D. 1974. *Similarity Methods for Differential Equations*, vol. 13. Berlin: Springer.

Bobenko, A. I. 1999. Discrete conformal maps and surfaces. In P. Clarkson and F. W. Nijhoff (eds), *Symmetries and Integrability of Difference Equations*. London Mathematical Society Lecture Note Series, vol. 255, Cambridge, UK: Cambridge University Press, 97–108.

Bobenko, A. I., and Pinkall, U. 1996a. Discrete isothermic surfaces. *J. Reine Angew. Math.*, **475**, 187–208.

Bobenko, A. I., and Pinkall, U. 1996b. Discrete surfaces with constant negative Gaussian curvature and the Hirota equation. *J. Differ. Geom.*, **43**(3), 527–611.

Bobenko, A. I. and Suris, Yu. B. 2002. Integrable systems on quad-graphs. *Int. Math. Res. Not.*, **2002**(11), 573–611.

Bobenko, A. I. and Suris, Yu. B. 2008. *Discrete Differential Geometry*. Graduate Studies in Mathematics, vol. 98. Providence, RI: American Mathematical Society.

Bobenko, A. I. and Suris, Yu. B. 2010. On the Lagrangian structure of integrable quad-equations. *Lett. Math. Phys.*, **92**, 17–31.

Bobenko, A. I. and Suris, Yu. B. 2015. Discrete pluriharmonic functions as solutions of linear pluri-Lagrangian systems. *Commun. Math. Phys.*, **336**(1), 199–215.

Bobenko, A. I., Mercat, C., and Suris, Yu. B. 2005. Linear and nonlinear theories of discrete analytic functions. Integrable structure and isomonodromic Green's function. *J. Reine Angew. Math.*, **2005**(583), 117–161.

Boelen, L. and van Assche, W. 2010. Discrete Painlevé equations for recurrence coefficients of semiclassical Laguerre polynomials. *Proc. Am. Math. Soc.*, **138**(4), 1317–1331.

Boelen, L., Smet, C. and van Assche, W. 2010. q-discrete Painlevé equations for recurrence coefficients of modified q-Freud orthogonal polynomials. *J. Differ. Equ. Appl.*, **16**(1), 37–53.

Bogdanov, L. V. and Konopelchenko, B. G. 1998a. Analytic-bilinear approach to integrable hierarchies. I. Generalized KP hierarchy. *J. Math. Phys.*, **39**(9), 4683–4700.

Bogdanov, L. V. and Konopelchenko, B. G. 1998b. Analytic-bilinear approach to integrable hierarchies. II. Multicomponent KP and 2D Toda lattice hierarchies. *J. Math. Phys.*, **39**(9), 4701–4728.

Boiti, M., Pempinelli, F., Prinari, B. and Spire, A. 2001. An integrable discretization of KdV at large times. *Inverse Prob.*, **17**(3), 515–526.

Boll, R. 2011. Classification of 3D consistent quad-equations. *J. Nonlinear Math. Phys.*, **18**(3), 337–365.

Boll, R. 2012. On multidimensional consistent systems of asymmetric quad-equations. *arXiv e-prints arXiv*:1201.1203.

Boll, R., Petrera, M. and Suris, Yu. B. 2013a. What is integrability of discrete variational systems? *Proc. R. Soc. London, Ser. A*, **470**, 30550.

Boll, R., Petrera, M. and Suris, Yu. B. 2013b. Multi-time Lagrangian 1-forms for families of Bäcklund transformations: Toda-type systems. *J. Phys. A: Math. Theor.*, **46**(27), 275204.

Boll, R., Petrera, M., and Suris, Yu. B. 2015. Multi-time Lagrangian 1-forms for families of Bäcklund transformations. Relativistic Toda-type systems. *J. Phys. A: Math. Theor.*, **48**(8), 085203.

Boll, R., Petrera, M. and Suris, Yu. B. 2016. On integrability of discrete variational systems. Octahedron relations. *Int. Math. Res. Not.*, **2016**(3), 645–668.

Boole, G. 1860. A Treatise on the Calculus of Finite Differences. London: McMillan & Co.

Borodin, A. 2004. Isomonodromy transformations of linear systems of difference equations. *Ann. Math.*, 1141–1182.

Boukraa, S., and Maillard, J.-M. 1995. Factorization properties of birational mappings. *Physica A*, **220**(3), 403–470.

Boukraa, S., Maillard, J.-M. and Rollet, G. 1994. Integrable mappings and polynomial growth. *Physica A*, **209**(1), 162–222.

Boussinesq, J. 1877. *Essai sur la théorie des eaux courantes*. Mémoires présentés par divers savants à l'Académie des sciences de l'Institut national de France, vol. XXIII. Imprimerie Nationale.

Brezin, E. and Kazakov, V. A. 1990. Exactly solvable field theories of closed strings. *Phys. Lett.*, **236**(2), 144–150.

Brezinski, C. 1977. *Accélération de la Convergence en Analyse Numérique*. Lecture Notes in Mathematics, vol. 584. Berlin Heidelberg: Springer-Verlag.

Brezinski, C. 2013. *Padé-Type Approximation and General Orthogonal Polynomials*. International Series of Numerical Mathematics. Basel: Birkhäuser.

Brezinski, C., He, Y., Hu, X. B., Redivo-Zaglia, M. and Sun, J. Q. 2012. Multistep ε-algorithm, Shanks' transformation, and the Lotka-Volterra system by Hirota's method. *Math. Comput.*, **81**(279), 1527–1549.

Brualdi, R. A. and Schneider, H. 1983. Determinantal identities: Gauss, Schur, Cauchy, Sylvester, Kronecker, Jacobi, Binet, Laplace, Muir, and Cayley. *Linear Algebra Appl.*, **52–53**(0), 769–791.

Bruschi, M., Ragnisco, O., Santini, P. M. and Tu, G. Z. 1991. Integrable symplectic maps. *Physica D*, **49**(3), 273–294.

Bustamante, M. D. and Hojman, S. A. 2003. Multi-Lagrangians, hereditary operators and Lax pairs for the Korteweg–de Vries positive and negative hierarchies. *J. Math. Phys.*, **44**(10), 4652–4671.

Butler, S. 2012a. A discrete inverse scattering transform for $Q3_\delta$. *arXiv preprint arXiv:1210.1869*.

Butler, S. 2012b. Multidimensional inverse scattering of integrable lattice equations. *Nonlinearity*, **25**(6), 1613.

Butler, S. and Joshi, N. 2010. An inverse scattering transform for the lattice potential KdV equation. *Inverse Prob.*, **26**(11), 115012.

Cadzow, J. A. 1970. Discrete calculus of variations. *J. Control*, **11**(3), 393–407.

Calogero, F. 1969. Solution of a three-body problem in one dimension. *J. Math. Phys.*, **10**(12), 2191–2196.

Calogero, F. and Degasperis, A. 1982. *Spectral Transform and Solitons*, vol. 1. Amsterdam: North-Holland Publ.

Cao, C., and Xu, X. 2012. A finite genus solution of the H1 model. *J. Phys. A: Math. Theor.*, **45**(5), 055213.

Cao, C. and Zhang, G. 2012. A finite genus solution of the Hirota equation via integrable symplectic maps. *J. Phys. A: Math. Theor.*, **45**(9), 095203.

Capel, H. W. and Sahadevan, R. 2001. A new family of four-dimensional symplectic and integrable mappings. *Physica A*, **289**(1–2), 86–106.

Capel, H. W., Nijhoff, F. W. and Papageorgiou, V. G. 1991. Complete integrability of Lagrangian mappings and lattices of KdV type. *Phys. Lett. A*, **155**(6–7), 377–387.

Carmichael, R. D. 1912. The general theory of linear q-difference equations. *Am. J. Math.*, **34**, 147–168.

Chihara, T. S. 2014. *An Introduction to Orthogonal Polynomials*. New York: Dover Publications.

Choodnovsky, D. V. and Choodnovsky, G. V. 1977. Pole expansions of nonlinear partial differential equations. *Il Nuovo Cimento B Series 11*, **40**(2), 339–353.

Clarkson, P. A., Hone, A. N. W. and Joshi, N. 2003. Hierarchies of difference equations and Bäcklund transformations. *J. Nonlinear Math. Phys.*, **10**, Supp 2, 13–26.

Conte, R. M. 1999. *The Painlevé Property: One Century Later*. CRM Series in Mathematical Physics. New York: Springer.

Conte, R. M. and Musette, M. 2008. *The Painlevé Handbook*. Netherlands: Springer.

Courant, R., Friedrichs, K. and Lewy, H. 1928. Über die partiellen Differenzengleichungen der mathematischen Physik. *Math. Ann.*, **100**(1), 32–74.

Cresswell, C. and Joshi, N. 1999a. The discrete first, second and thirty-fourth Painlevé hierarchies. *J. Phys. A: Math. Gen.*, **32**(4), 655.

Cresswell, C. and Joshi, N. 1999b. The discrete Painlevé I hierarchy. In P. Clarkson and F. W. Nijhoff (eds), *Symmetries and Integrability of Difference Equations*. London Mathematical Society Lecture Note Series, vol. 255, Cambridge, UK: Cambridge University Press, 197–205.

Darboux, G. 1914. *Leçons sur la Théorie Générale des Surfaces et les Applications Géométriques du Calcul Infinitésimal. Première partie: Généralités. Coordonnées curvilignes. Surfaces minima*. Paris: Gauthier-Villars.

Date, E., Kashiwara, M. and Miwa, T. 1981. Vertex operators and τ functions transformation groups for soliton equations, II. *Proc. Japan Acad. Ser. A Math. Sci.*, **57**(8), 387–392.

Date, E., Jimbo, M. and Miwa, T. 1982a. Method for generating discrete soliton equations. I. *J. Phys. Soc. Japan*, **51**(12), 4116–4124.

Date, E., Jimnbo, M. and Miwa, T. 1982b. Method for generating discrete soliton equations. II. *J. Phys. Soc. Japan*, **51**(12), 4125–4131.

Date, E., Jimbo, M. and Miwa, T. 1983a. Method for generating discrete soliton equations. III. *J. Phys. Soc. Japan*, **52**(2), 388–393.

Date, E., Jimbo, M. and Miwa, T. 1983b. Method for generating discrete soliton equations. IV. *J. Phys. Soc. Japan*, **52**(3), 761–765.

Date, E., Jimbo, M. and Miwa, T. 1983c. Method for generating discrete soliton equations. V. *J. Phys. Soc. Japan*, **52**(3), 766–771.

David Jr., E. E., et al. 1984. Renewing US mathematics: critical resource for the future. *Not. Am. Math. Soc.*, **31**(5), 435–466.

Deift, P., Li, L. C. and Tomei, C. 1989. Matrix factorizations and integrable systems. *Commun. Pure Appl. Math.*, **42**(4), 443–521.

Di Vizio, L., Ramis, J.-P., Sauloy, J. and Zhang, C. 2003. Équations aux q-différences. *Gaz. Math.*, **96**, 20–49.

Dickey, L. A. 1990. General Zakharov–Shabat equations, multi-time Hamiltonian formalism, and constants of motion. *Commun. Math. Phys.*, **132**(3), 485–497.

Dickey, L. A. 2003. *Soliton Equations and Hamiltonian Systems*, 2nd ed. Advanced Series in Mathematical Physics, vol. 26. River Edge, NJ: World Scientific Publishing Co. Inc.

Diller, J. and Favre, C. 2001. Dynamics of bimeromorphic maps of surfaces. *Amer. J. Math.* **123**(6), 1135–1169.

DLMF. 2010. *NIST Digital Library of Mathematical Functions*. http://dlmf.nist.gov/, Release 1.0.9 of 201408-29. Online companion to Olver et al. (2010).

Doliwa, A. 2005. On τ-function of conjugate nets. *J. Nonlinear Math. Phys.*, **12**, Supp 1, 244–252.

Doliwa, A. 2009. The τ-function of the quadrilateral lattice. *J. Phys. A: Math. Theor.*, **42**(40), 404008, 9pp.

Doliwa, A. 2010. Desargues maps and the Hirota–Miwa equation. *Proc. R. Soc. London Ser. A Math. Phys. Eng. Sci.*, **466**(2116), 1177–1200.

Doliwa, A. and Santini, P. M. 1997. Multidimensional quadrilateral lattices are integrable. *Phys. Lett. A*, **233**(4–6), 365–372.

Doliwa, A. and Santini, P. M. 1999. Planarity and integrability. In Degasperis, A. and Gaeta, G. (eds), *Symmetry and Perturbation Theory (Rome, 1998)*. River Edge, NJ: World Scientific Publishing Co. Inc., 167–177.

Doliwa, A. and Santini, P. M. 2000. Integrable discrete geometry: the quadrilateral lattice, its transformations and reductions. In Levi, D. and Ragnisco, O. (eds), *SIDE III – Symmetries and Integrability of Difference Equations (Sabaudia, 1998)*. CRM Proc. Lecture Notes, vol. 25. Providence, RI: American Mathematical Society, 101–119.

Doliwa, A., Mañas, M. and Alonso, L. M. 1999. Generating quadrilateral and circular lattices in KP theory. *Phys. Lett. A*, **262**(4), 330–343.

Doliwa, A., Santini, P. M. and Mañas, M. 2000. Transformations of quadrilateral lattices. *J. Math. Phys.*, **41**(2), 944–990.

Doliwa, A., Nieszporski, M. and Santini, P. M. 2004. Geometric discretization of the Bianchi system. *J. Geom. Phys.*, **52**(3), 217–240.

Doliwa, A., Grinevich, P., Nieszporski, M. and Santini, P. M. 2007. Integrable lattices and their sublattices: from the discrete Moutard (discrete Cauchy–Riemann) 4-point equation to the self-adjoint 5-point scheme. *J. Math. Phys.*, **48**(1), 013513, 28.

Dorfman, I. Ya. and Nijhoff, F. W. 1991. On a $(2+1)$-dimensional version of the Krichever-Novikov equation. *Phys. Lett. A*, **157**(2–3), 107–112.

Dragović, V. and Gajić, B. 2008. Hirota-Kimura type discretization of the classical nonholonomic Suslov problem. *Regul. Chaotic Dyn.*, **13**(4), 250–256.

Dragović, V. and Radnović, M. 2011. *Poncelet Porisms and Beyond: Integrable Billiards, Hyperelliptic Jacobians and Pencils of Quadrics*. Frontiers in Mathematics. Basel: Springer.

Drazin, P. G. and Johnson, R. S. 1989. *Solitons: An Introduction*. Cambridge Texts in Applied Mathematics. Cambridge, UK: Cambridge University Press.

Drinfeld, V. G. 1992. On some unsolved problems in quantum group theory. In P. P. Kulish (ed), *Quantum Groups*. Lecture Notes in Mathematics, vol. 1510. Berlin Heidelberg: Springer, 1–8.

Drinfel'd, V. G. and Sokolov, V. V. 1985. Lie algebras and equations of Korteweg–de Vries type. *J. Sov. Math.*, **30**(2), 1975–2036.

Duistermaat, J. J. 2010. *Discrete Integrable Systems: QRT and Elliptic Surfaces*. New York: Springer.

Duistermaat, J. J. and Joshi, N. 2011. Okamoto's space for the first Painlevé equation in Boutroux coordinates. *Arch. Ration. Mech. Anal.*, **202**(Dec.), 707–785.

Etingof, P., Schedler, T. and Soloviev, A. 1999. Set-theoretical solutions to the quantum Yang–Baxter equation. *Duke Math. J.*, **100**(2), 169–209.

Euler, L. 1761. De integratione aequationis differentialis $\frac{m\,dx}{\sqrt{1-x^4}} = \frac{n\,dy}{\sqrt{1-y^4}}$. *Nov. Comm. Acad. Sci. Petr.*, **6**, 35–57.

Faddeev, L. D. and Takhtajan, L. A. 2007. *Hamiltonian Methods in the Theory of Solitons*. English ed. Classics in Mathematics. Berlin: Springer. Translated from the 1986 Russian original by A. G. Reyman.

Falqui, G. and Viallet, C.-M. 1993. Singularity, complexity, and quasi-integrability of rational mappings. *Commun. Math. Phys.*, **154**(1), 111–125.

Ferapontov, E. V., Novikov, V. S. and Roustemoglou, I. 2014. On the classification of discrete Hirota-type equations in 3D. *Int. Math. Res. Not.*, **2015**(13): 4933–4974.

Field, C. M. and Nijhoff, F. W. 2003. A note on modified Hamiltonians for numerical integrations admitting an exact invariant. *Nonlinearity*, **16**(5), 1673.

Field C. M., Joshi N. and Nijhoff, F. W. 2008. q-Difference equations of KdV type and Chazy-type second-degree difference equations. *J. Phys. A: Math. Theor.*, **41**(33), 332005,13pp.

Flanders, H. 1963. *Differential Forms with Applications to the Physical Sciences*. New York: Dover Publications.

Flaschka, H. 1974. The Toda lattice. II. Existence of integrals. *Phys. Rev. B*, **9**(Feb), 1924–1925.

Flaschka, H. and Newell, A. C. 1980. Monodromy- and spectrum-preserving deformations I. *Commun. Math. Phys.*, **76**(1), 65–116.

Flaschka, H. and Newell, A. C. 1981. Multiphase similarity solutions of integrable evolution equations. *Physica D*, **3**(1), 203–221.

Fokas, A. S. and Ablowitz, M. J. 1981. Linearization of the Korteweg–de Vries and Painlevé II equations. *Phys. Rev. Lett.*, **47**(Oct), 1096–1100.

Fokas, A. S. and Ablowitz, M. J. 1982. On a unified approach to transformations and elementary solutions of Painlevé equations. *J. Math. Phys.*, **23**(11), 2033–2042.

Fokas, A. S. and Ablowitz, M. J. 1983. The inverse scattering transform for multidimensional (2 + 1) problems. In Wolf, K. B. (ed), *Nonlinear Phenomena (Oaxtepec, 1982)*. Lecture Notes in Physics, vol. 189. Berlin: Springer, 137–183.

Fokas, A. S., Its, A. R. and Zhou, X. 1992a. Continuous and discrete Painlevé equations. In Levi, D. and Winternitz, P. (eds), *Painlevé Transcendents, Their Asymptotics and Physical Applications*. New York: Springer Science+Business Media, 33–47.

Fokas, A. S., Its, A. R. and Kitaev, A. V. 1992b. The isomonodromy approach to matrix models in 2D quantum gravity. *Commun. Math. Phys.*, **147**(2), 395–430.

Fokas, A. S., Grammaticos, B. and Ramani, A. 1993. From continuous to discrete Painlevé equations. *J. Math. Anal. Appl.*, **180**(2), 342–360.

Fokas, A. S., Its, A. R., Kapaev, A. A. and Novokshenov, V. Yu. 2006. *Painlevé Transcendents: A Riemann–Hilbert Approach*. Providence, RI: American Mathematical Society.

Fomin, S. and Zelevinsky, A. 2002a. Cluster algebras I: foundations. *J. Am. Math. Soc.*, **15**(2), 497–529.

Fomin, S. and Zelevinsky, A. 2002b. The Laurent Phenomenon. *Adv. Appl. Math.*, **28**(2), 119–144.

Fomin, S. and Zelevinsky, A. 2003. Cluster algebras II: finite type classification. *Invent. Math.*, **154**(1), 63–121.

Fordy, A. P. and Hone, A. N. W. 2014. Discrete integrable systems and Poisson algebras from cluster maps. *Commun. Math. Phys.*, **325**(2), 527–584.

Fordy, A. P. and Kassotakis, P. G. 2006. Multidimensional maps of QRT type. *J. Phys. A: Math. Gen.*, **39**(34), 10773.

Fordy, A. P. and Marsh, R. J. 2011. Cluster mutation-periodic quivers and associated Laurent sequences. *J. Alg. Comb.*, **34**, 19–66.

Forrester, P. J. 2003. Growth models, random matrices and Painlevé transcendents. *Nonlinearity*, **16**(6), R27.

Forrester, P. J. 2010. *Log-Gases and Random Matrices (LMS-34)*. Princeton, NJ: Princeton University Press.

Forrester, P. J. and Witte, N. S. 2004. Discrete Painlevé equations, orthogonal polynomials on the unit circle, and N-recurrences for averages over $U(N)$ – $P_{\mathrm{III}'}$ and P_{V} τ-functions. *Int. Math. Res. Not.*, **2004**(4), 159–183.

Forrester, P. J. and Witte, N. S. 2006. Random matrix theory and the sixth Painlevé equation. *J. Phys. A: Math. Gen.*, **39**(39), 12211.

Freeman, N. C. and Nimmo, J. J. C. 1983. Soliton solutions of the Korteweg–de Vries and Kadomtsev–Petviashvili equations: The wronskian technique. *Phys. Lett. A*, **95**(1), 1–3.

Freud, G. 1976. On the coefficients in the recursion formulae of orthogonal polynomials. *Proc. R. Irish Acad. Sec. A*, **76**, 1–6.

Frobenius, G. 1881. Ueber Relationen zwischen den Näherungsbrüchen von Potenzreihen. *J. Reine Angew. Math.*, **90**, 1–17.

Fuchs, R. 1905. Sur quelques équations différentielles linéaires du second ordre. *C. R. Acad. Sci. Prais*, **141**, 555–558.

Fuchs, R. 1907. Über lineare homogene Differentialgleichungen zweiter Ordnung mit drei im Endlichen gelegenen wesentlich singulären Stellen. *Math Ann.*, **63**, 301–321.

Fukuda, A., Ishiwata, E., Iwasaki, M. and Nakamura, Y. 2009. The discrete hungry Lotka–Volterra system and a new algorithm for computing matrix eigenvalues. *Inverse Prob.*, **25**(1), 015007, 17.

Fukuda, A., Ishiwata, E., Yamamoto, Y., Iwasaki, M. and Nakamura, Y. 2013. Integrable discrete hungry systems and their related matrix eigenvalues. *Ann. Mat. Pura Appl.*, **192**(3), 423–445.

Gaeta, G. 2012. *Nonlinear Symmetries and Nonlinear Equations*. Mathematics and Its Applications. Netherlands: Springer.

Gambier, B. 1906a. Sur les équations différentielles dont l'intégrale générale est uniforme, *C.R. Acad. Sc. Paris*, **142**, 1403–1406.

Gambier, B. 1906b. Sur les équations différentielles du deuxième ordre et du premier degré dont l'intégrale générale est uniforme, *C.R. Acad. Sc. Paris*, **142**, 1497–1500.

Gambier, B. 1910. Sur les équations différentielles du second ordre et du premier degré dont l'intégrale générale est a points critiques fixes. *Acta Math.*, **33**(1), 1–55.

Garabedian, P. R. 1964. *Partial Differential Equations*. New York: John Wiley & Sons Inc.

Gardner, C. S., Greene, J. M., Kruskal, M. D. and Miura, R. M. 1967. Method for solving the Korteweg-deVries equation. *Phys. Rev. Lett.*, **19**(Nov), 1095–1097.

Garnier, R. 1912. Sur des équations différentielles du troisième ordre dont l'intégrale générale est uniforme et sur une classe d'équations nouvelles d'ordre supérieur dont l'intégrale générale a ses points critiques fixes. *Ann. Sci. Éc. Norm. Supér. (3)*, **29**, 1–126.

Gasper, G. and Rahman, M. 2004. *Basic Hypergeometric Series*. Encyclopedia of Mathematics and Its Applications. Cambridge, UK: Cambridge University Press.

Gelfand, I. M. and Fomin, S. V. 2000. *Calculus of Variations*. Dover Books on Mathematics. New York: Dover Publications.

Gelfand, I. M., Kapranov, M. and Zelevinsky, A. 2008. *Discriminants, Resultants, and Multidimensional Determinants*. New York: Springer Science & Business Media.

Gilson, C. R., Nimmo, J. J. C. and Ohta, Y. 2007. Quasideterminant solutions of a non-Abelian Hirota–Miwa equation. *J. Phys. A: Math. Theor.*, **40**(42), 12607.

Glasser, M. L., Papageorgiou, V. G. and Bountis, T. C. 1989. Mel'nikov's function for two-dimensional mappings. *SIAM J. Appl. Math.*, **49**(3), 692–703.

Goldstein, H. 1950. *Classical Mechanics*. Boston: Addison-Wesley.

Goncharenko, V. M. and Veselov, A. P. 2004. Yang-Baxter maps and matrix solitons. In Shabat, A.B., González-López, A., Mañas, M., Martínez Alonso, L. and Rodríguez, M.A. (eds), *New Trends in Integrability and Partial Solvability*. Nato Science Series II, vol. 132. US: Springer, 191–197.

Gragg, W. B. 1972. The Padé table and its relation to certain algorithms of numerical analysis. *SIAM Rev.*, **14**, 1–16.

Grammaticos, B. and Ramani, A. 1996. The Gambier mapping. *Physica A*, **223**(1–2), 125–136.

Grammaticos, B. and Ramani, A. 1999. On a novel q-discrete analogue of the Painlevé VI equation. *Phys. Lett. A*, **257**(5–6), 288–292.

Grammaticos, B. and Ramani, A. 2004. Discrete Painlevé equations: a review. In Grammaticos, B., Tamizhmani, T., and Kosmann-Schwarzbach, Y. (eds), *Discrete Integrable Systems*, Lecture Notes in Physics, Vol. 644, Berlin: Springer, 245–321.

Grammaticos, B., Ramani, A. and Papageorgiou, V. G. 1991. Do integrable mappings have the Painlevé property? *Phys. Rev. Lett.*, **67**(14), 1825–1828.

Grammaticos B., Ramani A. and Moreira, I. 1993. Delay-differential equations and the Painlevé transcendents. *Physica A*, **196**(4), 574–590.

Grammaticos, B., Ramani, A. and Hietarinta, J. 1994. Multilinear operators: the natural extension of Hirota's bilinear formalism. *Phys. Lett. A*, **190**(1), 65–70.

Grammaticos, B., Ohta, Y., Ramani, A. and Sakai, H. 1998. Degeneration through coalescence of the q-Painlevé VI equation. *J. Phys. A: Math. Gen.*, **31**(15), 3545.

Grammaticos, B., Nijhoff, F. W. and Ramani, A. 1999. Discrete Painlevé equations. In Conte, R. M. (ed), *The Painlevé Property: One Century Later*. CRM Series in Mathematical Physics. New York: Springer, 413–516.

Grammaticos, B., Ramani, A. and Tamizhmani, K. M. 2000. Deautonomisation of differential-difference equations and discrete Painlevé equations. *Chaos, Solitons Fractals*, **11**(5), 757–764.

Grammaticos, B., Kosmann-Schwarzbach, Y. and Tamizhmani, T. 2004. *Discrete Integrable Systems*. Lecture Notes in Physics, vol. 644. Berlin Heidelberg: Springer.

Grammaticos, B., Ramani, A., Satsuma, J., Willox, R. and Carstea, A. S. 2005. Reductions of integrable lattices. *J. Nonlinear Math. Phys.*, **12**, Supp 1, 363–371.

Grammaticos, B., Halburd, R. G., Ramani, A. and Viallet, C.-M. 2009. How to detect the integrability of discrete systems. *J. Phys. A: Math. Theor.*, **42**(45), 454002.

Grammaticos, B., Ramani, A., Satsuma, J. and Willox, R. 2012. Discretising the Painlevé equations à la Hirota–Mickens. *J. Math. Phys.*, **53**(2), 023506.

Grammaticos, B., Ramani, A., Willox, R., Mase, T. and Satsuma, J. 2015. Singularity confinement and full-deautonomisation: A discrete integrability criterion. *Phys. D*, **313**, 11–25.

Graves-Morris, P. R. and Roberts, D. E. 1997. Problems and progress in vector Padé approximation. *J. Comput. Appl. Math.*, **77**(1–2), 173–200.

Griffiths, D. J. 2005. *Introduction to Quantum Mechanics*, 2nd ed. Harlow, UK: Pearson Education.

Gromak, V. I. and Tsegel'nik, V. V. 1994. Functional relations between solutions of equations of P-type. *Differ. Uravn.*, **30**(7), 1118–1124, 1284.

Gromak, V. I., Laine, I. and Shimomura, S. 2002. *Painlevé Differential Equations in the Complex Plane*. De Gruyter Studies in Mathematics. Berlin: De Gruyter.

Gutknecht, M. H. and Parlett, B. N. 2011. From qd to LR, or, how were the qd and LR algorithms discovered? *IMA J. Numer. Anal.*, **31**(3), 741–754.

Hairer, E., Lubich, C. and Wanner, G. 2006. *Geometric Numerical Integration: Structure-Preserving Algorithms for Ordinary Differential Equations*. Springer Series in Computational Mathematics. Berlin Heidelberg: Springer.

Halburd, R. G. 2005. Diophantine integrability. *J. Phys. A: Math. Gen.*, **38**(16), L263–L269.

Halburd, R. G. and Korhonen, R. J. 2006. Existence of finite-order meromorphic solutions as a detector of integrability in difference equations. *Phys. D*, **218**(2), 191–203.

Halburd, R. G. and Korhonen, R. J. 2007a. Finite-order meromorphic solutions and the discrete Painlevé equations. *Proc. London Math. Soc. (3)*, **94**(2), 443–474.

Halburd, R. G. and Korhonen, R. J. 2007b. Meromorphic solutions of difference equations, integrability and the discrete Painlevé equations. *J. Phys. A: Math. Theor.*, **40**(6), R1–R38.

Hartshorne, R. 1977. *Algebraic Geometry*, Graduate texts in mathematics: 52. New York: Springer.

Hasselblatt, B. and Propp, J. 2007. Degree-growth of monomial maps. *Ergod. Theor. Dyn. Syst.*, **27**(05), 1375–1397.

Hay, M., Hietarinta, J., Joshi, N. and Nijhoff, F. W. 2007. A Lax pair for a lattice modified KdV equation, reductions to q-Painlevé equations and associated Lax pairs. *J. Phys. A: Math. Theor.*, **40**(2), F61–F73.

Hay, M., Kajiwara, K. and Masuda, T. 2011. Bilinearization and special solutions to the discrete Schwarzian KdV equation. *Journal of Math-for-Industry*, **3** (2011A-5), 53–62.

He, Y., Hu, X.-B., Sun, J. Q. and Weniger, E. J. 2011. Convergence acceleration algorithm via an equation related to the lattice boussinesq equation. *SIAM J. Sci. Comput.*, **33**(3), 1234–1245.

Herbst, B. M. and Ablowitz, M. J. 1989. Numerically induced chaos in the nonlinear Schrödinger equation. *Phys. Rev. Lett.*, **62**(18), 2065.

Herbst, B M., Varadi, F. and Ablowitz, M. J. 1994. Symplectic methods for the nonlinear Schrödinger equation. *Math. Comput. Simul.*, **37**(4), 353–369.

Hertrich-Jeromin, U., McIntosh, I., Norman, P. and Pedit, F. 2001. Periodic discrete conformal maps. *J. Reine Angew Math.*, **534**, 129–153.

Hietarinta, J. 1987a. A search for bilinear equations passing Hirota's three-soliton condition. I. KdV-type bilinear equations. *J. Math. Phys.*, **28**(8), 1732–1742.

Hietarinta, J. 1987b. A search for bilinear equations passing Hirota's three-soliton condition. II. mKdV-type bilinear equations. *J. Math. Phys.*, **28**(9), 2094–2101.

Hietarinta, J. 1987c. A search for bilinear equations passing Hirota's three-soliton condition. III. Sine-Gordon-type bilinear equations. *J. Math. Phys.*, **28**(11), 2586–2592.

Hietarinta, J. 1997. Permutation-type solutions to the Yang-Baxter and other n-simplex equations. *J. Phys. A: Math. Gen.*, **30**(13), 4757–4771.

Hietarinta, J. 2004. A new two-dimensional lattice model that is 'consistent around a cube'. *J. Phys. A: Math. Gen.*, **37**(6), L67–L73.

Hietarinta, J. 2005. Searching for CAC-maps. *J. Nonlinear Math. Phys.*, **12**, Supp 2, 223–230.

Hietarinta, J. 2011. Boussinesq-like multi-component lattice equations and multi-dimensional consistency. *J. Phys. A: Math. Theor.*, **44**(16), 165204, 22pp.

Hietarinta, J. and Kruskal, M. D. 1992. Hirota forms for the six Painlevé equations from singularity analysis. In Levi, D. and Winternitz, P. (eds), *Painlevé Transcendents, Their Asymptotics and Physical Applications*. New York: Springer Science+Business Media, 175–185.

Hietarinta, J. and Viallet, C. 1998. Singularity confinement and chaos in discrete systems. *Phys. Rev. Lett.*, **81**(Jul), 325–328.
Hietarinta, J. and Viallet, C. 2000. Discrete Painlevé I and singularity confinement in projective space. *Chaos, Solitons Fractals*, **11**(1–3), 29–32.
Hietarinta, J. and Viallet, C. 2012. Weak Lax pairs for lattice equations. *Nonlinearity*, **25**(7), 1955–1966.
Hietarinta, J. and Zhang, D. J. 2009. Soliton solutions for ABS lattice equations. II. Casoratians and bilinearization. *J. Phys. A: Math. Theor.*, **42**(40), 404006, 30pp.
Hietarinta, J. and Zhang, D. J. 2011. Soliton taxonomy for a modification of the lattice Boussinesq equation. *SIGMA* **7**, 061, 14pp.
Hietarinta, J. and Zhang, D. J. 2013. Hirota's method and the search for integrable partial difference equations. 1. Equations on a 3x3 stencil. *J. Differ. Equ. Appl.*, **19**(8), 1292–1316.
Hildebrand, F. B. 1968. *Finite-Difference Equations and Simulations*. Englewood Cliffs, NJ: Prentice-Hall Inc.
Hille, E. 1976. *Ordinary Differential Equations in the Complex Domain*. New York: Dover.
Hirota, R. 1971. Exact solution of the Korteweg–de Vries equation for multiple collisions of solitons. *Phys. Rev. Lett.*, **27**(Nov), 1192–1194.
Hirota, R. 1973. Exact three-soliton solution of the two-dimensional sine-gordon equation. *J. Phys. Soc. Japan*, **35**(5), 1566–1566.
Hirota, R. 1974. A new form of Bäcklund transformations and its relation to the inverse scattering problem. *Prog. Theor. Phys.*, **52**(5), 1498–1512.
Hirota, R. 1977a. Nonlinear partial difference equations. I. A difference analogue of the Korteweg-de Vries equation. *J. Phys. Soc. Japan*, **43**(4), 1424–1433.
Hirota, R. 1977b. Nonlinear partial difference equations. II. Discrete-time Toda equation. *J. Phys. Soc. Japan*, **43**(6), 2074–2078.
Hirota, R. 1977c. Nonlinear partial difference equations. III. Discrete sine-Gordon equation. *J. Phys. Soc. Japan*, **43**(6), 2079–2086.
Hirota, R. 1981. Discrete analogue of a generalized Toda equation. *J. Phys. Soc. Japan*, **50**(11), 3785–3791.
Hirota, R. 1986. Reduction of soliton equations in bilinear form. *Physica D*, **18**(1–3), 161–170.
Hirota, R. 2004. *The Direct Method in Soliton Theory*. Cambridge Tracts in Mathematics, vol. 155. Cambridge, UK: Cambridge University Press. Translated from the 1992 Japanese original and edited by A. Nagai, J. Nimmo and C. Gilson.
Hirota, R. and Ito, M. 1981. A direct approach to multi-periodic wave solutions to nonlinear evolution equations. *J. Phys. Soc. Japan*, **50**(1), 338–342.
Hirota, R. and Kimura, K. 2000. Discretization of the Euler top. *J. Phys. Soc. Japan*, **69**(3), 627–630.
Hirota, R. and Satsuma, J. 1976. A variety of nonlinear network equations generated from the Bäcklund transformation for the toda lattice. *Prog. Theor. Phys. Supp.*, **59**, 64–100.
Hirota, R. Tsujimoto, S. and Imai, T. 1993. Difference scheme of soliton equations. In Christiansen, P. L., Eilbeck, J. C., and Parmentier, R. D. (eds), *Future Directions of Nonlinear Dynamics in Physical and Biological Systems*. NATO ASI Series, vol. 312. US: Springer, 7–15.
Hirota, R., Kimura, K. and Yahagi, H. 2001. How to find the conserved quantities of nonlinear discrete equations. *J. Phys. A: Math. Gen.*, **34**(48), 10377–10386.
Hone, A. N. W. 2007. Sigma function solution of the initial value problem for Somos 5 sequences. *Trans. Am. Math. Soc.*, **359**(10), 5019–5034.

Hone, A. N. W. and Petrera, M. 2009. Three-dimensional discrete systems of Hirota-Kimura type and deformed Lie-Poisson algebras. *J. Geom. Mech.*, **1**(1), 55–85.

Hone, A. N. W. and Swart, C. 2008. Integrality and the Laurent phenomenon for Somos 4 and Somos 5 sequences. *Math. Proc. Cambridge Philos. Soc.*, **145**(7), 65–85.

Humphreys, J. E. 1992. *Reflection Groups and Coxeter Groups*. Cambridge Studies in Advanced Mathematics. Cambridge, UK: Cambridge University Press.

Hydon, P. E. 2014. *Difference Equations by Differential Equations Methods*. Cambridge Monographs on Applied and Computational Mathematics, vol. 149. Cambridge, UK: Cambridge University Press.

Hydon, P. E. and Viallet, C.-M. 2010. Asymmetric integrable quad-graph equations. *Appl. Anal.*, **89**(4), 493–506.

Iatrou, A. and Roberts, J. A. G. 2001. Integrable mappings of the plane preserving biquadratic invariant curves. *J. Phys. A: Math. Gen.*, **34**(34), 6617–6636.

Ince, E. L. 1956. *Ordinary Differential Equations*. Dover Books on Science. New York: Dover Publications Incorporated.

Ismail, M. 2005. *Classical and Quantum Orthogonal Polynomials in One Variable*. Encyclopedia of Mathematics and Its Applications. vol. 98. Cambridge, UK: Cambridge University Press.

Its, A. R., Kitaev, A. V. and Fokas, A. S. 1990. The isomonodromy approach in the theory of two-dimensional quantum gravitation. *Russ. Math. Surv.*, **45**(6), 155–157.

Iwasaki, K., Kimura, H., Shimemura, S. and Yoshida, M. 1991. *From Gauss to Painlevé: A Modern Theory of Special Functions*. Aspects of Mathematics. Braunschweig: Vieweg+Teubner Verlag.

Jacobi, C. G. J. 1846. Uber die Darstellung einer Reihe gegebener Werthe durch eine gebrochene rationale Function. *J. Reine Angew. Math.*, **30**, 127–156.

Jimbo, M. and Miwa, T. 1981. Monodromy perserving deformation of linear ordinary differential equations with rational coefficients. *Physica D*, **2**, 407–448.

Jimbo, M. and Miwa, T. 1983. Solitons and infinite-dimensional Lie algebras. *Publ. Res. Inst. Math. Sci.*, **19**(3), 943–1001.

Jimbo, M. and Sakai, H. 1996. A q-analog of the sixth Painlevé equation. *Lett. Math. Phys.*, **38**(2), 145–154.

Joshi, N. 1997. A local asymptotic analysis of the discrete first Painlevé equation as the discrete independent variable approaches infinity. *Methods Appl. Anal.*, **4**(2), 124–133.

Joshi, N. 2009. Direct "delay" reductions of the Toda equation. *J. Phys. A: Math. Theor.*, **42**(2), 022001.

Joshi, N. 2015. Quicksilver solutions of a q-difference first Painlevé equation. *Stud. Appl. Math*, **134**(2), 233–251.

Joshi, N. and Kruskal, M. D. 1994. A direct proof that solutions of the six Painlevé equations have no movable singularities except poles. *Stud. Appl. Math.*, **93**(3), 187–207.

Joshi, N. and Lobb, S. B. 2016. Singular dynamics of a q-difference Painlevé equation in its initial-value space. *J. Phys. A: Math. Theor*, **49**(1), 014002.

Joshi, N. and Lustri, C. J. 2015. Stokes phenomena in discrete Painlevé I. *Proc. R. Soc. London A Math. Phys. Eng. Sci.*, **471**(2177), 20140874.

Joshi, N. and Shi, Y. 2011. Exact solutions of a q-discrete second Painlevé equation from its iso-monodromy deformation problem: I. Rational solutions. *Proc. R. Soc. London A Math. Phys. Eng. Sci.*, rspa20110167.

Joshi, N. and Shi, Y. 2012. Exact solutions of a q-discrete second Painlevé equation from its iso-monodromy deformation problem. II. Hypergeometric solutions. *Proc. R. Soc. London A Math. Phys. Eng. Sci.*, **468**(2146), 3247–3264.

Joshi, N., Burtonclay, D. and Halburd, R. G. 1992. Nonlinear nonautonomous discrete dynamical systems from a general discrete isomonodromy problem. *Lett. Math. Phys.*, **26**(2), 123–131.

Joshi, N., Ramani, A. and Grammaticos, B. 1998. A bilinear approach to discrete Miura transformations. *Phys. Lett. A*, **249**(1–2), 59–62.

Joshi, N., Grammaticos, B., Tamizhmani, T. and Ramani, A. 2006. From integrable lattices to non-QRT mappings. *Lett. Math. Phys.*, **78**(1), 27–37.

Joshi, N., Nakazono, N. and Shi, Y. 2014. Geometric reductions of ABS equations on an n-cube to discrete Painlevé systems *J. Phys A: Math. Theor.*, **47**(50), 505201.

Joshi, N., Nakazono, N., and Shi, Y. 2015. Lattice equations arising from discrete Painlevé systems. I. $(A_2 + A_1)^{(1)}$ and $(A_1 + A_1')^{(1)}$ cases. *J. Math. Phys.*, **56**(9), 092705.

Kac, M. and van Moerbeke, P. 1975. On an explictly soluble system fo nonlinear differential equations related to certain Toda lattices. *Adv. Math.*, **16**, 160–169.

Kac, V. G. 1994. *Infinite-Dimensional Lie Algebras*. Cambridge, UK: Cambridge university Press.

Kahan, W. 1993. Unconventional numerical methods for trajectory calculations. *Lecture notes, CS Division, Department of EECS.*

Kajiwara, K. and Ohta, Y. 2008. Bilinearization and casorati determinant solution to the non-autonomous discrete KdV equation. *J. Phys. Soc. Japan*, **77**(5), 054004.

Kajiwara, K., Maruno, K. and Oikawa, M. 2000. Bilinearization of discrete soliton equations through the singularity confinement test. *Chaos, Solitons & Fractals*, **11**(1–3), 33–39.

Kajiwara, K., Noumi, M. and Yamada, Y. 2001. A study on the fourth q-Painlevé equation. *J. Phys. A: Math. Gen.*, **34**(41), 8563–8581.

Kajiwara, K., Noumi, M. and Yamada, Y. 2002. q-Painlevé systems arising from q-KP hierarchy. *Lett. Math. Phys.*, **62**(3), 259–268.

Kajiwara, K., Masuda, T. Noumi, M. Ohta, Y. and Yamada, Y. 2003. ${}_{10}E_9$ solution to the elliptic Painlevé equation. *J. Phys. A: Math. Gen.*, **36**(17), L263–L272.

Kajiwara, K., Masuda, T., Noumi, M., Ohta, Y. and Yamada, Y. 2004. Hypergeometric solutions to the q-Painlevé equations. *Int. Math. Res. Not.*, **2004**(47), 2497–2521.

Kajiwara, K., Masuda, T., Noumi, M., Ohta, Y. and Yamada, Y. 2005a. Construction of hypergeometric solutions to the q-Painlevé equations. *Int. Math. Res. Not.*, **2005**(24), 1439–1463.

Kajiwara, K., Masuda, T., Noumi, M., Ohta, Y. and Yamada, Y. 2005b. A geometric description of the elliptic Painleve equation. *Rokko Lect. Math.*, **18**, 43–48.

Kajiwara, K., Masuda, T., Noumi, M., Ohta, Y. and Yamada, Y. 2006. Point configurations, Cremona transformations and the elliptic difference Painlevé equation. In Delabaere, E. and Lodauy-Richaud, M. (ed), *Théories asymptotiques et équations de Painlevé, Seminaires et Congres Sémin. Congr*, vol. 14. Paris: SMF, 169–198.

Kajiwara, K., Noumi, M. and Yamada, Y. 2015. Geometric aspects of Painlevé equations. *arXiv:1509.08186*, 168pp.

Kakei, S., Nimmo, J. J. C. and Willox, R. 2009. Yang–Baxter maps and the discrete KP hierarchy. *Glasg. Math. J.*, **51**(2), 107–119.

Kakei, S., Nimmo, J. J. C. and Willox, R. 2010. Yang–Baxter maps from the discrete BKP equation. *SIGMA* **6**, 028, 11pp.

Kanki, M. 2013. *Studies on the Discrete Integrable Equations over Finite Fields*. Ph.D. thesis, University of Tokyo.

Kassotakis, P. and Nieszporski, M. 2011. Families of integrable equations. *SIGMA* **7**, 100, 14pp.

Kassotakis, P. and Nieszporski, M. 2012. On non-multiaffine consistent-around-the-cube lattice equations. *Phys. Lett. A*, **376**(45), 3135–3140.

Kay, I. and Moses, H. E. 1956. Reflectionless transmission through dielectrics and scattering potentials. *J. Appl. Phys.*, **27**(12), 1503–1508.

Kelley, W. G., and Peterson, A. C. 2001. *Difference Equations: An Introduction with Applications*. San Diego: Harcourt/Academic Press.

Kimura, K. and Hirota, R. 2000. Discretization of the Lagrange Top. *J. Phys. Soc. Japan*, **69**(10), 3193–3199.

Kimura, K., Yahagi, H., Hirota, R., Ramani, A., Grammaticos, B. and Ohta, Y. 2002. A new class of integrable discrete systems, *J. Phys. A: Math. Gen.*, **35**(43), 9205–3212.

King, A. D. and Schief, W. K. 2003. Tetrahedra, octahedra and cubo-octahedra: integrable geometry of multi-ratios. *J. Phys. A: Math. Gen.*, **36**(3), 785–802.

King, A. D. and Schief, W. K. 2006. Application of an incidence theorem for conics: Cauchy problem and integrability of the dCKP equation. *J. Phys. A: Math. Gen.*, **39**(8), 1899–1913.

Koekoek, R., Lesky, P. A. and Swarttouw, R. F. 2010. *Hypergeometric Orthogonal Polynomials and Their q-Analogues*. Springer Monographs in Mathematics. Berlin: Springer-Verlag.

Koenigs, G. 1891. Sur les systèmes conjugués à invariants égaux. *C. R. Acad. Sci. Paris.*, **113**, 1022–1024.

Koenigs, G. 1892. Sur les réseaux plans à invariants égaux et les lignes asymptotiques. *C. R. Acad. Sci. Paris.*, **114**, 55–57.

Kolchin, E. R. 1973. *Differential Algebra & Algebraic Groups*. Pure and Applied Mathematics. New York and London: Academic Press.

Konopelchenko, B. G. and Schief, W. K. 1998. Three-dimensional integrable lattices in Euclidean spaces: conjugacy and orthogonality. *R. Soc. London Proc. Ser. A Math. Phys. Eng. Sci.*, **454**(1980), 3075–3104.

Konopelchenko, B. G. and Schief, W. K. 2002a. Menelaus' theorem, Clifford configurations and inversive geometry of the Schwarzian KP hierarchy. *J. Phys. A: Math. Gen.*, **35**(29), 6125–6144.

Konopelchenko, B. G. and Schief, W. K. 2002b. Reciprocal figures, graphical statics, and inversive geometry of the Schwarzian BKP hierarchy. *Stud. Appl. Math.*, **109**(2), 89–124.

Konstantinou-Rizos, S. 2014. Darboux transformations, discrete integrable systems and related Yang–Baxter maps. *arXiv preprint arXiv:1410.5013*.

Konstantinou-Rizos, S., and Mikhailov, A. V. 2013. Darboux transformations, finite reduction groups and related Yang–Baxter maps. *J. Phys. A: Math. Theor.*, **46**(42), 425201.

Konstantinou-Rizos, S., Mikhailov, A. V. and Xenitidis, P. 2015. Reduction groups and related integrable difference systems of NLS type. *arXiv:1503.06406*.

Korepin, V. E., Bogoliubov, N. M. and Izergin, A.G. 1997. *Quantum Inverse Scattering Method and Correlation Functions*. Cambridge, UK: Cambridge University Press.

Korteweg, D. J. and de Vries, G. 1895. On the change of form of long waves advancing in a rectangular canal, and on a new type of long stationary waves. *Philos. Mag. Ser. 5*, **39**(240), 422–443.

Kouloukas, T. E. and Papageorgiou, V. G. 2009. Yang–Baxter maps with first-degree-polynomial 2×2 Lax matrices. *J. Phys. A: Math. Theor.*, **42**(40), 404012.

Kouloukas, T. E., and Papageorgiou, V. G. 2012. 3D compatible ternary systems and Yang–Baxter maps. *J. Phys. A: Math. Theor.*, **45**(34), 345204.

Kowalevski, S. 1889. Sur le probleme de la rotation d'un corps solide autour d'un point fixe. *Acta Math.*, **12**(1), 177–232.

Kowalevski, S. 1890. Sur une propriété du système d'équations différentielles qui définit la rotation d'un corps solide autour d'un point fixe. *Acta Math.*, **14**(1), 81–93.

Krichever, I. M. 1978. Rational solutions of the Kadomtsev–Petviashvili equation and integrable systems of N particles on a line. *Funct. Anal. Appl.*, **12**(1), 59–61.

Krichever, I. M. 1980. Elliptic solutions of the Kadomtsev-Petviashvili equation and integrable systems of particles. *Funct. Anal. Appl.*, **14**(4), 282–290.

Krichever, I. M. and Novikov, S. P. 1979. Holomorphic fiberings and nonlinear equations. Finite zone solutions of rank 2. *Sov. Math. Dokl*, **20**(4), 650–654.

Krichever, I. M. and Novikov, S. P. 1980. Holomorphic bundles over algebraic curves and non-linear equations. *Russ. Math. Surv.*, **35**(6), 53–79.

Krichever, I. M., Wiegmann, P. and Zabrodin, A. 1998. Elliptic solutions to difference non-linear equations and related many-body problems. *Commun. Math. Phys.*, **193**(2), 373–396.

Kruskal, M. D. and Clarkson, P. A. 1992. The Painlevé-Kowalevski and poly-Painlevé tests for integrability. *Stud. Appl. Math.*, **86**(2), 87–165.

Kupershmidt, B. A. 2000. *KP or mKP: Noncommutative Mathematics of Lagrangian, Hamiltonian, and Integrable Systems*. Mathematical surveys and monographs. Providence, RI: American Mathematical Society.

Lafortune, S., Ramani, A., Ohta, Y., Grammaticos, B., and Tamizhmani, K. M. 2001. Blending two discrete integrability criteria: singularity confinement and algebraic entropy. In Coley, A. A., Levi, D., Milson, R., Rogers, C. and Winternitz, P. (eds) *Bäcklund and Darboux Transformations. The Geometry of Solitons*. CRM Proceedings & Lecture Notes, vol. 29. Providence, RI: American Mathematical Society, 299–311.

Lamb, G. L. 1980. *Elements of Soliton Theory*. Pure and Applied Mathematics. New York: John Wiley & Sons.

Lax, P. D. 1968. Integrals of nonlinear equations of evolution and solitary waves. *Comm. Pure Applied Math.* **21**, 467–490.

LeCaine, J. 1943. The linear q-difference equation of the second order. *Am. J. Math.*, **65**, 585–600.

Levi, D. 1981. Nonlinear differential difference equations as Backlund transformations. *J. Phys. A: Math. Gen.*, **14**(5), 1083.

Levi, D. and Benguria, R. 1980. Bäcklund transformations and nonlinear differential difference equations. *Proc. Natl. Acad. Sci. U.S.A.*, **77**(9), 5025–5027.

Levi, D. and Petrera, M. 2007. Continuous symmetries of the lattice potential KdV equation. *J. Phys. A: Math. Theor.*, **40**(15), 4141.

Levi, D. and Winternitz, P. 1993. Symmetries and conditional symmetries of differential-difference equations., *J. Math. Phys.* **34**, 3713–3730.

Levi, D., and Winternitz, P. 2006. Continuous symmetries of difference equations. *J. Phys. A: Math. Gen.*, **39**(2), R1–R63.

Levi, D. and Yamilov, R. 1997. Conditions for the existence of higher symmetries of evolutionary equations on the lattice. *J. Math. Phys.*, **38**, 6648.

Levi, D. and Yamilov, R. I. 2009a. The generalized symmetry method for discrete equations. *J. Phys. A: Math. Theor.*, **42**(45), 454012, 18pp.

Levi, D. and Yamilov, R. I. 2009b. On a nonlinear integrable difference equation on the square. *Ufimsk. Mat. Zh.*, **1**, 101–105.

Levi, D., and Yamilov, R. I. 2011. Generalized symmetry integrability test for discrete equations on the square lattice. *J. Phys. A: Math. Theor.*, **44**(14), 145207, 22pp.

Levi, D. Pilloni, L. and Santini, P. M. 1981. Integrable three-dimensional lattices. *J. Phys. A: Math. Gen.*, **14**(7), 1567.
Levi, D. Ragnisco, O. and Rodriguez, M. A. 1992. On non-isospectral flows, Painlevé equations, and symmetries of differential and difference equations. *Theor. Math. Phys.*, **93**(3), 1409–1414.
Levi, D., Petrera, M. and Scimiterna, C. 2007. The lattice Schwarzian KdV equation and its symmetries. *J. Phys. A: Math. Theor.*, **40**(42), 12753.
Levi, D., Olver, P., Thomova, Z. and Winternitz, P. 2011. *Symmetries and Integrability of Difference Equations*. London Mathematical Society Lecture Note Series, vol. 381. Cambridge, UK: Cambridge University Press.
Levin, A. 2008. *Difference Algebra*. Algebra and Applications, vol. 8. Berlin: Springer.
Lobb, S. B. 2010. *Lagrangian Structures and Multidimensional Consistency*. Ph.D. thesis, University of Leeds.
Lobb, S. B. and Nijhoff, F. W. 2009. Lagrangian multiforms and multidimensional consistency. *J. Phys. A: Math. Theor.*, **42**(45), 454013.
Lobb, S. B. and Nijhoff, F. W. 2010. Lagrangian multiform structure for the lattice Gel'fand-Dikii hierarchy. *J. Phys. A: Math. Theor.*, **43**(7), 072003, 11pp.
Lobb, S. B. and Nijhoff, F. W. 2013. A variational principle for discrete integrable systems. *arXiv preprint arXiv:1312.1440.*
Lobb, S. B., Nijhoff, F. W. and Quispel, G. R. W. 2009. Lagrangian multiform structure for the lattice KP system. *J. Phys. A: Math. Theor.*, **42**(47), 472002.
Logan, J. D. 1973. First integrals in the discrete variational calculus. *Aequationes Math.*, **9**, 210–220.
Maeda, S. 1981. Extension of discrete Noether theorem. *Math. Japon.*, **26**(1), 85–90.
Magnus, A. P. 1995. Painlevé-type differential equations for the recurrence coefficients of semi-classical orthogonal polynomials. *J. Comput. Appl. Math.*, **57**(1), 215–237.
Magnus, A. P. 2009. Rational interpolation to solutions of Riccati difference equations on elliptic lattices. *J. Comput. Appl. Math.*, **233**(3), 793–801.
Maillet, J.-M. and Nijhoff, F. W. 1989. Integrability for multidimensional lattice models. *Phys. Lett. B*, **224**(4), 389–396.
Mano, T. 2010. Asymptotic behaviour around a boundary point of the q-Painlevé VI equation and its connection problem. *Nonlinearity*, **23**(7), 1585.
Marsh, R. J. 2013. *Lecture Notes on Cluster Algebras*. Zurich Lectures in Advanced Mathematics. European Mathematical Society.
Maruno, K. and Quispel, G. R. W. 2006. Construction of integrals of higher-order mappings. *J. Phys. Soc. Jpn.*, **75**, 123001–123005.
Maruno, K., Kajiwara, K., Nakao, S. and Oikawa, M. 1997. Bilinearization of discrete soliton equations and singularity confinement. *Phys. Lett. A*, **229**(3), 173–182.
Matveev, V. B. and Salle, M. A. 1991. *Darboux Transformations and Solitons*. Nonlinear Dynamics Series. Berlin: Springer.
McMillan, E. M. 1971. A problem in the stability of periodic systems. Brittin, W. E. and Odabasi, H. (eds), *Topics in Modern Physics. A Tribute to E.U. Condon*. Colorado Associated Univ. Press, 219–44.
McMullen, C. T. 2007. Dynamics on blowups of the projective plane. *Publ. Math. IHES* **105**(1), 49–89.
Mercat, C. 2001. Discrete Riemann surfaces and the Ising model. *Commun. Math. Phys.*, **218**(1), 177–216.
Mercat, C. 2004. Exponentials form a basis of discrete holomorphic functions on a compact. *Bull. Soc. Math. France*, **132**(2), 305–326.

Mercat, C. 2008. Discrete Riemann surfaces, linear and non-linear. *arXiv preprint arXiv:0802.1448.*

Mikhailov, A. V., Wang, J. P. and Xenitidis, P. 2011a. Cosymmetries and Nijenhuis recursion operators for difference equations. *Nonlinearity*, **24**(7), 2079–2097.

Mikhailov, A. V., Wang, J. P. and Xenitidis, P. 2011b. Recursion operators, conservation laws, and integrability conditions for difference equations. *Theor. Math. Phys.*, **167**(1), 421–443.

Milne-Thomson, L. M. 2000. *The Calculus of Finite Differences*. Providence, RI: AMS Chelsea Publishing (reprint of the 1933 edition).

Miura, R. M. 1968. Korteweg-de Vries equation and generalizations. I. A remarkable explicit nonlinear transformation. *J. Math. Phys.*, **9**, 1202–1204.

Miura, R. M., Gardner, C. S. and Kruskal, M. D. 1968. Korteweg–de vries equation and generalizations. II. Existence of conservation laws and constants of motion. *J. Math. Phys.*, **9**(8), 1204–1209.

Miwa, T. 1982. On Hirota's difference equations. *Proc. Japan Acad. Ser. A Math. Sci.*, **58**(1), 9–12.

Miwa, T., Jimbo, M. and Date, E. 2000. *Solitons: Differential Equations, Symmetries and Infinite Dimensional Algebras*. Cambridge Tracts in Mathematics, vol. 135. Cambridge, UK: Cambridge University Press. Translated from the 1993 Japanese original by M. Reid.

Moser, J. 1975. Three integrable Hamiltonian systems connected with isospectral deformations. *Adv. Math.*, **16**(2), 197–220.

Moser, J. and Veselov, A. P. 1991. Discrete versions of some classical integrable systems and factorization of matrix polynomials. *Commun. Math. Phys.*, **139**(2), 217–243.

Muğan, U., and Santini, P. M. 2011. On the solvability of the discrete second Painlevé equation. *J. Phys. A: Math. Theor.*, **44**(18), 185204.

Muir, T. 1890. *The Theory of Determinants in the Historical Order of Its Development. vol. I. Determinants in General. Leibniz (1693) to Cayley (1841)*. London: Macmillan and Co.

Murata, M. 2009. Lax forms of the q-Painlevé equations. *J. Phys. A: Math. Theor.*, **42**(11), 115201.

Murata, M., Sakai, H. and Yoneda, J. 2003. Riccati solutions of discrete Painlevé equations with Weyl group symmetry of type $E_8^{(1)}$. *J. Math. Phys.*, **44**(3), 1396–1414.

Nagai, A. and Satsuma, J. 1995. Discrete soliton equations and convergence acceleration algorithms. *Phys. Lett. A*, **209**(5–6), 305–312.

Nagai, A., Tokihiro, T. and Satsuma, J. 1998. The Toda molecule equation and the ϵ-algorithm. *Math. Comp.*, **67**(224), 1565–1575.

Nakamura, A. 1979. A direct method of calculating periodic wave solutions to nonlinear evolution equations. I. Exact two-periodic wave solution. *J. Phys. Soc. Japan*, **47**(5), 1701–1705.

Nevai, P. 1986. Géza Freud, orthogonal polynomials and Christoffel functions. A case study. *J. Approx. Theory*, **48**(1), 3–167.

Newell, A. C. 1985. *Solitons in Mathematics and Physics*. CBMS-NSF Regional Conference Series in Applied Mathematics, vol. 48. Philadelphia, PA: Society for Industrial and Applied Mathematics (SIAM).

Nieszporski, M. 2007. Darboux transformations for a 6-point scheme. *J. Phys. A: Math. Theor.*, **40**(15), 4193.

Nieszporski, M. and Santini, P. M. 2005. The self-adjoint 5-point and 7-point difference operators, the associated Dirichlet problems, Darboux transformations and Lelieuvre formulae. *Glasg. Math. J.*, **47**(A), 133–147.

Nijhoff, F. W. 1985. Theory of integrable three-dimensional nonlinear lattice equations. *Lett. Math. Phys.*, **9**(3), 235–241.

Nijhoff, F. W. 1987. Integrable hierarches, Lagrangian structures and non-commuting Flows. In M. Ablowitz, B. Fuchssteiner and M. D. Kruskal (eds), *Topics in Soliton Theory and Exactly Solvable Nonlinear Equations*. World Scientific, 150–181.

Nijhoff, F. W. 1997. On some "Schwarzian" equations and their discrete analogues. In Fokas, A. S. and Gelfand, I. M. (eds), *Algebraic Aspects of Integrable Systems: In Memory of Irene Dorfman*. Progress in nonlinear differential equations and their applications, vol. 26. Boston, MA: Birkhäuser, 237–260.

Nijhoff, F. W. 1999. Discrete Painlevé equations and symmetry reduction on the lattice. In Bobenko, A. I. and Seiler, R. (eds), *Discrete Integrable Geometry and Physics (Vienna, 1996)*. Oxford Lecture Ser. Math. Appl., vol. 16. New York: Oxford University Press, 209–234.

Nijhoff, F. W. 2002. Lax pair for the Adler (lattice Krichever-Novikov) system. *Phys. Lett. A*, **297**(1–2), 49–58.

Nijhoff, F. W. and Atkinson, J. 2010. Elliptic N-soliton solutions of ABS lattice equations. *Int. Math. Res. Not.*, **2010**(20), 3837–3895.

Nijhoff, F. W. and Capel, H. W. 1990. The direct linearisation approach to hierarchies of integrable PDEs in 2 + 1 dimensions: I. Lattice equations and the differential-difference hierarchies. *Inverse Prob.*, **6**(4), 567.

Nijhoff, F. W. and Capel, H. 1995. The discrete Korteweg-de Vries equation. *Acta Appl. Math.*, **39**(1–3), 133–158.

Nijhoff, F. W. and Enolskii, V. Z. 1999. Integrable mappings of KdV type and hyperelliptic addition theorems. In Clarkson, P. A. and Nijhoff, F. W. (eds), *Symmetries and Integrability of Difference Equations*. LMS Lecture Note Series 255, 64–78.

Nijhoff, F. W. and Pang, G. D. 1994. A time-discretized version of the Calogero-Moser model. *Phys. Lett. A*, **191**(1–2), 101–107.

Nijhoff, F. W. and Pang, G. D. 1996. Discrete-time Calogero-Moser model and lattice KP equations. In Levi, D., Vinet, L. and Winternitz, P. (eds), *Symmetries and Integrability of Difference Equations*. CRM Proceedings & Lecture Notes, vol. 9. Providence, RI: American Mathematical Society, 253–264.

Nijhoff, F. W. and Papageorgiou, V. G. 1991. Similarity reductions of integrable lattices and discrete analogues of the Painlevé II equation. *Phys. Lett. A*, **153**(6–7), 337–344.

Nijhoff, F. W. and Puttock, S. E. 2003. On a two-parameter extension of the lattice KdV system associated with an elliptic curve. *J. Nonlinear Math. Phys.*, **10**, Supp 1, 107–123.

Nijhoff, F. W. and Walker, A. J. 2001. The discrete and continuous Painlevé VI hierarchy and the Garnier systems. *Glasg. Math. J.*, **43A**, 109–123.

Nijhoff, F. W., Quispel, G. R. W. and Capel, H. W. 1983a. Direct linearization of nonlinear difference-difference equations. *Phys. Lett. A*, **97**(4), 125–128.

Nijhoff, F. W., Capel, H. W. and Quispel, G. R. W. 1983b. Integrable lattice version of the massive Thirring model and its linearization. *Phys. Lett. A*, **98**(3), 83–86.

Nijhoff, F. W., Capel, H. W., Wiersma, G. L. and Quispel, G. R. W. 1984. Bäcklund transformations and three-dimensional lattice equations. *Phys. Lett. A*, **105**(6), 267–272.

Nijhoff, F. W., Capel, H. W. and Wiersma, G. L. 1985. Integrable lattice systems in two and three dimensions. In Martini, R. (ed), *Geometric Aspects of the Einstein*

Equations and Integrable Systems (Scheveningen, 1984). Lecture Notes in Phys., vol. 239. Berlin: Springer, 263–302.

Nijhoff, F. W., Papageorgiou, V. G., Capel, H. W. and Quispel, G. R. W. 1992. The lattice Gel'fand-Dikii hierarchy. *Inverse Prob.*, **8**(4), 597.

Nijhoff, F. W., Ragnisco, O. and Kuznetsov, V. B. 1996. Integrable time-discretisation of the Ruijsenaars-Schneider model. *Commun. Math. Phys.*, **176**(3), 681–700.

Nijhoff, F. W., Hone, A. N. W. and Joshi, N. 2000a. On a Schwarzian PDE associated with the KdV hierarchy. *Phys. Lett. A*, **267**(2–3), 147–156.

Nijhoff, F. W., Joshi, N. and Hone, A. N. W. 2000b. On the discrete and continuous Miura chain associated with the sixth Painlevé equation. *Phys. Lett. A*, **264**(5), 396–406.

Nijhoff, F. W., Ramani, A., Grammaticos, B. and Ohta, Y. 2001. On discrete Painlevé equations associated with the lattice KdV systems and the Painlevé VI equation. *Stud. Appl. Math.*, **106**(3), 261–314.

Nijhoff, F. W. Atkinson, J. and Hietarinta, J. 2009. Soliton solutions for ABS lattice equations. I. Cauchy matrix approach. *J. Phys. A: Math. Theor.*, **42**(40), 404005, 34pp.

Nimmo, J. J. C. 2006. On a non-Abelian Hirota–Miwa equation. *J. Phys. A: Math. Gen.*, **39**(18), 5053.

Nimmo, J. J. C. and Schief, W. K. 1997. Superposition principles associated with the Moutard transformation: an integrable discretization of a $(2 + 1)$-dimensional sine-Gordon system. *Proc. R. Soc. London Ser. A*, **453**(1957), 255–279.

Nishioka, S. 2010. Transcendence of solutions of q-Painlevé equation of type $A_7^{(1)}$. *Aequa. Math.*, **79**(1–2), 1–12.

Nishioka, S. 2012. Irreducibility of q-Painlevé equation of type $A_6^{(1)}$ in the sense of order. *J. Differ. Equ. Appl.*, **18**(2), 313–333.

Noether, E. 1918. Variationsprobleme. *Nachr. Kong. Gesell. Wissensch. (Göttingen), Math.-Phys. Kl.*, **2**, 235–257.

Nörlund, N. E. 1954. *Vorlesungen über Differenzenrechnung*. Providence, RI: AMS Chelsea Publishing (reprint of the 1924 edition from Springer, Berlin).

Noumi, M. 2004. *Painlevé Equations Through Symmetry*. Translations of Mathematical Monographs. Providence, RI: American Mathematical Society.

Noumi, M., Tsujimoto, S. and Yamada, Y. 2013. Padé interpolation for elliptic Painlevé equation. In Iohara, K., Morier-Genoud, S. and Rémy, B.(eds), *Symmetries, Integrable Systems and Representations*. Springer, 463–482.

Ohta, Y., Satsuma, J., Takahashi, D. and Tokihiro, T. 1988. An elementary introduction to Sato theory. *Prog. Theor. Phys. Suppl.*, **94**, 210–241.

Ohta, Y. Hirota, R. Tsujimoto, S. and Imai, T. 1993. Casorati and discrete Gram type determinant representations of solutions to the discrete KP hierarchy. *J. Phys. Soc. Japan*, **62**(6), 1872–1886.

Ohta, Y., Tamizhmani, K. M., Grammaticos, B. and Ramani, A. 1999. Singularity confinement and algebraic entropy: the case of the discrete Painlevé equations. *Phys. Lett. A*, **262**(2–3), 152–157.

Ohta, Y., Ramani, A. and Grammaticos, B. 2001. An affine Weyl group approach to the eight-parameter discrete Painlevé equation. *J. Phys. A: Math. Gen.*, **34**(48), 10523.

Okamoto, K. 1979. Sur les feuilletages associés aux équation du second ordre à points critiques fixes de P. Painlevé. Espaces des conditions initiales. *Jpn. J. Math.*, **5**(1), 1–79.

Okamoto, K. 1981. On the τ-function of the Painlevé equations. *Physica D*, **2**(3), 525–535.

Okamoto, K. 1986. Studies on the Painlevé equations. *Annali di Matematica Pura ed Applicata*, **146**(1), 337–381.

Okamoto, K. 1999. The Hamiltonians associated to the Painlevé equations. In Conte, R. M. (ed), *The Painlevé Property: One Century Later*. CRM Series in Mathematical Physics. New York: Springer, 735–787.

Olshanetsky, M. A. and Perelomov, A. M. 1981. Classical integrable finite-dimensional systems related to Lie algebras. *Phys. Rep.*, **71**(5), 313–400.

Olver, F. W. J., Lozier, D. W., Boisvert, R. F. and Clark, C. W. (eds). 2010. *NIST Handbook of Mathematical Functions*. New York, NY: Cambridge University Press. Print companion to DLMF (2010).

Olver, P. J. 2000. *Applications of Lie Groups to Differential Equations*. Graduate Texts in Mathematics. New York: Springer.

Orfanidis, S. J. 1978. Sine-Gordon equation and nonlinear σ model on a lattice. *Phys. Rev. D (3)*, **18**(10), 3828–3832.

Ormerod, C. M. 2011. A study of the associated linear problem for q-PV. *J. Phys. A: Math. Theor.*, **44**(2), 025201.

Ormerod, C. M. 2012. Reductions of lattice mKdV to q-PVI. *Phys. Lett. A*, **376**(45), 2855–2859.

Ormerod, C. M. and Rains, E. M. 2016. A symmetric difference-differential Lax pair for Painlevé VI. *arXiv:1603.04393*.

Ormerod, C. M., Witte, N. S. and Forrester, P. J. 2011. Connection preserving deformations and q-semi-classical orthogonal polynomials. *Nonlinearity*, **24**(9), 2405.

Ormerod, C. M., Van der Kamp, P. H. and Quispel, G. R. W. 2013. Discrete Painlevé equations and their Lax pairs as reductions of integrable lattice equations. *J. Phys. A: Math. Theor.*, **46**(9), 095204.

Padé, H. 1892. Sur la représentation approchée d'une fonction par des fractions rationnelles. *Ann. Sci. Éc. Norm. Supér. (3)*, **9**, 3–93.

Painlevé, P. 1902. Sur les équations différentielles du second ordre et d'ordre supérieur dont l'intégrale générale est uniforme. *Acta Math.*, **25**(1), 1–85.

Painlevé, P. 1906. Sur les équations différentielles du second ordre á points critiques fixes. *C. R. Acad. Sci.*, **143**, 1111–1117.

Papageorgiou, V. G. and Tongas, A. G. 2007. Yang-Baxter maps and multi-field integrable lattice equations. *J. Phys. A: Math. Theor.*, **40**(42), 12677–12690.

Papageorgiou, V. G. and Tongas, A. G. 2009. Yang-Baxter maps associated to elliptic curves. *arXiv preprint arXiv:0906.3258*.

Papageorgiou, V. G., Nijhoff, F. W. and Capel, H. W. 1990. Integrable mappings and nonlinear integrable lattice equations. *Phys. Lett. A*, **147**(2–3), 106–114.

Papageorgiou, V. G., Nijhoff, F. W., Grammaticos, B. and Ramani, A. 1992. Isomonodromic deformation problems for discrete analogues of Painlevé equations. *Phys. Lett. A*, **164**(6), 57–64.

Papageorgiou, V. G., Grammaticos, B. and Ramani, A. 1993. Integrable lattices and convergence acceleration algorithms. *Phys. Lett. A*, **179**(2), 111–115.

Papageorgiou, V. G., Grammaticos, B. and Ramani, A. 1995. Orthogonal polynomial approach to discrete Lax pairs for initial-boundary value problems of the QD algorithm. *Phys. Lett. A*, **34**(2), 91–101.

Papageorgiou, V. G., Grammaticos, B. and Ramani, A. 1996. Integrable difference equations and numerical analysis algorithms. In Levi, D., Vinet, L. and Winternitz, P. (eds), *Symmetries and Integrability of Difference Equations*. CRM Proceedings & Lecture Notes, vol. 9. Providence, RI: American Mathematical Society, 269–280.

Papageorgiou, V. G., Tongas, A. G. and Veselov, A. P. 2006. Yang–Baxter maps and symmetries of integrable equations on quad-graphs. *J. Math. Phys.*, **47**(8), 083502, 16.

Papageorgiou, V. G., Suris, Yu. B., Tongas, A. G. and Veselov, A. P. 2010. On quadrirational Yang-Baxter maps. *SIGMA* **6**, 033, 9pp.

Periwal, V. and Shevitz, D. 1990. Unitary-matrix models as exactly solvable string theories. *Phys. Rev. Lett.*, **64**(12), 1326.

Perk, J. H. H. 1980. Quadratic identities for Ising model correlations. *Phys. Lett. A*, **79**(1), 3–5.

Petrera, M. and Suris, Yu. B. 2010. On the Hamiltonian structure of Hirota-Kimura discretization of the Euler top. *Math. Nachr.*, **283**(11), 1654–1663.

Petrera, M., Pfadler, A. and Suris, Yu. B. 2009. On integrability of Hirota–Kimura-type discretizations: experimental study of the discrete Clebsch system. *Experiment. Math.*, **18**(2), 223–247.

Petrera, M., Pfadler, A. and Suris, Yu. B. 2011. On integrability of Hirota-Kimura type discretizations. *Regu. Chaotic Dyn.*, **16**(3–4), 245–289.

Picard, É. 1887. Sur une classe d'équations différentielles. *C. R. Acad. Sci. Paris*, **104**, 41–43.

Picard, É. 1890. Sur une classe d'équations différentielles dont l'intégrale générale est uniforme. *C. R. Acad. Sci. Paris*, **110**, 877–880.

Quispel, G. R. W. 2003. An alternating integrable map whose square is the QRT map. *Phys. Lett. A*, **307**(1), 50–54.

Quispel, G. R. W., Nijhoff, F. W., Capel, H. W. and van ver Linden, J. 1984. Linear integral equations and nonlinear difference-difference equations. *Physica A*, **125**(2–3), 344–380.

Quispel, G. R. W., Roberts, J. A. G. and Thompson, C. J. 1988. Integrable mappings and soliton equations. *Phys. Lett. A*, **126**(7), 419–421.

Quispel, G. R. W., Roberts, J. A. G. and Thompson, C. J. 1989. Integrable mappings and soliton equations. II. *Phys. D*, **34**(1–2), 183–192.

Quispel, G. R. W., Capel, H. W., Papageorgiou, V. G. and Nijhoff, F. W. 1991. Integrable mappings derived from soliton equations. *Physica A*, **173**(1–2), 243–266.

Quispel, G. R. W., Capel, H. W. and Sahadevan, R. 1992. Continuous symmetries of differential-difference equations: the Kac-van Moerbeke equation and Painlevé reduction. *Phys. Lett. A*, **170**(5), 379–383.

Rains, E. M. 2008. An isomonodromy interpretation of the elliptic Painlevé equation. I. *arXiv preprint arXiv:0807.0258*.

Rains, E. M. 2011. An isomonodromy interpretation of the hypergeometric solution of the elliptic Painlevé equation (and generalizations). *SIGMA* **7**, 088, 24pp.

Ramani, A. and Grammaticos, B. 1996. Discrete Painlevé equations: coalescences, limits and degeneracies. *Physica A*, **228**(1–4), 160–171.

Ramani, A., Dorizzi, B. and Grammaticos, B. 1982. Painlevé conjecture revisited. *Phys. Rev. Lett.*, **49**(Nov), 1539–1541.

Ramani, A., Grammaticos, B. and Hietarinta, J. 1991. Discrete versions of the Painlevé equations. *Phys. Rev. Lett.*, **67**(14), 1829–1832.

Ramani, A., Grammaticos, B. and Satsuma, J. 1992. Integrability of multidimensional discrete systems. *Phys. Lett. A*, **169**(5), 323–328.

Ramani, A., Grammaticos, B. and Satsuma, J. 1995. Bilinear discrete Painlevé equations. *J. Phys. A: Math. Gen.*, **28**(16), 4655.

Ramani, A., Grammaticos, B. and Papageorgiou, V. G. 1996. Singularity confinement. In Levi, D., Vinet, L. and Winternitz, P. (eds), *Symmetries and Integrability of Difference*

Equations. CRM Proceedings & Lecture Notes, vol. 9. Providence, RI: American Mathematical Society, 303–318.

Ramani, A., Grammaticos, B., Lafortune, S. and Ohta, Y. 2000. Linearizable mappings and the low-growth criterion. *J. Phys. A: Math. Gen.*, **33**(31), L287.

Ramani, A., Carstea, A. S., Grammaticos, B. and Ohta, Y. 2002. On the autonomous limit of discrete Painlevé equations. *Physica A*, **305**(3–4), 437–444.

Ramani, A., Grammaticos, B., Tamizhmani, T. and Tamizhmani, K. M. 2003. The road to the discrete analogue of the Painlevé property: Nevanlinna meets singularity confinement. *Comput. Math. Appl.*, **45**(6–9), 1001–1012.

Ramani, A., Joshi, N., Grammaticos, B. and Tamizhmani, T. 2006. Deconstructing an integrable lattice equation. *J. Phys. A: Math. Gen.*, **39**(8), L145–L149.

Ramani, A., Grammaticos, B., Willox, R., Mase, T. and Kanki, M. 2015. The redemption of singularity confinement. *J. Phys. A: Math. Theor.*, **48**(11), 11FT02, 8pp.

Ramis, J.-P., Sauloy, J. and Zhang, C. 2013. *Local Analytic Classification of q-Difference Equations*. Providence, **355**. RI: American Mathematical Society.

Rasin, A. G. 2010. Infinitely many symmetries and conservation laws for quad-graph equations via the Gardner method. *J. Phys. A: Math. Theor.*, **43**(23), 235201.

Rasin, O. G. and Hydon, P. E. 2005. Conservation laws of discrete Korteweg-de Vries equation. *SIGMA* **1**, 026, 6pp.

Rasin, O. G. and Hydon, P. E. 2006. Conservation laws of NQC-type difference equations. *J. Phys. A: Math. Gen.*, **39**(45), 14055–14066.

Rasin, O. G. and Hydon, P. E. 2007a. Conservation laws for integrable difference equations. *J. Phys. A: Math. Theor.*, **40**(42), 12763–12773.

Rasin, O. G. and Hydon, P. E. 2007b. Symmetries of integrable difference equations on the quad-graph. *Stud. Appl. Math.*, **119**(3), 253–269.

Rasin, A. G. and Schiff, J. 2009. Infinitely many conservation laws for the discrete KdV equation. *J. Phys. A: Math. Theor.*, **42**(17), 175205.

Reid, W. T. 1972. *Riccati Differential Equations*. New York: Academic Press.

Ritt, J. F. 1934. Algebraic difference equations. *Bull. Am. Math. Soc.*, **40**(4), 303–308.

Ritt, J. F. and Doob, J. L. 1933. Systems of algebraic difference equations. *Am. J. Math.*, **55**, 505–514.

Roberts, J. A. G. 1990. *Order and Chaos in Reversible Dynamical Systems*. Ph.D. thesis, University of Melbourne.

Roberts, J. A. G. and Quispel, G. R. W. 1992. Chaos and time-reversal symmetry. Order and chaos in reversible dynamical systems. *Phys. Rep.*, **216**(2–3), 63–177.

Roberts, J. A. G. and Quispel, G. R. W. 2006. Creating and relating three-dimensional integrable maps. *J. Phys. A: Math. Gen.*, **39**(42), L605–L615.

Roberts, J. A. G. and Vivaldi, F. 2003. Arithmetical method to detect integrability in maps. *Phys. Rev. Lett.*, **90**(Jan), 034102.

Rogers, C. and Schief, W. K. 2002. *Bäcklund and Darboux Transformations*. Cambridge, UK: Cambridge University Press.

Ruijsenaars, S. N. M. 2002. A new class of reflectionless second-order AΔOs and its relation to nonlocal solitons. *Regul. Chaotic Dyn.*, **7**(4), 351–391.

Ruijsenaars, S. N. M. and Schneider, H. 1986. A new class of integrable systems and its relation to solitons. *Ann. Phys.*, **170**(2), 370–405.

Rutishauser, H. 1954a. Der Quotienten-Differenzen-Algorithmus. *Zeitschrift für angewandte Mathematik und Physik ZAMP*, **5**(3), 233–251.

Rutishauser, H. 1954b. Ein infinitesimales Analogon zum Quotienten-Differenzen-Algorithmus. *Arch. Math. (Basel)*, **5**, 132–137.

Sahadevan, R. and Capel, H. W. 2003. Complete integrability and singularity confinement of nonautonomous modified Korteweg-de Vries and sine-Gordon mappings. *Physica A*, **330**(3–4), 373–390.

Sakai, H. 2001. Rational surfaces associated with affine root systems and geometry of the Painlevé equations. *Commun. Math. Phys.*, **220**(1), 165–229.

Sakai, H. 2005. A q-analog of the Garnier system. *Funkcial. Ekvac.*, **48**(2), 273–297.

Sanz-Serna, J. M. and Calvo, P. 1994. *Numerical Hamiltonian Problems*. Applied mathematics and mathematical computation. London: Chapman & Hall.

Sauer, R. 1970. *Differenzengeometrie*. Berlin Heidelberg: Springer.

Sauloy, J. 1999. *Théorie de Galois des equations aux q-differences fuchsiennes*. Ph.D. thesis, Université de Toulouse 3, Toulouse.

Schief, W. K. 1996. The Tzitzeica equation: a Bäcklund transformation interpreted as truncated Painlevé expansion. *J. Phys. A: Math. Gen.*, **29**(16), 5153–5155.

Schief, W. K. 2003. Lattice geometry of the discrete Darboux, KP, BKP and CKP equations. Menelaus' and Carnot's theorems. *J. Nonlinear Math. Phys.*, **10**, Supp 2, 194–208.

Schief, W. K. 2006. On a maximum principle for minimal surfaces and their integrable discrete counterparts. *J. Geom. Phys.*, **56**(9), 1484–1495.

Schief, W. K. 2007. Discrete Chebyshev nets and a universal permutability theorem. *J. Phys. A: Math. Theor.*, **40**(18), 4775–4801.

Schief, W. K. and Rogers, C. 1999. Binormal motion of curves of constant curvature and torsion. Generation of soliton surfaces. *R. Soc. London Proc. Ser. A Math. Phys. Eng. Sci.*, **455**(1988), 3163–3188.

Schiff, L. I. 1968. *Quantum Mechanics*, 3rd ed. New York: McGraw-Hill.

Schmidt, R. J. 1941. On the numerical solution of linear simultaneous equations by an iterative method. *Philos. Mag. (7)*, **32**, 369–383.

Scott, R. F. 1904. *The Theory of Determinants and Their Applications*. Cambridge, UK: Cambridge University Press.

Shabat, A. B. 1992. The infinite-dimensional dressing dynamical system. *Inverse Prob.*, **8**(2), 303.

Shabat, A. B. 2002. Discretization of the Schrödinger spectral problem. *Inverse Prob.*, **18**(4), 1003.

Shabat, A. B. and Yamilov, R. I. 1990. Symmetries of nonlinear lattices. *Algebra i Analiz*, **2**(2), 183–208.

Shanks, D. 1955. Non-linear transformations of divergent and slowly convergent sequences. *J. Math. and Phys.*, **34**, 1–42.

Shi, Y., Zhang D. J. and Zhao, S. L. 2014. Solutions to the non-autonomous ABS lattice equations – Casoratians and bilinearization. *Sci. Sin. Math.*, **44**(1), 37.

Shohat, J. A. 1939. A differential equation for orthogonal polynomials. *Duke Math. J.*, **5**(2), 401–417.

Shohat, J. A. and Tamarkin, J. D. 1943. *The Problem of Moments*. Mathematical Surveys and Monographs. Providence, RI: American Mathematical Society.

Smirnov, S. 2010. Discrete complex analysis and probability. *arXiv preprint arXiv:1009.6077*.

Sogo, K. 1993. Toda molecule equation and quotient-difference method. *J. Phys. Soc. Jpn.*, **62**(4), 1081–1084.

Spicer, P. E., Nijhoff, F. W. and van der Kamp, P. H. 2011. Higher analogues of the discrete-time Toda equation and the quotient-difference algorithm. *Nonlinearity*, **24**(8), 2229–2263.

Spiegel, M. R. 1971. *Schaum's Outline of Theory and Problems of Calculus of Finite Differences and Difference Equations*. Schaum's outline series. New York: McGraw-Hill.

Spiridonov, V. P. and Zhedanov, A. S. 2007. Elliptic grids, rational functions, and the Padé interpolation. *Ramanujan J.*, **13**(1-3), 285–310.

Stephani, H. 1989. *Differential Equations: Their Solution Using Symmetries*. Cambridge, UK: Cambridge University Press.

Suris, Yu. B. 1989. Integrable mappings of standard type. *Funktsional. Anal. i Prilozhen.*, **23**(1), 84–85.

Suris, Yu. B. 1997a. A note on an integrable discretization of the nonlinear Schrödinger equation. *Inverse Prob.*, **13**(4), 1121.

Suris, Yu. B. 1997b. On an integrable discretization of the modified Korteweg–de Vries equation. *Phys. Lett. A*, **234**(2), 91–102.

Suris, Yu. B. 2003. *The Problem of Integrable Discretization: Hamiltonian Approach*. Progress in Mathematics, vol. 219. Basel: Birkhäuser Verlag.

Suris, Yu. B. 2012. Variational formulation of commuting Hamiltonian flows: multi-time Lagrangian 1-forms. *arXiv preprint arXiv:1212.3314*.

Suris, Yu. B. 2013. Variational symmetries and pluri-Lagrangian systems. *arXiv preprint arXiv:1307.2639*.

Suris, Yu. B. and Veselov, A. P. 2003. Lax matrices for Yang–Baxter maps. *J. Nonlinear Math. Phys.*, **10**, Supp 2, 223–230.

Symes, W. W. 1982. The QR algorithm and scattering for the finite nonperiodic Toda lattice. *Physica D*, **4**(2), 275–280.

Szegö, G. 1939. *Orthogonal Polynomials*. American Mathematical Society Colloquium Publications, no. nid. 23. Providence, RI: American Mathematical Society.

Taha, T. R. 1991. A numerical scheme for the nonlinear Schrödinger equation. *Comput. Math. Appl.*, **22**(9), 77–84.

Taha, T. R. and Ablowitz, M. J. 1984a. Analytical and numerical aspects of certain nonlinear evolution equations. I. Analytical. *J. Comput. Phys.*, **55**(2), 192–202.

Taha, T. R. and Ablowitz, M. J. 1984b. Analytical and numerical aspects of certain nonlinear evolution equations. II. Numerical, nonlinear Schrödinger equation. *J. Comput. Phys.*, **55**(2), 203–230.

Takenawa, T. 2001a. Algebraic entropy and the space of initial values for discrete dynamical systems. *J. Phys. A: Math. Gen.*, **34**(48), 10533.

Takenawa, T. 2001b. A geometric approach to singularity confinement and algebraic entropy. *J. Phys. A: Math. Gen.*, **34**(10), L95.

Takenawa, T. 2004. Discrete dynamical systems associated with the configuration space of 8 points in $P^3(C)$. *Commun. Math. Phys.*, **246**(1), 19–42.

Takenawa, T., Eguchi, M., Grammaticos, B., Ohta, Y., Ramani, A. and Satsuma, J. 2003. The space of initial conditions for linearizable mappings. *Nonlinearity*, **16**(2), 457.

Tamizhmani, K. M., Tamizhmani, T., Grammaticos, B. and Ramani, A. 2004. Special Solutions for Discrete Painlevé equations. In Grammaticos, B., Tamizhmani, T., and Kosmann-Schwarzbach, Y. (eds), *Discrete Integrable Systems*, Lecture Notes in Physics, vol. 644, Berlin: Springer. 323–381.

Toda, M. 1967. Vibration of a chain with nonlinear interaction. *J. Phys. Soc. Japan*, **22**(2), 431–436.

Toda, M. 1989. *Theory of Nonlinear Lattices*. Springer Series in Solid-State Sciences, vol. 20. Berlin: Springer-Verlag.

Tongas, A. 2007. On the symmetries of integrable partial difference equations. In Elaydi, S., Cushing, J., Lasser, R., Papageorgiou, V. G., Ruffing, A. and van Assche, W. (eds), Difference Equations, Special Functions and Orthogonal Polynomials, World Scientific, 654–663.

Tongas, A. and Nijhoff, F. W. 2005a. The Boussinesq integrable system: compatible lattice and continuum structures. *Glasg. Math. J.*, **47**(A), 205–219.

Tongas, A. and Nijhoff, F. W. 2005b. Generalized hyperbolic Ernst equations for an Einstein–Maxwell–Weyl field. *J. Phys. A: Math. Gen.*, **38**(4), 895.

Tongas, A. and Nijhoff, F. W. 2006. A discrete Garnier type system from symmetry reduction on the lattice. *J. Phys. A: Math. Gen.*, **39**(39), 12191.

Tongas, A., Tsoubelis, D. and Xenitidis, P. 2001a. A family of integrable nonlinear equations of hyperbolic type. *J. Math. Phys.*, **42**(12), 5762–5784.

Tongas, A., Tsoubelis, D. and Xenitidis, P. 2001b. Integrability aspects of a Schwarzian PDE. *Phys. Lett. A*, **284**(6), 266–274.

Tongas, A., Tsoubelis, D. and Papageorgiou, V. G. 2005. Symmetries and group invariant reductions of integrable partial difference equations. In Ibrahimov, N. H., Sophocleous, C. and P. A. Damianou, P. A. (eds), *Proceedings of the 10th International Conference in Modern Group Analysis (MOGRAN X)*, Nicosia: University of Cyprus, 222–230.

Tongas, A., Tsoubelis, D. and Xenitidis, P. 2007. Affine linear and D4 symmetric lattice equations: symmetry analysis and reductions. *J. Phys. A: Math. Theor.*, **40**(44), 13353.

Tremblay, S., Grammaticos, B. and Ramani, A. 2001. Integrable lattice equations and their growth properties. *Phys. Lett. A*, **278**(6), 319–324.

Trjitzinsky, W. J. 1933. Analytic theory of linear q-difference equations. *Acta Math.*, **61**(1), 1–38.

Tsoubelis, D. and Xenitidis, P. 2009. Continuous symmetric reductions of the Adler–Bobenko–Suris equations. *J. Phys. A: Math. Theor.*, **42**(16), 165203.

Tsuda, T. 2004. Integrable mappings via rational elliptic surfaces. *J. Phys. A: Math. Gen.*, **37**(7), 2721.

van Assche, W. and Foupouagnigni, M. 2003. Analysis of non-linear recurrence relations for the recurrence coefficients of generalized Charlier polynomials. *J. Nonlinear Math. Phys.*, **10**, Supp 2, 231–237.

van der Kamp, P. H. 2009. Initial value problems for lattice equations. *J. Phys. A: Math. Theor.*, **42**(40), 404019, 16pp.

van der Kamp, P. H. 2015. Initial value problems for quad equations. *J. Phys. A: Math. Theor.*, **48**(6), 065204.

van der Kamp, P. H. and Quispel, G. R. W. 2010. The staircase method: integrals for periodic reductions of integrable lattice equations. *J. Phys. A: Math. Theor.*, **43**(46), 465207, 34pp.

van der Put, M. and Singer, M. F. 1997. *Galois Theory of Difference Equations*. Lecture Notes in Mathematics, vol. 1666. Berlin: Springer-Verlag.

Vein, R. and Dale, P. 1999. *Determinants and Their Applications in Mathematical Physics*. Applied Mathematical Sciences, vol. 134. New York: Springer-Verlag.

Vereschagin, V. L. 1996. Asymptotics of solutions of the discrete string equation. *Physica D*, **95**(3), 268–282.

Vermeeren, M. 2014. *On the Pluri-Lagrangian Structure of the KdV Hierarchy*. M.Phil. thesis, Technische Universität Berlin.

Veselov, A. P. 1987. Integrable mappings and Lie algebras. *Dokl. Akad. Nauk SSSR*, **292**(6), 1289–1291.

Veselov, A. P. 1991a. Integrable maps. *Russ. Math. Surv.*, **46**(5), 1–51.

Veselov, A. P. 1991b. What is an integrable mapping? In Zakharov, V. E. (ed), *What Is Integrability?* Springer Series in Nonlinear Dynamics. Berlin Heidelberg: Springer, 251–272.

Veselov, A. P. 1992. Growth and integrability in the dynamics of mappings. *Commun. Math. Phys.*, **145**(1), 181–193.

Veselov, A. P. 2003. Yang–Baxter maps and integrable dynamics. *Phys. Lett. A*, **314**(3), 214–221.

Veselov, A. P. and Shabat, A. B. 1993. Dressing chains and spectral theory of the schroödinger operator. *Funkts. Anali. Prilozh.*, **27**(2), 1–21.

Viallet, C.-M. 2006. Algebraic entropy for lattice equations. *arXiv:math-ph/0609043v2.*

Viallet, C.-M. 2009. Integrable lattice maps: Q_V, a rational version of Q_4. *Glasg. Math. J.*, **51**(A), 157–163.

Von Westenholz, C. 2009. *Differential Forms in Mathematical Physics*, vol. 3. Amsterdam: Elsevier (reprint of the 1978 edition).

Wadati, M. and Toda, M. 1972. The exact N-soliton solution of the Korteweg-de Vries equation. *J. Phys. Soc. Japan.*, **32**(5), 1403–1411.

Wahlquist, H. D. and Estabrook, F. B. 1973. Bäcklund transformation for solutions of the Korteweg-de Vries equation. *Phys. Rev. Lett.*, **31**, 1386–1390.

Walker, A. J. 2001. *Similarity Reductions and Integrable Lattice Equations*. Ph.D. thesis, University of Leeds.

Wallenberg, G. 1917. Über Riccatische Differenzengleichungen. *Stzber. Berliner math. Ges.*, **16**, 17–28.

Wallenberg, G. 1918. Zur Theorie der Riccatische und Schwarzschen Differenzengleichungen. *Stzber. Berliner math. Ges.*, **17**, 14–22.

Whittaker, E. T. and Watson, G. N. 1902. *A Course of Modern Analysis*. Cambridge, UK: Cambridge University Press.

Wiersma, G. L. and Capel, H. W. 1987a. Lattice equations, hierarchies and Hamiltonian structures. *Physica A*, **142**(1–3), 199–244.

Wiersma, G. L. and Capel, H. W. 1987b. Lattice equations, hierarchies and hamiltonian structures: The Kadomtsev-Petviashvili equation. *Phys. Lett. A*, **124**(3), 124–130.

Wiersma, G. L. and Capel, H. W. 1988a. Lattice equations, hierarchies and Hamiltonian structures: II. KP-type of hierarchies on 2D lattices. *Physica A*, **149**(1–2), 49–74.

Wiersma, G. L. and Capel, H. W. 1988b. Lattice equations, hierarchies and Hamiltonian structures: III. The 2D toda and KP hierarchies. *Physica A*, **149**(1–2), 75–106.

Willox, R. 2000. A direct linearization of the KP hierarchy and an initial value problem for tau functions. *RIMS Kokyuroku*, **1170**, 111–118.

Willox, R. and Satsuma, J. 2004. Sato theory and transformation groups. a unified approach to integrable systems. In Grammaticos, B., Tamizhmani, T., and Kosmann-Schwarzbach, Y. (eds), *Discrete Integrable Systems*. Lecture Notes in Physics, vol. 644. Berlin Heidelberg: Springer, 17–55.

Willox, R., Tokihiro, T. and Satsuma, J. 1997. Darboux and binary Darboux transformations for the nonautonomous discrete KP equation. *J. Math. Phys.*, **38**(12), 6455–6469.

Witte, N. S. 2012. Semi-classical orthogonal polynomial systems on non-uniform lattices, deformations of the Askey table and analogs of isomonodromy. *arXiv preprint arXiv:1204.2328.*

Wojciechowski, S. 1982. The analogue of the Backlund transformation for integrable many-body systems. *J. Phys. A: Math. Gen.*, **15**(12), L653.

Wunderlich, W. 1951. Zur Differenzengeometrie der Flächen konstanter negativer Krümmung. *Österreich. Akad. Wiss. Math.-Nat. Kl. S.-B. IIa.*, **160**, 39–77.

Wynn, P. 1961. The epsilon algorithm and operational formulas of numerical analysis. *Math. Comp.*, **15**, 151–158.

Xenitidis, P. 2009. Integrability and symmetries of difference equations: the Adler-Bobenko-Suris case. *arXiv preprint arXiv:0902.3954.*

Xenitidis, P. 2011. Symmetries and conservation laws of the ABS equations and corresponding differential-difference equations of Volterra type. *J. Phys. A: Math. Theor.*, **44**(43), 435201, 22pp.

Xenitidis, P. and Nijhoff, F. W. 2012. Symmetries and conservation laws of lattice Boussinesq equations. *Phys. Lett. A*, **376**(35), 2394–2401.

Xenitidis, P. and Papageorgiou, V. G. 2009. Symmetries and integrability of discrete equations defined on a black-white lattice. *J. Phys. A: Math. Theor.*, **42**(45), 454025, 13pp.

Xenitidis, P., Nijhoff, F. W. and Lobb, S. B. 2011. On the Lagrangian formulation of multidimensionally consistent systems. *Proc. R. Soc. London Ser. A Math. Phys. Eng. Sci.*, **467**(2135), 3295–3317.

Xu, D. D., Zhang, D. J. and Zhao, S. L. 2014. The Sylvester equation and integrable equations: I. The Korteweg-de Vries system and sine-Gordon equation. *J. Nonlinear Math. Phys.*, **21**(3), 382–406.

Yamada, Y. 2009. A Lax formalism for the elliptic difference Painlevé equation. *SIGMA* **5**, 042, 15pp.

Yamilov, R. 2006. Symmetries as integrability criteria for differential difference equations. *J. Phys. A: Math. Gen.*, **39**(45), R541.

Yoo-Kong, S. 2011. *Calogero–Moser Type Systems, Associated KP Systems, and Lagrangian Structures*. Ph.D. thesis, University of Leeds.

Yoo-Kong, S. and Nijhoff, F. W. 2011. Discrete-time Ruijsenaars–Schneider system and Lagrangian 1-form structure. *arXiv preprint arXiv:1112.4576.*

Yoo-Kong, S. and Nijhoff, F. W. 2013. Elliptic (N, N')-soliton solutions of the lattice Kadomtsev-Petviashvili equation. *J. Math. Phys.*, **54**(4), 043511.

Yoo-Kong, S., Lobb, S. B. and Nijhoff, F. W. 2011. Discrete-time Calogero–Moser system and Lagrangian 1-form structure. *J. Phys. A: Math. Theor.*, **44**(36), 365203.

Yoshida, H. 1990. Construction of higher order symplectic integrators. *Phys. Lett. A*, **150**(5–7), 262–268.

Zabrodin, A. 1997a. Hirota difference equations. *Teoret. Mat. Fiz.*, **113**(2), 179–230.

Zabrodin, A. 1997b. Zero curvature representation for classical lattice sine-Gordon model via quantum R matrix. *J. Exp. Theor. Phys. Lett.*, **66**(9), 653–659.

Zabrodin, A. 1998. Tau-function for discrete sine-Gordon equation and quantum R-matrix. *arXiv preprint solv-int/9810003.*

Zabusky, N. J. and Kruskal, M. D. 1965. Interaction of "solitons" in a collisionless plasma and the recurrence of initial states. *Phys. Rev. Lett.*, **15**(Aug), 240–243.

Zakharov, V. E. and Manakov, S. V. 1985. Construction of higher-dimensional nonlinear integrable systems and of their solutions. *Funct. Anal. Appl.*, **19**(2), 89–101.

Zakharov, V. E. and Mikhailov, A. V. 1980. Variational principle for equations integrable by the inverse problem method. *Funct. Anal. Appl.*, **14**(1), 43–44.

Zhang, D. J. and Zhao, S. L. 2013. Solutions to ABS lattice equations via generalized Cauchy matrix approach, *Stud. Appl. Math.*, Wiley Online Library, **131**(1), 72–103.

Zhang, D. J., Zhao, S. L. and Nijhoff, F. W. 2012. Direct linearization of extended lattice BSQ systems. *Stud. Appl. Math.*, **129**(2), 220–248.

Zhang, D. J., Cheng, J. W. and Sun, Y. Y. 2013. Deriving conservation laws for ABS lattice equations from Lax pairs, *J. Phys. A: Math. Theor.*, **46**(26), 265202, 19pp.

Index

For EU product safety concerns, contact us at Calle de José Abascal, 56–1°, 28003 Madrid, Spain or eugpsr@cambridge.org.

www.ingramcontent.com/pod-product-compliance
Ingram Content Group UK Ltd.
Pitfield, Milton Keynes, MK11 3LW, UK
UKHW051510240426
470322UK00008B/111